CRYSTALLINE BEAMS AND RELATED ISSUES

THE SCIENCE AND CULTURE SERIES — PHYSICS

Series editor: A. Zichichi, *European Physical Society, Geneva, Switzerland*

1. Perspectives for New Detectors in Future Supercolliders, 1991
2. Data Structures for Particle Physics Experiments: Evolution or Revolution?, 1991
3. Image Processing for Future High-Energy Physics Detectors, 1992
4. GaAs Detectors and Electronics for High-Energy Physics, 1992
5. Supercolliders and Superdetectors, 1993
6. Properties of SUSY Particles, 1993
7. From Superstrings to Supergravity, 1994
8. Probing the Nuclear Paradigm with Heavy Ion Reactions, 1994
9. Quantum-Like Models and Coherent Effects, 1995
10. Quantum Gravity, 1996
11. Crystalline Beams and Related Issues, 1996

THE SCIENCE AND CULTURE SERIES — PHYSICS

CRYSTALLINE BEAMS AND RELATED ISSUES

Proceedings of the 31st Workshop of the INFN Eloisatron Project

Erice, Italy 12 – 21 November 1995

Editors

D. M. Maletić and A. G. Ruggiero

Series Editor

A. Zichichi

World Scientific

Singapore • New Jersey • London • Hong Kong

Published by

World Scientific Publishing Co. Pte. Ltd.
5 Toh Tuck Link, Singapore 596224
USA office: 27 Warren Street, Suite 401-402, Hackensack, NJ 07601
UK office: 57 Shelton Street, Covent Garden, London WC2H 9HE

British Library Cataloguing-in-Publication Data
A catalogue record for this book is available from the British Library.

CRYSTALLINE BEAMS AND RELATED ISSUES

ISBN-13 978-981-02-2785-2
ISBN-10 981-02-2785-X

CONTENTS

Preface ix

Introductory Talk

The Challenge of High Phase Space Density 3
I. Hofmann

I. SPACE CHARGE EFFECTS

Lattices with Minimal Space Charge Effects for Crystalline Beams 13
P. J. Channell and F. R. Neri

Anomalous Behavior of Schottky Noise for Cooled Beams 33
V. A. Lebedev, J. S. Hangst, J. S. Nielsen and N. Madsen

Longitudinal Instabilities and Crystallization 51
A. Burov and V. G. Vaccaro

Transverse Coherent Oscillations of Intense Coasting Beams 61
R. W. Hasse and I. Hofmann

Beam Instabilities and Their Cures 71
M. Chanel

II. COOLING TECHNIQUES

Methods of Beam Cooling 93
A. M. Sessler

Stochastic Cooling 111
F. Caspers

Electron Cooling 129
I. N. Meshkov

Spectroscopy of Trapped Ions 149
H. Walther

Three-Dimensional Laser Cooling by Synchrobetatron Coupling 173
H. Okamoto, S. Machida and A. M. Sessler

Ions in Electron Plasmas at Strong Coupling: Stopping Power and Electron–Ion Recombinations 185
G. Zwicknagel, Q. Spreiter, M. Miller, C. Toepffer and P. G. Reinhard

Electron Cooler Impedances 205
A. V. Burov

III. CRYSTALLINE BEAMS

The Physics of Crystalline Beams 217
J. P. Schiffer

Diverse Topics in Crystalline Beams 229
J. Wei, A. Draeseke, A. M. Sessler and X. P. Li

Analytical Studies of Crystalline Beams 253
A. F. Haffmans, D. Maletić and A. G. Ruggiero

Schottky Spectra and Crystalline Beams 275
D. V. Pestrikov

Studies of Crystalline Beams at Legnaro 295
A. Pisent and A. Burov

Intrabeam Scattering in Dense and Cold Beams 311
M. Seurer, Q. Spreiter and C. Toepffer

Crystalline Beams in the CRYSTAL Storage Ring at Legnaro 329
A. F. Haffmans, D. Maletić and A. G. Ruggiero

Space Charge Instability in Storage Rings with Circular Closed Orbits 343
D. V. Pestrikov

Particle Trapping with an RF Cavity 347
A. Chao

IV. EXPERIMENTS

Working Group on Experiments and Special Session on Electron Cooling 355
D. Reistad

The Legnaro CRYSTAL Proposal 365
L. Tecchio et al.

A Thermal Wave Model for Electromagnetic Traps 393
R. Fedele, R. Poggiani and G. Torelli

The Status of Laser Cooling Experiments in ASTRID 403
J. S. Hangst et al.

Review of Electron Cooling Investigation at the Novosibirsk INP 409
V. Parkhomchuk

Measurements of the Nonlinear Electron Cooling Force at the IUCF Cooler Ring 421
S. Y. Lee et al.

Review of Recent Work on Electron Cooling at LEAR 435
J. Bosser

Cooling Experiments at CRYRING 453
H. Danared, G. Andler and A. Källberg

Electron Cooling at the ESR 459
M. Steck et al.

Conclusive Talk

Are We Approaching the Ultimate Limit of Particle Accelerators? 473
G. H. Rees

List of Participants 485

PREFACE

Despite the great progress made during this century in the development and operation of particle accelerators, we are, at the moment, facing severe limitations which impede the attainment of very cold and dense beams of charged particles. These limitations are due to the interparticle Coulomb interactions: intrabeam scattering and space-charge effects.

There has been recent interest in the feasibility of producing a Crystalline Beam. This is a spatially ordered ensemble of circulating particles having very small velocity fluctuations relative to each other. Such a state may be obtainable from an ordinary warm beam through the application of intense and continuous phase-space cooling (e.g., stochastic, electron, laser).

Crystalline Beams represent the ultimate in the control of space-charge effects. The study, and eventual attainment, of Crystalline Beams may lead to a variety of other applications, but will certainly provide a deeper insight into the fundamental limitations of storage rings and particle colliders.

The 31st Workshop of the INFN ELOISATRON Project was held in Erice (Sicily) during the period 12-21 November 1995. About 50 scientists from Europe, USA, Japan and Russia participated. The Workshop was divided in two parts. During the first part, presentations were given on the most advanced topics concerning Crystalline Beams and Related Issues. There were four sessions which took place in sequence: Space Charge Effects, Cooling Techniques, Crystalline Beams, and Experiments. During the second part, four working groups were formed to discuss issues and eventually new concepts, also related to the subject, for each of the four topics given above. The summaries of each of the working groups were then given the last day of the Workshop.

The starting point, it was agreed, was to understand the present limitations due to space-charge effects of ordinary "warm" beams of charged particles. The presentations and discussions reported on the performance of low-energy storage rings mostly for beams of heavy ions. The common case of interest are storage rings operating below the transition energy, so that a variety of effects proper to very relativistic accelerators were not relevant. Cooling techniques are the method to decrease the temperature and the density of the ion beam to a level where transient effects are noticeable, caused by the balance of the space-charge forces and the cooling forces. It was pointed out that electron cooling is needed to cool the beam to a sufficiently low temperature which would then make the final step of laser cooling more effective. During the transition, the behavior of intrabeam scattering may at first sight be described as anomalous; but its complete understanding requires a refinement of the theory with a re-evaluation of the Coulomb log and of the range of the impact parameters. Space-charge forces themselves can no longer be described by a continuous charge distribution; but the corpuscolar aspect of the beam has to be taken into account. Finally, if the cooling rate is sufficiently high the beam may collapse toward an equilibrium configuration which would be the result of the particle-particle Coulomb repulsive forces balanced by the confinement forces of the storage ring magnets. This configuration would yield Crystalline Beams, the shape and feasibility of which depend on the properties of the storage ring magnet lattice, that is focussing and bending.

Brookhaven, Long Island, NY July 1996

Alessandro G. Ruggiero
Director of the Workshop

INTRODUCTORY TALK

The Challenge of High Phase Space Density

I. Hofmann

GSI Darmstadt, Planckstr. 1, D-64291 Darmstadt, Germany

Abstract

Space charge plays a crucial role in the transition from the more common "hot-beam" (emittance-dominated) regime to the "cold-beam" (space-charge-dominated) regime and to the ultimate state of "ordered" or "crystalline" beams. The study of such transitions is within reach due to advanced cooling techniques. What is learnt about space charge in this context is of quite general interest in beam physics whenever the phase space density is pushed to its limits. Beyond single-particle effects collective space charge effects are discussed, which play an important role as they evolve on the fast time-scale of the beam plasma frequency and may compete effectively with the cooling process and intrabeam scattering.

1. Introduction

Space charge effects are of interest in a number of projects involving protons or ions of non-relativistic energies and high phase-space-density, like cooler rings, high-intensity synchrotrons, heavy ion fusion storage rings, high-current linacs for various applications like fusion drivers, spallation neutron sources, radioactive waste transmutation linacs, and others. Recently important experimental progress on space-charge-dominated beams has been achieved in storage rings employing electron or laser cooling. It seems that these techniques allow fundamental studies of space charge phenomena, which so far have been studied only on the computer. These findings may also be useful to bridge the gap to the study of intense linac bunches close to the space charge limit, where crucial issues are the emittance growth or exchange, and the development of halos around the beam core. It is assumed that such halos are a result of nonlinear resonances and coherent oscillations in connection with space charge.

The space charge model we consider is that of a continuous, but not necessarily uniform charge distribution ("self-consistent field model"), which bridges the gap between conventional beams assuming uncorrelated particles and the exotic realm of ordered or crystalline beams. In the collective regime the approach to the space charge limit is described by the ratio of the oscillation frequencies (betatron or synchrotron tune) with and without space charge, ν/ν_0, which may be different in x, y or z. Generally speaking we deal with space-charge-dominated beams if ν/ν_0 is below roughly 0.7. In this overview we first discuss the relation between space-charge dominated beams and ordered beams; we continue with the problem of defining a self-consistent equilibrium; then we discuss departures from equilibrium by resonance effects of single particles as well as by collective effects; finally we compare collective effects with intrabeam scattering in terms of the coupling parameter Γ, and conclude with an outlook.

2. Equilibrium Solutions

For *coasting beams* a characteristic feature of the space charge limit are the collectively distorted *longitudinal* Schottky spectra observed in cooler rings. They are a consequence of the Debye shielding of longitudinal fluctuations by space charge. Shielding here means that particles around the ring are phase correlated, in contrast with the completely uncorrelated phases of particles in "hot" beams. In the limit of very cold beams the spectrum shows for each harmonic two lines that indicate the coherent beam plasma frequencies of the "fast" and "slow" wave. The transition to longitudinally ordered beams is expected to reveal new features if the strong short-range and long-range correlations are taken into account. Noise spectra of such strongly correlated beams adopt new features, which have been studied for instance in Ref. [1]. *Transversely* space charge dominated beams with small ν/ν_0 have not been observed yet in storage rings, but in high-current linear transport channels [2]. Transverse shielding means that the density distribution in the beam adopts a uniform profile, and the transverse focusing is balanced by the space charge defocusing force. Measurements of the beam profile and of the dependence of the beam radius on current would give direct evidence of the proximity to the space charge limit. For an emittance-dominated ("hot") beam one generally has

$$a = (\epsilon\beta)^{1/2}, \tag{1}$$

where $\beta(m)$ is given by the lattice. For a space-charge-dominated beam, instead, we obtain a limiting current density given by the balance with the focusing gradient, hence

$$a \propto I^{1/2}. \tag{2}$$

The equilibrium of *bunched* beams is a considerably more complicated subject, if space charge forces have to be taken into account in all three spatial directions. The simpler case of only longitudinally space-charge-dominated cooled bunches has been investigated in the frequency domain in Ref. [3]. It was shown there that the longitudinal space charge limit can be identified by measuring the incoherent (single particle) synchrotron satellite frequency shift, which vanishes at the space charge limit. Easier to measure is the bunch length, which adopts a limiting value for given intensity and progressive cooling as was shown in Ref. [4], where the longitudinal space charge limit was approached very closely.

For bunches in real 3-d an exactly self-consistent equilibrium can be written, for constant focusing, in terms of the total Hamiltonian. The most adequate case in the presence of some type of thermal relaxation, like cooling, appears to be a Maxwell-Boltzmann distribution:

$$f = f_0 \exp(-H/k_B T) \tag{3}$$

The independence of the transverse temperature from the position along the bunch (i.e. time) is a characteristic feature of the Maxwell-Boltzmann distribution and follows directly from Eq. 3. For the emittance-dominated beam the spatial profile is cylindrical, in accordance with Eq. 1; for the space-charge-dominated beam Eq. 2 results in a more ellipsoidal spatial shape as shown schematically in Fig. 1. Numerical solutions including the self-consistent potential by space charge have been calculated in Ref. [5]. Due to image charges on the conducting beam pipe the limiting line charge densities actually look somewhat different for bunches that are short or long bunches with respect to the beam pipe. For a short bunch in a quadratic RF potential one thus finds a parabolic line charge density and a uniformly charged ellipsoid at the space charge limit. For long bunches, however, image charges result in a logarithmic deviation from the ellipsoidal boundary given by (see Ref. [5]):

$$(z/z_m)^2 = 1 - (r/r_m)^2 + 1/g_m (r/r_m)^2 \ln (r/r_m)^2, \tag{4}$$

with the geometry factor $g_m = 1 + \ln(b/r_m)$, b the pipe radius and z_m, r_m the bunch half-axii.

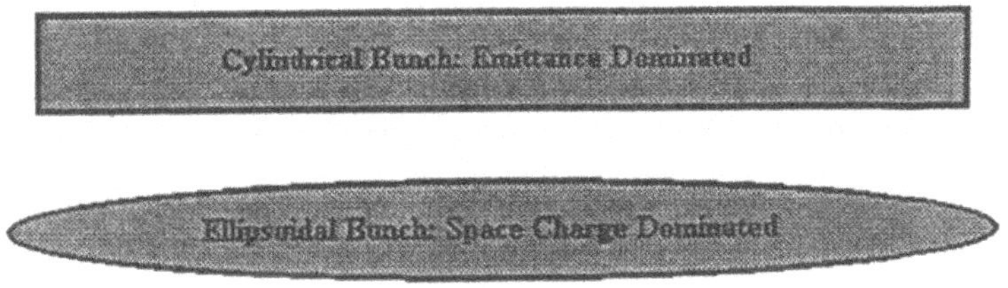

Figure 1: Bunch models (schematic).

The isotropic "thermal" distribution is, however, not necessarily realistic, if cooling or heating forces are anisotropic. This is the case for electron cooling, which is generally more efficient longitudinally; it is also true for laser cooling, which acts directly only in the longitudinal direction; and it applies to intrabeam scattering, which is in general anisotropic in storage rings due to the curvature rsp. dispersion effects. The experimental verification of such anisotropy effects is generally not trivial. In a recent experiment at the ESR heavy ion cooler storage ring at GSI it was shown recently by means of a streak camera that for weak space charge effects a bi-Maxwellian model with different temperatures longitudinally and transversely fits the data quite well [6]. The electron cooled bunch was extracted and focused on a scintillator, the light of which was recorded by a streak camera. This measurement allowed to resolve the bunch width vs. time. While the equidensity contours seemed to suggest an ellipsoidal spatial boundary, a quantitative evaluation of the horizontal rms size confirmed a "cylindrical bunch" model for which the transverse rms size is constant along the bunch. Based on Eq. 1 these measurements gave the emittance or temperature vs. time, which is consistent with the bi-Maxwellian distribution

$$f = f_0 \exp\left(-H_{lo}/k_B T_{lo}\right) \exp\left(-H_{tr}/k_B T_{tr}\right) \tag{5}$$

assuming different temperatures in the longitudinal and transverse degrees of freedom ("anisotropic bunches").

For space-charge-dominated bunches such measurements have not been performed yet. Analytical expressions for realistic anisotropic bunches with space charge are not known. What remains is computer simulation, where one can try to apply anisotropic cooling forces to relax an initially isotropic bunch distribution towards anisotropy. According to simulation and theoretical work it must be questioned, however, that in general stable equilibria can exist for space-charge-dominated bunches with temperature anisotropy, unless they are enforced by the continuous action of a sufficiently strong cooling force. Nonlinear space charge forces and collective effects can be a source of such loss of stability. The theoretical modelling of 3-d bunched beam equilibria close to the space charge limit is thus a challenging issue even in the absence of ordering efects.

3. Single-Particle Resonance Effects

The next step to consider are resonance effects of single particles with respect to the space charge force by the beam (coasting or bunched) itself, which is assumed to be in equilibrium. Such resonances may have to be considered even in the absence of machine resonances.

- Intrinsic space charge resonances - as opposed to resonances induced by nonlinearities of the external focusing - have been studied recently as mechanisms to explain the formation of halos in high current linac beams (for a review see Ref. [7]). Although most models consider the oscillating forces from strongly mismatched beams as driving terms (see section 4) it cannot be excluded that non-linear resonances driven by an otherwise stationary or perfectly matched charge distribution are a source of continuous beam loss close to the space charge limit.

- A resonance of the type $4\nu_y = 29$ driven by a space charge nonuniformity in connection with a lattice imperfection has been suggested to explain observed beam loss in a high-current synchrotron [8].

- Crossing of systematic machine resonances due to a space-charge-depressed single-particle tune can be an unsurmountable barrier even for super-fast cooling. Such a systematic resonance would occur, for instance, in a ring with a super-period of phase advance $\geq 180^0$. This has been studied by simulation in Ref. [9].

4. Collective Effects

Here we deal with coherent modes, which are deformations of the initially stationary equilibrium distribution. These deformations may become unstable under certain circumstances. Transversely they can be categorized as pure displacement (dipole mode, $m = 1$), envelope mismatch (quadrupole mode, $m = 2$; similarly a purely breathing mode exists for round beams), sextupole mode ($m = 3$), octupole mode ($m = 4$) etc. (see Fig. 2). For coasting beams one is generally concerned about transverse dipole modes getting

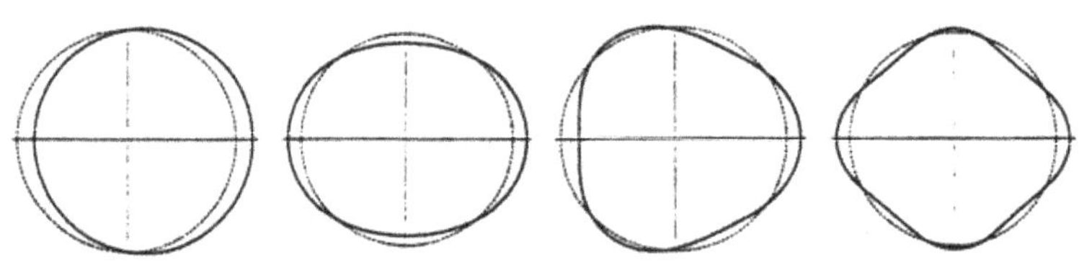

Figure 2: Coherent modes of transverse oscillation: $m = 1, 2, 3, 4$ (from left to right).

unstable at some longitudinal wave-length. For bunched beams one may have to consider multipolar modes in 3-d.

We need to take into account several types of collective effects that are of potential concern when approaching the space charge limit and the transition to ordered beams:

- In initially envelope mismatched beams ($m = 2$) the periodic oscillations of the linear or nonlinear space charge force may excite resonant behaviour of particles. This occurs predominantly for a 1:2 ("parametric resonance") relationship between the single-particle and the mismatched beam core oscillation frequency. In the presence of cooling forces it may be assumed, however, that such initial mismatch is practically absent.

- Large machine impedances at certain frequencies may drive modes with the corresponding wave-lengths unstable, in particular the purely longitudinal instability, or the transverse dipole mode. With sufficiently fast cooling one may again expect that these impedance driven instabilities are damped.

- In periodic focusing lattices the transverse modes $m = 2, 3, 4...$ may get in resonance with the periodic focusing and become exponentially unstable. The most dangerous case is the "envelope instability", which appears in a lattice with the zero-current phase advance per focusing cell exceeding 90^0 (see Ref. [10]). In general there is a finite width resonance band which, however, extends down to the space charge limit (i.e. zero phase advance) if the zero-current phase advance exceeds 127^0. To avoid this mode one has to keep the zero-intensity phase advance below 90^0. Higher order mode resonances have also been discussed, but they have practically negligible effect on the beam.

- For beams that are not in thermal equilibrium, but with pronounced anisotropy of temperature, mechanisms leading to equal temperatures have been explored. A possible candidate is an electrostatic instability similar to the "Harris mode" in uniform, but anisotropic plasmas [11]. The crucial point for beams is that for sufficiently strong anisotropy and near the space charge limit an internal multipole oscillation can become unstable and develop a coupling nonlinear field, which is responsible for the exchange of temperatures [12]. This "equipartitioning" mode has been considered for high-current linac bunches [13]. It can result in an "exchange" of temperatures and a final nearly isotropic distribution.

- Another deviation from thermal equilibrium would be a hollow distribution in synchrotron phase space, which could result from bunched beam laser cooling with a narrow band laser collecting particles on a particular synchrotron orbit. At a given position in the bunch this would correspond to a double-humped velocity distribution, which might be subject to the two-stream instability known in plasma physics. This coherent mode would fill up the central phase space part. One may speculate that this mode actually could support the cooling process.

The envelope instability and the anisotropy instability are *fast instabilities* in the sense that their growth times scale with the beam plasma period, which is comparable with the betatron tune for beams with ν/ν_0 below 0.5. Except for ν/ν_0 approaching 0, when these collective growth rates also vanish, they must be expected to be faster than intrabeam scattering effects, which are another candidate to cause heating and temperature isotropy. More studies are needed to compare these processes under realistic conditions, in particular when ordering effects are taken into account.

It is conceivable that the anisotropy instability could play an interesting role in laser cooling, which acts only longitudinally and therefore has a tendency to create a significant temperature difference. The instability would then present a desirable "collective sympathetic" cooling of the transverse degree of freedom. If confirmed by experiment it could be much more efficient than the presently discussed sympathetic cooling by intrabeam scattering.

5. Collective Effects vs. Ordering Effects

It may be assumed that most of the phenomena in space-charge-dominated beams with continuous charge distribution ("collective" regime) have some analogue in crystalline beams ("ordered" regime). It is therefore appropriate to characterize the border line between both. It may be noted that the square of the "tune depression" ν/ν_0, which characterizes space-charge-dominated beams, is the ratio of the effective restoring force over the externally applied force, which may adopt different values in different degrees of freedom.

In the "ordered" regime the appropriate quantity is the coupling parameter Γ, which is the ratio of the average inter-particle Coulomb energy over the thermal energy and a measure for inter-particle correlations:

$$\Gamma = \frac{e^2 Z^2/(4\pi\epsilon_0\rho)}{k_B T} \tag{6}$$

where ρ is the inter-particle distance and T the temperature. In general it must be assumed that Γ is also anisotropic with respect to different degrees of freedom. In a regime of small inter-particle correlations, where the character of periodic betatron oscillations is still preserved, a formal relationship between these parameters can be written as:

$$\Gamma = \frac{4r_p R^2 Z^2/A}{\beta^2 \rho a^2 \nu_0^2}(\nu/\nu_0)^{-2}, \tag{7}$$

Here we have used that for constant focusing the betatron oscillation follows $x = x_0 e^{i\nu s/R}$ and introduced the classical proton radius r_p (1.53×10^{-18} m) as well as the machine radius R. For small storage rings the zero-space-charge tune ν_0 is between 1...2. We note that for small ν/ν_0 (in practice ≤ 0.5) the density approaches a limiting value, which is described by the balance of the applied focusing force and the space charge defocusing and results as:

$$n_L = \frac{\beta^2\gamma^3}{4\pi Z^2/A r_p R^2}(\nu_0)^2. \tag{8}$$

This determines the limiting inter-particle distance ρ_L as $(3/(4\pi n_L))^{1/3}$.

To give an example we take a coasting beam of Mg^+ ions (A=24), $R = 10$, $\beta = 0.07$, $\nu_0 = 1.5$, which results in $\rho_L \approx 10$ μm. It is also convenient to express the beam radius a by the fictitious number of cylindrical crystalline shells that would fit into it (assuming that Γ could be made sufficiently large to lead to crystallization): $a^2 \approx \rho_L^2 N_{shells}$. This results readily in a relationship for the (transverse) coupling parameter:

$$\Gamma \approx \frac{2}{N_{shells}^2(\nu/\nu_0)^2}. \tag{9}$$

Equivalently N_{shells}^2 could be replaced by $N/(6 \times 10^6)$, where N is the total number of particles. Eq. 9 quantifies the connection between the "weak coupling" regime $\Gamma \approx 1$ and the space charge parameter ν/ν_0. For very large number of particles it corresponds to extremely small values of ν/ν_0; if the number of particles corresponds to three shells only (5×10^7 particles), we obtain $\nu/\nu_0 \approx 0.5$!

The practical significance of this result could lie in the relative importance of intra-beam scattering growth rates (found to be maximum around $\Gamma \approx 1$) versus growth rates from collective effects, which might be still faster as long as ν/ν_0 is not too small. To give an example: The maximum growth rate of the envelope instability (normalized by the plasma frequency ω_p) is found ≈ 0.2 for $\nu/\nu_0 \approx 0.3$ (Ref. [10]), which is half an order of magnitude larger than the maximum of the intrabeam scattering rate found at $\Gamma \approx 1$ (see Ref. [14]). This overlap region between ordering and collective effects will require more detailed studies in the future.

Outlook

We have discussed the beam physics issues that may be expected to be important close to the space charge limit within the model of a locally smeared out charge distribution ("collective" regime). Some of the processes considered may actually turn out to hinder the approach to the actual space charge limit, even in the presence of very efficient cooling forces. This could partly be due to their fast time-scale, which is given by the plasma period. In the collective realm analytical models exist under certain conditions, which allows a comparison with simulation. Unfortunately these conditions are not general enough - we don't even know how to write down a sufficiently general 3-d bunched beam equilibrium solution.

If one enters the "ordered" regime and includes Coulomb correlations between individual particles, one depends entirely on simulation. It will be necessary to show to what extent the collective space charge phenomena survive in the ordering and crystallization regime. One of the questions to be raised is: Can a crystalline beam be anisotropic?

So far we have mainly disussed cooling as means to provide the low temperature needed to reach the space charge limit. In reality one needs to look at the beams as being in an equilibrium between cooling and heating forces. This means that the cooling force itself introduces correlations among particles, which leads to some kind of dynamical equilibrium. Anomalous effects in Schottky noise spectra may reflect such dynamical effects. To obtain a better understanding extensive simulation work may be required, which also generates diagnostics output that can be directly compared with the experiment, like the fluctuation spectra.

References

[1] V.V. Avilov and I. Hofmann, Phys. Rev. **E 47**, 2019 (1993)
[2] M.G. Tiefenbach and D. Keefe, IEEE Trans. Nucl. Sci. **NS-32**, 2483 (1985)
[3] I. Hofmann and G. Kalisch, Phys. Rev. **E 53**, 2807 (1996)
[4] J.S. Hangst et al., Phys. Rev. Lett. **74**, 4432 (1995)
[5] N. Brown and M. Reiser, Phys. Plasmas **2**, 965 (1995)
[6] I. Hofmann et al., Phys. Rev. Lett. **75**, 3842 (1995)
[7] J.-M. Lagniel, Proc. 8th ICFA Advanced Beam Dynamics Workshop, Bloomington, Indiana, Oct. 11-13, 1995
[8] S. Machida, Nucl. Instr. Meth. **A 309**, 43 (1991)
[9] R.W. Hasse and I. Hofmann, these Proceedings
[10] I. Hofmann, L.J. Laslett, L. Smith and I. Haber, Part. Acc. **13**, 145 (1983)
[11] E.G. Harris, Phys. Rev. Lett. **2**, 34 (1959) and J. Nucl. Energy Part C, **2**, 138 (1961)
[12] I. Hofmann, IEEE Trans. Nucl. Sci. **NS-28**, 2399 (1981)
[13] R.A. Jameson, IEEE Trans. Nucl. Sci. **NS-28**, 2408 (1981)
[14] M. Seurer et al., J. Wei et al., these proceedings

I. SPACE CHARGE EFFECTS

Lattices With Minimal Space Charge Effects for Crystalline Beams

Paul J. Channell and Filippo R. Neri
Los Alamos National Laboratory
Los Alamos, NM 87545

1 Introduction

There are numerous techniques for cooling beams of charged particles including stochastic cooling, electron beam cooling, ionization (foil) cooling (for lepton beams only), and laser cooling which works only with ions with some electrons still attached. The successful implementation of laser cooling at Aarhus, [5], has led to interest in crystalline beams (see [6] and references contained therein), and it certainly seems that crystallization of small numbers of stored particles should be possible. There are limits, however, that may restrict the total number of charged particles stored; these include the limit on the space-charge tune shift, $|\Delta\nu| < 0.25$ (though the precise number is subject to debate) and intrabeam scattering.

In this paper we will be concerned with the possibility of intense crystalline beams; for simplicity we treat only the nonrelativistic case, though the relativistic case is a simple extension of this work. In the next section we review the limits on the number of particles stored and observe that the beam size scaling with beam temperature is the important dependence that determines the limits on the stored current as a function of beam temperature. In section 3 we use a general formalism to determine the beam size scaling and apply it to various kinds of focusing lattices and determine the relevant limits. In section 4 we use simulations that include lattice elements, a cooling model, and an N-body space-charge model to confirm the predictions of section 3 and to explore the details of various schemes. In the final section we summarize and discuss our results.

Note: Though we are interested ultimately in crystalline beams in this paper, the word *lattice* will refer throughout to an arrangement of magnetic elements in a beam line or storage ring and not to the arrangement of beam particles.

2 Limits

2.1 Space Charge

In general it is believed that one limit on the charge density in a storage ring arises from the space-charge tune shift induced by the collective interaction of the beam particles with themselves and the resulting crossing of lattice imperfection resonances by individual particles. To discuss this phenomenon, let us consider a simple one-dimensional model of the beam. Let us assume that the particle beam is an infinitely long cylinder of charge with a uniform particle density, n_0, (a charge density $\rho_0 = en_0$), out to a radius r_0. Note that we are assuming that there are enough particles at each azimuthal location that the beam can be described by a continuous distribution function; in some cases the total number of particles in a ring is so small that this assumption can be violated. With this assumption, Poisson's equation is

$$\nabla^2\phi = \begin{cases} -4\pi\rho_0 & \text{if } 0 \leq r \leq r_0 \\ 0 & \text{if } r \geq r_0 \end{cases} \tag{1}$$

In cylindrical coordinates the solution is clearly

$$\phi = \begin{cases} -\pi\rho_0 r^2 & \text{if } 0 < r \leq r_0 \\ -\pi\rho_0 r_0^2(1 + 2\ln(\frac{r}{r_0})) & \text{if } r \geq r_0 \end{cases} \tag{2}$$

The equation of motion for a nonrelativistic beam particle in the resulting field and in constant, linear, external focusing forces is

$$\frac{d^2x}{dt^2} + \omega_{\beta 0}^2 x = \frac{2\pi e^2 n_0 x}{m}, \tag{3}$$

where m is the mass of the particle, $\omega_{\beta 0}$ is the angular betatron frequency due to external focusing, and a similar equation holds in y. This can be rewritten as

$$\frac{d^2x}{dt^2} + (\omega_{\beta 0}^2 - \frac{\omega_p^2}{2})x = 0, \tag{4}$$

where $\omega_p^2 = 4\pi n_0 e^2/m$ is the beam plasma frequency squared. We thus see that in this simple model particles undergo simple harmonic oscillations with a reduced betatron frequency given by

$$\omega_\beta = \omega_{\beta 0}\sqrt{1 - \frac{\omega_p^2}{2\omega_{\beta 0}^2}}. \tag{5}$$

If we write the betatron frequencies in terms of the angular circulation frequency, ω_0, using the betatron tunes, ν_0, ν, as

$$\omega_{\beta 0} = \nu_0\omega_0, \tag{6}$$

$$\omega_\beta = \nu\omega_0, \tag{7}$$

then the tunes satisfy

$$\nu = \nu_0\sqrt{1 - \frac{\omega_p^2}{2\nu_0^2\omega_0^2}}. \tag{8}$$

If the external force is strong compared to the space-charge defocusing (the usual case, but not necessarily the one we are interested in) then we can expand the square root to get

$$\nu \approx \nu_0 - \frac{\omega_p^2}{4\nu_0\omega_0^2}, \tag{9}$$

with a resulting 'tune shift' given by

$$\Delta\nu \approx -\frac{\omega_p^2}{4\nu_0\omega_0^2}. \tag{10}$$

The plasma frequency can be expressed in terms of the number of particles in the ring, N, the ring radius, R, and the beam radius, r_0, as

$$\omega_p^2 = \frac{2Ne^2}{\pi m R r_0^2}. \tag{11}$$

The circulation frequency, ω_0, can be expressed in terms of the particle velocity, v, and the ring radius as

$$\omega_0 = \frac{v}{R}. \tag{12}$$

Combining equations 10, 11, and 12 we find that the tune shift can be reexpressed as

$$\Delta\nu = -\frac{Ne^2R}{2\pi\nu_0 r_0^2 m v^2}, \tag{13}$$

which is easily seen to be identical to the expression usually quoted, [1], in the limit of nonrelativistic beams.

Obviously this simple model has omitted a number of effects. For higher energy beams than we are considering in this paper it is essential to include relativistic effects; in the same regime the influence of the beam generated magnetic field must be included. Together these effects add a factor of γ^3 to the denominator of 13; one factor of γ comes from the mass increase and two factors of gamma come from the cancellation of electric and magnetic fields (if there is no neutralization). In addition, the electromagnetic boundary conditions used above are obviously not realistic; one can include the effects of image charges and image currents in the walls, the fact that the walls are not perfectly conducting (field penetration), and the presence of structures such as cavities in the walls.

The beam model used above is also too simple. Real beams have different sizes at different azimuthal locations in the ring; the above model should thus be considered some sort of azimuthal 'average' of the true situation. In addition, real beams are usually asymmetric, sometimes extremely so, in the two transverse directions, leading to the need to include geometric factors. Finally, beams are often bunched, i.e. nonuniform in the azimuthal direction; note that for bunched beams one must correctly take into account the time dependent penetration of magnetic field into the wall in order to have a completely accurate model. The paper by Bryant, [2], gives formulae that include most of these effects. Nevertheless, equation 13 is, in general, the largest term in the tune shift and certainly dominates for the nonrelativistic ion beams we consider in this paper.

In almost all cases beams are not of uniform density; this implies that the electric and magnetic fields do not vary linearly and also that the equation of motion is not adequately represented by equation 4, nonlinear terms must be added. In fact, particles near the 'core' of most beams will nearly obey equation 4, while particles near the periphery will experience a much reduced average space-charge field; thus the 'effective' tune shift will be given approximately by equation 10 for particles with low transverse energy and will be close to 0 for particles with high transverse energy. Thus, the formula, 10, for the tune shift is more accurately described as a formula for the tune *spread* in realistic beams; thus, individual particles will have tunes that range from ν_0 to $\nu_0 + \Delta\nu$. However, particles that resonate with one of the imperfection resonances of the machine lattice are usually rapidly lost, and since low order machine resonances occur at least at every quarter, half, and full integer tune, the usual limit cited for the allowable tune shift is $|\Delta\nu| < 0.25$, though the precise value is debatable.

However, resonances cause particle loss only when there is a driving term that excites the resonance; there is the possibility that such driving terms might be measured and corrected so that resonance crossing is possible. Some very nice experimental work showing that this is possible has been done at the AGS, see reference [4]. Nevertheless, it becomes much more difficult to cross successive resonances and crossing of the very strong integer resonances has not been accomplished. Thus, it is reasonable, for now, to say that the maximum allowable tune shift is given by

$$|\Delta\nu| < 1.0. \tag{14}$$

If we combine equation 14 with equation 13 we see that the space-charge limit can be written as

$$N < \frac{4\pi\nu_0 r_0^2 \mathcal{E}}{e^2 R} \tag{15}$$

where $\mathcal{E} = mv^2/2$ is the beam kinetic energy. As one would expect, one can confine more particles at higher energy, but the strong dependence in equation 15 is on the beam radius; as r_0 becomes smaller the number of particles that can be confined decreases quadratically in a linear lattice. It is thus important to determine the dependence of the beam size on the beam temperature if one is interested in intense crystalline beams; we return to this question in section 3.

2.2 Intrabeam scattering

It is not our intention to present even a barely adequate treatment of intrabeam scattering; we mention it only because it is one of the fundamental limits on stored beam intensity and because of its relevance to some of the simulation results to be presented later. A much better treatment can be found in Sørensen, [8], and in the references contained therein.

Intrabeam scattering is the process of two body scattering that occurs because the beam is, in fact, a collection of discrete particles. The most dramatic result of this occurs for bunched beams when two particle scattering transfers transverse momentum into longitudinal momentum shifts so large that the particles fall outside the longitudinal bucket and are lost from the machine; this is the Touschek effect. Even when this effect is negligible, two body scattering results in a diffusion of particles in phase space and effectively acts as a heating mechanism that tends to counteract externally imposed cooling.

We will not present a detailed model of intrabeam scattering but will simply extract some of the relevant parameter dependencies by examining the scattering rate. The rate of two body scattering is

$$R_{sc} \approx n\sigma v_{th}, \tag{16}$$

where n is the beam density, σ is the scattering cross section, and v_{th} is the magnitude of the thermal velocity in the beam frame of reference. The density, n, can be expressed in terms of the number of particles in the ring, N, the ring radius, R, and the beam radius, r_0, as

$$n = \frac{N}{2\pi^2 R r_0^2}. \tag{17}$$

The cross section in equation 16 is a complicated function of v_{th}. At higher thermal energies (> 1 eV) there are numerous atomic resonances. Below 1 eV the cross section is nearly the Rutherford cross section which increases rapidly as v_{th} decreases until the temperature becomes low enough that screening effects, i.e. multiparticle scattering, dominate and the cross section drops again. Nonetheless, even in the resonance region of the cross section the dependence on beam density is unchanged and so we can conclude that at least

$$R_{sc} \propto \frac{1}{r_0^2}. \tag{18}$$

Thus the heating rate due to intrabeam scattering increases quadratically with a decrease in beam size. We see that both beam space-charge tune shift limits and intrabeam scattering heating rate depend crucially on the beam size and its scaling as a function of beam temperature.

3 Derivation of Beam Sizes

Though for linear lattices the dependence of beam size on beam temperature is well known we will rederive it using a technique that is general enough to allow us to treat nonlinear lattices as well.

3.1 Multitime Scale Approximation

Particles in a periodic transport lattice or periodic storage ring typically traverse several periods of the external focusing force in the course of a single betatron oscillation, i.e. the period of the external force is small compared to the betatron period; of course this is not true for very strong focusing lattices (ones where the phase advance per cell exceeds 90°) so we will not consider such lattices in this paper. If we do have this separation of periods then it is natural to use the method of averaging due to Krylov-Bogoliubov-Mitropolsky (KBM) [7]. For linear lattices the result of this analysis will just be the well known 'smooth' approximation.

Let us consider the equations of motion in the beam frame of reference; for the nonrelativistic beams we are considering the transformation back to laboratory coordinates is straightforward. If the equation of motion can be written as

$$\frac{d^2\vec{x}}{dt^2} + \vec{f}(\vec{x}, t) = 0, \tag{19}$$

where $\vec{f}(\vec{x}, t)$ includes the spatial dependence of the forces and is a 'rapidly' oscillating function of t, then we can apply the KBM averaging technique. Thus, let $\vec{x} \equiv \vec{X} + \vec{x}_r$, where $\vec{X}$ is a slowly varying term (that is, slow compared to the variation of the forces) and $\vec{x}_r$ is a rapidly oscillating term and where we assume that

$$|\vec{X}| > |\vec{x}_r|. \tag{20}$$

Note that because $\vec{x}_r$ is rapidly oscillating

$$|\frac{d\vec{x}_r}{dt}| \sim |\frac{d\vec{X}}{dt}|$$

If we plug these definitions into equation 19 we find

$$\frac{d^2\vec{X}}{dt^2} + \frac{d^2\vec{x}_r}{dt^2} + \vec{f}(\vec{X} + \vec{x}_r, t) = 0, \tag{21}$$

Expanding this equation in a Taylor series, using equation 20, and keeping the zeroth and linear terms we find

$$\frac{d^2\vec{X}}{dt^2} + \frac{d^2\vec{x}_r}{dt^2} + \vec{f}(\vec{X}, t) + \vec{x}_r \cdot \nabla\vec{f}(\vec{X}, t) \approx 0. \tag{22}$$

If we now split this equation into a slowly varying part and a rapidly varying part we get

$$\frac{d^2\vec{X}}{dt^2} + \langle\vec{x}_r \cdot \nabla\vec{f}(\vec{X}, t)\rangle \approx 0, \tag{23}$$

and

$$\frac{d^2\vec{x}_r}{dt^2} + \vec{f}(\vec{X}, t) \approx 0, \tag{24}$$

where the angle brackets denote an average over the rapidly oscillating period and where we have dropped a small term that oscillates at twice the fast period. It is straightforward to solve equation 24 for $\vec{x}_r$ since $\vec{X}$ is constant on the fast time scale; integrating once we find

$$\vec{x}'_r \approx \frac{d\vec{x}_r}{dt}(0) - \int_0^t dt' \vec{f}(\vec{X}, t'). \tag{25}$$

The initial value term can be determined by requiring that the rapidly oscillating $\vec{x}_r$ have no secular term, i.e. that

$$\langle\vec{x}'_r\rangle = \frac{1}{T}\int_0^T dt\left(\frac{d\vec{x}_r}{dt}(0) - \int_0^t dt' \vec{f}(\vec{X}, t')\right) = 0, \tag{26}$$

This gives

$$\frac{d\vec{x}_r}{dt}(0) = \frac{1}{T}\int_0^T dt \int_0^t dt' \vec{f}(\vec{X}, t'), \tag{27}$$

Using equation 27 and integrating equation 25 once more we find

$$\vec{x}_r \approx \frac{t}{T}\int_0^T dt \int_0^t dt' \vec{f}(\vec{X}, t') - \int_0^t dt' \int_0^{t'} dt'' \vec{f}(\vec{X}, t''). \tag{28}$$

Given the solution, 28, it is now easy to evaluate the term involving $\vec{x}_r$ in equation 23. Note that with this KBM approximation technique there is no need to assume that the forces are linear; we will, in fact, examine some nonlinear examples after we establish that the technique gives the usual well known results in the linear case.

3.2 Application to AG Quadrupole

Let us begin by considering an alternating gradient (AG) quadrupole system. The equations of motion for $\vec{x} = (x, y)$ are given by

$$\frac{d^2x}{ds^2} + g(s)x = 0, \tag{29}$$

and

$$\frac{d^2y}{ds^2} - g(s)y = 0, \tag{30}$$

where we now use as independent variable s, the path length, and where $g(s) = eB_0'(s)/m\beta$ is related to the magnetic field gradient, $B_0'(s)$, and is a rapidly oscillating function of s. Comparing equations 29 and 30 with equation 19 and using equation 28 we find that

$$x_r \approx X\mathcal{R}(s). \tag{31}$$

and

$$y_r \approx -Y\mathcal{R}(s). \tag{32}$$

where

$$\mathcal{R}(s) \equiv (\frac{s}{S}\int_0^S ds \int_0^s ds' g(s') - \int_0^s ds' \int_0^{s'} ds'' g(s'')). \tag{33}$$

If we use these solutions in equation 23 we find that the 'slow' part of the solution satisfies

$$\frac{d^2X}{dt^2} + X <g(s)\mathcal{R}(s)>= 0, \tag{34}$$

and

$$\frac{d^2Y}{dt^2} + Y <g(s)\mathcal{R}(s)>= 0, \tag{35}$$

We thus see that the slow variables evolve according to the Hamiltonian given by

$$H = \frac{\vec{P}^2}{2} + <g(s)\mathcal{R}(s)> \frac{\vec{X}^2}{2}, \tag{36}$$

As a particular example, suppose we have a thin lens model of an AG quadrupole system with g given by

$$g(s) = G(\delta_P(s) - \delta_P(s - \frac{L}{2})), \tag{37}$$

where

$$G = \frac{eB_0' L_Q}{m\beta c^2} \tag{38}$$

with L_Q the quadrupole length and L the period, and where δ_P denotes the periodic delta function of period L. We then find that

$$<g(s)\mathcal{R}(s)> = \frac{G^2}{4}. \tag{39}$$

This result is identical to that usually found for thin-lens AG quadrupole lines in the limit of small phase shift per cell.

From equation 36 we can extract the scaling of the beam size with the transverse thermal velocity. A particle with a given transverse energy $\mathcal{E}_\perp$ has a maximum transverse momentum, P ($\approx x'$), given by

$$P_{\rm max} = \sqrt{\frac{\mathcal{E}_\perp}{\mathcal{E}}}. \tag{40}$$

The maximum spatial excursion for this same particle is

$$X_{\rm max} = \frac{P_{\rm max}}{\sqrt{<g(s)\mathcal{R}(s)>}}, \tag{41}$$

which implies that $r_0 \propto v_{\rm th}$. This scaling is exactly what one would expect. Simulations of a number of particles with MARYLIE of a thick-lens AG quadrupole line, (i.e. tracking 12 particles for 2048 periods, recording the maximum excursions in momentum and position, and fitting the resulting data with a nonlinear fit routine) confirm this scaling to several decimal places.

If we now examine the formula, 15, for the space-charge limit in a ring and use the scaling $r_0 \propto v_{\rm th}$, we see that the maximum number of particles that can be confined in a ring decreases linearly with the beam temperature. Of course, when the number of particles has decreased to the point where there are only on or two particles at a given azimuthal location, the continuous density model used in deriving the limit, 15, fails and one must resort to simulations to establish the real limits; we are interested, however, in confining considerably more particles than result in this case and will not consider this limit. As a particular example, let us consider the parameters of the quadrupole ring to be considered in section 4.2, i.e.

$$R = 13.18 \text{ m}, \tag{42}$$

$$\beta = 3.736 * 10^{-3}, \tag{43}$$

$$B_0' = 20 \text{ T/m}, \tag{44}$$

$$L_Q = 0.2 \text{ m}, \tag{45}$$

$$mc^2 = 21.415 \text{ GeV}, \tag{46}$$

and a tune

$$\nu_0 = 4.2. \tag{47}$$

We find, using equations 38, 40, 41, and 15, that the space charge limit is

$$N < 4.9 * 10^7 \mathcal{E}_\perp, \tag{48}$$

where the transverse energy in the beam frame, $\mathcal{E}_\perp$, is measured in eV. In other words, a beam with a transverse temperature of 1 K ($\approx 10^{-4}$eV) could only hold about 5000 particles; clearly in this case the space charge limit no longer applies as there are only one or two particles at each azimuthal location.

If we use the scaling $r_0 \propto v_{\rm th}$ in equation 18 we find that the intrabeam scattering heating rate is proportional to $1/v_{\rm th}^2$. Overcoming this heating with external cooling thus becomes more difficult as the beam temperature decreases.

It thus seems that both the heating rate due to intrabeam scattering and the space-charge tune shift increase as the beam is cooled in a linear focusing lattice and that these effects will severely limit the number of stored particles in such a cooling ring.

3.3 Application to AG Octupole

Because the application of KBM averaging does not require the original equations of motion to be linear, we can apply the results, equations 23 and 28, to AG nonlinear systems. Thus, let us now consider an AG octupole system; the equations of motion are given by

$$\frac{d^2x}{ds^2} + g(s)x(x^2 - 3y^2) = 0, \tag{49}$$

and

$$\frac{d^2y}{ds^2} - g(s)y(3x^2 - y^2) = 0, \tag{50}$$

where now

$$g(s) = \frac{eB_0'''(s)}{m\beta}, \tag{51}$$

is related to the third derivative of the magnetic field on axis. The solution in equation 28 is immediately applicable and, upon plugging the result into equation 23, a tedious but straightforward calculation shows that the slow variables evolve according to the Hamiltonian

$$H = \frac{\vec{P}^2}{2} + \langle g(s)\mathcal{R}(s)\rangle \frac{(\vec{X}^2)^3}{2}. \tag{52}$$

Repeating the analysis of the previous subsection we now find that

$$X_{\max} = \frac{P_{\max}^{1/3}}{(\langle g(s)\mathcal{R}(s)\rangle)^{1/6}}, \tag{53}$$

Note that the scaling of size with thermal velocity given by this formula is considerably different from that given by equation 41. Equation 53 indicates that a considerable change in the temperature, say by a factor of 10^6, which will give a change of a factor of 10^3 in the thermal velocity, will give a change of only a factor of 10 in the beam size; but this weaker dependence on the beam temperature is *exactly* what is needed to ameliorate the effects of space charge and intrabeam scattering because these effects depend crucially on the beam size, not the thermal velocity! The limit on the allowed number of particles, equation 15, now scales as the cube root of the beam temperature rather than linearly; in fact the applicability of equation 15 to such a nonlinear lattice is highly questionable and, in fact, in the next section we will see that when space charge is included the beam can be cooled to essentially zero temperature with no deleterious effects from space charge or intrabeam scattering.

Of course, the conclusion that AG octupole lattices can confine extremely cold beams rests on the validity of equation 53. To test the validity of this scaling we again ran simulations with MARYLIE of an AG octupole lattice. Again we tracked 12 particles for 2048 periods, recorded the maximum excursions in momentum and position, and fit the resulting data with a nonlinear fit routine; the results are shown in figure 1

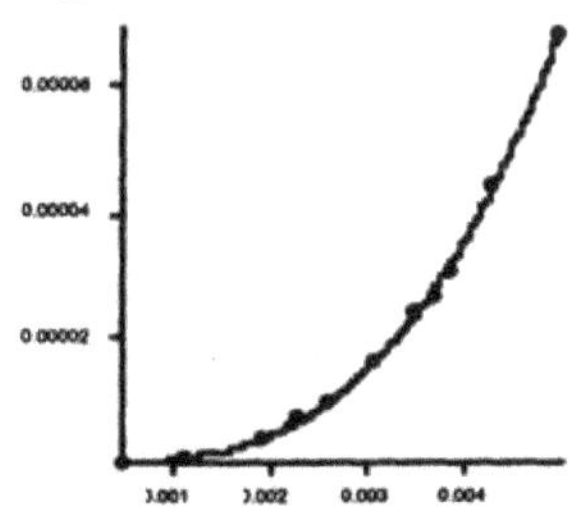

Figure 1. Maximum transverse velocity vs. maximum transverse displacement for an AG octupole line. The best fit curve ($X^{3.04}$) is also shown. Lengths are in meters, momenta in radians.

where the results are plotted along with the fit curve, where the best fit was found to be

$$P_{\max} \propto X_{\max}^{3.04}, \tag{54}$$

i.e. very nearly the scaling of equation 53. The KBM averaging technique again was successful.

As a final note in this subsection, observe that the dependence of the beam size on the strength of the machine focusing parameters given by equation 53 is also considerably weaker than the dependence on the machine parameters given by equation 41 for AG quadrupoles. One can thus expect that machine strength errors in an AG octupole machine will have less effect than equivalent errors in an AG quadrupole machine.

3.4 Application to AG Duodecapole

Let us consider in this section an even more nonlinear system, an alternating duodecapole line. Note that simulations will show that an AG octupole system is already adequate to store ultracold beams when space charge is taken into account, so that the considerations of this section are unnecessary; nevertheless, the analysis does serve to emphasize the conclusions of the previous section rather nicely. The equations of motion in an AG duodecapole line are given by

$$\frac{d^2x}{ds^2} + g(s)\frac{x}{3}(x^4 - 10x^2y^2 + 5y^4) = 0, \tag{55}$$

and

$$\frac{d^2y}{ds^2} - g(s)\frac{y}{3}(5x^4 - 10x^2y^2 + y^4) = 0, \tag{56}$$

where

$$g(s) = \frac{eB_0'''''(s)}{m\beta}, \tag{57}$$

is now related to the fifth derivative of the magnetic field on axis. The solution in equation 28 is again immediately applicable and, upon plugging the result into equation 23, a very tedious but straightforward calculation shows that the slow variables evolve according to the Hamiltonian

$$H = \frac{\vec{P}^2}{2} + <g(s)\mathcal{R}(s)> \frac{(\vec{X}^2)^5}{2}. \tag{58}$$

The scaling analysis of this equation now gives

$$X_{\max} = \frac{P_{\max}^{1/5}}{(<g(s)\mathcal{R}(s)>)^{1/10}}, \tag{59}$$

The dependence of the beam size on temperature is now so weak that to get a change in the beam size by a factor of 10 one must change the beam temperature by a factor of 10^{10}. Clearly the remarks made about space charge and intrabeam scattering for AG octupole systems apply to this case as well, though, as remarked above, it will turn out not to be necessary to consider AG duodecapoles.

4 Simulations

The prediction that AG octupole systems should have greatly relaxed space charge limits at very low temperature must be checked with simulations in order to correctly include space charge effects in a realistic setting; in this section we will carry this out and develop realistic designs for nonlinear rings with enhanced limits. The simulations we present were carried out with the code MARYLIE, [3], a third order particle tracking code that can simulate fairly realistic models of drifts, dipoles, quadrupoles, and octupoles. In addition, the ability to add user written subroutines to MARYLIE was employed to include models of both cooling and space charge effects.

To be specific, throughout this section we will investigate designs suitable for an 80 MeV singly charged sodium beam, Na^+; the energy and ion were chosen so that the beam is nonrelativistic but with a significant $B\rho$ so that the lattice design is definitely nontrivial.

4.1 Simulation Model

In addition to the simulation of standard lattice elements in MARYLIE, user subroutines were used to add models of two physical effects, namely cooling and space charge. The cooling routine was quite simple; each time the cooling routine was called the transverse momenta were decreased by an amount proportional to the transverse momentum, i.e.

$$\delta p_i = -\epsilon_i p_i \quad i = (x, y), \tag{60}$$

so that the transverse energy was damped exponentially by this routine. In principle, the damping in the two transverse directions could have been different, but in all cases actually run the two damping rates were taken to be the same.

Space charge was modeled in two dimensions using N-body techniques, i.e. the space-charge force on a particle was just the sum of the space-charge forces due to all the other particles; in all the simulations reported here, 100 particles were used. However, in order to avoid the usual singularities associated with point particle collisions, each particle was modeled as a charge cloud with a potential given by

$$\phi(\vec{x}) \propto -\ln(\frac{|\vec{x}-\vec{x}_0|^2}{\lambda^2} + \exp(-\frac{1}{2}\frac{|\vec{x}-\vec{x}_0|^2)}{\lambda^2})), \tag{61}$$

where $\vec{x}_0$ is the location of the particle, and λ is an adjustable cutoff parameter that can be used to control the amount of 'smoothing'. This model for the single particle potential has a couple of advantages over other smooth models such as

Gaussian potentials or Gaussian densities with potentials given by 'erf' functions, namely, 1) the continuous particle density associated with this potential is everywhere positive, and 2) the potential and associated fields can be evaluated in terms of functions that are a part of standard FORTRAN and can thus be evaluated quickly. The logarithm appears in equation 61 because we are using a two dimensional space-charge model rather than a three dimensional one. Note that MARYLIE is fully three dimensional; our restriction to a two dimensional model arises because of the computing time that would be required to simulate tens of thousands of particles in order to achieve a reasonable resolution for a three dimensional model. Simulating even 100 particles using an N-body model required hours of CPU time, almost all of which were spent in the space-charge routine. Clearly further work in three dimensions will require the development of a much more efficient space-charge routine, probably a particle-in-cell (PIC) model.

4.2 Simulation Results

Numerous runs with the code were done to ensure that the results obtained were reasonable, i.e. runs with all combinations of (no damping - damping) and (space charge - no space charge). In this paper we will only present results that included all lattice effects, full symmetric damping, and full space charge.

We first consider two lattices that consisted of drifts and either 1) AG finite length quadrupoles, or 2) AG finite length octupoles. Element lengths, damping coefficients, and space-charge intensity were taken to be identical in both cases.

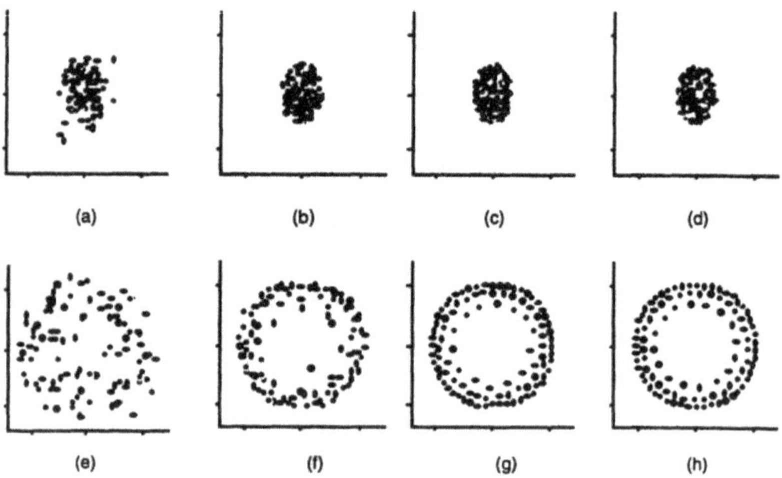

Figure 2. Simulation results for an AG quadrupole line (a-d) and an AG octupole line (e-h) with both cooling and space charge. The results are shown after 1000 periods (a and e), after 4000 periods (b and f), after 7000 periods (c and g), and after 10000 periods (d and h).

The results are shown in figure 2. Figure 2-h is reproduced at higher resolution in figure 3, where the (x, x') transverse phase space is also shown; note that the crystallization is essentially complete.

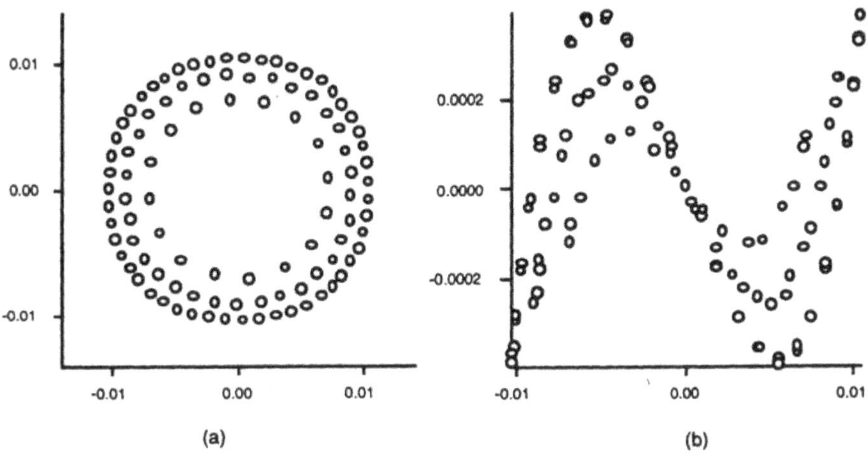

Figure 3. Simulation results for an AG octupole line with both cooling and space charge. Figure 3a is the same as figure 2h and figure 3b is the associated (x, x') phase space.

As one can see from figure 2, the quadrupole lattice results in a beam that is significantly smaller than the beam that results from an octupole latice. On the other hand, the beam in the octupole lattice has 'frozen' into a crystal configuration while the beam in the quadrupole lattice has not settled down even after 10 cooling times.

In fact, the beam in the quadrupole lattice cannot cool further because of the heating effect from numerical intrabeam scattering. Note that the intrabeam scattering present in the simulation is stronger than the real intrabeam scattering for the same charge density; this is because each simulation particle has the total charge of hundreds or thousands of real particles, i.e. the simulation beam is more discrete than the real beam. Nonetheless, the behavior of the quadrupole lattice beam in the simulation does represent qualitatively the behavior of a real beam, even though the real beam would encounter the intrabeam scattering limit at a lower temperature. The important point to note, however, is that the octupole lattice beam is not limited even by the artificially large numerical intrabeam scattering heating because it does not get smaller; the tendency to become slightly smaller in the octupole line is more than offset by the space-charge defocusing.

Though the intrabeam scattering limit for the quadrupole lattice beam occurs at too high a temperature for this simulation, this quadrupole lattice beam could not have become smaller than it is anyway because of the need to limit the tune shift in a real lattice. These simulations did not include errors so that in the simulation the temperature could have been lowered further in the absence of intrabeam scattering, but in a real lattice structure resonances would have driven the beam to the wall. Again, the important point to note is that the space-charge defocusing tune shift for the octupole lattice beam is strictly limited because the beam does not get smaller.

In general, a linear lattice must have linear external forces larger than the linear term from space-charge defocusing in order for the beam to be confined, i.e. the total Hamiltonian must have a quadratic term that is net focusing; as a result, for cool beams the quadratic terms will dominate and the space-charge tune shift and intrabeam scattering must increase as the beam is cooled. On the other hand, a nonlinear lattice will have a Hamiltonian with quadratic terms that are net defocusing (due to space charge), and as a result the beam cannot continue to shrink as it is cooled; confinement is due instead to the nonlinear terms. The simulations above dramatically confirm this picture.

Of course, one must worry about the effects of errors in the octupole lattice as well. As we remarked earlier, strength errors should be less of a problem with an octupole lattice than with a quadrupole lattice, not only beacuse the size is more insensitive to the strength than in a quadrupole lattice, but also because there are no significant linear resonances in an octupole lattice to cause losses; nonlinearly created resonances tend to be smaller and to cause less particle loss than linear resonances. However, one might worry about roll errors.

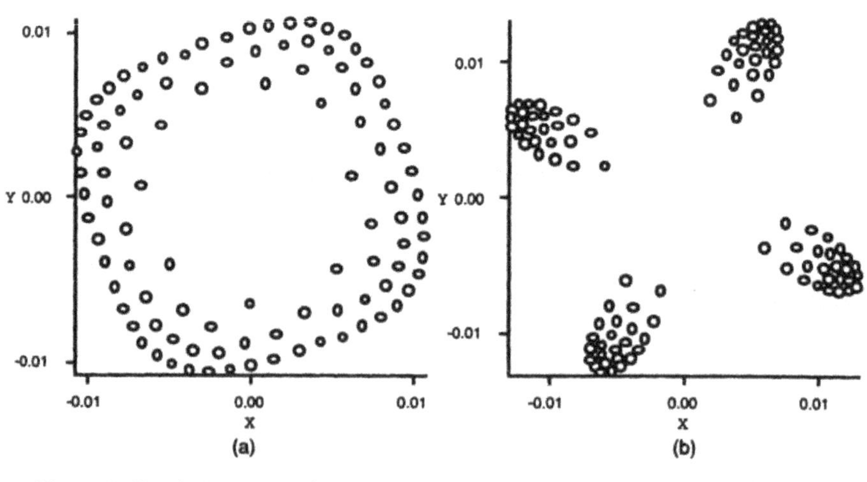

Figure 4. Simulation results for an AG octupole line with a 0.5° roll error, (a), and with a 3.0° roll error, (b).

In figure 4 we show the results of introducing roll errors of 0.5° and 3.0° in one octupole into the simulation above. As can be seen, in neither case does the error interfere with crystallization, though the larger error does dramatically affect the shape of the crystallized beam.

Up to now the lattices we have discussed have consisted of focusing elements and drifts; a realistic ring design must, of course, also include bends, i.e. dipoles.

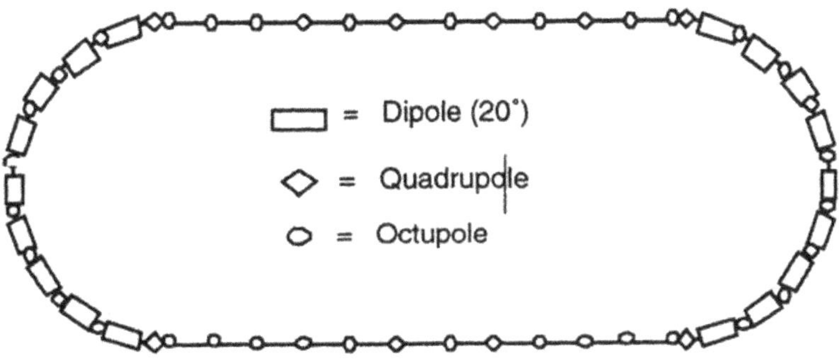

Figure 5. AG octupole ring. The lattice is a regular array of alternating octupoles except for the bend magnets and the quadrupoles at each end of the 180° bends that make the linear terms defocusing.

In the vertical direction a bend acts like a drift and thus does not interfere with the overall focusing properties of the lattice. However, in the horizontal (bend) plane a dipole generally acts like a focusing element and thus will give overall linear focusing which can make the beam shrink with lower temperatures and exacerbate the problems of space charge and intrabeam scattering. Note, though, that a 180° normal entry bend has a horizontal effect that is a negative identity, i.e. acts like there is no net focusing. As a result if we consider a lattice such as that shown in figure 5, where a 180° bend is perturbed only slightly by a periodic array of focusing elements in the bend and by the effects of finite dipole gaps, the behavior should be the same as found above; figure 6, which compares the results from the lattice of figure 5 with the results from a lattice with the octupoles replaced by quadrupoles, confirms this.

Note that the weak quadrupoles introduced at the ends of the 180° bend for the lattice of figure 5 were inserted to slightly decrease the vertical defocusing of the dipoles due to finite gaps and to introduce an equal amount of horizontal defocusing. Thus, the resulting lattice of figure 5 has linear tunes that are small and nearly equal, but imaginary; the net confinement is provided by the nonlinear elements, of course.

Though, as pointed out before, the space charge limit doesn't strictly apply to the AG octupole lattice because of the lack of linear driving resonances, let

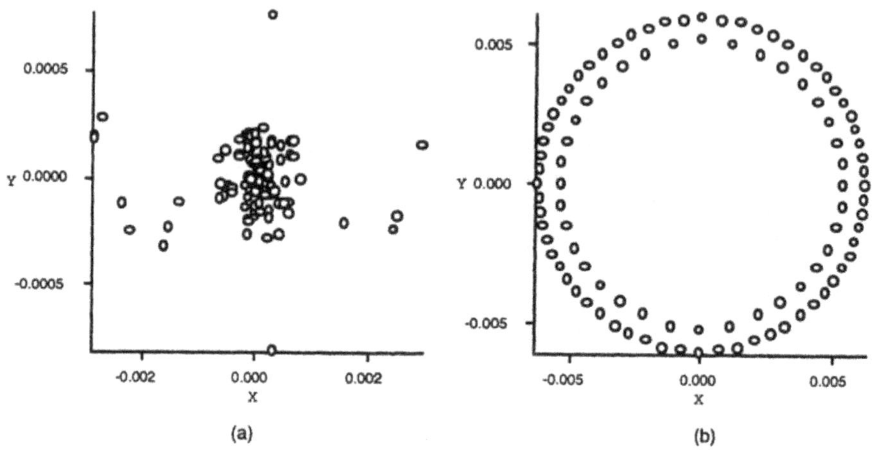

Figure 6. Simulation with cooling and space charge of a beam in the lattice of figure 5 with quadrupoles instead of octupoles (a) and with octupoles as in figure 5 (b). The simulation was run for 70 damping times; the quadrupole beam has clearly not cooled down.

us apply equation 15 to this lattice anyway, using the beam radius of figure 6 ($r_0 \approx 0.006$), a beam energy of 80 MeV, machine radius given by equation 42, a particle mass given by equation 46, and a machine tune given by equation 47; we find that the space charge limit in this case, 15, is

$$N < 8.0 * 10^{12}, \tag{62}$$

considerably larger than the limit found for the equivalent quadrupole ring even though the octupole ring beam is considerably colder than the quadrupole ring beam.

5 Summary and Discussion

We have shown that theory predicts a drastically different scaling of beam size with beam temperature for AG octupole lattices as compared with AG quadrupole lattices. As a result, the effects of the space-charge tune shift and of intrabeam scattering are greatly ameliorated; in fact, simulations show that these effects are negligible for AG octupole lattices and beams can completely crystallize at high intensity. We have also seen that AG octupole lattices are relatively insensitive to errors and that in a realistic ring the effects of dipoles need not interfere with crystallization.

By far, the most glaring omission in this paper is the lack of realistic three-dimensional space charge simulations of beams in AG octupole lattices. Though

3D space-charge effects cannot significantly change the transverse scaling of the beam size with beam temperature and thus cannot change the conclusion that space charge and intrabeam scattering effects will be strictly limited in an AG octupole lattice, 3D space charge might introduce enough coupling of the beam with the lattice, particularly in the bends, that the regular beam structure repeatedly 'breaks' and the beam fails to completely crystallize. Developing and benchmarking a 3D PIC code to accurately model these effects and investigate the extent of the crystallization is a task for future development.

References

[1] H. Bruck, *Accéléateurs Circulaires de Particules, Introduction a la Theorie*, Saclay, Institut National des Sciences et Techniques Nuclearies, 1966.

[2] P.J. Bryant, *Betatron Frequency Shifts Due to Self and Image Fields*, CERN Internal Report 88-10 (1988), pp 62-78.

[3] A. Dragt and D. Douglas, *MARYLIE, the Maryland Lie Algebraic Beam Transport and Tracking Program*, IEEE Trans. Nucl. Sci. **NS-30**, p.2442 (1983).

[4] C. Gardner, Y. Shoji, L. Ahrens, J.W. Glenn, Y.Y. Lee, T. Roser, A. Soukas, W. van Asselt, and W.T. Weng, *Observation and Correction of Resonance Stopbands in the AGS Booster*, Proc. of the 1993 Part. Acc. Conf., IEEE (1993), pp.3633- 3635.

[5] J.S. Hangst, M. Kristensen, J.S. Nielsen, O. Poulsen, J.P. Shiffer, and P. Shi, *Laser Cooling of a Stored Ion Beam to 1mK*, Phys. Rev. Lett., Vol 67, no. 10, (1991), pp 1238-1241.

[6] Jie Wei, Xiao-Ping Li, and Andrew M. Sessler, *Low-Energy States of Circulating Stored Ion Beams: Crystalline Beams*, Phys. Rev. Lett., Vol 73, no. 23, (1994), pp 3089-3092.

[7] P. Lochak and C. Meunier, *Multiphase Averaging for Classical Systems With Applications to Adiabatic Theorems*, Springer-Verlag, New York (1988).

[8] A.H. Sørensen, *Introduction to Intrabeam Scattering*, CERN Internal Report 88-10 (1988), pp 135-151.

ANOMALOUS BEHAVIOR OF SCHOTTKY NOISE FOR COOLED BEAMS

V. A. Lebedev
CEBAF, 12000 Jefferson Ave., Newport News, VA 23606, USA
and
J.S.Hangst, J.S.Nielsen, N.Madsen
Aarhus University, Aarhus C. 8000, Denmark

ABSTRACT

The properties of the thermal noise in a beam subjected to cooling in a storage ring are considered. It is shown that a non-Gaussian tail in the beam distribution function creates anomalous behavior of the noise. The effect is important for both laser and electron coolings but it is much more powerful for laser cooling where the noise of a cooled beam is significantly amplified. Good agreement between the theoretical predictions and experimental results obtained at the ASTRID storage ring has been found. The possibility of extracting dynamical information from the Schottky spectra is also discussed.

1. Introduction

The study of beam thermal noise (Schottky noise) is one of the most commonly used methods for measuring properties of a continuous beam. For uncooled (hot) beam the noise spectral power is directly connected with distribution function but it becomes distorted by the particle interaction with beam temperature decrease what makes difficult to extract distribution function from Schottky spectra. In this article we will analyze the case of longitudinal particle motion, where the fluctuations of the total number of particles inside a pick-up electrode are measured. The signal from a single particle has the form of a short pulse that is repeated at the rotation frequency; thus, the spectral density of the beam noise consists of narrow peaks located at harmonics of the revolution frequency.

If the beam density is small or if the beam has sufficiently large momentum spread, so that the collective beam interaction is small enough, one can consider the motion of separate particles to be independent. Then the noise spectral density around each harmonic, $S_n(\delta\omega_n)$, is directly related to the particle distribution function in revolution frequency[1,2]

$$S_n(\delta\omega_n) = \frac{N}{n} f_\omega(\delta\omega_n / n), \qquad \delta\omega_n = \omega - n\omega_0, \tag{1}$$

and one can easily measure the momentum distribution function and calculate the rms momentum spread in the beam. Here $f_\omega(\delta\omega)$ is the particle distribution function in revolution frequency normalized so that $\int f_\omega(\delta\omega)d\delta\omega=1$, N is the number of particles in the beam and n is the harmonic number. One can see from Eq.1 that the integral of the spectral density around each peak is equal to the number of particles and does not depend on other parameters.

In a dense and cold beam the motion of each particle is strongly influenced by the motion of other particles; thus the noise spectral density should be determined by the particle interactions. The general theory of such fluctuations was developed in plasma physics[3] and successfully applied to accelerators in Ref.2, which we will follow here. For the most frequently encountered case, when the beam energy is below the transition energy and the main contribution to the ring impedance is from Coulomb repulsion, the noise spectral density should be strongly suppressed by particle interactions. In this case the spectrum around each harmonic usually consists of two peaks located symmetrically around this harmonic. These peaks are associated with charge density waves propagating up and down stream along the beam. It was shown[2] that in the thermodynamic equilibrium, when the velocity distribution function is Gaussian, the integral of the spectral density around each harmonic is equal to

$$\left\langle A_n^{\,2}\right\rangle = \int\limits_{\substack{around\\harmonic}} S_n(\omega)d\omega = N\frac{\overline{\delta\omega^{\,2}}}{\overline{\delta\omega^{\,2}}+\Omega^2}. \tag{2}$$

Here $\overline{\delta\omega^{\,2}}$ is the rms spread of the particle revolution frequencies, Ω is the so-called coherent shift and we use the same normalization of the longitudinal beam density as in Ref.2,

$$\rho(\theta,t)=\frac{1}{2\pi}\sum_{n=-\infty}^{n}A_n(t)e^{in(\omega_0 t-\theta)}, \qquad \int_0^{2\pi}\rho(\theta,t)d\theta = N . \tag{3}$$

For purely capacitive impedance (Coulomb particle repulsion) Ω is equal to the frequency in the beam frame at which these coherent waves propagate around the ring. One can see that for a cold beam, when $\overline{\delta\omega^{\,2}} < \Omega^2$, the fluctuations are strongly suppressed by the particle interaction.

This noise suppression was studied theoretically and experimentally for the first time in Novosibirsk for an electron cooled beam[2,4]. Good agreement between theory and experiment was found. But recently, experiments with laser cooling have exhibited a very strong discrepancy between the developed theory and experimental results[5,6]. In particular, it was found that after switching on the cooling the noise spectral density, instead of decreasing, increased by two orders of magnitude. As will be seen below, the main reason of this effect is that the distribution function in laser cooling experiments is far away from the Gaussian one. For electron cooling the approximation of Gaussian distribution function is more accurate but very often is not fully adequate. Although the agreement between theory and experiment was found in electron cooling experiments there was not another method allowing to measure the momentum spread and verifying the obtained results, Thus, this agreement should be rather considered as qualitative than quantitative. Laser cooling allows to measure the longitudinal distribution function directly, what, consequently, allows to do more accurate test of the theory. This article is devoted to an analysis of the effect of non-Gaussian distribution on the Schottky noise for both laser and electron cooling.

2. Review of the Theory

We will follow Ref.2 modified only, as necessary, to correct obsolete definitions and minor inconsistencies. There it was shown that the noise spectral density around the n-th harmonic of the revolution frequency is equal to

$$S_n(\delta\omega_n) = \frac{N}{n}\frac{f_\omega(\delta\omega_n/n)}{\left|\varepsilon_n(\delta\omega_n)^2\right|}, \qquad \delta\omega_n = \omega - n\omega_0, \tag{4}$$

where $\varepsilon_n(\omega)$ is the analog of the plasma dielectric permitivity. For the case of a beam without collisions it can be expressed through the integral of the distribution function

$$\varepsilon_n(\omega) = 1 + \frac{\Omega_n^{\ 2}}{n}\int_{-\infty}^{\infty}\frac{1}{\omega - n\omega' - i0}\frac{\partial f_\omega(\omega')}{\partial\omega'}d\omega' . \tag{5}$$

The contour used for the integral includes the real axis and is deformed around the pole at $\omega'=\omega/n$ above the real axis. Here the coherent shift for n-th harmonic is determined by

$$\Omega_n^{\ 2} = i\frac{NZ^2e^2\omega_0}{4\pi^2\overline{R}}\frac{d\omega}{dp}nZ_n, \qquad \frac{d\omega}{dp} = \frac{\omega_0}{p_0}\left(\frac{1-\alpha\gamma^2}{\gamma^2}\right), \tag{6}$$

where N is number of particles in the beam and eZ is their charge, $\overline{R}$ is the average ring radius, α is the momentum compaction factor, and Z_n is the longitudinal ring impedance for harmonic n. The coherent shift Ω_n is the characteristic frequency describing the time evolution of a density perturbation having wavelength $2\pi\overline{R}/n$. In general, the impedance Z_n has both real (resistive) and imaginary (capacitive or inductive) parts.

One can see that in the case of a hot and weak beam the beam permitivity is close to 1, $\varepsilon \approx 1$, and Eq.(4) coincides with Eq.(1). It should be noted that Eq.(4) can be used only if a beam is stable, i.e., all roots of the dispersion equation,

$$\varepsilon_n(\omega) = 0, \tag{7}$$

have positive imaginary parts.

Usually, for low energy beams, Coulomb repulsion produces the main contribution to the storage ring impedance. In this case the impedance is purely capacitive and is given by[7]

$$\frac{Z_n}{n} = \frac{4\pi}{c\beta\gamma^2 i}\left(\ln\frac{r_c}{r_b} + \frac{1}{2}\right) = \frac{Z_0}{\beta\gamma^2 i}\left(\ln\frac{r_c}{r_b} + \frac{1}{2}\right), \tag{8}$$

where β and γ are the relativistic factors, c is the velocity of light, $Z_0=4\pi/c$ is the impedance of the free space ($Z_0\approx377\ \Omega$), and r_c and r_b are the radii of the vacuum chamber and the beam, respectively. Thus, for beams below transition energy, $\alpha\gamma^2 < 1$, the square of the coherent shift becomes real and positive

$$\Omega^2 \equiv \frac{\Omega_n^{\ 2}}{n^2} = \frac{NZ^2e^2(1-\alpha\gamma^2)}{\pi M\overline{R}^3\gamma^5}\left(\ln\frac{r_c}{r_b} + \frac{1}{2}\right). \tag{9}$$

For a Gaussian distribution function

$$f_\omega(\omega) = \frac{1}{\sqrt{2\pi\overline{\delta\omega^2}}}\exp\left(-\frac{\omega^2}{2\overline{\delta\omega^2}}\right) \tag{10}$$

the integral (5) can be reduced to the following expression[8]

$$\varepsilon_n(\omega) = 1 + \frac{\Omega_n^2}{n^2\overline{\delta\omega^2}}\left(1 - 2ye^{-y^2}\int_0^y e^{x^2}dx - i\sqrt{\pi}\,ye^{-y^2}\right), \qquad y = \frac{\omega}{n\sqrt{2\overline{\delta\omega^2}}}. \tag{11}$$

The results of numerical calculations for a fixed number of particles and different revolution frequency spreads (longitudinal temperatures) are shown in Figure 1. One can see that with decreasing beam temperature, two peaks, corresponding to two waves, appear (only the right-hand part of the spectrum are shown in Figure 1). The position of the peaks has a rather weak dependence on the beam temperature. With decreasing temperature the peaks become sharper and approach the positions determined by the value of the coherent shift.

Although the described above theory has been used at many installations there was not a direct prove that it allows to extract a correct value of longitudinal momentum spread for cold beam. The first direct test, when longitudinal momentum spread of a cold beam (but with cooling switched off) was independently determined by a technique another than Schottky noise, was done at ASTRID[5,6]. The Schottky spectra, which were strongly distorted by particle interactions, were analyzed using Eq.11 above. The extracted velocity spreads were then compared to direct measurements of the velocity distribution, carried out using a laser induced fluorescence (LIF) technique[5]. There was found a good agreement of both measurements for $n\sqrt{\overline{\delta\omega^2}}/\Omega_n \geq 0.7$. These measurements confirm the validity of the model described above for Gaussian velocity distribution. Note, that the beam temperature in these experiments was not sufficiently small to observe the effect of non-Gaussian tails discussed below.

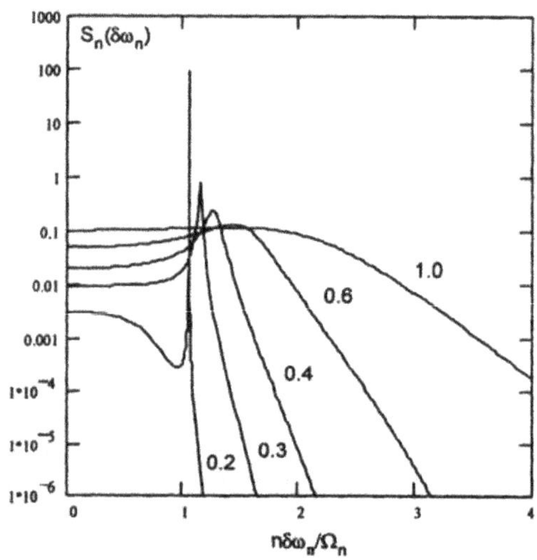

Figure 1. Dependence of the spectral density on frequency for different particle momentum spreads. The numbers on the plot are values of $n\sqrt{\overline{\delta\omega^2}}/\Omega_n$

3. Experimental Evidence of Anomalous Behavior of Schottky Noise in a Laser Cooled Beam

The analysis given below is based on the experimental results obtained in the laser cooling experiments[5,6] carried out at the ASTRID storage ring. The relevant parameters for the cooling experiments are presented in Table I. All measurements of the Schottky noise were done at the 23-rd revolution frequency harmonic of about 514 kHz.

Table I. ASTRID storage ring parameters

Ion kinetic energy	100 keV
Ions	$^{24}Mg^{+}$
Beam current	2-4 μA
Number of particles	$(5.6\text{-}11.2)\cdot 10^{8}$
Circumference	40 m
Revolution frequency	22.3 kHz
Betatron tunes, ν_x/ν_z	2.21/2.64
Momentum compaction factor, α	0.07
Laser wave lengths,co-/counter-propagating	≈278.8/280.4 nm
Laser power	20-40 mW

To understand the difference between the behavior of Schottky noise in laser and electron cooling, we need to consider the process of laser cooling in more detail. The cooling force is created by two counter-propagating laser beams[5]. Because the natural width of the absorption line is usually very small compared to the laser detuning, the cooling force usually consists of two peaks associated with two lasers. The distance between the peaks is determined by the detuning of the laser frequencies. An example of such a cooling force is shown in Figure 2.

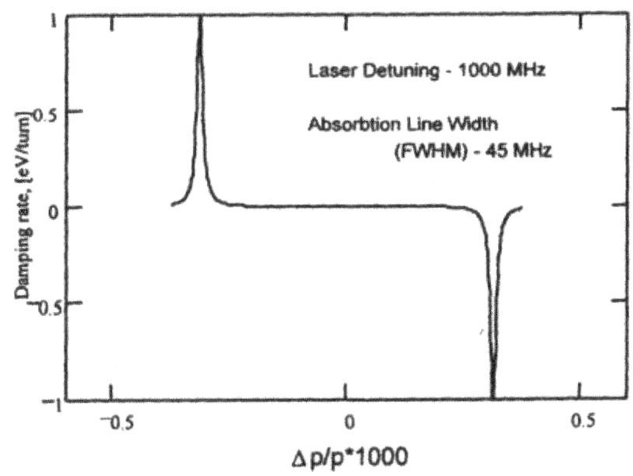

Figure 2. The calculated friction force for laser detunings of 1000 MHz.

To cool the beam, the laser detuning is initially set so that the force peaks are located outside of particle distribution function. Then as the lasers are scanned in frequency the particles are pushed by the cooling force to the center of the distribution function. During scanning, such a cooling process produces a particle distribution function which has very sharp edges and is distinct from the Gaussian one in that it has virtually no tails. Examples of experimental measurements of particle distribution functions and Schottky spectra at different stages of cooling are shown in Figure 3. As one can see from Figure 3, the behavior of Schottky noise in a laser cooled beam is significantly different than described in the above theory. Although, as expected, two peaks appear in the spectra during beam cooling, there are three important distinctions. First, just after switching on the beam cooling, the integral of the spectral density around the 23-rd revolution harmonic, $\left\langle A_n^{\,2}\right\rangle\Big|_{n=23}$, increases rapidly with decreasing momentum spread and reaches a value higher by two orders of magnitude than that observed without cooling (Figure 4). As scanning proceeds, its value decreases and has roughly the same value as without cooling at the very end of the process, when the beam momentum spread is more than 10 times smaller. Secondly, the splitting between peaks depends strongly on the momentum spread (Figure 5), while from the theory for a Gaussian beam this dependence is expected to be rather weak. Finally, the heights of the two peaks are widely different. Figures 3-5 demonstrate these described distinctions.

The non-Gaussian tails in the distribution function is the clue to the understanding of the strange behavior of Schottky signal described above. As will be seen, the reason for the great amplification of the noise is the disappearance of Landau damping.

4. Beam Permitivity without Friction

In order to simplify calculations we will first neglect the cooling force. This is justified by the fact that most of particles experience virtually no force when the scan of lasers is stopped. The particles have been just pushed by the cooling force towards the center of the distribution function, where they stay without friction. Their velocities will not change in the absence of diffusion. To estimate the effect of the drag force we can compare the damping decrement at the center of the distribution function (zero relative velocity), $\lambda = \dfrac{1}{M}\dfrac{\partial F}{\partial v}\bigg|_{v=0}$,

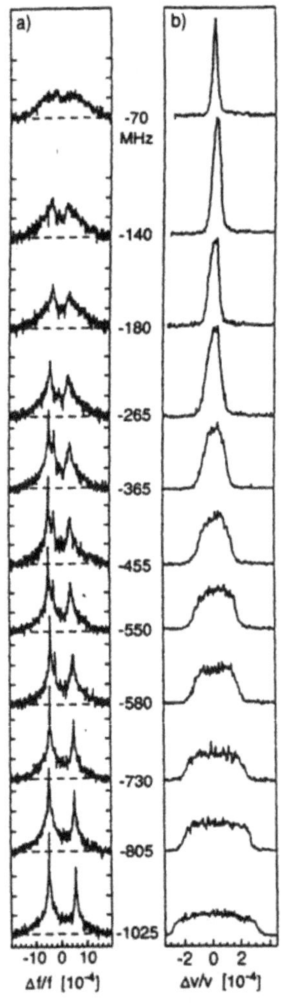

Figure 3. a) Schottky spectra measured at various detunings (shown between figures) during laser cooling. Dashed lines indicate the noise floor of the measurement; each vertical tick mark is one decade in relative power (10 dB). b) The corresponding velocity distribution, measured by laser induced fluorescence. Each spectrum contains about 10^3 fluorescence counts.

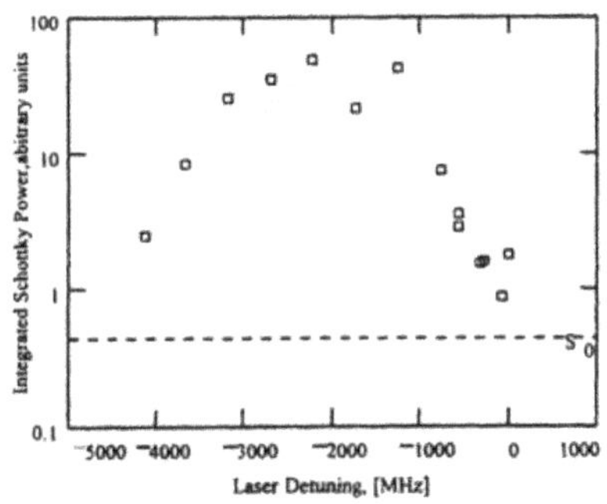

Figure 4. Dependence of the integral of spectral density around the 23-rd harmonic on the laser detuning. Dotted line shows value without laser cooling.

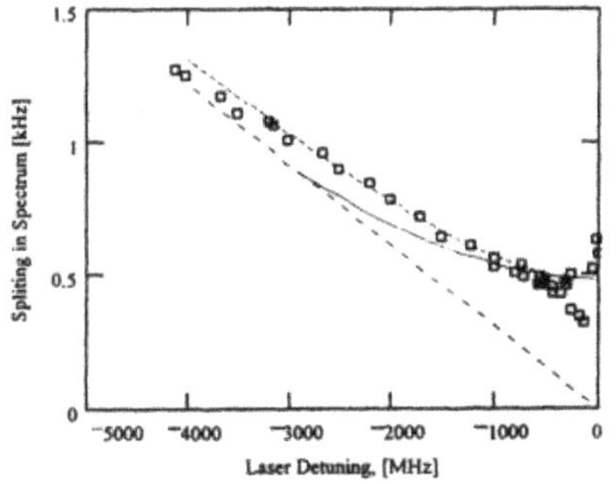

Figure 5. Dependence of splitting between the peaks in the noise spectra on the detuning of the lasers. The dashed line shows the value of the laser detuning expressed in terms of the shift of the revolution harmonic which was used for measurements. Dotted and solid lines show the theory predictions for the rectangular distribution function and for the distribution function described by Eq.(12), respectively.

and the coherent shift. The results of such a comparison for the parameters of the experiment are shown in Figure 6. One can see that when the detuning of the lasers is more than ~500 MHz, the decrement is at least two orders of magnitude smaller than the coherent shift, and the cooling force can be neglected in the first approximation.

Thus we consider a model where we neglect the friction force and use a simple expression for the distribution function which should approximate the results of the measurements better than the Gaussian distribution. To calculate the beam permitivity we will use the following normalized distribution function

$$f_{\omega}(\omega)=\begin{cases}\dfrac{45}{64\omega_b}\left(1-2\left(\dfrac{\omega}{\omega_b}\right)^4+\left(\dfrac{\omega}{\omega_b}\right)^8\right) & , \ |\omega|\le\omega_b, \\ 0 & , \ |\omega|\ge\omega_b. \end{cases} \tag{12}$$

Here $2\omega_b$ is equal to the full width of the distribution function. This function has much steeper edges than a Gaussian and represents a much better approximation to the results of the measurements. Figure 7 shows fits of this function and the Gaussian distribution to the experimentally measured distribution for a laser detuning of 1025 MHz. As one can see, for this particular measurement there is very good agreement with the suggested functional shape, although for most cases the measured distribution functions have even steeper edges and the agreement is not quite as good. The vertical lines in the Figure show positions where particles are in resonance with the lasers. Substituting Eq.12 into

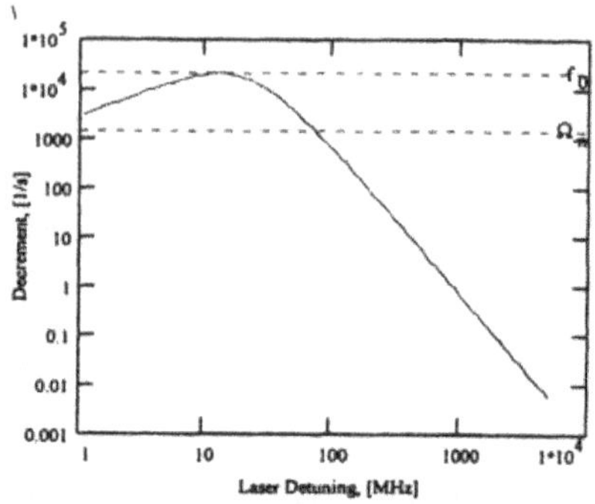

Figure 6. Dependence of the damping decrement at zero relative velocity on the detuning of lasers for the parameters of the ASTRID storage ring. The revolution frequency and the coherent shift for the 23-rd harmonic are shown by dashed lines.

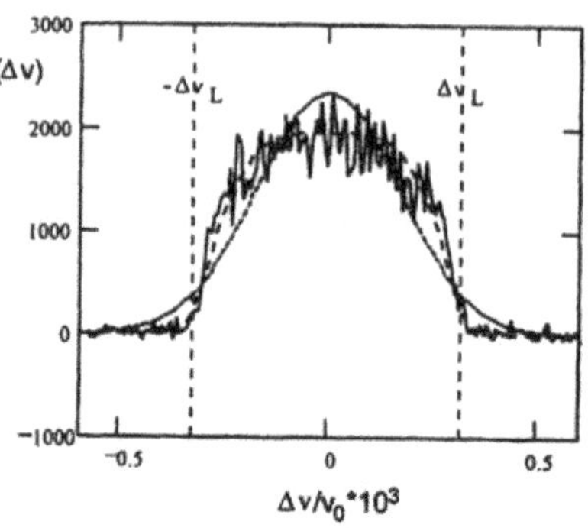

Figure 7. The measured velocity distribution function for laser detuning of 1025 MHz. Also shown are fits to this function: Gaussian distribution (dashed line) and the distribution described in Eq.(12) (dotted line). Vertical dotted lines show the velocities where ions are in resonance with laser radiation. Relative velocity and revolution spreads are related by $\delta\omega_n/(n\omega_0) = (1-\alpha)\Delta v/v_0$.

Eq.5 and performing the calculation, we finally find that the permitivity is given by

$$\varepsilon_n(\omega) = 1 + \frac{45\Omega_n^{\,2}}{8n^2\omega_b^{\,2}}\left(\frac{8}{21} + \frac{8}{5}y^2 - \frac{2}{3}y^4 - 2y^6 - y^3(y^4-1)\ln\frac{y-1}{y+1}\right), \quad y = \frac{\omega}{n\omega_b}. \quad (13)$$

For $\omega_b \le 2.13\Omega_n/n$ the equation, $\mathrm{Re}(\varepsilon_n(\omega_r)) = 0$ has two real positive roots (see Figure 8). The solution for the larger root can be approximated as

$$\omega_r = n\omega_b\left(\frac{\Omega_n}{n\omega_b} + .282\left(\frac{n\omega_b}{\Omega_n}\right)^{.8}\right), \quad n\omega_b \le 2.13\Omega_n \,. \quad (14)$$

to an accuracy of about 1%. Taking into account that, for $n\omega_b \le 1.87\Omega_n$, the imaginary part of the permitivity is also equal to zero at these points, we find that ε becomes zero too, *i.e.*, the dispersion equation, $\varepsilon_n(\omega_r) = 0$, has real roots for $n\omega_b \le 1.87\Omega_n$. These frequencies, where ε becomes zero, determine the positions of peaks in the spectra. The solid line in Figure 5 shows the splitting between peaks calculated from Eq.14. As one can see the peak positions approach the coherent shift only for the case when $n\omega_b \ll \Omega_n$. To test how strongly this result depends on a particular form of the distribution function we also calculated the beam permitivity for a rectangular distribution function[1]. In this case one can get that

[1]The use of this function is not fully correct for frequencies near the distribution function edge where ε goes to infinity. We can neglect this in our consideration because the frequencies in which we are interested are well outside of this point.

$$\varepsilon_n(\omega) = 1 - \frac{\Omega_n^2}{\omega^2 - (n\omega_b)^2}, \quad \omega_r = \sqrt{\Omega_n^2 + (n\omega_b)^2}\,. \tag{14a}$$

The results are shown by the dotted line in Figure 5. As one can see the behavior of the curve is in better agreement with the experimental results for detuning larger than 3 GHz

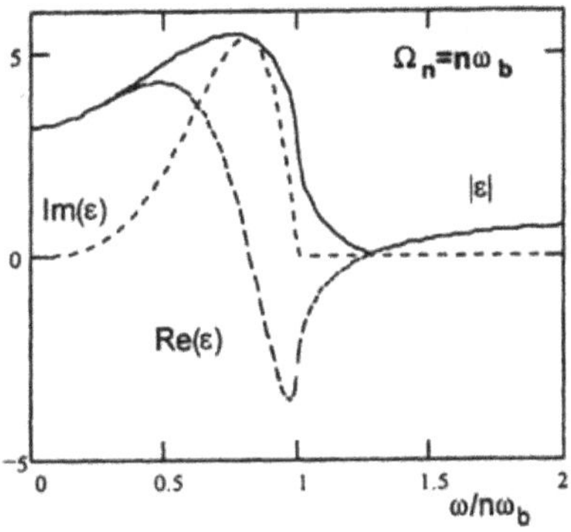

Figure 8. Dependence of beam permitivity on frequency for the distribution described in Eq.(12), $\Omega_n/n\omega_b$=1.

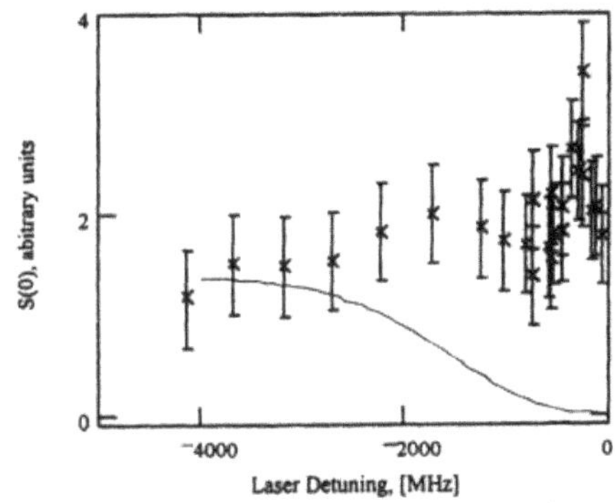

Figure 9. Dependence of the noise spectral density at zero relative frequency on laser detuning.

where Eq.14 predicts the disappearance of real roots for dispersion equation but, apparently, the approximation of rectangular distribution is more adequate for large detuning. Thus, we can conclude that, in contrast to a Gaussian distribution function, the distribution functions with steep edges reproduce the dependence of the splitting between peaks which has been observed in experiments.

At zero relative frequency we can neglect the influence of the tails. According to Eq.13 the beam permitivity at zero frequency is

$$\varepsilon(0) \approx 1 + \frac{45}{21}\frac{\Omega_n^2}{n^2\omega_b^2}\,. \tag{15}$$

Then the noise spectral density at zero frequency is equal to

$$S(0) \approx \frac{45N}{64n\omega_b}\frac{1}{\left(1+\dfrac{45}{21}\dfrac{\Omega_n^2}{n^2\omega_b^2}\right)^2}\,. \tag{16}$$

A comparison of the measured data and the predictions of Eq.16 is shown in Figure 9. Taking into account that, the height of the peaks changes by more than 1000 times over the range of laser detunings, one can appreciate that there is good agreement between theory and experiment, at least for large laser detuning (more than 1000 MHz) where we can neglect the cooling force. As can be seen below, the finite resolution of measurements also contributes to an increase of the spectral density at zero frequency because of large heights of the peaks in the spectra.

5. Influence of Tails on Schottky spectra

Although we obtain the right prediction for the splitting between the zeros of the permitivity, the model does not yet predict the existence of peaks in the noise spectrum, since the distribution function is equal to zero at the frequencies corresponding to the zeros of $\varepsilon_n(\omega)$. In reality, there must be small tails in the distribution function which will supply particles to these regions. Although there is a small number of particles, the noise spectral power becomes very large because ε is very small.

We can consider that the particle distribution function consists of two parts: the functions for the central part and for the tails,

$$f_s = f + \delta f. \tag{17}$$

Because only a small number of particles are located in the tails, one can neglect their contribution to ε for all frequencies except when ε becomes zero. Here one can also neglect the contribution of the tails to the real part of ε because it slightly changes the positions of the peaks but does not change their heights. Thus we can let

$$\varepsilon_n(\omega) = 1 + \frac{\Omega_n^{\,2}}{n} \int_{-\infty}^{\infty} \frac{\partial(f + \delta f)/\partial\omega'}{\omega - n\omega' - i\Delta} d\omega' \approx \varepsilon_{0n}(\omega) + \pi i \frac{\Omega_n^{\,2}}{n^2} \left.\frac{\partial \delta f}{\partial \omega}\right|_{\omega' = \omega/n} . \tag{18}$$

where $\varepsilon_{0n}(\omega)$ is the beam permitivity without the tails. Let ω_r be the real root of the dispersion equation, $\varepsilon_{0n}(\omega_r) = 0$, then we can expand ε in a Taylor series in the vicinity of ω_r.

$$\varepsilon_n(\omega) \approx \left.\frac{\partial \varepsilon_{0n}}{\partial \omega}\right|_{\omega=\omega_r} \cdot (\omega - \omega_r) + \pi i \frac{\Omega_n^{\,2}}{n^2} \left.\frac{\partial \delta f}{\partial \omega}\right|_{\omega = \omega_r/n} . \tag{19}$$

If $\left.\partial \delta f / \partial\omega\right|_{\omega=\omega_r} < 0$, then the system is stable and we can use Eq.4. Substituting Eq.19 into Eq.4 we get an equation describing the spectral density near the peak

$$S_{peak_n}(\delta\omega) = \frac{N}{n} \frac{\delta f(\omega_r / n)}{\left(\left.\frac{\partial \varepsilon_{0n}}{\partial \omega}\right|_{\omega=\omega_r} \delta\omega \right)^2 + \left(\pi \frac{\Omega_n^{\,2}}{n^2} \left.\frac{\partial \delta f}{\partial \omega}\right|_{\omega=\omega_r/n} \right)^2}, \quad \delta\omega = \omega - n\omega_0 \pm \omega_r . \tag{20}$$

For a cold beam, $\omega_b << \Omega_n$, these peaks produce the main contribution to the harmonic amplitude. Then, to find the total integral around n-th harmonic we use,

$$\left\langle A_n^{\,2} \right\rangle \approx 2 \int_{-\infty}^{\infty} S_{peak_n}(\omega) d\omega = \frac{2Nn\delta f(\omega_r / n)}{\Omega_n^{\,2} \left(\frac{\partial \varepsilon_{0n}}{\partial \omega} \right)^2_{\omega=\omega_r} \left(\frac{\partial \delta f}{\partial \omega} \right)_{\omega=\omega_r/n}} . \tag{21}$$

The factor of 2 appears in Eq.21 because there are two peaks in the spectrum. One can see that, if the derivative of the distribution function in the tails goes to zero, the Landau damping disappears and the spectral density goes to infinity. For a positive derivative the system becomes unstable.

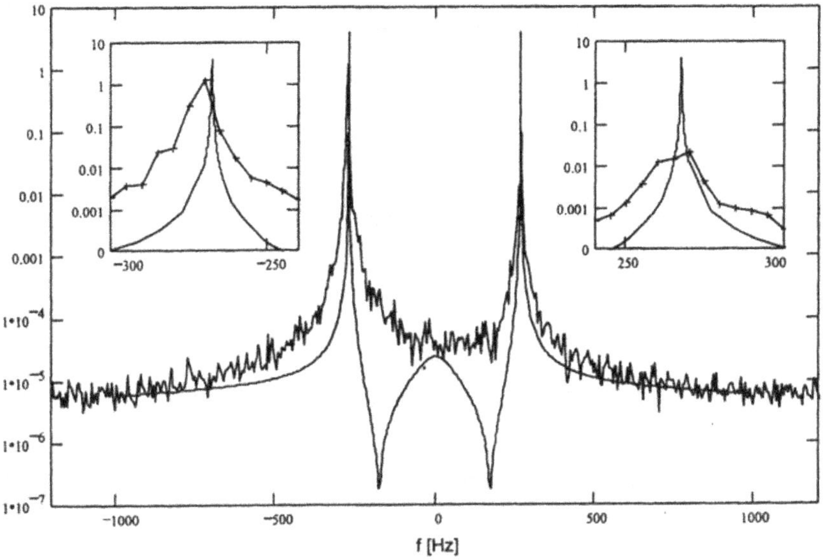

Figure 10. Comparison of the experimentally measured Schottky spectrum for laser detuning of 1025 MHz and the theory prediction. Insets show the peaks on a smaller scale.

Figure 10 demonstrates the calculated and measured spectra. The distribution function for calculation is linearly combined from two parts: the core and the tails. The core is described by Eq.(12) with $\omega_b/2\pi$=175 Hz (see Figure 7). The tails include 15% of the particles and are described by Gaussian distribution with temperature equal to the transverse beam temperature, so that $\sqrt{\overline{\delta v^2}}\,/\,v_0 = 2.7\cdot 10^{-3}$ ($n\sqrt{\overline{\delta\omega^2}}$ = 1340 Hz). A plot of the corresponding distribution function is shown in Figure 11. The friction force was neglected in the theory. One can see that predicted width of the peaks is smaller than experimentally measured. It can be related to effects of both: the friction force and the final resolution of the Fast Fourie Transformation (FFT). Numerical experiments carried out on the FFT of an ideal sinusoid exhibit almost identical behavior with the measured spectral density in the peak. Thus, the peak width should be first related to the FFT resolution. The noise

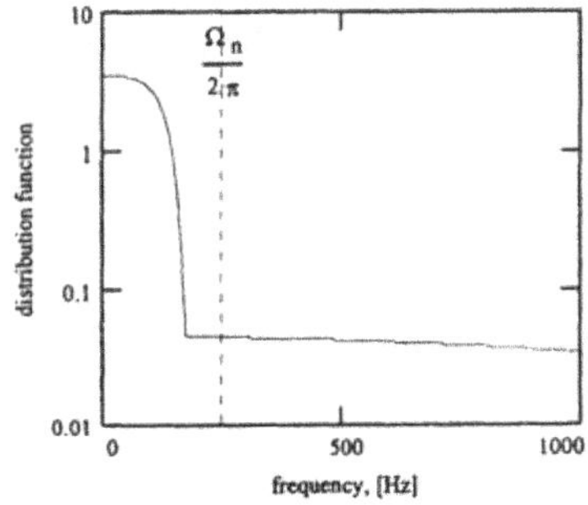

Figure 11. Distribution function for spectrum of Figure 10.

floor is determined by noises of electronic and also limits an accuracy of the measurements.

6. Diffusion and the Origination of the Tails

From the general point of view a beam in a storage ring cannot be not in thermodynamic equilibrium because of its motion. It means that the distribution function is not Gaussian and the approximation of the Gaussian distribution has to be used precautiosly. There are two main mechanisms which produce non-Gaussian tails. They are intrabeam scattering and collisions with residual gas. For intense beam the first usually produces the main contribution and we will consider it here in more details. In this case a decrease of the transverse temperature is usually limited by non-linear resonance due to Coulomb tune shift (generally, the tune shift does not exceed 0.1-0.3) and transverse temperature is much larger than the longitudinal one. Then, a single (Touschek) scattering can change particle velocity by amount much larger than the longitudinal temperature what creates non-Gaussian tails.

To calculate the longitudinal distribution function we will consider a single dimension model. It is justified by the fact that the transverse temperature is stabilized by non-linear resonance and is much larger than the longitudinal one. Then, the dynamic of the system can be described by the generalized Vlasov equation for distribution function

$$\frac{\partial f}{\partial t}+\frac{\partial}{\partial p}\left(\left(F_{beam}(p)+F_{cool}(p)\right)f\right)=\frac{1}{2}\frac{\partial}{\partial p}\left(D(p)\frac{\partial f}{\partial p}\right)+\int_{-\infty}^{\infty}W(p-p')f(p')dp' \,. \qquad (22)$$

Here $F_{cool}(p)$ and $F_{beam}(p)$ are the friction forces acting on particle due to cooling (laser or electron) and due to friction in the cooled beam, $D(p)$ is the diffusion and $W(p)$ is the probability for particle to be scattered with a longitudinal momentum change of p. Strictly speaking, $F_{beam}(p)$ depends on distribution function but for the pan-cake distribution this friction force can be calculated with good accuracy everywhere except a region of small velocities which is not important for calculations of the distribution function in tails.

For round beam the following expression can be used to calculate the number of particles scattered from the beam core with longitudinal velocity more than v (Touschek lifetime)[10]:

$$\frac{1}{N}\frac{dN}{dt}=\frac{e^4 N}{4\sqrt{\pi}M^2 v^2 \bar{v}_x \bar{R}\sigma^2}\left[\frac{\sqrt{\pi}}{2}(1+2\Delta^2)\left(1-\frac{2}{\sqrt{\pi}}\int_0^{\Delta}e^{-x^2}dx\right)-\Delta e^{-\Delta^2}\right], \qquad (23)$$

Here N is the number of particles in the beam, e and M are the charge and mass of the ion, $\bar{v}_x$ is the rms transverse velocity, σ is the rms beam radius, $\bar{R}$ is the mean radius of the storage ring and $\Delta = v/\bar{v}_x$. Differentiating Eq.23 with respect to v, one can calculate the probability

$$W(p) = \frac{1}{2}\frac{\partial}{\partial p}\left(\frac{1}{N}\frac{dN}{dt}\right), \tag{24}$$

Figure 12 shows the results of numerical solution of Eq.22 for parameters of ASTRID storage ring. Initially narrow distribution without tails was placed between peaks of laser cooling force. One can see that after 0.1 s a large fraction of the particles is already outside of cooling region. The simulation also showed that although particles leave the cooling region they are not lost and the tails shape are determined by balance of the diffusion and the friction force originated from beam itself. So that, the length of the tail is about transverse velocity of the ions. Experimental measurements of the distribution function on ASTRID show good coincidence with numerical simulations. Simulation also showed that with increase of the cooling force the derivative of the distribution function changes its sign outside of cooling region (see Figure 13). It is associated with stronger attraction of the scattered particles to the peaks of the friction force. It overwhelms diffusion which

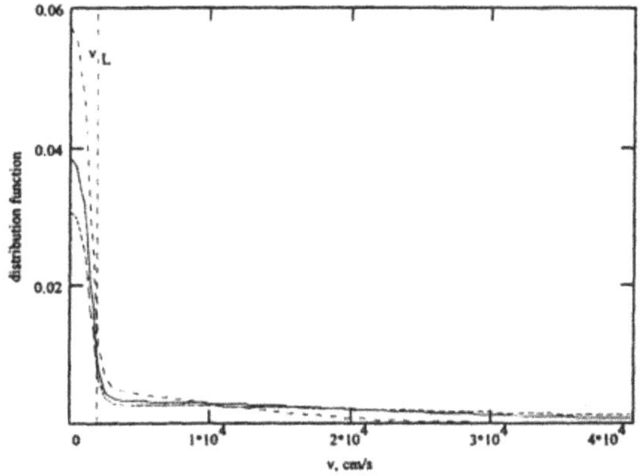

Figure 12. Dependence of the distribution function on velocity; laser detuning of 70 MHz, power of lasers of 20 mW, beam current of 2 μA, beam emittance of $8.7 \cdot 10^{-4}$ cm; dashed, solid and dotted lines correspond to the time after injection of 0.1, 0.5 and 1 s respectively; vertical dotted line shows a position of the peak of laser cooling force.

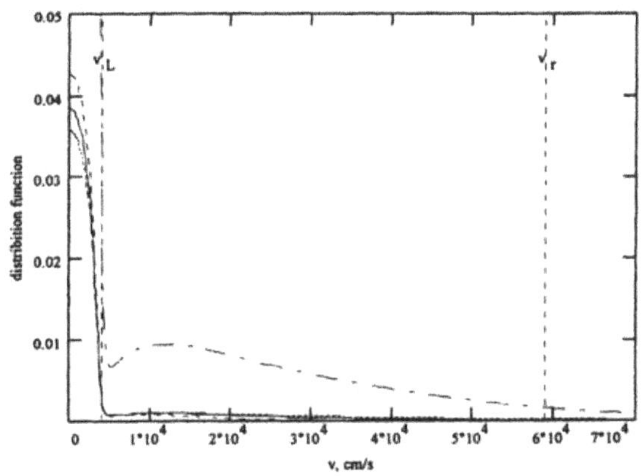

Figure 13. Dependence of the distribution function on velocity; laser detuning of 150 MHz, power of lasers of 36 mW, other parameters are the same as in Figure 10; dashed, solid and dotted lines correspond to the time after injection of 0.1, 0.5 and 1 s, respectively; the dash-dotted line is the same as dotted line (1 s) but its vertical scale is increased in 10 times to see tails.

smoothened the distribution function for smaller laser power. As it was shown in previous sections this change of the derivative produces longitudinal instability which amplifies Schottky noise. This small or even negative derivative explains the strongest amplification of the noise for the case when velocity of the wave propagating in a beam, $v_r = R\Omega_n/n$, is in the range $\approx(1\text{-}1.5)v_L$, where v_L is the velocity when particle is in resonance with laser radiation ($v_L \approx v_b$).

Although the effect of the tails does not so clearly manifest itself in electron cooling it also distorts the Schottky spectra. It becomes very important for deeply cooled beam when the particle population in tails becomes much larger than it should be for Gaussian distribution. Figure 14 shows the distribution function when the rms longitudinal velocity spread is 10 times smaller than the transverse one and is mainly determined by intrabeam scattering. One can see that for velocity of about $3\sigma_v$ tail starts to be different from the Gaussian distribution and for 5 σ_v it is larger more then 2 orders of magnitude. The linear dependence of the friction force on velocity was used in this simulations what is justified by the small longitudinal velocity spread. The effect will be stronger if one takes into account decreasing friction force with increasing velocity.

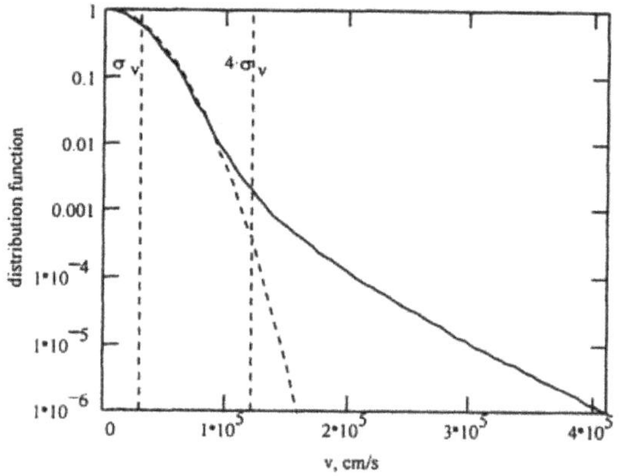

Figure 14. Longitudinal distribution function for electron cooling - solid line and the idealized Gaussian distribution - dashed line. Longitudinal rms velocity spread is 10 times smaller than the transverse one and is mainly determined by intrabeam scattering.

7. Influence of the Friction force on the Schottky spectrum

The cooling force changes the beam permitivity and Schottky spectrum. The theory for the case of a linear friction force and a constant diffusion was built in Ref.2. It was shown that the noise spectral density is equal to[11]

$$S_n(\delta\omega_n) = \frac{N}{\pi\delta\omega_n \left|\varepsilon_n(\delta\omega_n)^2\right|} \mathrm{Im}\left(\frac{n^2\overline{\delta\omega^2}}{\Omega_n^2}\left(1-\varepsilon_n(\delta\omega_n)\right)\right), \qquad \delta\omega_n = \omega - n\omega_0 . \quad (25)$$

Eqs.4 and 25 coincide for the Gaussian distribution only, when

$$\mathrm{Im}\left(\varepsilon_n(\delta\omega_n)\right) = -i\pi\frac{\Omega_n^2}{n^2\overline{\delta\omega^2}} f\left(\frac{\delta\omega_n}{n}\right), \quad \frac{\partial f}{\partial\omega} = -\frac{\omega}{\overline{\omega^2}} f \quad (26)$$

and, consequently, this theory is not applicable for the general case of non-Gaussian distribution. Although, there is not a correct theory a simple formula can be derived to estimate the spectral density in vicinity of peaks.

If the damping decrement is sufficiently small we can write for frequencies $\delta\omega_n \approx \Omega_n$,

$$\varepsilon_n(\delta\omega_n) \approx 1 - \frac{\Omega_n^{\ 2}}{\delta\omega_n(\delta\omega_n - i\lambda)} + \pi i \frac{\Omega_n^{\ 2}}{n^2} \frac{\partial f}{\partial\omega}\bigg|_{\delta\omega_n/n} \approx \frac{2\tilde{\omega} - i\lambda_s}{\Omega_n}, \quad \tilde{\omega} = \delta\omega_n - \Omega_n \ , \tag{27}$$

where

$$\lambda_s = \lambda - \pi \frac{\Omega_n^{\ 3}}{n^2} \frac{\partial f}{\partial\omega}\bigg|_{\Omega_n/n} \ . \tag{28}$$

and λ is a single particle decrement averaged over the particle distribution function

$$\lambda = -\frac{1}{M} \int_{-\infty}^{\infty} \frac{F(\mathrm{v})}{\mathrm{v}} f(\mathrm{v}) d\mathrm{v} \ . \tag{29}$$

The last term in Eq.28 corresponds to the contribution from Landau damping while the first one is related to external damping.

From the other hand, for strong damping when the beam permitivity can be expressed by Eq.27 we can rewrite Eq.25 as

$$S_n(\tilde{\omega}) \approx \frac{N}{\pi\Omega_n |\varepsilon_n(\tilde{\omega})^2|} \mathrm{Im}\left(\frac{n^2 \overline{\delta\omega^2}}{\Omega_n^{\ 2}} \left(\frac{\Omega_n^{\ 2}}{\delta\omega_n(\delta\omega_n - i\lambda)} \right) \right) \approx \frac{ND}{2\pi n^2 (\Omega_n / n)^4 |\varepsilon_n(\tilde{\omega})^2|}, \tag{30}$$

where we took into account that rms frequency spread is determined by an equilibrium between diffusion and friction, $\overline{\delta\omega^2} = D/(2\lambda)$. Combining Eqs. 4 and 30 we obtain,

$$S_n(\tilde{\omega}) \approx \frac{N}{n|\varepsilon_n(\tilde{\omega})^2|} \left(f\left(\frac{\Omega_n}{n}\right) + \frac{D}{2\pi n (\Omega_n / n)^4} \right) \ . \tag{31}$$

This formula is correct for both limiting cases of zero and strong external dampings and interpolates the spectral density for intermediate case where it predicts quite different noise behavior for the cases of Gaussian and non-Gaussian distributions. From general point of view one can consider that there are two sources of the noise. The first is the free motion of the particles when each particle produces δ-function in spectrum; and the second is the diffusion which significantly widens spectral density of each particle and amplifies its relative contribution to the spectral density of the beam. One can see that in the case of very strong damping the noise spectral density near peaks does not depend directly on the distribution function and is determined by diffusion D, *i.e* one can neglect effect of non-Gaussian tails of the distribution function on Schottky spectra in this case.

One can see from Eqs. 27, 28 and 31 that on different harmonics the external and Landau dampings have different relative contributions into beam permitivity and spectral density. The relative contribution of the Landau damping into the total damping decrement and the relative contribution of tails in spectra grow linearly with increasing *n*. Thus, the measurements at different harmonics would allow an identification of the separate contributions from Landau damping and external (laser or electron) damping and allows one to estimate amount of particles in tails. Generally speaking, the spectral density of fluctuations of a cold beam is determined by four parameters. They are the coherent shift, Ω_n, the coherent decrement, λ_s, the value of the distribution function at the resonant frequency, $f(\Omega_n/n)$. Although an accurate analysis allows, in principal, extract this parameters from measurements they have indirect character and, strictly speaking, the beam distribution function and temperature cannot be extracted from measurements of the Schottky noise.

As can be seen from Eqs. 27 and 31 the width of peaks in a spectrum are determined by damping decrement λ_s. The dependence of the damping decrement on laser detuning calculated from measurements is shown in Figure 15. At large detunings the decrement is small and is mainly bound up with Landau damping for small detuning it grows fast and is mainly determined by the laser cooling.

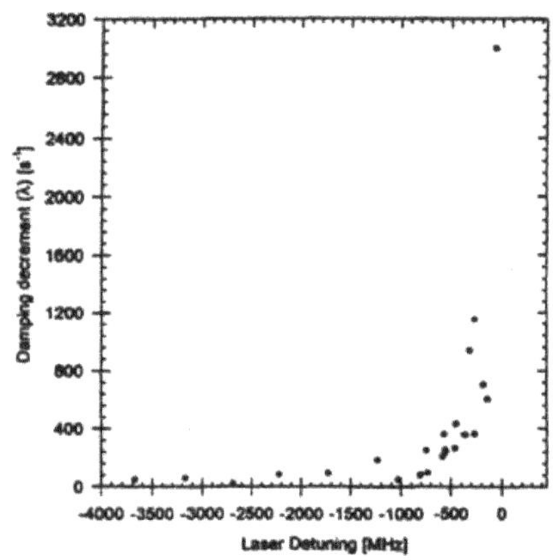

Figure 15. Damping decrement as function of laser detuning extracted from Schottky measurements.

Taking into account that in Touschek scattering a particle cannot get longitudinal velocity much larger than the rms transverse velocity, and assuming that the particles in the tails are evenly distributed in a velocity range determined by the transverse velocity spread, $-\bar{v}_x \le v \le \bar{v}_x$ we can estimate the total number of particles in the tails. The results of such an estimate are in agreement with calculations of the particle distribution in tails and experimental measurements for small detunings. Unfortunately, the accuracy of the tails measurements is not currently sufficient to see tails for large detuning.

8. Discussion

The ASTRID data which are analyzed in this article represent the first simultaneous measurements of both Schottky spectra and velocity distributions in a very cold beam[5,10]. In general, good agreement was obtained between experimental measurements and the proposed theoretical model.

The experimental results from ASTRID illustrate that the model for Schottky noise based upon a Gaussian velocity distribution is not applicable to a laser cooled beam during cooling. The model presented here emphasizes the importance of the detailed shape of the velocity distribution in determining the fluctuation spectrum. In particular, the tails of the velocity distribution play an important role. In general, particle beams in storage rings often have an asymmetric velocity distribution, due to energy loss in the rest gas, and they may also have long, non-Gaussian tails. The analysis presented here may thus have wider application, for example to electron cooled beams. It was also shown that intrabeam scattering in electron cooled beam also creates non-Gaussian tails which can significantly change Schottky spectra.

In previous works, the difference in peak heights in the Schottky spectra of electron cooled beams was attributed to the real part of the storage ring impedance[9,11], which causes an increase of the lower peak and a decrease of the upper peak. Such a difference has been observed in many experiments, but it should be noted that in the case of the laser cooled beam in ASTRID, the intensity of both peaks was strongly increased (compared to a Gaussian beam) which implies that the effect of the tails determines the main contribution to Schottky noise. The effect of the tails may be also responsible for different heights of the peaks in the case of electron cooling, because the inelastic scattering on residual gas atoms produces an asymmetry in the tails of the velocity distribution; it creates a tail at smaller energy but can not influence the tail at higher energy. Like the real part of the ring impedance, it causes an increase of the lower peak height. The effects of these two possible mechanisms could be easily distinguished by measurements at different gas pressures, which should have a strong effect on the tails but cannot, of course, change the ring impedance. It was demonstrated that for experiments on ASTRID an asymmetry arose due to differences in laser powers or spatial overlap of the laser and ion beams.

Although the Schottky spectra of laser cooled beams appear to be very complicated, this article demonstrates that quantitative information about beam parameters can be extracted from them. In the future, it is very important to do measurements on different harmonics of the revolution frequency, because both the ring impedance from the vacuum chamber walls and the effect of the drag force depend strongly on harmonic number. It is hoped that the theoretical model presented here will prove valuable in diagnosing the behavior of very cold beams which are not in equilibrium.

The authors are grateful to professor J.P.Schiffer for careful reading of the manuscript and useful remarks. This work was supported by the Danish National Research Foundation through the Aarhus Center for Advanced Physics.

References

1. H.Koziol, Beam diagnostics for accelerators, in CERN accelerator school: 5-th General Accel.Phys.Course, Geneva (1994).
2. V.V.Parkhomchuk and D.V.Pestrikov, Sov.Phys.Zh.Tekh.Fiz. **50**, 1411(1980).
3. Yu.L.Klimontovich, Statistical Theory of Non equilibrium Processes in Plasma [in Russian], Moscow State Univ., Moscow (1964).
4. E.N.Dementiev, et al, Sov.Phys.Zh.Tekh.Fiz. **50**, 1717(1980).
5. J.S.Hangst, et al, Phys.Rev.Lett.**74**, 86 (1995).
6. V.A.Lebedev, et al, Phys.Rev.E. **52**, 4345 (1995).
7. J.-L. Laclare, Coasting beam longitudinal coherent instabilities, in CERN accelerator school: 5-th General Accel.Phys.Course, Geneva (1994).

8. A.I.Akhiezer, Plasma electrodynamics, Pergamon, 1975.

9. N.S.Dikansky, D.V.Pestrikov, Physics of the intense beams in storage rings, [in Russian], "Nauka", Novosibirsk (1989).

10. H.Brook, Accêlêrateurs circulaires de particules (Preses Universitaires de France, 1966).

11. D.V.Pestrikov, The longitudinal Schottky noise of intense beam, in Proceedings of the workshop "Electron Cooling and New Cooling Techniques", Legnaro, Padova - Italy, 15-17 May, 1990, World Scientific (1991); Preprint 90-48, BINP, Novosibirsk, 1990.

Longitudinal Instabilities and Crystallization

A. Burov *
Laboratori Nazionali di Legnaro, I-35020, Legnaro, Italy
V. G. Vaccaro
Dip. di Scienze Fisiche, Università di Napoli,
INFN Sezione di Napoli , I-80125, Napoli, Italy

May 9, 1996

Abstract

Longitudinal coherent motion of a coasting beam is discussed for parameters typical to the crystallization problem. Properties of the wall impedance of a storage ring are studied for a low relativism, $\gamma - 1 \leq 1$. Both broad and narrow band impedances are shown to be damped at low energies, the damping factors are found. The impedance of an electron cooler is discussed, it is concluded that it should likely to be the dominant contribution in the total impedance budget.

1 Introduction

Beams suggested for crystallization are typically assumed to be low-current, smaller or even much smaller than 1μA. For these beams, the coherent interactions could be significant only at very low temperatures, when the beam becomes a space-charge dominated and the Landau damping goes to zero. At this temperature region the cooling rate has to be high enough to stabilize the coherent motion driven by an impedance of the ring [1].

The dependence of the impedance on a beam velocity has to be taken into account here. In an ultrarelativistic limit, all the fields excited by a point charge lag behind it, which is referred to as the causality principle for wake fields, see e. g. [2]. This causality principle does not work for low and moderate energies, where the relativistic factor $\gamma \simeq 1$: the Coulomb field is not here a δ-function of a longitudinal coordinate, but smoothly increase and decrease during the time $\tau \simeq r/(\gamma\beta c)$, where r is an impact parameter, βc is the beam velocity. It follows that a broad band wall impedance $Z(\Omega)$ is exponentially depressed at frequencies $\Omega \geq \gamma\beta c/b$, where b is an aperture radius. A pure space charge impedance is the only remaining at these frequencies, but it does not lead to instabilities of itself, when the beam is below the transition energy. A narrow band impedance is also

*On leave from the Budker Institute of Nuclear Physics, 630090, Novosibirsk, Russia.

depressed in a nonrelativistic case, but to a lesser extent, then the broad band one, partly due to lower values of frequencies, which are below the cut-off. An electron beam of the electron cooler introduces its own impedance [6, 7, 12] which can be the main contribution in the total impedance of the machine for the case of nonrelativistic beams.

2 Dispersion Equation

Single-particle equations of motion can be presented in the following form:

$$\dot{z} = \eta\beta c\delta, \quad \dot{\delta} = qE_z/p, \tag{1}$$

where $z = R_0(\theta - \omega_0 t)$ is the displacement from the reference particle, $\delta = \Delta p/p$ is the relative momentum deviation, the dots mean time derivatives,

$$\eta = \frac{p}{\omega_0}\frac{d\omega_0}{dp} = 1/\gamma^2 - 1/\gamma_t^2$$

is the slippage factor, q is the charge of an ion and $E_z = U/(2\pi R_0)$ is an electric field. The kinetic equation for the phase-space density $\psi(z, \delta, t)$ follows:

$$\frac{\partial\psi}{\partial t} + \dot{z}\frac{\partial\psi}{\partial z} + \dot{\delta}\frac{\partial\psi}{\partial\delta} = \frac{\partial}{\partial\delta}\left(\lambda\delta\psi + d\frac{\partial\psi}{\partial\delta}\right), \tag{2}$$

where the Fokker-Planck term in the right hand side takes the cooling and diffusion into account; λ is the cooling rate and d is the diffusion coefficient.

Assuming the phase-space density to be a sum of the main equilibrium term (subscript 0) and a small perturbation (subscript 1),

$$\psi = N(g_0(\delta) + g_1(\delta)\exp(-i\Delta\Omega t + inz/R_0)), \quad \int g_0(\delta)d\delta = 1/(2\pi R_0), \quad \Delta\Omega = \Omega - n\omega_0$$

the kinetic equation (2) can be linearized over perturbational terms $\propto g_1, U$. The voltage U is induced by the current perturbation $I_1 = Nq\beta c \int d\delta g_1(\delta)$ by means of the ring impedance Z:

$$U = -Z(n, \Omega, \beta)I_1 \exp(-i\Delta\Omega t + inz/R_0)). \tag{3}$$

From here, the dispersion relation $\Delta\Omega = \Delta\Omega(n)$ can be found.

For low energy rings, the main part of the impedance is contributed by the space charge:

$$Z \approx Z_s = iZ_0 nL/(\gamma\beta^2), \tag{4}$$

where $L = 1/2 + \ln(r_{max}/a)$ is the logarithmic factor, with $r_{max} = \min(b, R_0/n)$, a stands for the beam radius and b denotes the radius of the aperture. Assuming the corrections introduced in the dispersion relation by the ring resistance $\mathcal{R} = \mathrm{Re}Z$, temperature (Landau damping) and the cooling to be small in the comparision with the main space-charge term, the dispersion relation can be written as [7]:

$$\begin{aligned}\Delta\Omega = & \pm n\omega_0 \frac{\beta_s}{\beta}\left\{1 + \frac{\pi i}{2}\left(\frac{\beta_s}{\beta\eta}\right)^2 \cdot 2\pi R_0 g_0'\left(\pm\frac{\beta_s}{\beta\eta}\right) - i\frac{\mathcal{R}}{2|Z|}\right\} - i\frac{\lambda}{2}. \\ \beta_s = & \sqrt{\frac{|Z/n|\eta I_0}{2\pi E/q}}, \quad E = \gamma M c^2, \ I_0 = qN\beta c/(2\pi R_0).\end{aligned} \tag{5}$$

According to Eq.(5), instabilities caused by the real part of the impedance $\mathcal{R}$ can be avoided due to the Landau damping or due to cooling. In the first case longitudinal temperature of the beam must be sufficiently high. Assuming the distribution function to be Gaussian,

$$2\pi R_0 g_0(\delta) = \frac{1}{\sqrt{2\pi}\Delta}\exp(-\delta^2/2\Delta^2)$$

,
the stability condition can be expressed as:

$$x_s^2 \leq 2\ln\left(\sqrt{\frac{\pi}{2}}\frac{|Z|}{\mathcal{R}}x_s^3\right), \quad x_s = \frac{\beta_s}{\beta\eta\Delta}. \tag{6}$$

The factor $|Z|/\mathcal{R}$ is usually pretty large. Therefore the stability condition (6) is almost independent on the impedance. The stability diagram (Eq.6) shown in Fig.1 demonstrates that for a wide range of the perturbative impedances the threshold dispersion Δ is almost constant:

$$\Delta_{th} \approx 0.2\beta_s/(\beta\eta).$$

In terms of a temperature $T = E\eta\beta^2\Delta^2$ it can be expressed as:

$$T_{th} \approx 0.04E\beta_s^2/\eta. \tag{7}$$

For instance, the oscillations in a 1μA beam of Li_7^+ with $\beta = 0.06$ [8] are Landau-damped when the temperature $T \geq T_{th} \approx 2$ K, assuming the logarithmic factor $L = 4$. This low temperature correspondes to the momentum dispersion $\delta = 3 \cdot 10^{-6}$. The frequency of ion-ion Coulomb collisions is pretty high here: $\nu_{ii} \simeq nr_0^2 cL/(\beta^3\Delta^3) \simeq 10^6 s^{-1}$, the same order as the revolution frequency, which approves the assumption of thermal equilibrium.

This limit can be removed if the cooling rate is high enough:

$$\lambda > 2\Lambda = n\omega_0 \frac{\beta_s}{\beta}\frac{\mathcal{R}}{|Z|} \tag{8}$$

If this is satisfied for all the wavenumbers n, the oscillations are stable even without Landau damping. For a given λ the restriction (8) can be treated as a safe condition imposed on the resistance $\mathcal{R}$. Taking into account that the cooling rate of electron cooler is normally already independent on the ion temperature at this low-temperature region [10] and assuming that the Li beam is cooled with the rate $\lambda = 10\,s^{-1}$, the threshold condition on the resistance is found as $\mathcal{R} < 10$ kΩ.

3 Impedance Damping Factors

3.1 Wall Resistivity

Assuming the field dependence on the longitudinal coordinate and time as $e^{ik(z-vt)}$, $k = n/R_0$, $v = \beta c$, Maxwell's equations reduce to a single equation for a longitudinal electric field E_z excited by a charge linear density perturbation. Applying Leontovich boundary condition at the resistive wall surface, [3], $E_z(b) = \sqrt{-ikv/(4\pi\sigma)}B_\theta$, $k = n/R_0$ the electric field E_z in a perfectly conducting tube and its perturbation $\tilde{E}_z$ due to a finite walls conductivity σ [4] can be found.

The fields E_z, $\tilde{E}_z$ and the current perturbation I_1 are connected by corresponding impedances, $E_zC = -I_1 Z$, $\tilde{E}_zC = -\tilde{Z}I_1$, where $C = 2\pi R_0$ is the ring circumference, which gives an imaginary and a real part of the total impedance $\mathrm{Im}(Z+\tilde{Z}) \approx \mathrm{Im}Z = |Z|$, $\mathrm{Re}(Z+\tilde{Z}) = \mathrm{Re}\tilde{Z} = \mathcal{R}$:

$$\frac{Z}{n} = \frac{iZ_0 L}{\gamma^2\beta}, \quad \frac{\mathcal{R}}{n} = \frac{Z_0}{2}\frac{\delta_s}{b}\beta f(\kappa) \tag{9}$$

where $\delta_s = c/\sqrt{2\pi\sigma kv}$ is the skin depth, $L = \ln(r_{max}/a) + 1/2$ is the logarithmic factor with $r_{max} = \min(b, 1/k)$, a is the beam radius, $Z_0 = 4\pi/c = 377\Omega$. Here a high frequency damping factor $f(\kappa)$, $\kappa = kb/\gamma$ has been introduced:

$$f(\kappa) = \kappa\frac{K_1(\kappa)I_0(\kappa) - K_0(\kappa)I_1(\kappa)}{I_0^2(\kappa)} = \begin{cases} 1, & \text{if } \kappa \ll 1 \\ \pi e^{-2\kappa}, & \text{if } \kappa \gg 1 \end{cases} \tag{10}$$

$K_m(\kappa)$, $I_m(\kappa)$ are modified Bessel functions. The plot of this broad-band damping factor is shown on Fig.2.

The resistance $\mathcal{R}$ achieves its maximum at the dimensionless wavenumber $\kappa = 0.43$,

$$\mathcal{R}_{max} = 0.24 Z_0 (R/b)\sqrt{\gamma v/(2\pi\sigma b)},$$

Assuming $\beta = 0.4$, $\sigma = 1.3\cdot 10^{16}s^{-1}$, $R = 10\,m$, $b = 5\,cm$, it gives $\mathcal{R}_{max} = 3\,\Omega$.

3.2 Broad Band Impedances

The resistive wall impedance $\mathcal{R}$ (Eq.9) can be represented in terms of its ultra relativistic value $\mathcal{R}_{ur}$ and the damping factor $f(\kappa)$ (Eq.10) :

$$\mathcal{R}(\Omega) = \mathcal{R}_{ur}(\Omega)f(\kappa), \quad \kappa = \Omega b/(\gamma v). \tag{11}$$

The factor $f(\kappa)$ reflects the strong decrease of the incident field energy flux at the distance $r = b$ as $\sim \exp(-2kb/\gamma)$. The incident field of the beam, damped near the walls as $f^{1/2} \sim \exp(-kb/\gamma)$, produces proportionally to itself a perturbation of the surface current density. The energy loss which is a product of the field $\tilde{E}_z$ and the current induced, is damped quadratically, as

$$f^{1/2}\cdot f^{1/2} = f \simeq \pi\exp(-2\kappa), \quad \kappa = kb/\gamma. \tag{12}$$

The energy loss is proportional to the real part of the impedance, so the last one is damped in the same way.

Thus, the asymptotics (12) of Eq.(11) follows from the general physical consideration, and can be applied to all kinds of the broad band wall impedance.

The real part of the broad band impedance has non-zero value only above the low frequency cut-off: $\Omega \geq \Omega_c$. For the circular cross-section of the vacuum chamber $\Omega_c = 2.4c/b$. Taking into account that $\Omega = kv$, the damping factor f near the lowest possible frequency (cut-off) occurs to be:

$$f \simeq \pi \exp\left(-4.8/(\gamma\beta)\right)$$

So, the broad band impedance of wall imperfections (discontinuities, shallow cavities, irises, etc.) cannot play any role for insufficiently relativistic particles. Even for $\beta = 0.4$, the damping factor $f = 2 \cdot 10^{-5}$, which makes the broad band impedance completely negligible.

3.3 Narrow Band Impedances

Another possibility for energy loss and for instabilities is connected with the radiation in low frequency ($\Omega < \Omega_c$) modes of some elements of the vacuum chamber. At the resonance, the impedance is pure real, it achieves here its maximum, the shunt impedance $R_s(v)$. The velocity dependence of the shunt impedance is determined by a transite time factor $T(v)$ (see, e. g. [5]):

$$R_s(v) = R_s(c)\left|\frac{T(v)}{T(c)}\right|^2, \quad T(v) = \int_{-\infty}^{\infty} E(z)e^{-i\omega z/v}dz, \tag{13}$$

where $E(z)$ is an arbitrary normalized eigenfield distribution, ω is the eigenfrequency. Assuming

$$E(z) \propto \frac{\cos(k_{\|}z)}{(1+\exp(-z/\Delta z))(1+\exp((z-g)/\Delta z)},$$

where the parameter $\Delta z = b/2.4$ takes into account the finit depth of the evanescent field, the time-transite factor (Eq.13) can be calculated analytically, which gives

$$T(v) = -\frac{\pi i \Delta z}{2}\left\{\frac{\exp(-i(\omega/v + k_{\|})g) - 1}{\sinh(\pi\Delta z(\omega/v + k_{\|}))} + \frac{\exp(-i(\omega/v - k_{\|})g) - 1}{\sinh(\pi\Delta z(\omega/v - k_{\|}))}\right\}.$$

The result for the harmonics without longitudinal variation ($k_{\|} = 0$) is more simple to read:

$$\frac{R_s(v)}{R_s(c)} = [\beta\sin(\omega g/(2v))/\sin(\omega g/(2c))]^2\left\{\frac{\delta\psi}{\sinh(\delta\psi)}\right\}^2,$$

where $\psi = \omega g/v$, g is a gap length, $\delta\psi = \pi\omega\Delta z/v$. The $\{...\}$- factor reflects an influence of a finit depth of the evanescent fields at the entrance and exit of the narrow band element and entails the scaling exponential factor to be $\propto \exp(-2.6\omega b/c)$.

3.4 An Electron Cooler Impedance

An electron beam of a cooler can be considered as a medium which responds to fields generated by an ion beam. This reaction can be described in terms of an impedance introduced in the ring by the electron beam [6, 7, 12]. The impedance can be calculated within the hydrodynamical approach if the dispersion of electron velocities Δv_e is small enough [12]: $k\Delta v_e \tau \ll 1$, where $\tau = l_c/v$ is the time of a pass through the cooler. Assuming no difference between electron and ion velocities, the dynamics of the magnetized electron medium can be described in a reference frame by the following set of equations [9]:

$$\begin{aligned} \frac{\partial \tilde{n}_e}{\partial t} + \bar{n}_e \frac{\partial \tilde{v}_e}{\partial z} &= 0 \\ \frac{\partial \tilde{v}_e}{\partial t} - \frac{e}{m}\frac{\partial \Phi}{\partial z} &= 0 \\ \frac{1}{r}\frac{\partial}{\partial r}\left(r\frac{\partial \Phi}{\partial r}\right) + \frac{\partial^2 \Phi}{\partial z^2} &= 4\pi e(\tilde{n}_e - \tilde{n}_i) \end{aligned} \tag{14}$$

Here $\bar{n}_e$ is an electron density, $\tilde{n}_e, \tilde{n}_i$ are electron and ion density perturbations, $\tilde{v}_e$ is a perturbation of electron velocity, Φ is an electrostatic potential. The solution of Eqs.(14) with zero boundary conditions in the entrance of the electron beam is a superposition of two waves: one is travelling together (+ wave) and another opposite (− wave) the beams. Frequencies $\omega_{\pm\mu}$ and wavenumbers $q_{\pm\mu}$ of the waves are connected with the longitudinal wavenumber of the ion perturbation k by means of the Doppler condition:

$$q_{\pm\mu} v \pm \omega_{\pm\mu} = kv$$

The subscript $\mu = 1, 2, \ldots$ counts the number of radial excitation. Due to the dispersion relation for the electron medium $\omega_{\pm\mu} = q_{\pm\mu} u_\mu$.

The solution of Eqs.(14) for the long-wave limit $ka \ll 1$ was found in [12]:

$$\mathrm{Re}Z = 2\sum_\mu \frac{(1-\cos(\omega_{\mu+}\tau)) - (1-\cos(\omega_{\mu-}\tau))}{\kappa_\mu^2 F_\mu^2 u_\mu}. \tag{15}$$

Here $\kappa_\mu = q_\mu a \omega_e/\omega_\mu = a\omega_e/u_\mu$ has to be found from matching conditions for the potential Φ at $r = a$ and zero boundary codition at $r = b$, where a is the electron beam radius, and b is the radius of the vacuum chamber. The radius of the ion beam is not significant when it is much less than one of the electron beam. This condition is well satisfied for the discussed region of the ion temperature. The mentioned boudary conditions for the potential Φ reduce to an equation for the eigennumbers κ_μ:

$$\begin{aligned} \kappa_\mu J_1(\kappa_\mu)/J_0(\kappa_\mu) &= 1/\ln(1/(qa)), \quad \text{if } qb \gg 1, \\ J_0(\kappa_\mu) &= 0, \quad \text{if } q(b-a) \ll 1. \end{aligned} \tag{16}$$

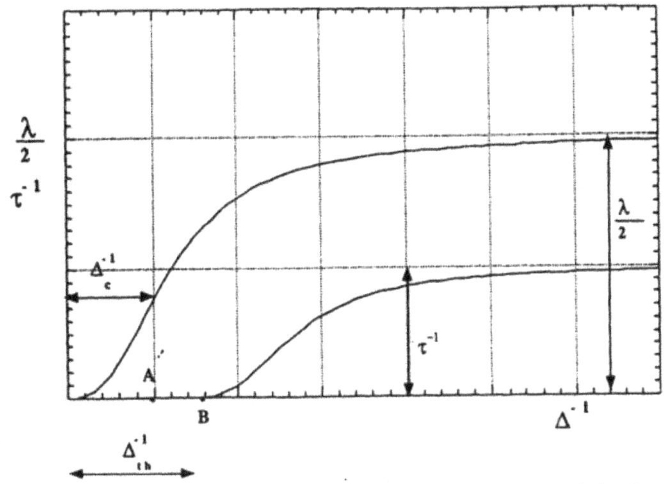

Figure 1: The stability diagram. The line shows threshold value of the impedance at a given dispersion of the velocities, or vice-versa.

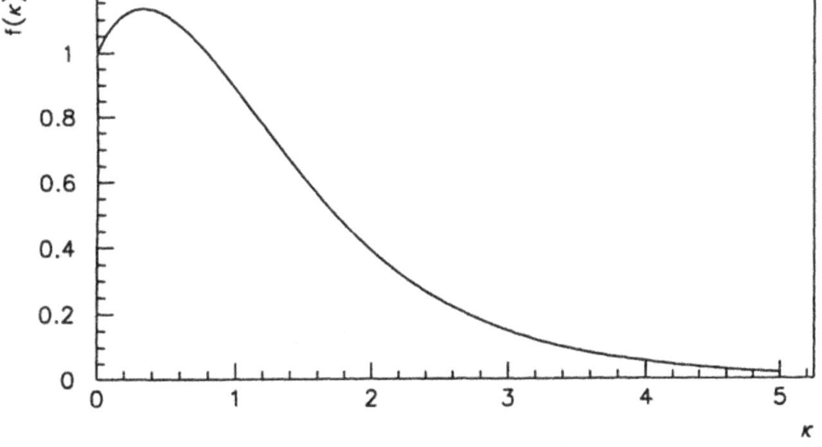

Figure 2: The damping factor of the broad-band impedance as it is for the resistive wall. The horisontal axis is for the parameter $\kappa = kb/\gamma$.

The factor F_μ in the deminator of Eq.(15) is a radial form-factor:

$$\int_0^a J_0(\kappa_\mu r/a)J_0(\kappa_\nu r/a)rdr = \delta_{\mu\nu}a^2F_\mu^2/2,$$
$$F_\mu = \begin{cases} J_0(\kappa_\mu) & \text{if } qb \gg 1 \\ -J_1(\kappa_\mu) & \text{if } q(b-a) \ll 1 \end{cases} \tag{17}$$

The contributions in the real part of the impedance (15) from the positive (+) and negative (−) waves can be seen to have opposite signs: an emission of the positive wave takes away a longitudinal momentum of ions, i. e. decelerates them; on the contrary, radiation of the negative wave with a negative momentum accelerates the ion beam. The negative sign of the impedance at $\omega\tau < \pi$ follows from the fact that the phase advance of the accelerating negative (−) wave is higher than one of the decelerating positive (+) wave. The absolute value of the real part of the impedance achieves its maximum at $ka \simeq 1$, this maximum value can be estimated as:

$$|\mathrm{Re}Z|_{\max} \simeq 0.1Z_0\frac{\omega_e\tau}{\beta}, \quad \omega\tau \gg 1. \tag{18}$$

Substituting, for example, $n_e = 3\cdot 10^7$ cm^{-3}, $g = 1$m, $\beta = v/c = 0.06$ (Li_7^{+1} in the CRYSTAL ring [8]), it gives the cooler impedance $\mathrm{Re}Z = 10K\Omega$, which is three orders of magnitude more than typical values of the resistive-wall impedance. Thus, the impedance of the electron cooler could be the main source of the longitudinal instability of the coasting ion beam.

4 Conclusions

The concept of the storage ring impedance, developed mainly for ultrarelativistic beams cannot be at once applied to low and moderate energy cases. The reason is that the causality principle for wake fields generally is not valid here, which cause an exponential damping of impedances above a certain threshold. The approach suggested gives the possibility to estimate the wall impedance for an arbitrary energy if its ultrarelativistic value is known. In a particular, at low energies, $\beta \leq 0.5 \div 0.8$, the broad-band wall impedance is shown to be exponentially damped and too weak to cause instabilities of a cooled beam. A narrow-band impedance could be dangerous if its eigenfrequency ω is rather low: $\pi\omega b/(2.4v) \leq 1$. In this case the eigenfield of the cavity-like element has effectively sharp boundaries and the time-transite factor has no exponential damping. The electron beam introduces its own impedance in the ring; the estimations show that this can be the main part in the total impedance budget.

5 Acknowledgments

The report has been done in the framework of the CRYSTAL project. The authors are particularly grateful to L. Tecchio, N. S. Dikanski and A. Aleksandrov.

References

[1] A. Sessler and V. Vaccaro, CERN Report ISR-RF/67-2 (1967).

[2] A. W. Chao *Physics of Collective Beam Instabilities in High Energy Accelerators*, Wiley, N.- Y. 1993.

[3] J. D. Jackson. *Classical Electrodynamics*, Wiley, N.- Y. and London, 1962.

[4] P. L. Morton, V. K. Neil, and A. M. Sessler, J. Appl. Phys. **37**, 3875 (1966).

[5] L. Palumbo, V. Vaccaro and M. Zobov *Wake Fields and Impedance*, LNF-INFN Report 94/041, 1994

[6] N. S. Dikansky, D. V. Pestrikov. *The Physics of Intense Beams and Storage Rings*, AIP, N.- Y. 1994.

[7] A. Burov, in *Proc. of the Workshop on Beam Cooling and Related Topics* , p.230, 1994.

[8] L. Tecchio et al. *CRYSTAL. A Storage Ring for Crystalline Beams*. LNL-INFN Int. Rep., 1994.

[9] L. S. Bogdankevich, A. A. Rukhadze, Sov. Uspekhi Phys. Nauk, **103**, 4, p. 609 (1971).

[10] I. N. Meshkov, Phys. Part. Nucl. (Rus., translated) **25**(6), pp. 631-661 (1994).

[11] A. Burov, Wall Impedances for Low and Moderate Energies, Proc. PAC'95, Dallas, USA, 1995.

[12] A. Burov, Electron Cooler Impedances, Proc. PAC'95, Dallas, USA, 1995.

TRANSVERSE COHERENT OSCILLATIONS OF INTENSE COASTING BEAMS [1]

R.W. HASSE and I. HOFMANN

GSI Darmstadt, P.O. Box 110552, D-64220 Darmstadt, Germany

ABSTRACT

In order to study space charge effects on the transverse coherent motion of a coasting beam we perform computer simulations of transverse single-particle and coherent oscillations with high space charge derived from Fourier spectra of the time evolution of the electric potentials. Results with solenoidal and periodic focusing with initial K-V and waterbag distributions are compared with those obtained from the envelope equation. An increase of space charge is simulated by cooling in one or both transverse directions.

1. Introduction

Heavy ion inertial confinement fusion requires beams of highest intensities. At high space charge densities single-particle (s.p.) motion and coherent oscillations of the beam might lead to instabilities and in circular machines to dangerous crossings of integer or half-integer resonances. For circular machines such space charge phenomena have been studied previously[1] for simplified lattices as well. A comprehensive treatment in a realistic ring by Machida[2] focuses on the emittance growth. There elliptical symmetry has been assumed for the oscillations of the charge density.

It is the purpose of this work to study the limits of space charge and the influence of space charge on the coherent and incoherent frequencies of quadrupole and higher order oscillations of a coasting beam without limitation to elliptical symmetry. To that end we perform computer simulations with a Particle-In-Cell (PIC) code with solenoidal focusing and in simplified lattices. Coasting beams with initially circular or elliptical cross sections are simulated by following the trajectories of N (up to 65536) macro-particles moving in a two dimensional $(x-y)$ transverse focusing field and in the self-consistent electric field obtained by solving the Poisson equation on a grid (up to 101×101) with a rectangular boundary condition. In case of constant solenoidal focusing the horizontal and vertical betatron tunes will be denoted by ν_{x0} and ν_{y0}, respectively.

The time varying self-consistent space charge potential is calculated for a number of (up to 1000) revolutions and is recorded on virtual horizontal and vertical pick-ups

[1] Proc. 31st Eloisatron Workshop on Crystalline Beams and Related Issues, November 11-21, 1995, Erice, Italy

which are located somewhere outside the beam, as well as on an (unphysical) central one. After many betatron oscillations it is Fourier transformed in order to get the Schottky spectra $\Phi(\omega)$. From them the frequency shifts and the s.p. and coherent tunes can be read off.

With solenoidal focusing mostly an initial K-V distribution is employed and the results are compared with those obtained from solutions of the envelope equation. With periodic lattices of type FODO, on the other hand, the space charge density is increased by cooling the macro-particles. This is done by decreasing the velocities per time step by a certain amount. Single-particle tunes are obtained by following the phase space angle of particle j, $\Phi_i^j = \mathrm{arctg}(v_i^j/\omega_r x_i^j)$, where ω_r is the revolution frequency, and measuring the time when Φ_i^j jumps by 2π.

As results, in particular the coherent even and odd frequencies split up if the focusing constants are different in horizontal and vertical directions and for beams with non-circular cross sections and the two branches can cross whereas the incoherent particle tunes do not play an influential role. Also, even with very strong cooling one cannot cross systematic stop-bands of periodic lattices.

2. Solenoidal focusing

Initially the envelopes of the particles with emittances ϵ_i and envelopes a_i are in a matched state obtained from a solution of the envelope equation with a K-V distribution,

$$\nu_{i0}^2 a_i - \frac{\epsilon_i^2}{a_i^3} - \frac{\nu_p^2 a_x a_y}{a_x + a_y} = 0, \qquad i = x, y \;, \tag{1}$$

where ν_p is the tune corresponding to the plasma frequency. In a K-V distribution all particles have the same transverse energy and are distributed randomly within the space (a_i) and velocity $(v_i = \epsilon_i/a_i)$ boundaries. By virtue of the random initial positions and momenta all types of coherent and incoherent modes are excited automatically.

The influence of space charge is controlled by the parameter η or the tune ν_0 or the plasma frequency ω_p derived from the beam current according to:

$$\eta = Q/2r_0^2\nu_0^2 = \omega_p^2/2\nu_0^2 = 1 - \nu^2/\nu_0^2, \tag{2}$$

where $Q = 4qI/Mv^3\gamma^3$ is an effective current, r_0 is the beam radius, I the current, M the mass, v the velocity, γ the relativistic factor. $\eta = 0$, thus, means no space charge and $\eta = 1$ corresponds to the limiting space charge where the incoherent frequency vanishes.

A K-V distribution of charged particles creates an average field which has an inverted harmonic potential. This partly compensates for the harmonic solenoidal focusing potential, thus effectively reducing the s.p frequency to $\nu = (\nu_0^2 - \omega_p^2/2)^{1/2}$ for a round beam.

2.1 Incoherent (single-particle) tunes

Elliptical focusing is introduced by the frequency asymmetry parameter τ, viz $\nu_{x0}=\nu_0/\tau$, $\nu_{y0}=\nu_0\tau$. Furthermore, non-cylindrical beams are characterized by the spatial asymmetry parameter σ, viz $a_x=r_0\sigma$, $a_y=r_0/\sigma$. Then the incoherent s.p. frequencies of a K-V distribution become

$$\nu_x^2=\omega_0^2/\tau^2-\omega_p^2/(1+\sigma^2), \qquad \nu_y^2=\omega_0^2\tau^2-\omega_p^2\sigma^2/(1+\sigma^2)\,. \tag{3}$$

A waterbag distribution, on the other hand, creates a non-harmonic self-consistent potential so that the s.p tunes are spread over a certain interval. However, if the initial waterbag envelopes are rms matched according to the K-V initial envelopes, the average tunes are expected to follow the same systematics of eq.(3). As can be seen from Fig.1, our results show that this is the case even for large space charges. Only for very high space charges larger than about 90% of the space charge limit there are deviations from eq.(3). These deviations are smaller if using a waterbag distribution thus indicating that the particles arrange more in a waterbag rather than in a K-V distribution.

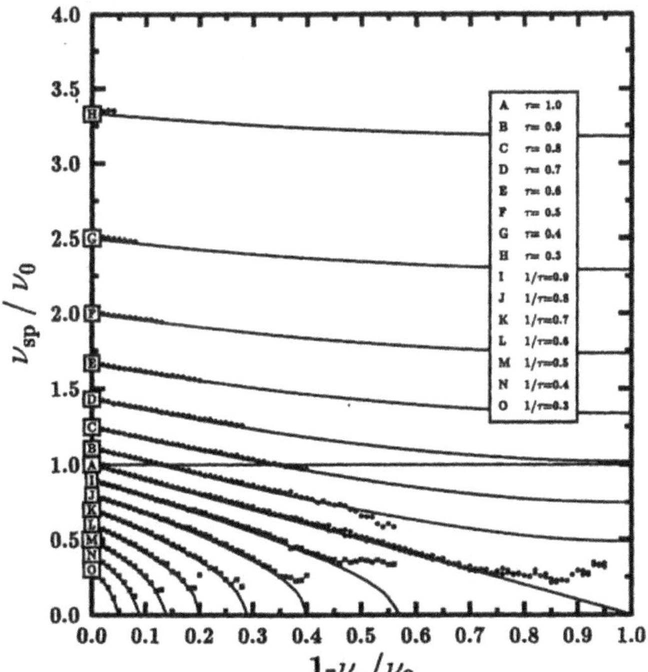

Figure 1: Single-particle tunes for various focusing asymmetries τ. Squares are results from simulations with initial waterbag distributions and lines are from eq.(3).

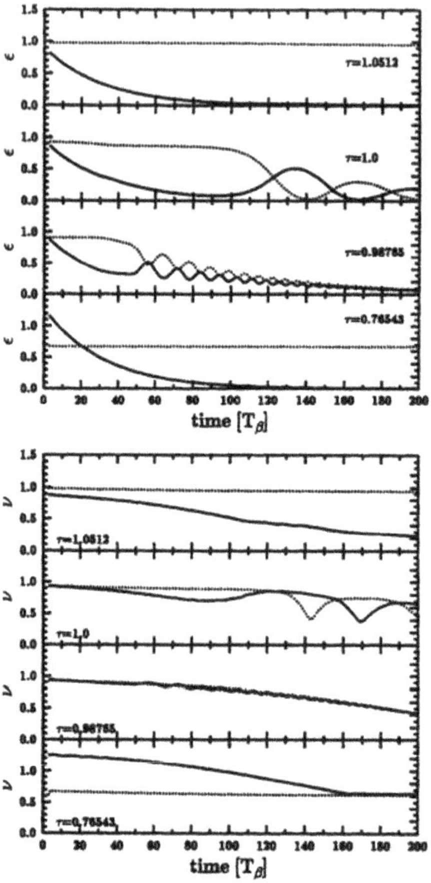

Figure 2: Horizontal (cooled, full lines) and vertical (uncooled, dotted lines) emittances (top) and s.p. tunes (bottom) vs. time in units of betatron periods. (For solenoidal fucusing and various focusing asymmetries τ).

In order to investigate this effect further, we increased the space charge in one (the horizontal) direction with a cooling mechanism which cools this emittance to half of its initial value after 20 betatron oscillations. For the various focusing asymmetries τ under consideration we observed that the cooling process did not give rise to instabilities when the s.p tunes did not cross each other, see Fig.2. However, when the tune in the cooled direction became almost equal to the other tune the particles interacted severely resulting in third order (pear-shaped) disturbances of the density distribution which tilted the emittance. In this way the tunes are equalized and they did not cross.

2.2 Coherent tunes

The coherent oscillation frequencies of a round K-V beam have been calculated by Gluckstern[3] by decomposing the electric field in the Vlasov equation according to the radial (j) and azimuthal (m) node numbers. This yields the characteristic equation

$$\frac{j!\,m!}{(m+j-1)!}\sum_{\mu=0}^{j}\frac{(m+j+\mu-1)!}{\mu!(j-\mu)!(m+\mu)!}\int_0^\infty ds\, i\lambda e^{i\lambda s}\cos^{m+2\mu}(s)+\delta_{j0}+\frac{2(-)^j\,m!\,j!\,\nu^2}{(m+j-1)!\,\omega_p^2}\equiv 0\,, \tag{4}$$

where $\lambda=\omega/\nu$. The integral in eq.(4) is solved analytically with the help of Mathematica in form of a polynomial. Note that $j=0, m=0$ does not exist and that the dipole mode $j=0, m=1$ is unshifted, $\omega_{\ell=1}=\nu_0$. The lowest nontrivial excitations branch down from $\omega=2\nu_0$,

$$j=1, m=0: \quad \omega_{\ell=0}^2 = 2\nu_0^2+2\nu^2 \text{ monopole (breathing)} \tag{5}$$

$$j=0, m=2: \quad \omega_{\ell=2}^2 = \nu_0^2+3\nu^2 \quad \text{quadrupole.} \tag{6}$$

However, also the modes $(j,m)=(4,2)$, $(1,2)$, $(2,0)$ have branches at $\omega=2\nu_0$. Its lowest one, $(2,0)$, results from

$$\omega_{\ell=3}^4-2\omega_{\ell=3}^2(\nu_0^2+9\nu^2)=4\nu_0^2(\nu_0^2-17\nu^2). \tag{7}$$

The coherent monopole ($\ell=0$ or even) and octupole ($\ell=3$) oscillations are only observed with the center signal because the field of a round beam with conserved charge is not influenced outside the beam.

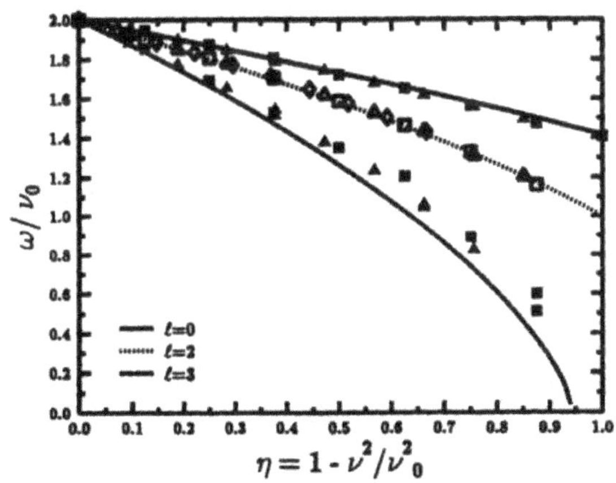

Figure 3: Coherent frequencies branching from $\omega=2\nu_0$. The lines are theoretical results from eq.(4).

These modes are shown in Fig.3. The monopole and quadrupole frequencies follow well the analytical result. The frequencies of the $\ell = 3$ mode, however, being very sensitive to the actual distribution[4] are less degressed than for a pure K-V distribution.

For elliptical beams the even ($\ell = 0$: monopole) and odd ($\ell = 2$: quadrupole) coherent frequencies are in linear approximation in η and for $\sigma \not\ll 1$:

$$\omega_{\ell=0}/\nu_0 \approx 2 - \tfrac{1}{2}\eta\ , \qquad \omega_{\ell=2}/\nu_0 \approx 2 - \tfrac{3}{4}\eta \tag{8}$$

They mix under deformation with the arithmetic mean $\omega_{\ell=02}$ and :

$$\omega_{\ell=02}/\nu_0 \approx 2 - \tfrac{5}{8}\eta\ , \qquad \omega_{\ell=0,2}/\nu_0 \approx 2 - \tfrac{1}{4}\left(3 - \tfrac{\sigma^2}{1+\sigma^4}\right)\eta. \tag{9}$$

Here we observe that eq.(9) is well confirmed by the simulations for arbitrary ν_0.

3. Periodic lattices

For non-solenoidal focusing as is the case in periodic lattices, the waterbag distribution is more realistic. The initial emittances, envelopes and velocity boundaries in the horizontal and vertical directions are calculated from the envelope equation and we carry out an rms matching. The initial conditions of the particles are then obtained by randomly choosing the positions and velocities within these rms boundaries. This always yields a well matched beam to start with.

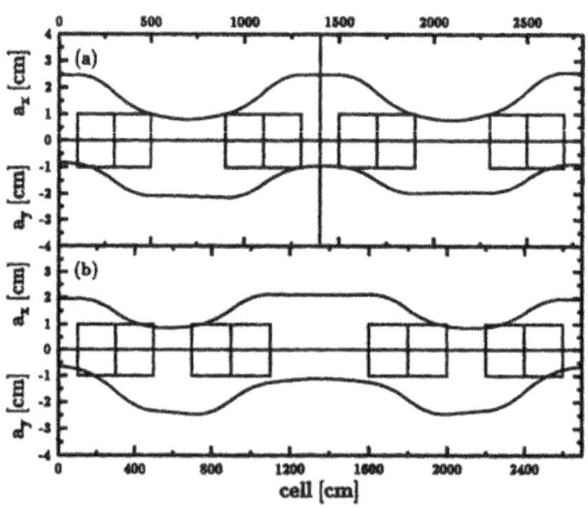

Figure 4: Lattice cells with horizontal (upper part) and vertical (lower) envelopes. (a): double cell with an envelope stop-band and (b): single cell similar to (a) but with a single-particle stop-band

We study two lattice cells with quadrupole magnets and free sections shown in Fig.4. Cell (a) has an envelope stop-band and cell (b) which is similar to cell (a) but has a s.p. stop-band instead. Cell (b) is similar to cell (a) but has shorter drift sections between the FD-magnets and longer drift sections between the two cells (a). A realistic lattice consists of four to eight such cells and tunes multiply accordingly. The beam properties are: charge $q = 1$, mass $A = 208$, energy $E = 48$ MeV/A, velocity $\beta = 0.3092$, and emittances $\epsilon_x = \epsilon_y = 6.25$ (π) mm mrad. Both cells have a stiffness of $B\rho = 2.1 \times 10^8$ G cm. Cell (a) has length 2.8 m, focusing gradient $B' =$ 3000 G/cm and defocusing gradient $B' = -3135$ G/cm and cell (b) has length 2.7 m, $B' = 3109$ G/cm and $B' = -3036$ G/cm. The current I, beta functions $\beta_{x,y}$, s.p. phase advances $\sigma_{x,y}$, and coherent even and odd phase advances $\phi_{0,2}$ in these cells are listed in the following table.

Table: Currents I, matched betatron functions $\beta_{x,y}$, s.p. tunes $\sigma_{x,y}$ and coherent even and odd tunes $\phi_{0,2}$ of lattices (a) and (b) as obtained from the envelope equation.

Cell (a): I [A]	$\beta_x[\frac{\text{cm}}{\text{rad}}]$	$\beta_y[\frac{\text{cm}}{\text{rad}}]$	σ_x	σ_y	ϕ_0	ϕ_2
0	2393.5	281.9	118.0°	118.0°	236.0°	236.0°
111.4	2423.8	357.7	102.4°	102.6°	231.6°	218.9°
243.1	2614.9	453.6	86.3°	86.9°	196.7°	184.3°

Cell (b): I [A]	$\beta_x[\frac{\text{cm}}{\text{rad}}]$	$\beta_y[\frac{\text{cm}}{\text{rad}}]$	σ_x	σ_y	ϕ_0	ϕ_2
0	1506.1	173.6	228.0°	228.0°	456.0°	456.0°
115.9	1453.4	140.9	198.8°	199.2°	415.9°	426.6°
228.8	1536.1	82.4	177.5°	186.4°	381.4°	422.0°

An increase in space charge density is simulated through cooling by reducing the kinetic energy of the relative motion of the particles by 2% per cell. The total time spectrum is recorded at the virtual pick-ups for 125 transits through the cells and is then Fourier transformed. As can be seen from Fig.5, after passage through about 50 cells in both cases the horizontal and vertical s.p. tunes become equal, thus indicating that the quarter-integer stop-band (90° per cell, tune 0.25 in Fig.5) and half-integer resonance (180° per super-cell, tune 0.5 in Fig.5), respectively, have been reached and that they cannot be crossed. The emittances, envelopes and average velocities, too, level off, however not at the same values in horizontal and vertical directions.

In Fig.6 are shown the frequency spectra in these two cells. In case (a) there is a pronounced coherent response at one half of the revolution frequency (and a multiple thereof) corresponding to twice the s.p. stop-band frequency. Here a typical envelope resonance[5] has been reached. For case (b), on the other hand, the envelope equation predicts vertical and horizontal coherent phase advances of about 380°($\omega/\nu_0 = 1.055$) and 420°($\omega/\nu_0 = 1.17$) according to the Table which also can be found in Fig.6(b).

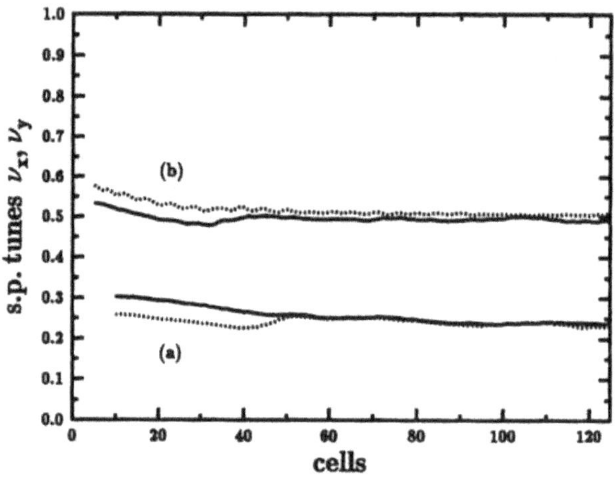

Figure 5: Time evolution of the horizontal (full lines) and vertical (dotted lines) single-particle tunes as fractions of the zero-intensity phase advances per single cell in cells (a) and (b) of Fig.4.

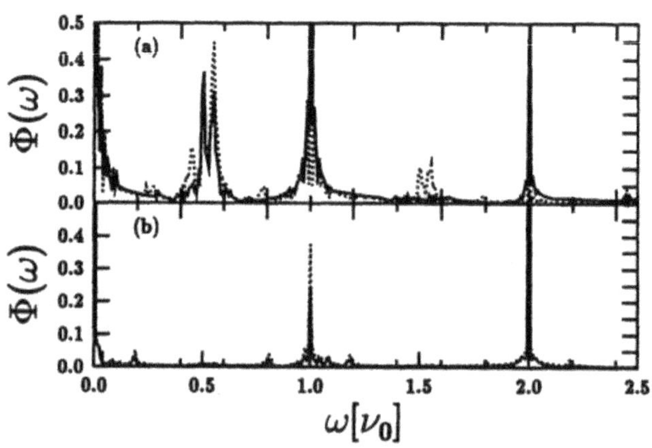

Figure 6: Horizontal (full lines) and vertical (dotted lines) frequency spectra after 125 transits through cells (a) and (b) of Fig.4.

4. Summary

We found that the single-particle tunes obtained in the simulations with solenoidal focusing agree well with those obtained from the envelope equation (with a K-V distribution) sufficiently far away from the space charge limit. Close to the space charge limit, however, a non-standard redistribution of the particles occurs. Increasing the space charge up to the space charge limit by cooling one degree of freedom we observed that the single-particle tunes did not cross. This calls for further investigation. The lowest coherent transverse oscillation frequencies also agree with those obtained from the characteristic equation derived with a K-V distribution. With periodic focusing we could not cross stop-bands by (even strong) cooling.

References

1. I. Hofmann and K. Beckert, *Proc. Part. Acc. Conf. Vancouver, 1985, IEEE Trans. Nucl. Sc. NS-32* (1985) p. 2264
2. S. Machida, *Nucl. Instr. Meth.* **A309** (1991) 43
3. R.L. Gluckstern, *Proc. Proton Lin. Acc. Conf., Batavia, 1970*, p.811
4. I. Hofmann, *Phys. Fluids* **23** (1980) 296
5. I. Hofmann, L.J. Laslett, L. Smith and I. Haber, *Part. Acc.* **13** (1983) 145

BEAM INSTABILITIES AND THEIR CURES

M. CHANEL
PS Division, CERN
1211 Geneva 23, Switzerland

ABSTRACT

A general review of the sources of instability on dense coasting beams obtained by cooling is given. The most useful formulae are derived concerning the impedances, the threshold of instabilities and their growing rate. The effect of cooling rate is analysed. The effect of active feedback is derived, the needed characteristics are indicated. Interesting results obtained in different cooler rings are shown to sustain the results of the theory.

Preliminary

Instability is the rule, stability is a constant fight of the human race.

1. Introduction

In the case of high-beam intensity or high-beam density, the motion of particles in an accelerator is not given only by the external electric and magnetic fields. As particles go through the ring, they interact with their surroundings via the electromagnetic fields, called the wake field, created by their own charge and current. This field extends for a certain distance in front and behind the particles that created it. In the case of high γ, there is no field in front of the particles. As an example of this interaction, the resistivity of the vacuum chamber causes ohmic losses caused by the wake field creating image current in the wall of the vacuum chamber. The wake field can act back on the particles and if certain conditions are satisfied, the initial disturbance gets amplified and hence an instability is generated. In some cases, the disturbance grows indefinitely causing beam losses; in others, the disturbance saturates and leads to a new equilibrium. In the later case, if external cooling forces are applied, the beam will return to the situation where instability can appear again and a relaxation phenomenon is observed.

In general, the main signature is an intensity dependence. For cooling rings the beam density plays an important role as it is always maximised even if it is limited by the equilibrium between cooling forces, intra-beam scattering and other blow-up mechanisms.

In this paper we outline the typical instabilities observed in cooler rings and the techniques used to avoid them. At the core of the discussion on instabilities is the concept of coupling impedance of the machine which is closely related to the wake field. Once the impedance is known it is possible to evaluate the thresholds and growth rates of the instabilities. The latter are related to the dispersion relations which indicate the amount of Landau damping present in the beam. The dispersion relations are related to the response of the beam to a small disturbance. When applying an external disturbance the dispersion relations can be determined. This technique is called the beam transfer function (BTF). All the characteristics of the system can be deduced from this type of measurement (e.g. impedance and damping).

In the case of cooler rings, the cooling time can be viewed as damping decrement for beam instabilities. The influence of this effect is included in the beam transfer function (BTF) measurements. Like in the electronic system, active feedback systems are used to stabilize the beam. A review of the characteristics of this system is proposed.

2. Wake Fields and Impedances

2.1. *Definitions*[1-3]

In the simplest version, the beam consists of a single particle of charge q travelling in a cylindric beam pipe of radius b. The beam trajectory is a straight line parallel to the axis but is offset transversely to it by x. We consider a test particle of charge e travelling parallel to the beam at distance z with some transverse position. From Maxwell's equations, one can calculate the transverse and longitudinal electromagnetic forces on the test particle. By integrating these forces along the test particle trajectory over a distance $L >> b$, we obtain the wake functions $W_{//}(z)$ and $W_\perp(z)$ which are defined as:

$$\int_0^L F_{//} ds = -eq\, W_{//}(z) \qquad \int_0^L F_\perp ds = -eq\, x\, W_\perp(z) \tag{1}$$

where higher order terms have been neglected.

The corresponding impedances for the distance L are defined by the Fourier transform of the wake functions and are given by:

$$Z_{//}(\omega) = \frac{1}{\beta c}\int_{-\infty}^{\infty} dz\, e^{-i\omega z/\beta c} W_{//}(z) \qquad Z_\perp(\omega) = \frac{1}{\beta c}\int_{-\infty}^{\infty} dz\, e^{-i\omega z/\beta c} W_\perp(z) \tag{2}$$

Of course, in a real machine, all particles play the role of particle generating wake field and test particles. Impedances summarize the electromagnetic effects from the environment traversed by the beam. Thus the vacuum chamber resistivity, rf cavities, bellows, discontinuities, vacuum chamber dimension changes, ports, curvature of the chamber, electron beam of the cooler all contribute to the impedance.

2.2. *Properties of Impedances*

As implied by the definitions, the wake functions are real functions. Therefore, considering the complex conjugate of Eqs. (2) we conclude that $Z_{//}(-\omega) = Z^*_{//}(\omega)$ and $Z_\perp(-\omega) = -Z^*_\perp(\omega)$. Then the real part of $Z_{//}(-\omega)$ is a symmetric function and the imaginary part is antisymmetric; the transverse impedance $Z_\perp(-\omega)$ has the opposite parity properties due to the extra factor i in its definition.

In calculating beam instabilities, the longitudinal impedances appear divided by ω. It is customary to deal with $Z_{//}/n$ where n is defined by $n = \omega/\omega_0$ (ω_0 being the revolution frequency). Then $Z_{//}(\omega)/n$ has the same parity as $Z_\perp(\omega)$.

For the resistive wall impedance of a smooth, infinitely long cylindrical pipe of radius b, there exists a relation between longitudinal and transverse impedance[1].

$$Z_\perp(\omega) = \frac{2L}{b^2\omega_0}\frac{Z_{//}(\omega)}{n} \tag{3}$$

This relation is approximately valid for other impedances generated by discontinuities like bellows, small cavity-like structures providing their size is small compared to b.

Equations (1) and (2) imply that $Z_{//}(\omega)/n$ is measured in Ω while $Z_\perp(\omega)$ is measured in Ω/m.

2. Estimations of Impedances[1-3]

2.1. *Space Charge*

For a beam having a uniform distribution of particles the space-charge impedance is capacitive and is given by

$$\frac{Z_{//}}{n} = -i\frac{Z_0}{2\beta_0\gamma_0^2}\left(1+2\ln\frac{b}{a}\right) \tag{4}$$

and

$$Z_\perp = -i\frac{RZ_0}{\beta_0^2\gamma_0^2}\left(\frac{1}{a^2}-\frac{1}{b^2}\right) \tag{5}$$

where $Z_0 = 377\ \Omega$ is the free-space impedance, β_0 and γ_0 the relativistic factors, b the vacuum chamber radius, a the beam radius and R the ring radius.

In a cooler ring these impedances are typically $Z_{//}/n \sim 2$ to 5 kΩ and $Z_\perp \sim 1$ GΩ/m for $\beta_0 = 0.3$.

2.2. *The Resistive Wall Impedance*

For the case of a vacuum chamber wall much thicker than the skin depth δ_0, the following expressions hold for the impedance of the walls:

$$\frac{Z_{//}(\omega)}{n} = (1+i)\frac{Z_0\beta_0}{2b}\delta_0\frac{1}{\sqrt{n}} \tag{6}$$

and

$$Z_\perp(\omega) = (1+i)\frac{Z_0R}{b^3}\delta_0\frac{1}{\sqrt{n}} \tag{7}$$

where $\delta_0 = \sqrt{2\sigma/\omega_0\mu_0}$, σ being the resistivity of the vacuum chamber and μ_0 the free-space permittivity. The resistive wall impedance is estimated to be $Z_{//}/n \sim 1\ \Omega$ and $Z_\perp \sim 1$ MΩ/m for $n = 1$.

2.3. *The Resonator Impedance Model*

In practical rings, it is impossible to know the impedances accurately and completely. However, impedances of individual components can be often calculated or measured at least for certain frequency ranges. A few parameters can model a given ring component. A simple and generally used model for the longitudinal impedance corresponds to the superposition of resonant structures like rf cavities.

$$Z_{//}(\omega) = \sum_r Z_{//,r}(\omega) \tag{8}$$

where $Z_{//,r}(\omega)$ is a single resonator impedance defined by:

$$Z_{//,r}(\omega) = \frac{R_{s,r}}{1 + iQ_r\left[(\omega/\omega_r) - (\omega_r/\omega)\right]} \tag{9}$$

The quality factor Q_r, the resonant frequency ω_r and the shunt impedance $R_{s,r}$ (at $\omega = \pm\omega_r$) define the model.

The real part of this resonator impedance has a peak at $\omega = \pm\omega_r$ and a width of ω_r/Q_r. The imaginary part changes sign as the frequency crosses the resonant value. By definition a broad-band resonator has a low Q, typically $Q \approx 1$, and therefore wide peaks. A narrow-band resonator has large Q and hence narrow peaks (usually rf cavity has $Q \sim$ 100...1000). Taking the Fourier transform of Eq. (9), the decay time of the excitation produced by the resonator is obtained as $\tau_r = 2Q_r/\omega_r$. Therefore, narrow resonance ring for a long time.

For a broad-band resonator with $Q = 1$, the imaginary part of $Z_{//,r}(\omega)/n$ is inductive in the range $-\omega_r < \omega < \omega_r$ and almost independent of ω. For the transverse impedance, the same properties are also true, using the relation (3).

An rf accelerating cavity can have a shunt impedance of typically 1 to 3 kΩ.

A broad-band resonator with ω_r and Q chosen to be $\omega \equiv \omega_c$ ($\omega_c = c/b$ being the cut-off frequency of the vacuum chamber) and $Q = 1$ is often used to model the whole impedance of a storage ring. The shunt impedance R_s is determined empirically to fit to the data.

2.4. *Other Sources of Impedances*

- At high frequencies ($\omega > \omega_c$), according to the diffraction model[1], the impedances $Z_{//}/n$ and $Z_\perp$ fall off as $\omega^{-3/2}$.

- The electron beam of the electron cooling can be regarded as an additional impedance. This is treated elsewhere in this conference[4],.

Figure 1 illustrates the impedance of a storage ring.

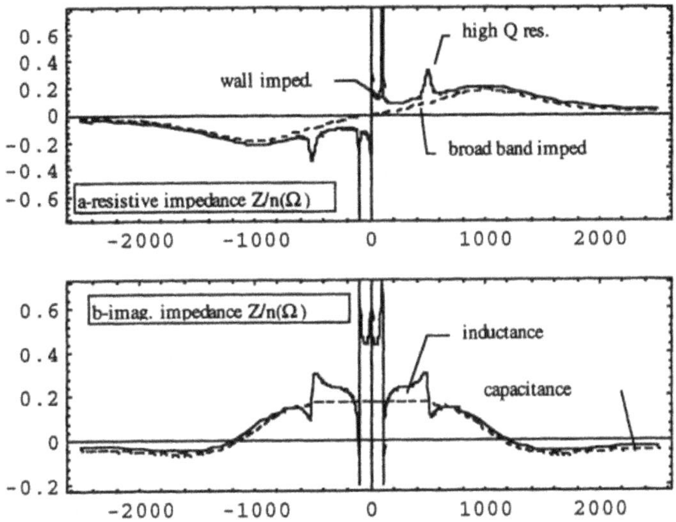

Figure 1 - Illustration of the impedance of a storage ring. The space-charge impedance which is reactive is not shown as it is very large compared to other sources of impedance.

3. Stability in the Transverse Plane[1-5]

3.1. *General Derivation*

In the transverse plane, each particle is performing an oscillation of frequency ω_i due to the focusing property of the ring. Take a collection of these harmonic oscillators perturbed by a common driving force which comes from their average displacement $<x>$. This driving force is linked to the wake field defined in section 2. The equation representing the movement of each particle is then:

$$\ddot{x}_i + \omega_i^2 x_i = c < x > \tag{10}$$

where c is some complex constant.

Let's assume that everything depends on time via $\exp(i\omega t)$ and look for values of ω, real or complex, that satisfy Eq. (10).

With $x_i = X_i e^{i\omega t}$ the solution of this equation is:

$$X_i = c < x > \frac{1}{\omega_i^2 - \omega^2}$$

For simplicity, assume that all ω_i and ω are close to some frequency $n\omega_0 + \omega_\beta$ ($\omega_\beta = Q\omega_0$ and $n\omega_0 + \omega_\beta$ is one of the different mode frequencies, n is positive or negative and in the range where the distributions around two consecutive modes are not overlapping). The final solution is obtained by averaging over-all particles which have a normalised betatron frequency distribution $\rho(\omega_i)$:

$$1 = \frac{c}{2\omega_\beta} \int \frac{\rho(\omega_i) d\omega_i}{\omega - \omega_i} \tag{11}$$

Replacing c by the wake-field forces, the constant $\xi_1 = c/2\omega_0$ becomes:

$$\xi_1 = -\frac{iqJ_c Z_\perp(\omega)}{2Q_0 m_0 c^2 \gamma} \tag{12}$$

where q is the charge of each particle, I the beam current, m_0 the mass of particle and c the velocity of light.

The integration of Eq. (10) is performed in the complex plane, taking into account the pole at $\omega = \omega_i$, and applying the residual theorem:

$$1 = +\xi_1 \left[PV \int \frac{\rho(\omega_i) d\omega_i}{\omega - |\omega_i|} + i\pi \ \text{sign}|\omega_i| \rho(\omega) \right] \tag{13}$$

The principal value PV is an integral in the real plane. To be practical we transform this dispersion integral to be close to the two modes (|n| and -|n|), and we use:

$$f(u) = \Delta\omega \ PV \int \frac{\rho(\omega_i) d\omega_i}{\omega - \omega_i}$$

and

$$g(u) = \pi \ \Delta\omega \ \rho(\omega) \tag{14}$$

$$u = \frac{\omega - \omega_e}{\Delta\omega}$$

$\Delta\omega$ is related to the frequency spread of the distribution and ω_e is the central frequency. We get:

$$+\frac{\xi_1}{\Delta\omega} = \frac{1}{f(u) \pm i g(u)} \tag{15}$$

The the right-hand side of Eq. (15) traced out as a function of u in the normalized complex D plane defines a stability boundary condition. The left-hand side is represented by a single point in the same D plane. If this point lies on the boundary curve, it means that the solution of ω for Eq. (15) is real and the beam is just at the limit of stability. If it lies outside the area enclosed by the curve (on the side), the beam is unstable.

3.1. Lorentz and Elliptical Distribution

The stability boundary depends strongly on the detailed shape of the distribution. We propose to analyse the Lorentz distribution which has large tails and the elliptical distribution which has no tail.

The Lorentz distribution is written:

$$\rho(\omega) = \frac{\Delta\omega}{\pi} \frac{1}{(\omega-\omega_c)^2 + \Delta\omega^2} \tag{16}$$

where $\omega_c = n\omega_0 + \omega_\beta$, and finally

$$\frac{\xi_1}{\Delta\omega} = \frac{\omega_c - \omega}{\Delta\omega} - i \tag{17}$$

The complex mode frequency is then:

$$\omega = n\omega_0 + \omega_\beta + \xi_1 - i\Delta\omega \tag{18}$$

The beam stability condition is then:

$$\text{Im}\,\xi_1 < \Delta\omega \tag{19a}$$

or

$$-k\ \text{Re}\left[Z_\perp\left(n\omega_0 + \omega_\beta\right)\right] < \Delta\omega \tag{19b}$$

where k is a positive constant.

One concludes the expected result that the frequency spread helps to stabilize the beam. The stability condition is always satisfied if $\text{Re}[Z_\perp(n\omega_0 + \omega_\beta)] > 0$ which is the case for $n\omega_0 + \omega_\beta > 0$. It follows that the fast waves are always stable. It is the stability of the slow waves ($n < 0$) which requires a large frequency spread for Landau damping.

In case the above condition (eq. 19) is not fulfilled, a perturbation in the beam will grow until the beam recovers some additional frequency spread due to non-linearities or the beam is lost. The growth rate is then:

$$\frac{1}{\tau} = \text{Im}(\Omega) = -k\ \text{Re}(Z_\perp) + \Delta\omega \tag{20}$$

Well above the instability threshold, $\Delta\omega$ can be neglected and the ultimate growing rate is:

$$\frac{1}{\tau_{inst}} = q\frac{I\ \text{Re}(Z_\perp(\omega))}{4\pi Q_0 m_0 \gamma_0 c\omega_0^2} \tag{21}$$

The elliptical distribution with $u = (\omega-\omega_c)/\Delta\omega$ is

$$\rho(u) = \frac{2}{\pi\,\Delta\omega}\sqrt{1-u^2}\ \mathrm{H}[1-|u|] \tag{22}$$

where H is the Heaviside function
$$\begin{Vmatrix} \mathrm{H}(x) = 1 & x > 0 \\ \mathrm{H}(x) = 0 & x < 0 \end{Vmatrix}$$

The stability is given by:

$$|\xi_1| < \frac{\Delta\omega}{2} \tag{23}$$

which implies the similar conditions as for the Lorentz distribution, with the difference that the imaginary part of impedance (space charge) is very important even in the presence of a small real part.

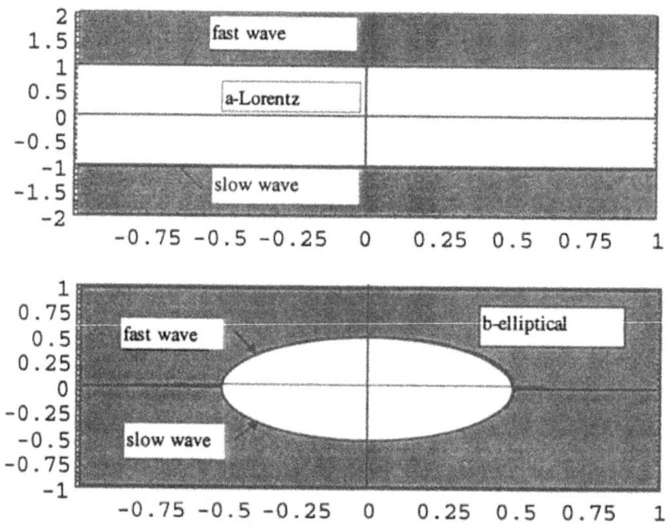

Figure 2 - The stability boundary diagrams for a Lorentz (a) and an elliptical distribution (b) for two frequencies corresponding to the same mode number n. In the upper D-plane is the fast wave and in the lower D-plane is the slow wave. Notice that a distribution with large tail gives larger stability limit.

3.3. Betatron Frequency Spread

One source of the frequency spread comes from the momentum spread which gives two contributions; one due to a dependence of the revolution frequecy ω_0 on $\Delta p/p$ through the slipping factor $\eta = (1/\gamma^2)-(1/\gamma_{tr}^2)$, and one due to the dependance of ω_β through the chromaticity ξ.

$$\Delta\omega = \left|n\omega_0\eta + \xi\omega_\beta\right| \frac{dp}{p} \tag{24}$$

For $\eta = 0$ (close to transition) the spread comes from the chromaticity, for $\xi = 0$ the spread comes from a spread in revolution frequency.

For the particular mode which has $(n\omega_0\eta + \xi\omega_\beta) = 0$, spreads have to be provided by other sources like the dependence of ω_β on amplitude introduced by octupoles. To avoid this requirement, it would be desirable to arrange the parameters in such a way that $(n\omega_0\eta + \xi\omega_\beta) = 0$ occurs in the fast wave regime. This means for $n > 0$, ξ must have the same sign as η which leads to $\xi > 0$ above transition and $\xi < 0$ below transition.

All the above implies that the spread of the negative mode frequencies of the particles $\Delta\omega_n = (-n+q)\delta\omega_0 + \omega_0\,\Delta q$ due to revolution and tune spread is large. Under this condition, particles respond differently enough to each other to prevent a coherent built-up.

In the transverse planes, the incoherent frequencies of the particles are shifted away from the coherent frequencies by $|\Delta Q_{ic}-\Delta Q_c|\omega_0$. At low energy, which is the case for cooling rings, the coherent frequency shift $\Delta Q_c \omega c_0$ is very small compared to the incoherent frequency shift $\Delta Q_{ic}\omega_0$.

As a rule of thumb[8], stabilization by Landau damping requires that the modulus $|\Delta\omega_n|$ of the frequency shift due to space charge is smaller than the total (FWHM) frequency spread by about a factor of π.

$$\left|\Delta\omega_n\right| < \frac{1}{\pi}\left|(n-Q)\omega_0 - \omega_0\,\Delta Q\right|_{FWHM} \tag{25}$$

The frequency-spread term coming from the dependence of tune on the octupoles has been neglected for at least two reasons:

- The octupole spread is constant and does not depend on the mode frequency.
- The external octupole power needed to stabilize the beam is very high due to small beam size, and this octupole power will give difficulties to inject a hot beam into the ring with good efficiency.

It emerges from Eq. (26) that cooled beams can easily lose Landau damping for two reasons: as the emittances shrink, the required tune spread increases, and as the momentum spread decreases, the actual spread diminishes.

The instability threshold has been computed for protons and lead ions (Fig. 3) in LEAR.

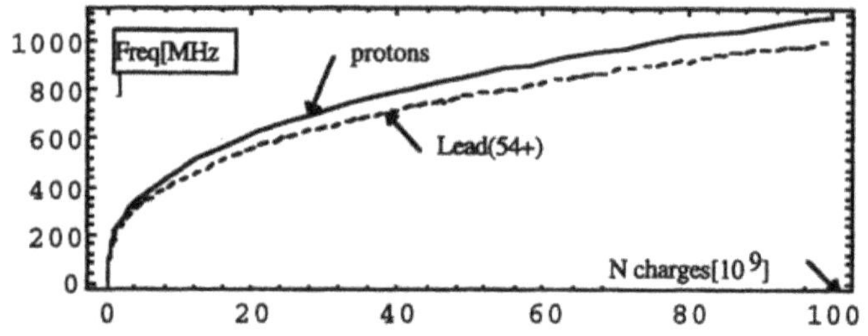

Figure 3 - The threshold for instability at LEAR for 310 MeV/c protons and 341 MeV/c/charge lead ions Pb^{54+} versus the number of charges. The tune shift, emittances and momentum spread used to obtain this curve are obtained by computation of the equilibrium between intra-beam scattering and electron cooling. It appears that for 5×10^{10} protons in LEAR, the beam instability can occur up to mode frequency 400.

3.4. Other Stabilizing Mechanisms

In the case of absence of Landau damping, is it possible that a strong cooling can compensate the growth rate ? Using again Eq. (10) and adding a damping term in the rhs proportional to

$$-\frac{b}{\tau_{cool}} < \dot{x} >, \tag{26}$$

after some computation, it appears that:

$$\tau_{cool} < \frac{\tau_{inst}}{2} \tag{27}$$

is the necessary condition to avoid instabilities[6].

The locus of the stability curve in the D-plane is shifted away from the horizontal axis, leaving room to accommodate a larger real impedance.

In general, observed growth rates are in the range of 10 to 100 ms. With the actual electron cooler, or even laser cooling with coupling to transverse planes, the cooling time constant in the same range is only obtained for high-charge state of the ions like gold or lead. Fast growth rates (in the ms range) are predicted and observed driven by trapped electrons from the ionisation of the residual gas. These can probably not be compensated by beam cooling. We note also that frequency shifts and growth rates for ion beams scale as Z^2/A, just as the optimistic scaling for the electron-cooling rate. Some recent measurements and theories[7] predict an electron-cooling rate as $Z^{1.5}/A$ in the ultimate beam cooling process.

Obviously, to avoid instability the first antidote consists of reducing as much as possible the real impedance of the machine. The resistive wall impedance implies that the tune is not close and below an integer and it is easily damped by a small tune spread. The vacuum chamber has to be smooth enough to avoid local resonances due to transitions and small cavities. A problem arises from the kicker magnet used for fast injection or extraction. Good improvements have been obtained at LEAR by just insuring the ground continuity between the vacuum chamber and the magnet itself, but no drastic changes were seen on the beam stability. In the cooler rings, there are a lot of electrostatic or Schottky pick-ups[8] and kickers used for beam observation. Great care has to be taken in building them in such a way that they do not add resistive impedance. Good matching to the head amplifiers and resistor termination are necessary.

Other antidotes come from active systems. For large incoherent tune shift, one can imagine creation in the momentum distribution a tail towards low frequencies[10]. This tail is such that the corresponding incoherent betatron frequency covers the coherent frequency with the same factor π, as discussed earlier. This requires for the stability of mode n a very large momentum tail extending to:

$$\left(\frac{\Delta p}{p}\right) = \frac{\pi}{|\eta| n}|\Delta Q_{ic}| \tag{28}$$

It is feasible for small $|\Delta Q_{ic}|$ to stabilize high-order modes, but the realisation is not very practical as it deteriorates the beam quality needed for physics purposes. It was tested at LEAR using two methods:

- A white noise was applied to the low-frequency side of the momentum distribution to diffuse particles in the tail, but the noise power needed is very high and it perturbs the cooling.
- A modulation was applied on the cathode voltage of the electron cooler to create the tail but this is detrimental to the cooling process.

All the potentially unstable modes can be stabilized by an active feedback system[1]. It consists of an electrostatic pick-up detecting the emerging coherent oscillation, and after the necessary amplification this signal is fed into a kicker to the beam with the right delay and phase advance. This resistive feedback can be modelled in the rhs of Eq. (10) by adding a term:

$$g < y > (t - \tau) \tag{29}$$

with $\tau = \pi/2\omega_\beta$.

The effect is the same as for the cooling forces, i.e. an additional resistive impedance with $Z_d(-\omega) = Z_d(\omega)$. This damping impedance has to be larger than $\mathrm{Re}(Z_\perp(\omega))$ for all the potentially unstable modes. It means that the bandwidth of the system must be very large, as predicted in Fig. 3. The damper in LEAR is composed of two systems; one for low-frequency from 50 kHz to 70 MHz, and the second for higher frequencies. The latter is just the transverse stochastic cooling system where the overall gain is decreased to 50 dB.

3.1. *Beam Transfer Function Measurements*[1,2,6,11]

In practical accelerator operation, it is often useful to find out how the beam responds to a sinusoidal driving force. This beam response, called the beam transfer function (BTF), contains valuable information about the beam and the storage ring. Taking again Eq. (10) and adding to the rhs an external driving force d $<\dot{x}>$, the BTF becomes[10]:

$$\frac{1}{BTF} = \frac{\text{excitation}}{\text{response}} = Z_{\perp}(\omega\pm) \pm \frac{ki}{[f(u) \pm i\,g(u)]} \tag{30}$$

with k a real constant, ± for fast and slow wave.

This 1/BTF corresponds to the locus of the dispersion relation rotated by 90° clockwise, amplified by k and finally shifted by the impedance vector Z_T. As seen previously, resistive damping coming from the active feedback or the electron cooling drifts the locus in the same way.

This is an important tool to measure the effects of the different impedances, or to analyse the effect of the active feedback (Fig. 4)

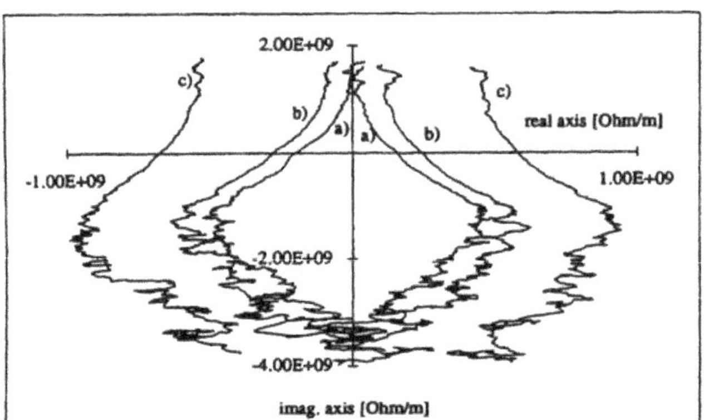

Figure 4 - Example of an horizontal stability diagram of a beam with 8×10^9 protons at LEAR (310 MeV/c) at the 70th harmonic (~84 MHz) with 50 dB as the gain for the low-frequency feedback system and different gains for the broad-band stochastic cooling system. a) 50 dB, b) 70 dB, c) 85 dB.

3.6. *Higher Order Modes*

We have treated only the dipole instability, i.e. modes where the beam moves as a rigid body. But higher-order modes ("throbbing beam") may become unstable, like coherent quadrupolar mode (oscillation of the beam envelope), or more generally coherent beam shape oscillations with frequencies,

$$\omega_{np} = \omega_0(n - pQ) \quad \text{with } p = 2, 3, \ldots \tag{32}$$

Some, but probably not all, potentially dangerous modes can be stabilized by an active feedback. Also the amount of frequency spread needed for stabilization is in general smaller than for dipole mode.[12]

4. Stability in the Longitudinal Plane

4.1. General Derivation[1,2,13,14]

Landau damping of longitudinal instability in a coasting beam is rather similar to the oscillator problem discussed in Section 3, but there are important differences which make it worthwhile handling the problem by use of Vlasov equation. This is:

$$\frac{\partial \psi}{\partial t} + \dot{\theta}\frac{\partial \psi}{\partial \theta} + \dot{\delta}\frac{\partial \psi}{\partial \delta} = 0 \tag{32}$$

where $\psi(\theta,\delta,t)$ is the normalised particle distribution in the longitudinal phase space, $\theta = s/R$ and $\delta = \Delta p/p$. For a coasting beam with a small perturbation around the mode n

$$\psi = \psi_0 + \psi_n e^{i(n\theta - \omega t)} \tag{33}$$

We get, after insertion into Eq. (33), integration overall momenta and changing from the variable δ to frequency ω, the dispersion relation:

$$1 = i\frac{I\eta q\omega_0^2}{2\pi m_0 c^2 \gamma_0 \beta_0^2}\frac{Z_{//}}{n}\int \frac{\partial \psi_0/\partial \omega}{(\omega_n/n) - \omega}d\omega \tag{34}$$

The variables ω_n/n and ω are taken around ω_0.

After integration in the complex plane, we obtain:

$$1 = \xi_2\left(PV\int \frac{\partial \psi_0/\partial \omega}{(\omega_n/n) - \omega}d\omega + i\pi\frac{\partial \psi_0(\omega_n/n)}{\partial \omega}\right) \tag{35}$$

with
$$\xi_2 = i\frac{I\eta q\omega_0^2}{2\pi m_0 c^2 \gamma_0 \beta_0^2}\frac{Z_{||}}{n}$$

As for the transverse plane, we denote the width of the spectrum by $\Delta\omega$ and the dispersion relation becomes:

$$\frac{\xi_2}{\Delta\omega^2} = \frac{1}{f(u) + ig(u)} \tag{36}$$

with
$$f(u)=\Delta\omega^2\ PV\int\frac{(\partial\psi_0/\partial\omega)d\omega}{(\omega_n/n)-\omega}$$

$$g(u)=\pi\ \Delta\omega^2\frac{\partial\psi_0}{\partial\omega}\left(\frac{\omega_n}{n}\right)$$

The rhs of Eq. (37) traced out as a function of ω in the normalised phase space D_2 defines a stability boundary condition. If the frequency ω_n approaches the harmonic of the revolution frequency, the curve comes from infinity, moves around the origin whilst ω_n moves through $n\omega_0$ and moves again to large values after. The central part is very often approximated by a circle of radius of about 1. This gives a very simple stability condition:

$$\left|\frac{Z}{n}\right|\leq\frac{2\pi\ \beta^2\gamma\ A m_0c^2\eta(\Delta p/p)_\sigma^2}{qI} \tag{37}$$

which is called the Keil-Schnell criterion, where $(\Delta p/p)_\sigma$ is the RMS width of the momentum distribution.

In the case of space-charge dominated beams (Im $Z_{//}<0$) and rings working below transition, the point corresponding to $\xi/\Delta\omega^2$ is located to the side where the locus goes to infinity. In particular, all cooler rings work below transition.

The equilibrium between cooling and diffusion (IBS) forces gives a Gaussian distribution of particles. In that case, the Keil-Schnell criterion can be overcome by some factors[13]. The low-frequency tail of the distribution is then very important to stabilize the beam (Figs. 5,6). One can ask the question whether a growing instability in this tail will lead to a diffusion process of the particles towards the low frequency tail restoring a sufficient amount of Landau damping. We have always observed at LEAR, after having created an uniform longitudinal distribution over a certain bandwidth prior to slow extraction, that coherent signals appear at the sharp edges of the distribution. These signals stay for a while but decrease slowly during the time the sharp edges are smoothed out.

The effect of the cooling rate ν has been calculated[6]. It gives an opening of the "thermometer shaft" of the stability diagram (Fig. 7). It ensures the possibility of having stable beam with large space-charge effect. In the case of laser cooling (ν>> 1000 s^{-1}) we can consider that the distribution is a Dirac function and the stability condition becomes:

$$\frac{\nu}{2}>\sqrt{\frac{I\eta q}{2\pi m_0c^2\beta_0^2\gamma}}\ \frac{\mathrm{Re}|Z_{//}/n|}{2\sqrt{\mathrm{Im}|Z_{//}/n|}}$$

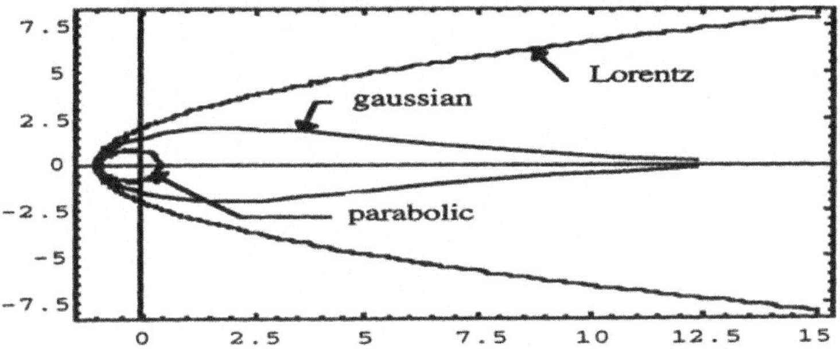

Fig. 5 - Locus of the rhs of the dispersion relation in D_2 plane for three distributions: a) Lorentz, b) Gaussian, e) Parabolic.

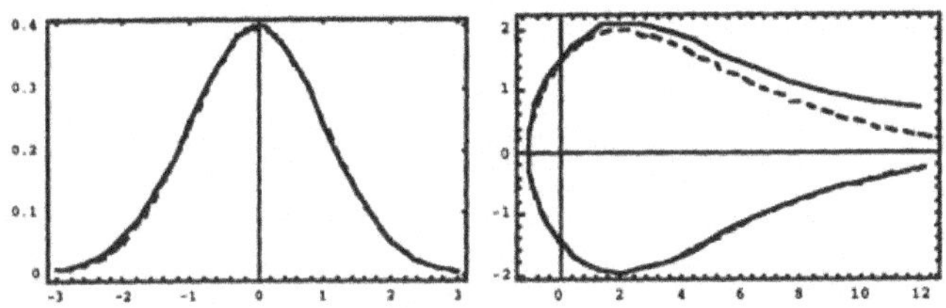

Fig. 6 - Effect of a small low-frequency tail onto the stability diagram. a) Gaussian distribution with (solid) and without (dash) tail: the tail contains 1% of the particles. b) Corresponding stability diagram; Notice the large effect of a very small tail.

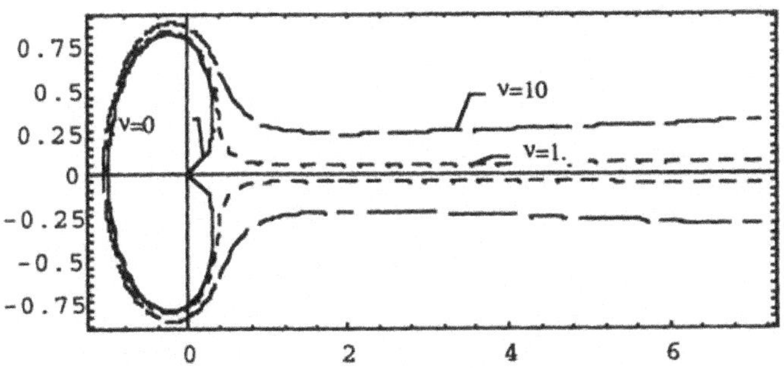

Figure 7 - Effect of the cooling on the stability diagram for a parabolic distribution. a) $v = 0$, b) $v = 1\ s^{-1}$, c) $v = 10\ s^{-1}$, LEAR, $\Delta p/p = 2 \times 10^{-4}$

4.2. Longitudinal Beam Transfer Function[14]

To know more about longitudinal impedances of the machine and the frequency spread of the beam in the longitudinal plane, it is useful to find out how the beam responds to a small sinusoidal excitation applied on a longitudinal "kicker". This beam response (BTF) has been computed and gives - with some normalisation - a trace of its locus in the impedance plane:

$$\frac{1}{BTF} = \frac{ik_2}{f(u)+ig(u)} + \frac{Z_{//}}{n}$$

This BTF^{-1} corresponds to the locus of the dispersion relation rotated by 90° anti-clockwise and shifted by the impedance vector $Z_{//}/n$. It has been used extensively at LEAR and ESR to find out the impedance versus frequency (Fig. 8, 9, 10).

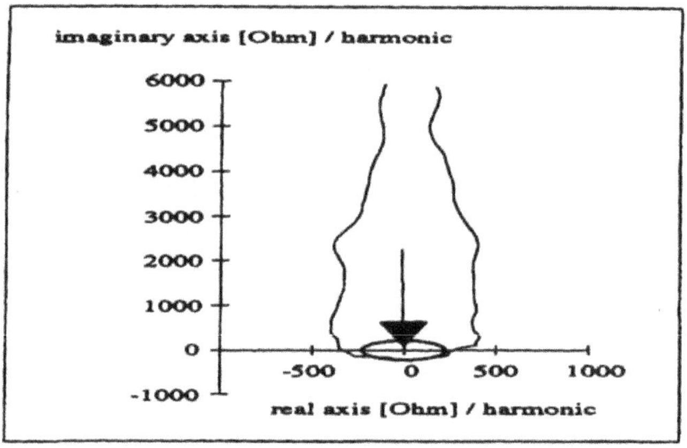

Figure 8 - A typical longitudinal BTF measurement at LEAR[15] for 5×10^{10} protons at 310 MeV/c. The arrow represents the impedance and the circle is the Keil-Schnell criterion.

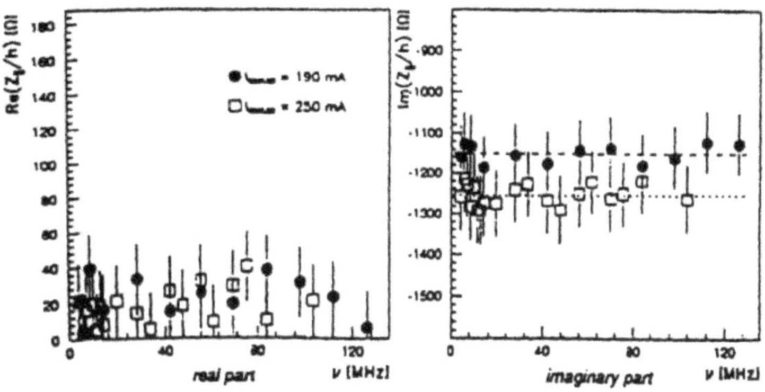

Figure 9 - $Z_{//}/n$ measured at ESR[15] for two electrons current of the electron cooling (190 and 250 mA). These figures show $Z_{//}/n \approx (20 - I\ 1200)\Omega$.

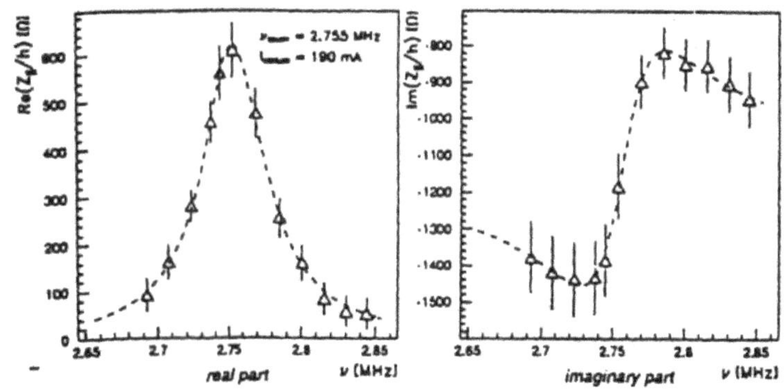

Figure 10- Impedances around the fundamental frequency of the rf cavity at ESR[16]. The imaginary part of the impedance is shifted by -1100 Ω corresponding to the space-charge impedance. The shunt impedance is 600 Ω.

The high shunt resistance of the rf cavity is a source of instability which appears to give a self bunching of the beam. This is important that the gap of the rf cavity is short-circuited during coasting beam operation. It is also recommended to decrease the shunt impedance by adding a feedback system to the rf power system of the cavity. (A reduction from 2500 Ω to 200 Ω has been obtained at LEAR).

In general it is much easier to obtain beam stability in the longitudinal plane. For coasting beam operation, to our knowledge, none of the cooler rings uses a longitudinal active feedback system.

5. Conclusion

The longitudinal stability of the beam is generally achieved for cooler rings working below transition. For transverse planes, there is a need for a powerful broad-band active feedback system. The cooling rate is a source of beam stabilization if it is greater than 1000 s^{-1}. This can be achieved with laser cooling and good coupling between longitudinal and transverse plane.

6. Acknowledgements

The author would like to thank D. Möhl for useful discussions during the preparation of this report.

7. References

1. A.W. Chao, *Physics of Collective Instabilities in High Energy Accelerators*, (John Wiley & Sons, Somerset, USA, 1990).
2. H. Wiedeman, *Part. Acc. Physics II*, (Springer-Verlag, Berlin-Heidelberg, 1995).
3. J.L. Laclare, in *Proc. of the CERN Accelerator School* (Gif-sur-Yvette, 1985), eds. P. Bryant and S. Turner (CERN 85-19, 1985) p. 377.
4. V. G. Vaccaro, this conference.
5. H.G. Hereward, in *Proc. of CERN Accelerator School* (Oxford, 1987), ed. S. Turner (CERN 87-03, 1987) p. 255.
6. S. Cocher, *Effets collectifs d'un faisceau refroidi d'ions lourds dans un anneau de stockage*, Thèse (Université de Paris-Sud, Centre d'Orsay, 1989).
7. a) I. Hofmann, in *Proc. of the Workshop on Beam Cooling and Related Topics* (Montreux, 1993), ed. J. Bosser (CERN 94-03, 1994) p. 330.
   b) A. Wolf, C. Ellert, M. Grieser, D. Habs, B. Hochadel, R. Repnow, and D. Schwalm, in *Proc. of the Workshop on Beam Cooling and Related Topics* (Montreux, 1993), ed. J. Bosser (CERN 94-03, 1994) p. 416.
   c) G. Zwicknagel, C. Toepffer and P.G. Reinhard, in *Proc. of the Workshop on Beam Cooling and Related Topics* (Montreux, 1993), ed. J. Bosser (CERN 94-03, 1994) p. 245.
8. a) K. Hübner and V.G. Vaccaro, *Dispersion Relations and the Stability of Coasting Particle Beams*, (CERN/ISR/TH 70-44, 1970).
   b) W. Schnell and B. Zotter, *A Simplified Criterium for Transverse Stability of a Coasting Beam*, (CERN/ISR/GS/RF/76-26, 1976)
9. K.Y. Ng, *Field, Impedances and Structures*, Fermilab report FN 443, (1987).

10. I. Hofmann, in *Proc. of the 3rd LEAR Workshop* (Tignes, 1985), eds. U. Gastaldi, R. Klapisch, J.M. Richard and J. Tran Tranh Van (Editions Frontières, Gif-sur-Yvette) p. 159.
11. D. Möhl, A. Sessler, *The Use of RF Knockout for Determination of the Characteristics of the Transverse Coherent Instabilities of Coasting Beams,* (Proc 1971 International Conference on High Energy Accelerators, CERN 1971, p.334)
12. E. Mustafin, *Self Stabilization of Higher-Multipole Modes of Coherent Transverse Oscillations Due to Nonlinear Coulomb Tune-Shift,* in Proc. of the fourth European Particle Conference, EPAC94, Vol.2, P.1174 World Scientific eds.
13. I. Hofmann, K. Beckert, S. Cocher, and U. Schaaf, in *Proc. of EPAC 90 (Nice),* **Vol. 1**, eds. P. Marin and P. Mandrillon (Editions Frontières, Gif.sur-Yvette, 1990) p. 229.
14. A. Hofmann, in *Proc. of the CERN Accelerator School* (Nordwijkerhout, 1991), ed. S. Turner (CERN 92-01, 1992) p. 88.
15. U. Oeftiger, *Measurement of Beam Properties and Beam Environment in LEAR and COSY Using RF Excitation Methods*, Thesis, (University of Bonn, 1994).
16. G. Kalisch, K. Beckert, B. Franzke, I. Hofmann, U. Schaef, in *High Energy Density in Matter*, (Annual Report 1991), GSI 92-27, GSI, Darmstadt, p. 26.

II. COOLING TECHNIQUES

Methods of Beam Cooling *

Andrew M. Sessler
Center for Beam Physics
Lawrence Berkeley Laboratory
Berkeley, California 94720

ABSTRACT

Diverse methods which are available for particle beam cooling are reviewed. They consist of some highly developed techniques such as radiation damping, electron cooling, stochastic cooling and the more recently developed, laser cooling. Methods which have been theoretically developed, but not yet achieved experimentally, are also reviewed. They consist of ionization cooling, laser cooling in three dimensions and stimulated radiation cooling.

Introduction

In particle beam physics, the concept of "cooling" has a different meaning than in the rest of physics. In, say, thermodynamics, "cooling" means the reduction of the material (gas, liquid, solid, plasma) temperature. In particle beam physics that is not adequate for "cooling". In particular, as one focuses and de-focuses beams, the transverse temperature can easily be changed. The "cooling" of particle beams is based upon the concept of phase space and a theorem first introduced by Liouville [1] (see Fig. 1).

Liouville's theorem states that in a conservative Hamiltonian system, such as a single particle in an external magnetic and electric field, the phase density of many non-interacting systems (having slightly different initial conditions) is preserved as one follows the motion of the system. Thus, this theorem is applicable to a beam of particles when the interaction between particles is negligible and can be ignored, which is often true — but not always true — for high energy beams.

Even within the restrictions of Liouville's theorem it is possible to arrange to interchange phase space (between, say, longitudinal and transverse degrees of freedom). Of course, other Hamiltonian requirements, summarized in the requirement that the transformation be symplectic, greatly restricts this process. Sometimes, although it is not possible, or convenient, to exchange phase space, it is possible to introduce correlations between the degrees of freedom.

---

* Work supported by the Director, Office of Energy Research, Office of High Energy and Nuclear Physics, Division of High Energy Physics of the U. S. Department of Energy, under Contract No. DE-AC03-76SF00098.

Figure 1 : Joseph Liouville (1809 - 1882)

A proof of Liouville's theorem can be found in many text books [2]. A simple way of looking at this theorem is to think of it as equivalent to the condition of incompressible flow in the phase space of a given system. Let the system be described by the N coordinates q_α and the N conjugate nomenta p_α, $\alpha=1,2,...,N$. The phase space is just the 2N-dimensional space with coordinates q_α and p_α, and the development in time of the state of the system is represented by the trajectory of a single point in phase space. Just as with fluid flow, there is a well-defined velocity field at each instant of time, which assigns to each point in phase space a definite velocity, with components q_α and p_α given as functions of the q's and p's by Hamilton's equations. For fluid flow in any number of dimensions the condition that volumes are preserved by the flow is equivalent to the vanishing of the divergence of the velocity field:

$$\nabla \cdot \mathbf{v}(x) = 0 \ . \tag{1}$$

For phase space the components of the velocity field are

$$\dot{q}_\alpha = v_\alpha(q, p, t) = \partial H / \partial p_\alpha \ , \tag{2}$$

$$\dot{p}_\alpha = F_\alpha(q, p, t) = -\partial H / \partial q_\alpha \ , \tag{3}$$

The divergence condition (1) then becomes

$$\Sigma_\alpha (\partial v_\alpha / \partial q_\alpha + \partial F_\alpha / \partial p_\alpha) = \Sigma_\alpha (\partial^2 H / \partial p_\alpha \partial q_\alpha - \partial^2 H / \partial q_\alpha \partial p_\alpha) = 0 \ , \tag{4}$$

which is automatically satisfied as a consequence of Hamilton's equations, and thus demonstrates the validity of Liouville's theorem. It is important to note that the Hamiltonian may depend explicitly on the time t.

Often a dynamical system is represented by an ensemble of possible states, with a distribution function f giving the number $n\Delta V$ of particles systems in a small volume ΔV of phase space:

$$n\Delta V = f(q, p, t)\Delta V \ . \tag{5}$$

If the q_i and p_i are Hamiltonian variables, then ΔV is conserved, and since the number of particles is clearly invariant, we deduce that moving with the particles (or set of systems) the density function f is constant:

$$\frac{df}{dt} = 0 \ . \tag{6}$$

Note that we can apply Liouville's theorem to an ensemble of non-interacting particles, but not to an ensemble of interacting particles.

However if one has a conservative fluid then Liouville's theorem exactly applies, as was shown in Ref. [3]. To a collection of particles interacting through their self-generated electric and magnetic fields Liouville's theorem again applies. (It doesn't work when "hard scattering" - - non-fluid-like behavior is included.)

It is, then, in the context of Liouville's theorem that "cooling" is defined. It means an actual reduction of phase space volume; that is, an increase in phase space density. This can only be achieved by violating the assumptions behind Liouville's theorem. Generally, this is not easy, but, nevertheless, a number of effective cooling methods have been devised. We review them in this paper. For each of the methods, we leave to the reader the task of determining how Liouville's theorem for a continuous conservative medium is circumvented.

Radiation Cooling

The radiation of accelerating electrons has been known since the last century, but it was only in 1956 that Kolomenski and Lebedev [4] realized that the radiation reaction has a damping effect upon the emitting electron.

The radiation of an electron moving in a circular accelerator is well-known. The power emitted is:

$$P = \frac{2}{3}\frac{e^2\beta^4\gamma^4 c}{R^2} , \tag{7}$$

where e is the electron charge, β is the electron speed divided by c, c the speed of light, γ the relativistic factor, and R the radius of the circular orbit.

Hence the radiated energy per turn is

$$\delta E = \frac{4}{3}\frac{\pi e^2}{R}\beta^3\gamma^4 , \tag{8}$$

or, in practical units (and taking β=1):

$$(\delta E)(\text{MeV}) = 8.85 \times 10^{-2}\frac{E^4(\text{GeV})}{R(\text{m})} . \tag{9}$$

The frequency spectrum of the radiation is complicated; for low frequencies it varies as $\omega^{2/3}$. The radiation drops off exponentially for ω larger than a critical frequency, ω_c, where

$$\omega_c = \frac{3\gamma^3 c}{R} . \tag{10}$$

Note that this frequency varies as γ^3 times the revolution frequency. Thus the radiation can extend up to very high frequencies.

The emission of radiation has an effect on the radiation particle [4-6]. The radiation reaction can cause either damping or undamping of the electrons' oscillations (transversely and in energy) about the equilibrium orbit. If we characterize this exponential damping rate by rate constants $\alpha_x, \alpha_y, \alpha_E$ then

$$\alpha_i\begin{pmatrix} x \\ y \\ E \end{pmatrix} = \frac{c\gamma^3 e^2 J_i}{3R^2 mc^2} , \tag{11}$$

where the subscript i stands for x, y or E, and m is the electron mass.

The damping partition numbers satisfy:

$$J_y = 1, J_x + J_E = 3 . \tag{12}$$

One can arrange by proper lattice design, as one must in a storage ring, to have damping in all three directions.

Thus, on the basis of the above analysis, an electron beam in a storage ring will just damp and damp so that its transverse size becomes smaller and smaller. This is approximately true, and beams become very small indeed, but they do not become arbitrarily small. Why not? Because quantum effects need to be taken into account; i.e. that electrons radiate discrete photons and that the hard photons, which are radiated statistically, kick the electron. In fact, the size of electron beams at equilibrium is determined by these quantum mechanical effects. The energy spread σ_E of the beam, which also damps to zero classically, is (in a uniform field),

$$\left(\frac{\sigma_E}{E}\right)^2 = \left(\frac{55}{32\sqrt{3}}\right)\left(\frac{\hbar}{mc}\right)\frac{\gamma^2}{J_E R} , \tag{13}$$

and one can see that the finiteness of σ_E is due to a quantum mechanical effect; i.e. to the non-zero nature of Planck's constant $\hbar$.

For transverse motion we have for the normalized emittances

$$\varepsilon_{ny} \geq \left(\frac{55}{32\sqrt{3}}\right)\left(\frac{\hbar}{mc}\right)\frac{\gamma}{2\sqrt{n}} , \tag{14}$$

$$\varepsilon_{nx} \geq \left(\frac{55}{32\sqrt{3}}\right)\left(\frac{\hbar}{mc}\right)\left[\frac{(1-2n)^2}{n(1-n)}\right]\gamma^3 , \tag{15}$$

for field gradient $0<n<3/4$, where the formulas have been given for a constant gradient, weak focusing machine.

Recently some insight into the process of radiation damping was provided through an analysis, by Huang, Chen and Ruth, who considered radiation damping in a uniform focusing channel [7]. Their analysis is quantum mechanical, but need not be. They show that an electron introduced into such a channel will experience damping of its transverse betatron oscillations at a rate

$$\alpha = -\tfrac{2}{3}\frac{r_e k}{mc} \,, \tag{16}$$

where r_e is the classical electron radius and k is the focusing strength of the channel ($V(x)=kx^2/2$). The rate of damping is very slow (no γ^3 in Eq. 16!) but, there is no quantum excitation from photon recoil as there is for motion in a circular orbit. This is because there is no continual, forced, emission of photons. Consequently, the final emittance reached is only determined by uncertainty in the lowest state of an oscillator and can be derived, easily, from the fact that $E \geq \hbar\,\omega/2$. Thus,

$$\varepsilon_n \geq \hbar / 2mc = \lambda c / 2 \,, \tag{17}$$

where λ_c is the Compton wavelength and ε_n is the normalized emittance.

The contrast between Eq. (17), and Eqs. (14 & 15) are most dramatic (γ dependence !) and very significant in practice (since damping rings usually have $\gamma \sim 2\times10^3$).

Electron Cooling

Electron cooling was first suggested by Budker [8] (see Fig. 2). A good discussion may be found in Ref. 9.

$$\tau_e = \frac{C}{L_e}\frac{F_1\gamma^2}{r_e r_i n_e c \ln\Lambda}\left[\left(\frac{k_B T_{be}}{m_e c^2}\right)^{3/2} + \left(\frac{k_B T_{bi}}{m_i c^2}\right)^{3/2}\right] . \tag{18}$$

Here n_e is the electron density, assumed to be the same as the ion density n_i, r_e and r_i are the classical electron and ion radii. F_1 is a constant that for a smooth focusing system has the value $F_1 = 3/4\sqrt{2\pi} \approx 0.3$ and γ is the relativistic energy factor (identical for both beams); the electron and ion temperatures are measured in the beam frame. L_e / C is the fraction of the storage ring occupied by the cooling section, and $\ln\Lambda$ is the Coulomb logarithm. When equilibrium is reached, the two beam temperatures are the same (i.e., $T_{bi} = T_{be}$). Assuming that both beams have identical transverse cross sections, one obtains an emittance ratio of $\varepsilon_i/\varepsilon_e \approx (m_e/m_i)^{1/2}$ that is, the ion-beam emittance is considerably smaller than that of the electron beam.

Figure 2 : Gersh I. Budker, (1918 - 1977)

Stochastic Cooling

The idea of stochastic cooling is due to Simon van der Meer [9] (see Fig. 3). The article by Bisognano presents a comprehensive treatment of the subject [10]. At the most elementary level, a pick-up takes a signal from a section of beam, amplifies it and sends it across the diameter of an accelerator so as to put an appropriate signal upon a beam kicker just as the same section of beam arrives on its circular route.

Suppose there are N particles in the ring and that $f = 1/T$ is the revolution frequency, and that W is the bandwidth of the electronics. Then the pick-up "sees" a number of particles

$$n = \frac{N}{2WT} \quad . \tag{19}$$

Under influence of pick-up and kicker the betatron amplitude of oscillation becomes

$$x_i \rightarrow x_i - g\sum_{j=1}^{n} x_j \quad , \tag{20}$$

where g is the effective gain of the system.

$$\Delta\left(x_i^2\right) \equiv \left(x_i - g\sum_{j=1}^{n} x_j\right)^2 - x_i^2 \quad , \tag{21}$$

$$\Delta x_i^2 = -2gx_i\left(\sum_{j=1}^{n} x_j\right) + g^2 \sum_{j=1}^{n}\sum_{k=1}^{n} x_j \cdot x_k \quad . \tag{22}$$

Initially there are no correlations so

$$\left\langle \Delta x_i^2 \right\rangle = -2g\left\langle x_i^2 \right\rangle + ng^2\left\langle x_i^2 \right\rangle \quad . \tag{23}$$

Optimum cooling is obtained at g=1/n and there

$$\left\langle \Delta x_i^2 \right\rangle = -\frac{1}{n}\left\langle x_i^2 \right\rangle \; . \tag{24}$$

So the rate of damping of rms betatron amplitudes is :

$$\frac{1}{\tau} = \frac{1}{2}\,\frac{1}{2}\,\frac{1}{n}\,\frac{1}{T} = \frac{W}{2N} \quad . \tag{25}$$

x not x' damped (first $\frac{1}{2}$); rms (second $\frac{1}{2}$); happens once a turn ($\frac{1}{T}$)

Figure 3 : Simon van der Meer, (Nobel Prize, 1984)

From this we see it is good to make W as big as possible and that it works best at small N.

If there is noise then

$$x_i \rightarrow x_i - g\left(\sum_{j=1}^{n} x_j + r \right) \quad . \tag{26}$$

↑ amplifer noise expressed as apparent x - amplitude at the pick-up

Now

$$\frac{1}{\tau} = \frac{W}{2N\left[1 + \frac{\langle r^2 \rangle}{\langle x_i^2 \rangle}\right]} \quad . \tag{27}$$

↑ $\left(\frac{\text{noise power}}{\text{signal power}}\right)$

Thus we see that amplifier noise only slows things down, doesn't stop cooling at some effective temperature (of course some — less important sources of noise — do introduce a temperature limit).

There have been efforts to raise the typical frequency employed in stochastic cooling (~ 5 GHz) to much higher frequencies (~ 90 GHz). This work, so far, has only been theoretical. Recently, there have been proposals by Mikhailichenko and Zolotoev and Zolotorev and Zholents to extend operation into the optical range [11,12]. For this purpose it is proposed to use wigglers for the pick-ups and kickers while wide-band lasers are employed as amplifiers. Experimental work on these ideas is being initiated, as well as theoretical analysis taking into account 3-D effects [13].

Laser Cooling

The cooling of not-fully-ionized ions by means of Doppler laser cooling was first introduced in 1975 by a number of atomic physicists [14]. It was applied to beams of ions in a storage ring by groups in Heideberg, Germany and Aarhus, Denmark [15]. A rather comprehensive treatment can be found in the thesis of J. Hangst [16].

Laser cooling is the result of velocity-selective transfer of photon momentum from a laser beam to a moving ion. In the most basic laser cooling scheme, known as Doppler cooling, particles have a closed transition (i.e. the population is confined to two levels) between internal energy levels. Those particles which are in resonance with a laser beam absorb photons. Each absorbed photon transfers momentum of magnitude $h\nu / c$ to the particle, which recoils in the direction of the laser beam propagation. When a photon is spontaneously emitted by the excited particle, the particle again recoils, but the average momentum transfer to a particle after many spontaneous emissions is negligible, because the angular distribution of the emission is symmetric. There is thus a net radiation pressure force, directed along the laser beam, on resonant particles. The process is shown in Fig. 4. Due to the small magnitude of the photon momentum (optical photons have at most a few eV/c), it is necessary for an ion to absorb many photons to achieve macroscopically significant acceleration. It is for this reason that the optical transition must be closed. The force is dependent on the velocity of the particle, by virtue of the Doppler shift. By tuning the frequencies of two counter-propagating lasers to accelerate slow particles and decelerates fast ones, it is possible to reduce the velocity spread (in one dimension) of a collection of particles; hence the name "cooling".

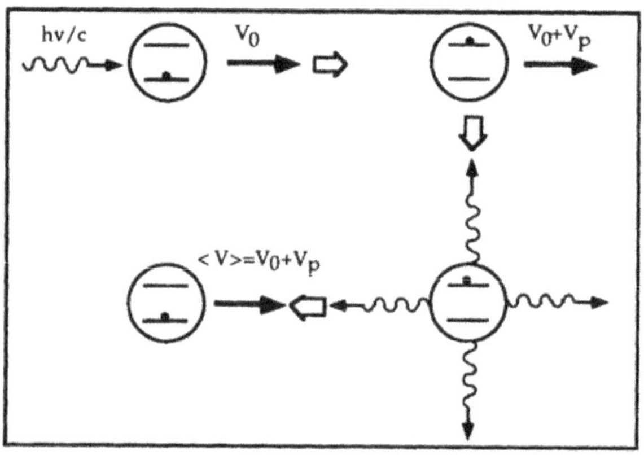

Figure 4 : The principle of laser cooling. The net effect of photon absorption and emissions is that the ion has a change in velocity.

The minimum temperature T_{min} achievable by Doppler cooling is given as

$$k_B T_{min} = \frac{\hbar\Gamma}{2} \quad , \tag{28}$$

where k_B is the Boltzmann constant, and Γ is the spontaneous decay rate of the upper level back to the lower level; $\Gamma/2$ is the radiative linewidth of the transition. The maximum laser cooling force

$$F_{LC}^{max} = \left(\frac{\Delta p}{\Delta t}\right)_{max} = \frac{\hbar k_0 \Gamma}{2} \quad , \tag{29}$$

where k_0 is the wave number.

In a storage ring, laser cooling is sometimes achieved using two laser beams, overlapping the ion beam for up to a few meters in the machine. One laser propagates opposite to the ions' direction of travel and can decelerate particles with velocities above the mean. The second laser is co-propagating and acts to accelerate slow particles. The initial energy spread in a stored beam is usually much higher than the linewidth of the cooling transition. The lasers must then be tunable over a wide frequency range in order to interact with all of the ions.

There are at least four different ways in which to achieve laser cooling, all of which have been used in practice. That is, it is not necessary to employ two laser beams if: (1) one laser can be swept in frequency, or an auxiliary force is employed which is (2) constant, (3) varies linearly, or (4) is produced by an RF bucket. Of course, much theoretical work, employing the Fokker-Planck equation, etc. has been done on this subject. The essence of the laser cooling, however, has been described here.

Physicists who do laser cooling generally work with the concept of temperature. In terms of particle velocities the longitudinal temperature, $T_\parallel$, and the transverse temperature, $T_\perp$, are given by

$$\frac{1}{2} k_B T_\parallel = \frac{1}{2} m \left(\delta v_\parallel\right)^2 \quad , \tag{30}$$

$$k_B T_\perp = \frac{1}{2} m \left(v_x^2 + v_y^2\right) \quad , \tag{31}$$

where k_B is Boltzman's constant.

For particle beams these are not very good quantities as they are not invariants as one moves about the storage ring. In practical units

$$T_\parallel(^\circ K) = 2.32 \times 10^4 \left(\frac{\delta p}{p_0}\right) E_0[eV] \quad , \tag{32}$$

$$T_{\perp}(^{\circ}K) = 2.32x10^{4} \frac{\varepsilon_{\perp}[m-rad]}{\langle\beta\rangle[m]} E_0[eV] \quad , \tag{33}$$

where E_0 is the kinetic energy of the reference particle, $\varepsilon_{\perp}$ is the transverse beam emittance, and $\langle\beta\rangle$ denotes the averaged value of the betatron function.

Experimental achievements have been $T_{\parallel} \sim 1mK$ for $^7Li^+$ and $^{24}Mg^+$ ions at 100 keV, and $T_{\parallel} \sim 30mK$ for $^7Be^+$ at 7.3 MeV. The achievements correspond to $\Delta p/p \sim 4x10^{-7}$ (to be contrasted with $\Delta p/p \sim 4x10^{-5}$ from electron cooling prior to the laser being used). Unfortunately, laser cooling is only effective in the longitudinal direction, and in these mK-experiments the transverse temperature remained at the 1000°K value (to which it had been brought by electron cooling).

Ionization Cooling

There are three methods of cooling, which are reviewed here, which have not yet been achieved experimentally, but have been subject to detailed theoretical analysis. Ionization cooling is special to muons, but is the basis of much recent work on Muon-Muon Colliders; it is the first of the three methods which we review. The possibility was first put forward by Sasha Skrinsky [17] and analyzed, in considerable detail, by Neuffer [18].

The concept is quite simple: muons passing through a material medium lose energy (and momentum) through ionization interaction. The losses are parallel to the particle motion, and therefore include transverse and longitudinal momentum losses; the transverse energy loses reduce (normalized) emittance. Re-acceleration of the beam (in rf cavities) restores only longitudinal energy. The combined process of ionization energy loss plus rf re-acceleration reduces transverse momentum and hence reduces transverse emittance. However, the random process of multiple scattering in the material medium increases the emittance. There can also be longitudinal cooling, under proper circumstances, as we discuss below. The equation for transverse cooling can be written in a differential-equation form as:

$$\frac{d\varepsilon_{\perp}}{dz} = -\frac{\frac{dE_{\mu}}{dz}}{E_{\mu}} E_{\perp} + \frac{\beta_0}{2} \frac{d\langle\theta_{rms}^2\rangle}{dz} \quad , \tag{34}$$

where $\varepsilon_{\perp}$ is the (unnormalized) transverse emittance, dE_{μ}/dz is the absorber energy loss per cooler transport length z, β_0 is the betatron function in the absorber and θ_{rms} is the mean accumulated multiple scattering angle in the absorber. Note that $dE_{\mu}/dz = f_A \, dE_{\mu}/ds$ where f_A is the fraction of the transport length occupied by the absorber, which has an energy absorption coefficient of dE_{μ}/ds. Also the multiple scattering can be estimated from:

$$\frac{d\left\langle \theta_{rms}^2 \right\rangle}{dz} \equiv \frac{f_A}{L_R}\left(\frac{0.014}{E_\mu}\right)^2 \quad , \tag{35}$$

where L_R is the material radiation length and E_μ is in GeV. (The differential-equation form assumes the cooling system is formed from small alternation absorber and reaccelerator section; a similar difference equation would be appropriate if individual sections are long).

If the parameters are constant, the equations may be combined to find a minimum cooled (unnormalized) emittance of

$$\varepsilon_\perp \rightarrow \frac{(0.014)^2}{2E_\mu} \frac{\beta_0}{L_R \frac{dE_\mu}{dz}} \quad , \tag{36}$$

or, when normalized

$$\varepsilon_N = \varepsilon_\perp \gamma \Rightarrow \frac{(0.014)^2}{2m_\mu c^2} \frac{\beta_0}{L_R \frac{dE_\mu}{dz}} \quad . \tag{37}$$

Longitudinal (energy-spread) cooling is also possible, if the energy loss increases with increasing energy. The energy loss function for muons, dE/ds, is rapidly decreasing (heating) with energy for $E_\mu < 0.3$ GeV, but is slightly increasing (cooling) for $E_\mu > 0.3$ GeV. This natural dependence can be enhanced by placing a wedge-shaped absorber at "non-zero dispersion" region where position is energy-dependent. Longitudinal cooling is limited by statistical fluctuations in the number and energy loss of muons (Landau straggling).

Muon colliders, using ionization cooling must observe the limit of Eq. (37) (and the corresponding limit on energy spread), but, nevertheless, appear to have an adequately small beam size to create an interesting luminosity.

Laser Cooling in Three Dimensions

As we saw earlier, laser Doppler cooling only works in the longituindal direction, but there it works very well indeed. At high temperature, intra-beam scattering will cause the transverse temperature to stay close to the longitudinal temperature and, therefore, laser cooling will cool in all three directions. As the temperature is reduced, however, intra-beam scattering becomes greatly reduced and laser cooling (given the finite lifetime of beams due to residual gas scattering) only works in the longitudinal direction.

Transverse laser illumination is not practical (the overlap with the particle beam is not adequate). The idea of a coupling cavity, or of an rf cavity operating in a location where there is dispersion, has recently been introduced [19].

Considering a coupling cavity as a cavity operating in the TM_{210} mode, and forming a Hamiltonian for the system which consists of an rf banding cavity (i.e., a normal rf cavity) and a linear lattice, one arrives at the equations of motion

$$\frac{d^2x}{d\theta^2} + \nu_x^2 x = 2\pi\Gamma_c \psi \delta_p(\theta - \theta_c) \quad , \tag{38}$$

$$\frac{d^2x}{d\theta^2} + 2\pi\nu_L^2 \psi \delta_p(\theta - \theta_b) = -2\pi\xi_0 \Gamma_c x \delta_p(\theta - \theta_c) \quad , \tag{39}$$

where x is the radial extent of the particle, ψ is its phase with respect to the ordinary cavity, θ is measured about the circumference, θ_b and θ_c are the location of the bunching and the coupling cavities. The quantity $\zeta_0 = \gamma - 1/\gamma^2$ where γ is the relativistic factor. The transverse tune is ν_x and the longitudinal tune is close to ν_L (equal to ν_L if $\nu_L << 1$). The periodic delta functions are $\delta_p(\theta)$ and the degree of coupling is given by Γ_c. For a rectangular cavity (surely not used in practice)

$$\Gamma_c = \frac{eV_c}{2\beta_0 cp0}\left(\frac{R}{a}\right) \quad , \tag{40}$$

where R is the radius of the machine, 2a is the transverse size of the cavity having voltage V_c. The particle momentum $p_0 = \beta_0 \gamma_0 mc$.

It is simple to solve, by matrix methods, Eqs. 38 and 39, after introducing a laser frictional force within a small region. One finds that if the coupling condition $\nu_x - \nu_L$ = integer is satisfied, then the damping rate per turn is:

$$\mathrm{Im}(\nu) = -\frac{1}{2}\,\mathrm{In}\left[1 + \frac{(2\pi)^3 \xi_0 \Gamma_c{}^2}{2\nu_x}\right]\coth\left(\frac{\Lambda_D}{2}\right) \quad , \tag{41}$$

$$\mathrm{Im}(\nu) \approx \frac{\Lambda_D}{2} + \frac{1}{2}\,\mathrm{In}\left[1 + \frac{(2\pi)^3 \xi_0 \Gamma_c{}^2}{2\nu_x}\right]\coth\left(\frac{\Lambda_D}{2}\right) \quad , \tag{42}$$

where $\frac{\Lambda_D}{2}$ is the laser damping per turn if there is no coupling.

It is easy to see that if the coupling, Γ_c, is zero then only the longitudinal motion is damped, but at large Γ_c the two damping rate become equal. Provided $\Gamma_c > \Gamma_{co}$ where

$$\Gamma_{co} = \frac{\Lambda_D}{4\pi^{1/2}}\left(\frac{v_x}{-\xi_0}\right)^{1/2} , \tag{43}$$

then essentially x and ψ are damped at the same rate (about 1/2 the rate for damping just the longitudinal motion).

With a skew quadrupole the vertical motion may also be damped. Ordinary rf cavities in regions of dispersion also can be used to couple motion. Much theoretical work with tracking of particles and study of how much one can be off resonance (a good bit) has been done, but no experimental work has yet been initiated.

Stimulated Radiation Cooling

Recently it has been proposed by Bessonov, and analyzed in detail by Bessonov and Kim [20], that one can stimulate non-fully-ionized ion beams and, thus, have effective radiation damping in circumstances where natural radiation of heavy ions is negligible. Essentially the bound electron is caused to radiate (transfer from one bound level to another) by means of a laser whose frequency is close to resonance. The laser beam itself acts as a "wiggler" in the radiation process.

If the laser, of wavelength λ in the laboratory, is exactly on resonance, then the cross-section is $\sigma = (2\gamma\lambda)^2 / 4\pi$, (Doppler shifted into ion frame), while off resonance it is $\sigma = \left(\frac{8}{3}\pi\right) r_e^{\,2}$ and (typically) 18 orders of magnitude smaller. The laser cannot, however, due to the temperature of the beam and the resulting Doppler shifts, be on resonance for all ions. In light of the many orders of magnitude, it is possible to have a broad band laser, of width $\Delta\omega$, so that the average cross-section

$$\sigma = \left(\frac{8}{3}\pi\right)\left(\frac{2\gamma\lambda}{r_e}\right)\left(\frac{\omega}{\Delta\omega}\right) r_e^{\,2} , \tag{44}$$

is still large enough to be used for beam cooling.

Note that in this process, as contrasted with laser cooling, an rf cavity is essential and the laser is very broad band. Unlike laser cooling, which only changes the ion energy by a small amount, here, in order to get damping by n e-foldings it is required to change the ion energy by $n(\gamma M_i c^2)$ Thus, as in ordinary radiative cooling, the rf cavity and energy replacement, are essential ingredients. On the other hand, as in laser cooling, it is the electron transition, and the transfer of momentum to the ion, which is basic to the mechanism.

The damping rate for vertical oscillations (similar formulas can be obtained for each degree of freedom) is (approximately):

$$\tau = \frac{SR}{c\lambda\ell\gamma}\left(\frac{\Delta\omega}{\omega}\right)\frac{P_A}{P} \quad , \tag{45}$$

where S is the cross-section of the laser beam, of frequency ω and spread $\Delta\omega$, and power P, R is the radius of the storage ring, ℓ is the length of the interaction region, and $P_A = m_e m_i c^5/e^2$. In a numerical example the transverse damping time, for a nitrogen like xenon ion beam at $\gamma = 97$, is 411 seconds.

Conclusion

We have seen that there are a number of ways to circumvent Louville's theorem. Attempts to do so were initiated by the MURA Group in 1956, but with no success. Physicists knew about radiation of electrons, but the MURA Group was interested in proton damping. The radiation damping of electrons was quickly developed, but it wasn't until ten years later that Budker with electron cooling, and then van der Meer, with stochastic cooling, devised effective methods to circumvent the theorem and, thus, damp protons and anti-protons.

In the last decades, the methods of Budker and of van der Meer have been greatly refined. Also, a number of new methods have been proposed, and one has even been demonstrated experimentally. The most effect (in terms of the lower temperature achieved) is the new method of laser cooling.

It is interesting to note that special cooling methods have been developed for different species. Thus radiation cooling works for electrons, stimulated radiation cooling works for relativistic non-fully ionized ions, laser cooling works on non-relativistic non-fully ionized ions, and ionization cooling works for mu mesons.

In the future, we may expect even colder beams useful, for a diversity of purposes, from collider applications to the achievement of crystalline beams.

Acknowledgments

This being a review paper the author has drawn freely on his earlier papers as well as upon the references cited. The section on Ionization Cooling follows closely the paper by D. Neuffer.

References

1. Liouville, J., Journ. de Math. **3**, 349 (1838).
2. Goldstein, H. Classical Mechanics, Addison Wesley, Reading, MA (1980), p. 426;
   A. J. Lichtenberg, Phase Space Dyamics of Particles, Wiley, New York, (1969).
3. Mills, R. L. and A. M. Sessler, "Liouville's Theorem and Phase Space Cooling", Proceedings of the Workshop on Beam Cooling and Related Topics, Montreaux, CERN Report 94-03, 4 (1994).
4. Kolomenski, A. and A. Lebedev, "The Effect of Radiation on the Motion of Relativistic Electrons in Synchrotrons", CERN Symposium, 447 (1956).
5. Robinson, K., Phys. Rev. 111, 373 (1958); Sands, M. Phys. Rev. **97**, 470 (1955).
6. Sands. M., "The Physics of Electron Storage Ring, An Introduction", Stanford Linear Accelerator Center, SLAC 121 (1970).
7. Huang, Z., P. Chen, and R. Ruth, Phys. Rev. Lett. **74**, 1759 (1995).
8. Budker, G. I., Atomnaya Energiya **22**, 346 (1967).
9. van der Meer, S., "Stochastic Damping of Betatron Oscillations in the ISR", CERN Internal Report CERN/ISR-PO/72-31 (1972).
10. Bisognano, J. J., Proc. of the Inter. Conf. on High Energy Accelerators, 772, (1980).
11. Mikhailichenko, A., and M. Zolotorev, Phys. Rev. Lett, **71**, 4146, (1993).
12. Zolotorev, M., and M. Zholentz, "Femtosecond Pulses of Synchrotron Radiation", Lawrence Berkeley Laboratory, LBL #37556, (1995).
13. Kim, K.-J., "Analysis of Optical Stochastic Cooling", presented at PAC, (1995).
14. Wineland, D. J. and H. Dehmelt, Bull. Am. Phys. Soc. **20**, 637 (1975); T. Hanchand, A. Schawlow, Opt. Comm. **13**, 68 (1975).
15. Schroder, S., et al., Phys. Rev. Lett. **64**, 2901 (1990); J. S. Hangst, et al., Phys. Rev. Lett, **67**, 1238 (1991).
16. Hangst, J., "Laser Cooling of a Stored Ion Beam - A First Step Towards Crystalline Beam", Argonne National Laboratory Report ANL/PHY-93/1 (1993).
17. Skrinsky, A. N., Proc. XXth Int. Conf. on High Energy Physics, AIP Conf. Proc. **68**, 1056 (1980).
18. Neuffer, D., Part. Acc. **14**, 75 (1983).
19. Okamoto, H. , and A. M. Sessler, and D. Mohl, Phys. Rev. Lett. **72**, 3977, (1994).
20. Bessonov, E.G. and K.-J. Kim, *Phys. Rev. Lett*. **76**, 431, (1996).

STOCHASTIC COOLING

F. CASPERS
PS Division, CERN
1211 Geneva 23, Switzerland

ABSTRACT

Examples of stochastic cooling (S.C.) systems which are operational in several accelerator laboratories are discussed. The status of new systems under construction (GSI Darmstadt, KFA Jülich) is outlined in the subsequent chapters. For bunched beam stochastic cooling (BBSC) the results of tests on large machines like the CERN SPS and FNAL TEVATRON are compared with data obtained from smaller rings (AC and LEAR). This technique (BBSC), if found to be applicable reliably to large rings, may be considered for future projects at BNL and DESY. Since technological issues play an important role a separate chapter gives a status of the UHV compatible materials, microwave amplifiers, signal treatment and the possible impact of high temperature superconductors for stochastic cooling. A chapter on theoretical aspects outlines aperture limitations of microwave pick-ups for high γ-beams. Projects focus in particular on S.C. in the RIKEN RI machine for special radioactive ions. A new technique, the "Optical Stochastic Cooling", has a potential for beam cooling provided it can be made operational and reliable.

Introduction

The stochastic cooling technique has been shown to be operational for nearly 20 years in the ISR and the experimental cooling ring ICE at CERN. Since then it has been sucessfully used in several machines at CERN, in the INS (Tokyo) and at FERMILAB and has contributed to the discovery of the W and Z particles by its ability to increase phase-space density beyond a factor of 10^9 in several steps. Originally applied to antiprotons where it really became one of the workhorses for $\bar{p}$ accumulation and storage, it can now be considered for a wider range of applications such as protons and heavy ions. The theoretical aspects for coasting beam stochastic cooling are rather well understood and in line with the technological progress. The bandwidth of the S.C. systems has increased by more than one order of magnitude to the several GHz we have now. Despite its complexity S.C. was found to be a reliable tool for phase-space reduction of "warm" particle beams. However, for cooling of "cool" beams to "cold", electron cooling appears more favorable.

1. Review of Existing Systems

1.1. Stochastic Cooling in the AAC

The peak performance and the operational values (1990) compared with the design data are shown in Table 1. The daily stacking rate is the average during stacking periods and takes into account the fluctuations of the PSB, PS and AAC machines[1] (AAC = Antiproton Accumulator Complex).

Table 1 - AAC Performance

		1988	1989 and afterwards	
	DESIGN	OPERATION	OPERATION	PEAK
Production beam (ppp)	1.0×10^{13}	1.35×10^{13}	1.45×10^{13}	
Repetition period	2.4 s	4.8 s	4.8 s	
Yield ($\bar{p}$/p)	10.0×10^{-6}	5.7×10^{-6}	5.4×10^{-6}	5.8×10^{-6}
$\bar{p}$/pulse	5.0×10^{7}	4.9×10^{7}	7.0×10^{7}	7.7×10^{7}
Stacking rate (10^{10}/h)	7.5	3.6	5.3	6.13 (13.10.90)
Daily production (10^{11})	10	6.0	8.5	11.5
Daily stack. rate (10^{10}/h)		3.3	4.4	5.16
Stack. intensity max.	1.0×10^{12}	0.85×10^{12}	1.03×10^{12}	1.31×10^{12} (9.10.89)
Transverse emittances 95% of beam	1-2π	2-3π	2-3π	
Total efficiency	50%	63%	91%	93%

The 1-3 GHz AC stochastic cooling system in which moving pick-up and kicker electrodes accompany the beam as it shrinks, was identified as a critical item in achieving the design performance of the AAC, even when using a 4.8 s instead of a 2.4 s cycle as originally planned[2]. This cooling system is subdivided into three frequency bands (1-1.65,1.65-2.4 and 2.4-3.1 GHz). It consists of 9 pick-up and kicker tanks each operating in one of the frequency bands and in $\Delta p/p$ as well as in the transverse plane.

Despite having cryo-pick-up structures and preamplifiers the available power (about 10 kW for CW signals) limits the gain to values below optimum during most of the cycle. Several improvements were introduced during 1988 and 1989.

- Additional cryo-cooling of the pick-up loops and combiner boards.
- Periodic filters (power saver filters) for certain betatron systems to reduce the noise power transmitted to the kickers.
- Two-stage momentum cooling filters.
- A dynamic phase correction as function of pick-up and kicker positions.
- Additional power for band 3 (2.4- 3.1 GHz).
- Improved pick-up and kicker movement versus time.

The first of these improvements was the most expensive and complex, but also the most effective. By lowering the temperature from 100 K to 30 K the thermal noise was reduced by 4 dB, and acceptable transverse emittances and efficiencies were obtained after 2.4 s. But also the 4.8 s cycle was improved (Table 2).

Table 2 - AC stochastic cooling performances

Date	4.8 s		2.4 s	
	$\varepsilon_{H,V}$ (mm.mrad)	ε (4 eVs)	$\varepsilon_{H,V}$ (mm.mrad)	ε (4 eVs)
May 88	$\sim 15\pi$	68%	30π	35%
May 89	4π	92%	13π	70%

Improvements were also needed for the AA precooling system (momentum and vertical):

- A second frequency band was added: 1.6-2.4 GHz (initially only 0.8 to 1.6 GHz).
- Preamplifiers with better noise figures.
- Damping of propagating TE-modes in the vacuum chamber by ferrite material (to avoid coupling to the stack)

Nevertheless the precooling and the stack-tail systems are not sufficiently fast to digest the $\bar{p}$ with adequate efficiency for the 2.4 s cycle, while the precooling and the stack-tail systems do not limit the efficiency with a 4.8 s cycle.

As for the AC, the AA precooling gain is limited below the optimum condition by thermal noise and available power. The cheapest way to obtain sufficient gain and speed would be to add cryogenic cooling to the AA precooling pick-ups. Even so, barely 10% could be gained in overall daily production rate, since the number of production cycles per hour available from the PS using 2.4 s cycles only increases by a factor of typically 1.5 to 1.7 and not 2 due to other users (LEAR, SPS, East HALL, LEP). For this reason (and others such as budget and manpower) the cryogenic pick-ups in the AA were abandoned and hence also the 2.4 s cycle operation.

In the years following the improvements (until end of 95) stacks around 10^{12} have regularly been obtained and stacking efficiencies beyond 80%. Presently the stacking rate is limited to 3.5×10^{10}/h due to modifications on the proton beam and due to a reduced yield (magnetic horn) of 4.5×10^{-6}.

1.2. Stochastic Cooling at LEAR

The LEAR stochastic cooling systems consist of two chains, each working in all three planes, to cover two ranges of particle momentum in the machine. The high momentum system can be adjusted for a momentum range from 200 MeV/c to 2 GeV/c (protons and antiprotons) corresponding to a variation of particle velocity $\beta = v/c$ by a factor of 4.35.

The low momentum system is adjustable from 105 MeV/c to 61 MeV/c[3]. Figure 1 illustrates the configuration used in 1993. Over the years (beginning in 1986) the position of stochastic cooling pick-ups and kickers has been changed together with with the type of coupling structure. Initially the longitudinal cooling systems had ferrite ring pick-ups (24 rings each) and kickers of the same type. These ferrites were outside the vacuum (outgassing) on a section of a ceramic vacuum chamber. With these units the usable frequency range extended from 30 MHz to 200 MHz showing a rather poor phase response (dispersion). However, at that time only a momentum cooling (high energy) range from $\Delta p/p = 0.6\%$ was required.

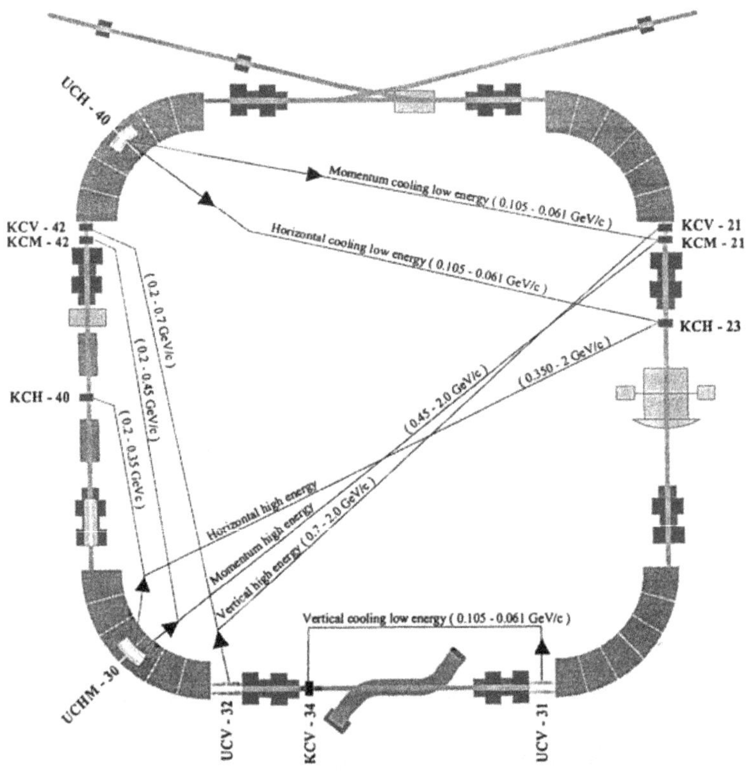

Fig. 1 - LEAR stochastic cooling systems

After the implementation of the "bunch-to-bucket" transfer from the PS into LEAR and the adiabatic debunching in LEAR, the longitudinal stochastic cooling after injection into LEAR turned out to be more important.

Later LEAR was operated at higher momenta than injection, so the increase in bandwidth for the stochastic cooling system became mandatory. Now loop couplers with variable external delays and a horizontal cooling system using the same loop couplers as the vertical one have been installed (BHN30). The usable bandwidth extends from 3 MHz to 1.1 GHz. In order to avoid undesired feedback between pick-ups and kickers due to imperfect shielding as well as microwave mode propagation in the beam pipe, pick-ups and kickers were placed as far apart from each other as possible.

As a particular feature the horizontal and the vertical stochastic cooling system is used as a broadband damper by adding 30 dB of attenuation during electron cooling at 310 MeV/c and 200 MeV/c.

Stochastic coooling is applied also between 105 MeV/c and 2 GeV/c to counteract diffusion mechanisms such as intrabeam scattering during the slow extraction process. The need of high fluxes of particles by the experiments and the use of long spills up to 3 h at high energy necessitates circulating beams of 4×10^{10} particles. In these cases intrabeam scattering becomes important and has to be compensated by stochastic cooling. Transverse emittances of 5π mm mrad are measured while the momentum spread is maintained below 0.3% at 4π mm mrad. As for the performance of the LEAR stochastic cooling systems Table 3 gives an indication about the values presently obtained in 1995.

Table 3 - LEAR Stochastic Cooling Performances

LEAR momentum	Particle intensity	Longitudinal emittances			Horizontal emittances		Cooling time	Vertical emittances		Cooling time
					π mm.mrad (95%)			π mm.mrad (95%)		
MeV/c	$N\times10^9$	$\Delta p/p_i$	$\Delta p/p_f$	T (s)	$\varepsilon_{H,i}$	$\varepsilon_{H,f}$	T (s)	$\varepsilon_{V,i}$	$\varepsilon_{V,f}$	T (s)
2000	1		0.001			2			2	
1500	1		0.001			2			2	
609	1	0.002	0.0005	70	40	15	70	20	10	70
310	1	0.001	0.001		30	15	100	20	10	100
200	1	0.002	0.001	100	30	15	100	20	10	100
105	1	0.002	0.001	100	30	20	100	20	15	100

1.3. Stochastic Cooling at FNAL

In terms of future developments there is a planned upgrading of FNAL debuncher pick-ups and kickers [4]. These new pick-ups will be cooled to 4 K and both pick-ups and kickers have movable planar loops. An upgrade of the accumulator system is also under consideration (in order to obtain a higher flux in the main injector area). In the case of accumulator stack-tail upgrade to 2-4 GHz , a redesign of the lattice will be required as well. For increased collider luminosity a three-fold increase (from presently 6×10^{10}/h) in $\bar{p}$ production rate to 2×10^{11}/h is discussed. As a new project the "recycler" will be built. This is a $\bar{p}$-storage ring with permanent magnets and it will be situated inside the main injector tunnel. Stochastic cooling in this machine is used to maintain beam size. However, electron cooling at 9 GeV/c is the main work horse in this machine for rapid beam cooling.

1.4. Stochastic Cooling at TARN1

Stochastic cooling experiments at TARN1 have been carried out from February 1984 until May 1985 when TARN 1 was closed. These experiments were done with 7 MeV protons and 28 MeV α-particles. (7 MeV/amu) with $N = 10^7$, $\beta = 0.12$, $f_{rev} = 1.13$ MHz and $\eta = 0.7$. The stochastic cooling frequency range extended from 20 MHz to 95 MHz. It was the first time that stochastic cooling was applied to ions other than protons. The pick-up and kicker structures used were helical travelling wave couplers for $\Delta p/p$ cooling only. The measured coupling impedance amounts 250 Ω at 40 MHz , 400 Ω at 80 MHz and 290 Ω at 120 MHz and was found to be close to theoretical values. In the course of the stochastic cooling experiments ($N = 1.2 \times 10^7$) an initial momentum spread of 1.2% (FWHM) was reduced to 0.06% in 420 s on α-paricles. For protons it took 85 s to cool to $\Delta p/p = 0.08\%$. In both cases the system was set to 99 dB.

2. Review of Stochastic Cooling Under Construction

2.1. Stochastic Cooling at the Experimental Storage Ring ESR (GSI Darmstadt)

The purpose of this stochastic cooling stage is the reduction of the phase-space volume of exotic fragments (heavy ions) prior to rf-stacking and subsequent electron cooling[5]. Note that due to the high charge state of the ions there will be an excellent signal-to-noise ratio at the pick-ups. For example, a fully stripped uranium ion yields the same signal power as 8464 (uncorrelated) protons. The momentum cooling range is $\Delta p/p = 0.35\%$ to 0.1% and the transverse cooling leads to an emittance reduction (H,V) from 20×10^{-6} m to 2.5×10^{-6} m. The frequency range of the cooling systems extends from 0.9 to 1.7 GHz with an installed CW power of 2 kW.

The ESR being a multi-purpose machine, all pick-ups and kickers have to be positioned on the precooling (injection) orbit inside dipoles and quadrupoles. This measure liberates straight sections for other purposes. All electrodes are situated in regions of large dispersion (for rf-stacking) and thus a longitudinal kick leads to a transverse deflection of the beam as well. A modified Palmer cooling method using longitudinal corrections to compensate for bad mixing is applied. Since the pick-up signals contain information on momentum deviation and on betatron motion, a signal subtraction scheme (Fig. 2) for disentangling the longitudinal and the the transverse plane has been implemented. Undesired transverse kicks linked to a longitudinal correction are eliminated in the same turn[6].

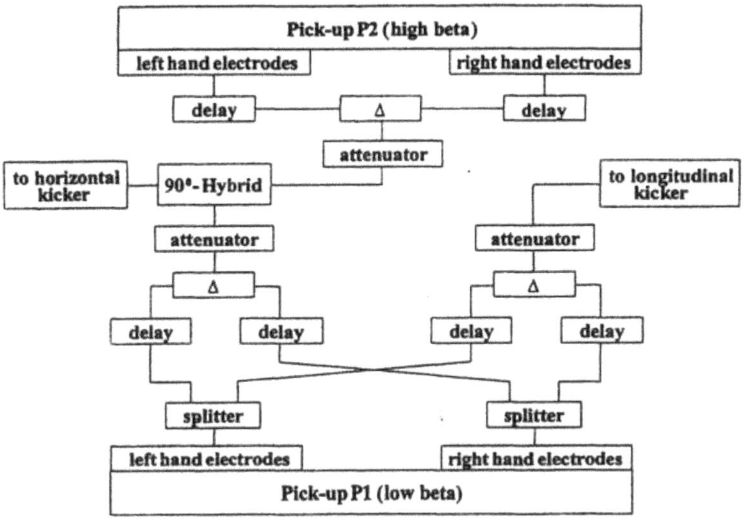

Fig. 2 - ESR stochastic cooling

2.2. Stochastic Cooling at COSY (KFA Jülich)

The COSY stochastic cooling system (1-1.8 GHz, 1.8-3 GHz) is at present in the installation phase[7-11]. The pick-ups are cryogenically cooled to 20 K. They are independently adjustable in their position and allow for closed orbit suppression. Uncooled preamplifiers with a noise temperature below 50 K are mounted outside the vacuum tanks. Programmable delays permit an energy adjustment range from 0.8 GeV to 2.5 GeV.

At present one horizontal and one vertical pick-up tank are in the machine. The beginning of 1996 is the scheduled installation time for two kicker tanks, to be followed later in 1996 by two more pick-ups.

All active rf components for stochastic cooling have been delivered, tested and fulfill the specifications.

3. Bunched Beam Stochastic Cooling

3.1. Bunched Beam Stochastic Cooling in the SPS at CERN

Tests were carried out on low intensity proton bunches (about 10^9 particles per bunch) using a corrugated wall pick-up with a bandwidth of 2 GHz (-10 dB) and 11 GHz centre frequency[12]. Measurements performed up to 1990 indicated the presence of coherent rf-related signals in the 11 GHz range. This contamination of the pure Schottky noise is related to a macro-structure in the bunch which depends itself on the past rf history and on intensity. As a consequence these tests have been discontinued after the measured results on the Schottky spectra.

3.2. Bunched Beam Stochastic Cooling Tests at Tevatron

A 4-8 GHz BBSC system has been installed in the Tevatron[13] aiming to reduce the emittance blow-up and to increase the luminosity lifetime during collider runs. The length between kicker and pick-up is 60 m ($=\lambda_\beta/4$). The pick-ups and kicker have movable plates each with 16 co-planar loops. The cooling systems use fiber optic delay lines for the notch filter in order to get pico-second timing stability and also to take advantage from the low fiber losses for implementation of a high Q-notch filter (a kind of photon storage ring). In this project for beam transfer function (BTF) measurements a "bucket gating" technique was applied for dynamic range improvements. The observed decrease of signal power at high revolution harmonics (Fig.3) is much slower than predicted for a smooth Gaussian bunch shape. Measurements carried out on a 6 GHz bandwidth wall current monitor using a 20 GHz sampling scope indicate a smooth Gaussian bunch shape with about 5 ns FWHM. The strong signal in the GHz range is thus probably due to coherent instabilities, which may cause a microstructure in the bunch but do not seem to lead to observable losses.

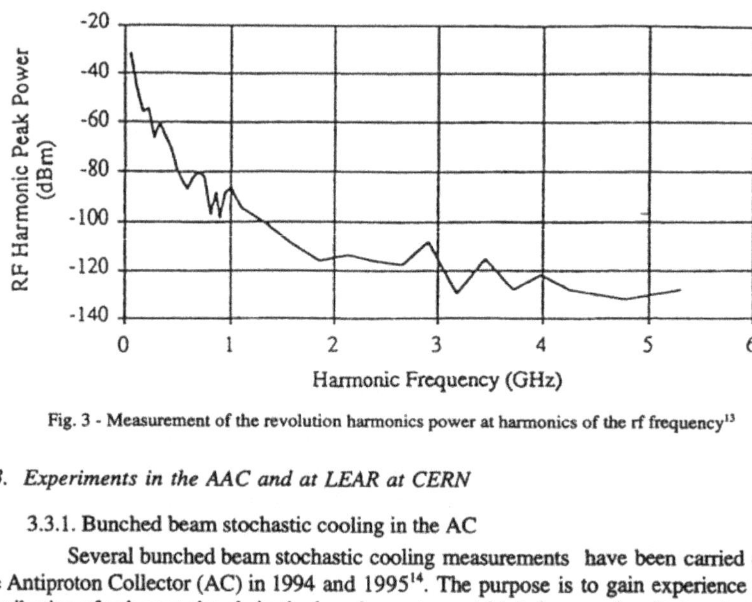

Fig. 3 - Measurement of the revolution harmonics power at harmonics of the rf frequency[13]

3.3. Experiments in the AAC and at LEAR at CERN

3.3.1. Bunched beam stochastic cooling in the AC

Several bunched beam stochastic cooling measurements have been carried out on the Antiproton Collector (AC) in 1994 and 1995[14]. The purpose is to gain experience on the distribution of coherent signals in the bunch spectrum at high frequencies. In bunched beam cooling experiments at LEAR it has been observed that the amplitude of the coherent signals for short bunches decays much more slowly towards higher frequencies than expected from the Fourier transform of a Gaussian shaped bunch. The LEAR measurements indicate that this deviation depends on the longitudinal density of the bunch. The AC experiments were carried out with antiproton beam of about 5×10^7 particles and rf bucket sizes (h=1) between 1 and 15 eVs, and $\Delta p/p$ values between 0.1% and 3%. The time structure of the bunch was observed by means of a fast scope connected via a 30 dB preamplifier to the AC longitudinal pick-ups. Spectral distributions have been monitored using the AC Schottky pick-up (non-resonant mode) as well as the stochastic cooling pick-ups (1-3 GHz). For a bunch length of about 50 ns in a 12 eVs bucket strong longitudinal coherent signals have been observed up to 3 GHz (observation limited by the pick-up bandwidth). Also the cooling process itself (on the bunched beam) could be well observed both in time and frequency domain. Furthermore a tendency to produce very fast longitudinal blow-ups on cold bunched beams was noted with a subsequent turbulence-like behaviour extending over several minutes. Leaving the longitudinal systems on has a stabilizing effect (comparable to a longitudinal damper) with respect to the instabilities mentioned above. Also bunched beam cooling in the transverse plane has been tested successfully. Future experiments in this context are aimed to further investigate coherent signal distribution beyond 1 GHz in the time domain.

3.3.2. Observation of bunched beam Schottky spectra at 309 MeV/c at LEAR

The results of these tests are compiled in Table 4. It can be seen that for the lower intensity the cut-off point of the frequency spectrum is approximately at $f_c = 6/\Delta t$, as expected for a smooth Gaussian bunch. However, at the higher intensity, the cut-off point is considerably higher as well, indicating a high frequency structure on the bunch and/or very sharp edges. Note that electron cooling has been applied prior to the measurement in order to achieve a sufficiently small $\Delta p/p$ for getting short bunches[15].

Table 4 - Measured bunch length and cut-off frequency as function of intensity (rounded numbers). For a smooth Gaussian bunch the product $f_c\,\Delta t$ is about 6 [15].

Intensity N [p]	Voltage V [volt]	Bunch length Δt [ns]	Cut-off frequency f_c [MHz]	Length × cut-off frequency $f_c \times \Delta t$ [GHz × ns]
10^9	100	180	30	5.4
10^9	1000	80	100	8
5×10^9	100	450	40	18
5×10^9	200	300	70	21
5×10^9	1000	150	300	45

3.4. RHIC, DESY (Possible Projects)

3.4.1. Bunched beam stochastic cooling activities at RHIC

Recently at BNL studies[16] have begun to fight luminosity reduction due to intra-beam scattering (IBS) of highly charged ions (Au^{79+}) by means of BBSC. Due to the high charge state, one can expect a very good signal/noise ratio (like at GSI Darmstadt). The cooling should operate at relatively low beam intensities (10^9 ions /bunch). From theoretical estimates a 4-8 GHz momentum and transverse cooling system could return an improvement of a factor of 3 in average luminosity. But there are still many remaining questions such as the effect of coherent signals at high revolution harmonics as seen in the SPS and at FNAL. Presently there are no firm plans for implementation, but BBSC is a good candidate for a future upgrade.

3.4.2. Bunched beam stochastic cooling activities at DESY

Luminosity lifetimes (run times) at HERA (820 GeV proton collisions with electrons) are typically limited to 24 h due to the emittance blow-up of the proton beam. Thus a BBSC system operating on the proton beam should be able to maintain at least constant transverse emittances[17]. There are short bunches (<2 ns FWHM with 5×10^{11} protons/bunch) in the machine using two rf-systems which operate at 52 MHz and 208 MHz respectively. First tests planned for the 95/96 shutdown are the installation of a 4-8 GHz FNAL-TEVATRON prototype pick-up (movable planar arrays) and subsequent Schottky signal measurements. As a next step the study of coherent and incoherent parts of the 4-8 GHz Schottky signal from the pick-up is foreseen. Depending on the results a general outline of a BBSC system may be developed.

4. Status of Technology

4.1. UHV-Compatible Materials

For the construction of stochastic cooling pick-ups and kickers requirements for rather special UHV (ultra-high vacuum) compatible materials occur. Examples of such materials are[18]:

- printed circuit substrates,
- resistive film microwave absorbers,
- ferrite based microwave absorbers,
- "plastic"-like materials (super-isolation, "VESPEL").

Typical storage rings need to work with vacuum in the order of 10^{-9} to 10^{-12} Torr. Thus low outgassing and possibly bakeable materials are mandatory. Depending on the application there may be an in situ bakeout (up to 300 °C) and cryo-temperatures for operation. Substrate materials for printed electrodes are alumina (used at CERN, GSI) and teflon (used at FNAL). Ferrites in the beam pipe (e.g. used as microwave mode absorbers) should be slightly conductive (bulk or surface conductivity) in order to avoid ion trapping due to surface charges.

4.2. Microwave Amplifiers

Room temperature low noise pre-amplifiers (>octave bandwidth) are now available from a few MHz to beyond 5 GHz with noise temperatures below 60 K. Cryogenic low-noise amplifiers exhibit noise temperatures from around 1 K to about 20 K depending on bandwidth and cryo-temperature within the frequency range from several MHz to 10 GHz. Class A solid state amplifiers with octave bandwidth in the range 1-3 GHz and >100 W CW power have been available for nearly 10 years and have shown a good reliability record (CERN-AC)[19].Their price per Watt installed power is competitive with that of TWTs. Note that in practical operation the rated CW power of such a class A amplifier cannot be used for stochastic cooling purposes. Due to the different amplitude density distribution of a noise-like signal as compared to a sine wave, severe nonlinear effects (saturation) already occur at about 40% of the rated power level.

4.3. Signal Treatment

Apart from the pick-up and kicker amplifiers critical elements in the stochastic cooling signal path include:

- notch filters,
- stable, low dispersion and broadband delays,
- high-isolation 180° hybrids,
- gain equalizers.

For high quality notch filters optical fibre delays have been successfully used at FNAL[13]. Bulk acoustic wave (BAW) devices and acoustic charge transport (ACT devices are also suitable for broadband delays and signal convolution.

A variable signal delay using RAM-based digital storage has been applied for LEAR (60 MHz bandwidth)[20].

4.4. Possible Impact of High-Temperature Superconductors (HTSC)

The evolution of the HTSC technology has led to the development of several HTSC-based microwave structures, which are typically (in a normal conducting version) part of a stochastic cooling system. Examples are low loss transmission lines and signal combiner networks as well as high-Q filters and resonators[21]. Since stochastic cooling pick-ups are often at cryo-temperatures (<70 K) for the reduction of thermal noise, a further improvement in sensitivity may be obtained by increasing signal combiner performance using HTSC elements. HTSC structures require single crystalline substrates (e.g. alumina with a buffer layer) and are available now as wafers with more than 100 mm diameter.

However, it must be mentioned that HTSCs do not appear very suitable for high power applications as they exhibit rather small inter-modulation levels compared to classic superconductors. Also these structures are sensitive to crystal defects. Presently the HTSC (often also referred to as HTS) technology is delicate but evolving fast.

5. Theoretical Limitations of Pick-Up and Kicker Apertures for High-γ Beams

- The field slice of a single particle moving on a straight line with $\gamma = E/(mc^2)$ has a (half) opening angle of $1/\gamma$(rad). This leads to a maximum admissible lateral distance r for some PU structure given by the condition $r/\gamma = \lambda/4$ thus $r_{max} = \gamma \cdot \lambda/4$ instead of $r_{max} = \lambda/4$. [22]
- Assuming a general PU structure and still working in the range of validity of the above condition the sensitivity would decay proportionally to $1/r$ (surface current density decreases as $1/r$, the number of PU elements on an azimuth increase with r, thus sensitivity decreases as $1/r$).
- Only for a kicker structure of infinite length is the synchronous wave remaining. The intensity of this synchronous wave (seen by the particle) decays very rapidly with r (evanescent field) and its phase velocity quickly approaches that of c with increasing r.

For structures of finite length individual calculation is required. E.g. a single small button PU with a diameter smaller than $\lambda/2$ will give a good response for $r < r_{max}$. If the output signals of an array of such button PUs are combined without back reaction to the beam (each PU has its own preamplifier) it should even be possible to obtain a good sensitivity for infinite long structures (arrays of button PUs) as each element of this array may be considered to work independently and virtually independent of the beam velocity, provided the signal combiner network is adapted to the beam velocity and the highest frequency of interest is still present in the image current modulation.

The configuration discussed above does not assume any change of beam trajectory (like in ondulators or wigglers etc.)

6. Studies for Future Projects

6.1. Stochastic Cooling at ADRIA

INFN Legnaro has proposed the heavy ion accelerator complex ADRIA. The main purpose of ADRIA is the acceleration of heavy ions to energies around 1 GeV/u[23]. For nuclear experiments investigating the Coulomb barrier, unstable nuclei produced in a thin target are decelerated to 10-20 MeV/u. The ADRIA complex consists of a heavy ion injector, a fast cycling synchrotron (Booster) and a slow cycling synchrotron used as a decelerator. This decelerator is foreseen to operate both with electron and stochastic cooling. During deceleration the beam momentum decreases by a factor of 10 as the momentum spread, and the transverse emittance accordingly. The combination of electron cooling and stochastic cooling is designed to produce a final momentum spread in the stack of 0.01% and a cooling time of 0.3 seconds. Table 5 gives an overview of the stochastic cooling parameters.

The stochastic cooling pick-up is to be made by a pair of striplines (operating for a $\beta = 0.8$) and situated in a zero dispersion region. The frequency range chosen (1-2 GHz) represents a compromise between optimized desired mixing, Schottky band overlap and technological constraints on the bandwidth of microwave amplifiers. As in all stochastic cooling applications for highly charged heavy ions the signal/noise ratio on the pick-up would be excellent.

Table 5 - Stochastic Cooling Parameters

Total number of ions	5×10^7	
Mass number	200	u
Charge state	80	
Specific kinetic energy	1.0	GeV/u
β relativistic factor	0.8	
Bandwidth	1 - 2	GHz
Method	Notch filter	
No. of pick-ups	16	
No. of kickers	32	
Schottky power (initial)	1.0	kW
Thermal power	negligible	
Amplifier gain	160	dB

6.2. RIKEN Radioactive Isotope (RI) Beam Factory

The aim of this proposed facility[24] is to provide high intensity beams of radioactive isotopes over a very large mass range with energies up to several hundred MeV/nucleon. The injector consists of an ECR ion source (18 GHz), followed by a tunable RFQ, another subsequent LINAC and two small cyclotrons. The beam produced by that configuation is sent to the Multi-Use Experimental Storage Rings (MUSES). MUSES may be subdivided into 3 parts, namely an Accumulator-Cooler Ring (ACR), a Booster Synchrotron Ring (BSR) and a Double Storage Ring (DSR). The DSR permits various types of experiments, such as head on collisions and ion-ion merging. The function of the ACR is exclusively the accumulation and cooling of RI beams prior to the acceleration in the BSR. With the ACR, which is intended to improve on beam quality (transverse emittance and momentum spread), acceptance requirements for the DSR are considerably reduced. With an rf-stacking repetition frequency of 10 Hz the longitudinal cooling time constant τ must be smaller than 100 ms (Table 6). In parallel to the stochastic cooling the electron cooling in the ACR operates at the same time until a stack of 10^5 paricles is reached with a momentum spread of less than 0.15% and 1π mm mrad transverse emittance.

For a comparison, both the stochastic and electron cooler parameters are reproduced in Tables 6 and 7.

Table 6 - Parameters of the RI Stochastic Cooler

Longitudinal cooling time, s	(number of ions 1×10^8)	
U^{92+}	(150 MeV/u)	0.06
C^{6+}	(400 MeV/u)	0.01
Bandwidth, W	(MHz)	2000
Ambient temperature, T_n	(K)	18
Total microwave power, P	(kW)	10
Pick-up sensitivity, Z_p (Ω)	300	

Table 7 - Parameters of the RI Electron Cooler

Maximum electron energy	(keV)	300
Maximum cooled ion energy	(MeV/nucleon)	500
Maximum electron current	(A)	10
Cathode diameter	(mm)	5.8
Electron beam diameter at cooling section	(mm)	50
Length of cooling section	(m)	3.0
Cooling time for 200 MeV/nucleon $^{132}Sn^{50+}$	(s)	5.4
for initial $\Delta p/p = 0.5\%$, emittance $\varepsilon = 125\pi$ mm·mrad		

7. Optical Stochastic Cooling (OSC)

The basic principle of the OSC consists in using an undulator as a pick-up which emits visible or infrared light that passes an optical amplifier and is then sent to an undulator which acts as a kicker[25,26]. This method could in principle be applied for longitudinal cooling (the particle receives an energy kick assuming correct delay settings) or betatron cooling (the particle gets a corresponding transverse kick). The basic layout of an OSC system is depicted in Fig. 4.

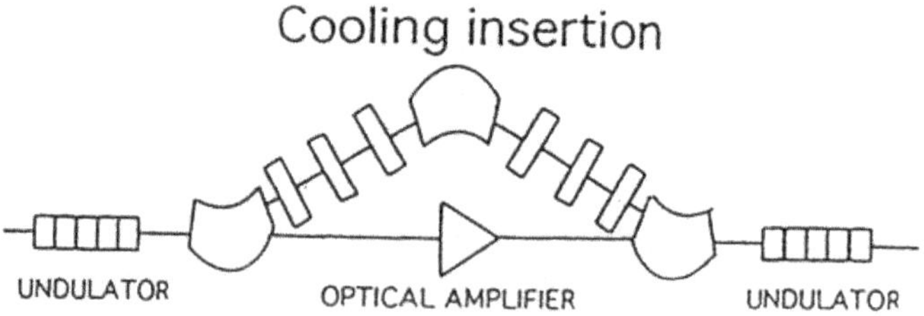

Fig. 4 - Principal layout of an Optical Stochastic Cooling system

In both cases a special lattice insertion is used which makes the time of flight dependent on the momentum error or the betatron amplitude respectively (for momentum and betatron cooling). One might envisage several applications which are facilitated by the extremely large bandwidth of OSC such as:

- Cooling in proton colliders (Tevatron, Hera , LHC)
- Cooling of electrons (positrons) in low energy storage rings
- Cooling of high energy heavy ions

Independent of these applications OSC imposes challenging requirements in advanced optical diagnostic techniques and handling femtosecond beam gymnastics.

Amongst concerns related to OSC the following points require particular attention:

- the tolerances in synchronisation,
- the preservation of the fluctuations in the electron beam during the propagation between pick-up and kicker (isochronicity or mixing)
- the preservation of fluctuations in the light beam from the pick-up passing the optical tract (optical amplifier chain)

These points are of special relevance since the bandwidth of an OSC system is of the order of 10 μm (assuming green light with a wavelength of 0.5 μm and 5% relative bandwidth).

The tolerances in this case in terms of isochronicity are of the order of 0.1 μm.

In the near future a test of the isochronicity will be carried out at the Advanced Light Source (ALS) booster synchrotron at LBL. This experiment uses two undulators installed in a special beam line on an extracted beam, which are located several meters apart, and act as pick-ups, and one will observe the superposition of both (delay corrected) light beam images on an image detector. The particle beam (electrons) parameters for this test configuration are given in Table 8.

Table 8 - Particle beam parameters for interference visibility test

BEAM PARAMETERS:		
Extraction energy	MeV	150
Horizontal emittance	m rad	1.5×10^{-7}
Vertical emittance	m rad	1×10^{-8}
Energy spread		7×10^{-4}
REQUIRED BEAM STABILITY:		
Horizontal position		$\delta x/\sigma_x = 0.2$
Horizontal angle		$\delta x'/\sigma_{x'} = 0.2$
Vertical position		$\delta y/\sigma_y = 1$
Vertical angle		$\delta y/\sigma_y = 1$
Energy		$\delta E/E = 3 \times 10^{-4}$

Apart from the questions mentioned above, the undesired mixing between pick-up and kicker may be worsened by beam turbulence, a parameter which is difficult to predict and to specify. Also in general, closed orbit and positioning errors are an issue for OSC. The optical amplifier chain is supposed to be (in analogy to microwave stochastic cooling) a "high fidelity" amplifier. Thus intermodulation and other nonlinear distortion effects should be small. The dispersion of the optical amplifier must be known (possibly be compensated) and should be stable in time and also against signal amplitude. Thus there are challenging problems to be solved before this method can be used reliably in a storage ring.

8. Conclusion

Coasting beam (microwave) stochastic cooling has become the work horse for $\bar{p}$ accumulation (CERN, FNAL) and for improved beam purity (LEAR). The physics and technology of this technique are well understood and also mastered in practical operation for a frequency range from a few MHz to about 10 GHz. Bunched beam stochastic cooling works well in small machines (AC, LEAR) and moderate number of particles ($<10^{10}$), but there are still considerable difficulties (beam dynamics) for big colliders.

Technological developments (active components) are heading towards higher frequencies, better sensitivities, cheaper power, easier signal treatment and improved reliability. UHV (ultra-high vacuum) compatible materials and technolgies permit improvements in pick-up and kicker design. Printed circuits are a challenging alternative to conventional loop couplers, in particular in the microwave range.

Apart from microwave mode interference, precise mechanical alignments and particle mixing problems, theoretical limitations do not prevent the extension of pick-up /kicker structures beyond 10 GHz for high-γbeams.

Optical stochastic cooling is a new idea which requires careful evaluation of theoretical, technological and practical limitations.

9. Acknowledgement

The author would like to thank D. Möhl, M. Chanel and R. Ley for many fruitful discussions, S. Maury for support, D.J. Warner for reading the manuscript, and A. Molat-Berbiers for typing and editing the manuscript.

8. References

1. B. Autin, G. Carron, F. Caspers, V. Chohan, C.D. Johnson, E. Jones, G. Le Dallic, S. Maury, C. Metzger, D. Möhl, Y. Orlov, F. Pedersen, A. Poncet, J.C. Schnuriger, T.R. Sherwood, C.S. Taylor, L. Thorndahl, S. van der Meer, and D.J. Williams, in *Proc. of EPAC 90 (Nice)*, **Vol. 1**, eds. P. Marin and P. Mandrillon (Editions Frontières, Gif.sur-Yvette, 1990) p. 614.
2. H. Koziol and S. Maury, *Parameter List for the Antiproton Accumulator Complex (AAC) (Ed. 1994)* (CERN/PS 95-15 (AR/BD), 1995).

3. F. Caspers, M. Chanel and J.C. Perrier, in *Proc. of the Workshop on Beam Cooling and Related Topics* (Montreux, 1993), ed. J. Bosser (CERN 94-03, 1994) p. 207.
4. R. Pasquinelli, Private communication.
N. Tokuda, H. Vonehara, A. Noda, S. Watanabe, M. Yoshizawa, N. Takahashi, M. Takanaka, K. Chida, A. Mizobuchi, T. Nakanishi, and T. Katayama, *IEEE Trans. Nucl. Sci.*, **NS-32**, 5 (1985) p. 2415.
5. F. Nolden, W. Bourgeois, H. Eickhoff, B. Franzke, P. Raabe, A. Scior, A. Schwimm, and G. Trageser, in *Proc. of the Workshop on Beam Cooling and Related Topics* (Montreux, 1993), ed. J. Bosser (CERN 94-03, 1994) p. 137.
6. B. Franzke, Private communication.
7. R. Maier, U. Bechstedt, J. Dietrich, S. Martin, D. Prasuhn, A. Schnase, H. Schneider, H. Stockhorst, and R. Tölle, in *Proc. of the Int. Workshop on Polarized Beams and Polarized Gas Targets (Cologne, 1995),* (Univ. of Cologne, to be published).
8. R. Maier, U. Bechstedt, J. Dietrich, U. Hacker, S. Martin, D. Prasuhn, P. v. Rossen, H. Stockhorst, and R. Tölle, in *Proc. of EPAC 94 (London)*, **Vol. 1**, eds. V. Suller and Ch. Petit-Jean-Genaz (World Scientific, Singapore, 1994) p. 165.
9. P. Brittner, H.U. Hacker, D. Prasuhn, G. Schug, H. Singer, W. Spieß, amd R. Stassen, in *Proc. of EPAC 94 (London)*, **Vol. 2**, , eds. V. Suller and Ch. Petit-Jean-Genaz (World Scientific, Singapore, 1994) p. 1203.
10. D. Prasuhn, Private communication.
11. R. Maier, U. Bechstedt, J. Dietrich, S. Martin, D. Prasuhn, A. Schnase, H. Schneider, H. Stockhorst, and R. Tölle, *Commissioning of the Cooler Synchrotron COSY*, (paper presented at the Cyclotron Conf., Capetown, Oct. 1995).
12. D. Boussard, *Proc. of the Joint US-CERN Accelerator School* (Texas 1986), eds. M. Month and S. Turner (*Lecture Notes in Physics*, **296**, Springer-Verlag Berlin Heidelberg, 1987) p. 289.
13. G. Jackson, in *Proc. of the Workshop on Beam Cooling and Related Topics* (Montreux, 1993), ed. J. Bosser (CERN 94-03, 1994) p. 127.
14. F. Caspers, S. Maury, C. Metzger, and D. Möhl, *Bunched Beam Stochastic Cooling* (Cooling Club Newsletter, Sept. 1994).
15. J. Bosser, F. Caspers, M. Chanel, M. Church, S. Jacobson, S. Maury, and D. Möhl, *LEAR MD Report: Bunched Beam Schottky Spectrum* (PS/AR/Note 94-15 (MD), CERN, 1994).
16. Jie Wei, in *Proc. of the Workshop on Beam Cooling and Related Topics* (Montreux, 1993), ed. J. Bosser (CERN 94-03, 1994) p. 132.
17. M. Wendt, Private communication.
18. F. Caspers, *Experience with UHV-Compatible Microwave Absorbing Materials at CERN*, (CERN/PS 93-10 (AR), 1993).
19. G. Carron, F. Caspers and L. Thorndahl, *Development of Power Amplifier Modules for the ACOL Stochastic Cooling Systems* (CERN 85-01, 1985).

20. M. Chanel, R. Maccaferri and J.C. Perrier, in *Proc. of EPAC 90 (Nice),)*, **Vol. 2**, eds. P. Marin and P. Mandrillon (Editions Frontières, Gif.sur-Yvette, 1990) p. 1574.
21. H. Chaloupka, M. Hein, G. Müller, and H. Piel, in *Proc. of the ESA ESTEC Workshop on Space Applications of High Temperature Superconductors (1993).*
22. F. Caspers and G. Dôme, in *Proc. of EPAC 94* (London), **Vol. 2**, eds. V. Suller and Ch. Petit-Jean-Genaz (World Scientific, Singapore, 1994) pp. 1208-1210.
23 A. Dainelle, A.G. Ruggiero and L. Tecchio, in *Proc. of the Workshop on Beam Cooling and Related Topics* (Montreux, 1993), ed. J. Bosser (CERN 94-03, 1994) p. 212.
24. Y. Yano, A. Goto, Y. Miyazawa, M. Hemmi, T. Chiba, T. Kageyama, K. Ikegami, M. Takanaka, T. Kawaguchi, S. Kohara, M. Kase, T. Tominaka, Y. Batygom, T. Kubo, T. Nakagawa, Y. Tanaka, E. Ikezawa, M. Wakasugi, T. Mitsumoto, N. Inabe, O. Kamigaito, H. Okuno, S. Fujishima, A. Ohtomo, and T. Ohkawa (RIKEN), T. Katayama, S. Watanabe, and M. Tomizawa (INS-Tokyo), J. Xia, Y. Rao, and Y. Yuan (IMP-Lanzhou*), RIKEN RI Beam Factory Project*, (paper presented at the Cyclotron Conf., Capetown, Oct. 1995).
25. A. Michailichenko and M. Zolotorev, *Phys. Rev. Lett.* **71** (1993) 4146.
26. M. Zolotorev and A. Zholents, *Phys. Rev.* ***E*** **50** (1994) 3087.

ELECTRON COOLING
(Theory, experiment, computer simulations)

I.N.MESHKOV

Joint Institute for Nuclear Researches,
141980, Dubna, Russia

ABSTRACT

Although electron cooling devices became now rather routine instruments of 10 heavy particle storage rings, some vague points still exist in understanding of its physics. First of all does the classical electron cooling theory fit numerically to experiment data? Then, how significant is a disagreement between different theoretical approaches and which one is the better approximation to the reality? An attempt to answer these questions and to give a look on the trends in the method development is made in present report.

1. Electron Cooling Theory

1.1. The Magnetised Electron Beam with Flattened Distribution of Electrons over Velocities. Three Areas of Particle Velocity

The peculiarity of electron distribution over velocities influences essentially on the efficiency of electron cooling process [1,2]. If the electron beam is obtained by acceleration in an electrostatic field, the velocity distribution of its electrons is significantly anisotropic in the electron frame: the spread in longitudinal velocities is much smaller than the spread in transverse ones. A sufficiently good approximation of this anisotropic distribution function is the Maxwellian function (see details in [3]):

$$f(\vec{v})d^3v = \left(\frac{m}{2\pi}\right)^{3/2} \frac{1}{T_\perp\sqrt{T_\parallel}} e^{-mv_\perp^2/2T_\perp - mv_\parallel^2/2T_\parallel} 2\pi v_\perp dv_\perp dv_\parallel ,$$

$$\frac{1}{2}m\langle v_\perp^2\rangle = \frac{1}{2}m\left(\langle v_x^2\rangle + \langle v_y^2\rangle\right) \equiv m\Delta_\perp^2 = T_\perp \approx T_c \geq 0{,}1\ \text{эB},.$$

$$m\langle v_\parallel^2\rangle \equiv m\Delta_\parallel^2 = T_\parallel \approx \frac{T_c^2}{\beta^2\gamma^2 mc^2} + e^2 n_e^{1/3} \approx 3\cdot10^{-7} + 1\cdot10^{-4}\ \text{эB} \qquad (1.1)$$

Here v, n_e - electron velocity and density in the particle rest frame, $T_\perp, T_\parallel$ - transverse and longitudinal electron temperature, $\Delta_\perp$, $\Delta_\parallel$ - the rms values of the electron velocity spread in the electron frame. In this case it is possible to distinguish three characteristics region of high, low and superlow velocities of the cooled particle (Fig.1):

$$
\begin{aligned}
&\Delta_\perp < V && \text{- High velocity (H)} \\
&\Delta_\parallel < V < \Delta_\perp && \text{- Low velocity (L)} \\
&V < \Delta_\parallel && \text{- Superlow velocity (S).}
\end{aligned} \tag{1.2}
$$

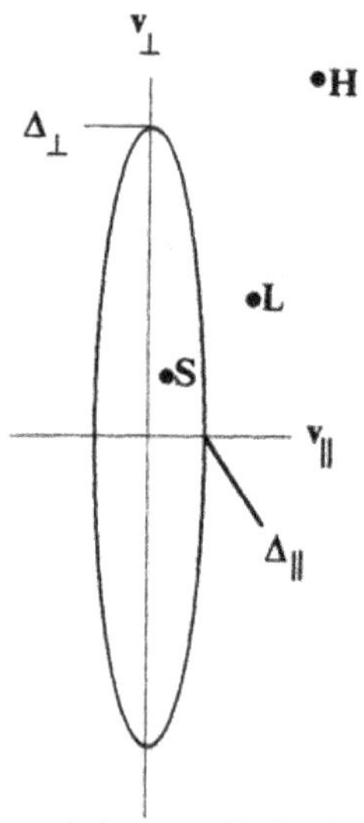

Fig. 1. Flattened electron velocity distribution. The line of constant density labelled H,L,S is the position of the particle in velocity space for the three cases in (1.2).

In the first region, that of *high velocities*, the cooling dynamics is essentially the same as that in the case of an isotropic Maxwellian distribution, which follows directly from the Coulomb analogy: the field of a "charged ellipse" coincides at large distances with the field of a point charge.

In the *low velocity* region the friction force is the analogy of the Coulomb force of a charged disk (a layer) near its axis. In the region of *superlow velocities* the particle falls inside the "disk", where the friction force has only a longitudinal component (again, the Coulomb analogy).

1.2.Binary collisions

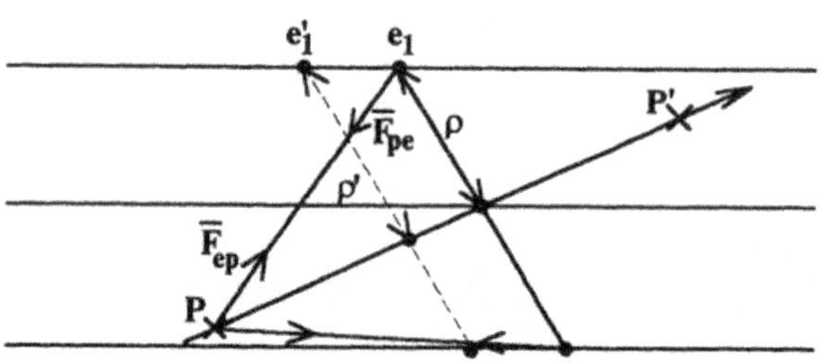

Fig.2. A collision of the particle (P) with magnetised electron (e) ee' - the electron trajectory, pp' - the particle trajectory. The impact parameters alters during the collision, this leads to variation of the Coulomb interaction force magnitude and, as result, gives some momentum transfer.

This is convenient to begin the analysis of electron-particle binary interaction with the case of so called magnetised collisions, when electron can be presented as a bead on a knitting-needle, which corresponds to a magnetic field line. Then during the collision the electron displaces along the line and the impact parameter changes (Fig.2), what leads to nonzero momentum transfer from the particle to the electron. Finally, such an interaction mechanism results in a finite magnitude of the interaction force, acting to the particle [1-3]. This is to be underlined, that all three components of the force have nonzero magnitudes even at infinite magnetic field $B \to 0$.

In a finite magnetic field the electron moves by a spiral trajectory, which axis coincides with a magnetic field line. Then the first dimension parameter appears - the radius of the Larmor circle

$$\rho_\perp = \frac{m v_\perp c}{eB} \le \frac{m \Delta_\perp c}{eB} \quad . \tag{1.3}$$

And now one can define three kinds of the impact parameter ρ (Fig.3).

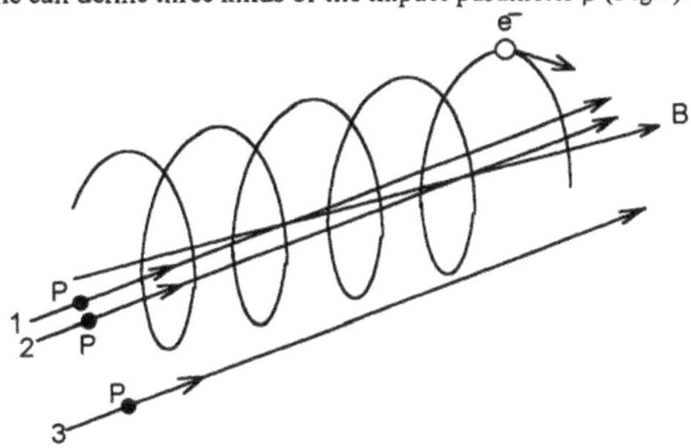

Fig.3.Three kinds of collisions: 1 - *F*ast, 2 - *A*diabatic, 3 - "*M*agnetised".

1. Small impact parameters, which correspond to so called *F*ast collisions, when the collision time is much less of cyclotron revolution frequency:

$$\tau_{collision} \sim \frac{\rho}{\left|\vec{V} - \vec{\Delta}_\parallel\right|} << \frac{1}{\omega_B} \equiv \frac{mc}{eB} \quad . \tag{1.4}$$

The small ρ lies in an interval

$$\rho_{\min} \le \rho \le \rho_F \quad , \tag{1.5}$$

where

$$\rho_{\min} = \frac{ze^2}{m\left|\vec{V} - \vec{\Delta}_\perp\right|}, \quad \rho_F = \frac{\left|\vec{V} - \vec{\Delta}_\parallel\right|}{\omega_B} \quad . \tag{1.6}$$

2. Intermediate impact parameters and *A*diabatic collisions are described with the following expressions:

$$\rho_F \le \rho \le 2\rho_\perp \ . \tag{1.7}$$

At such ρ the electron has a chance to collide with the particle few times, circulating around its trajectory:

$$N_{collision} \sim 1 + \frac{\Delta_\perp}{\pi\left|\vec{V} - \vec{\Delta}_\parallel\right|} \ . \tag{1.8}$$

3. Large impact parameters - *M*agnetised collisions. Correspondingly, we have the case of quasimagnetised collisions, when the electron looks like "the bead":

$$2\rho_\perp \le \rho \le \rho_{max} \ , \tag{1.9}$$

$$\rho_{max} = \max \begin{cases} a - \textit{electron beam radius}, \\ R_D - \textit{Debye radius}, \\ |\vec{V} - \vec{\Delta}_\perp| \cdot \tau_{travel} \ , \quad \tau_{travel} = l_{cool} / \beta c \, . \end{cases} \tag{1.10}$$

Each one of three kinds of collisions has own peculiarity in each of three region of the electron velocity space, defined above (Fig.1). In the "H" region only two kinds -the *F*ast and *M*agnetised collisions (no *A*diabatic ones!) are possible. In the "L", "S" regions all three kinds of collisions can occur. Some peculiarity of magnetised collisions at "S" (superlow) particle velocities take place: only longitudinal component of the friction force of magnetised collisions has here nonzero value. It depends also on the particle sign: additional friction for negatively charged particles appears.

The impact parameters have different magnitudes (for the same kind of collisions - F, L, or S) at different particle velocity values. As results *the Coulomb Logarithm "Zoo"* appears (Formulae (1.12)-(1.13) below).

The next problem - what is Debye radius R_D at Superlow velocity? The formal Debye sphere is poor of electrons there. One can solve this problem artificially, taking

$$R_D = (k/n_e)^{1/3}, \ k \sim 3z. \tag{1.11}$$

This means, that the "Debye" sphere has about $3z$ electrons for shielding of the particle charge.

1.3. The friction force (binary collisions)

The combination of all the conditions of electron-particle interactions leads to rather complicated formulae for the friction (cooling) force [3]:

$$F_\perp \approx -\frac{2\pi q}{m} V_\perp \begin{cases} \frac{1}{V^3}\left(2L_{FH} + \frac{V_\perp^2 - 2V_\parallel^2}{V^2} L_{MH}\right), & \text{H} \\ \frac{2}{\Delta_\perp^3}\left(L_{FL} + N_L L_{AL}\right) + \frac{V_\perp^2 - 2V_\parallel^2}{V^2}\frac{L_{ML}}{V^3}, & \text{L} \\ \frac{2}{\Delta_\perp^3}\left(L_{FS} + N_S L_{AS}\right) + \frac{L_{MS}}{\Delta_\parallel^3}, & \text{S} \end{cases}$$

$$F_{\parallel} \approx -\frac{2\pi q}{m} V_{\parallel} \begin{cases} \frac{1}{V^3}\left(2L_{FH} + \frac{3V_{\perp}^2}{V^2} L_{MH} + 2\right), & \text{H} \\ \frac{2}{\Delta_{\perp}^2 V_{\parallel}}\left(L_{FL} + N_L L_{AL}\right) + \left(\frac{3V_{\perp}^2}{V^2} L_{ML} + 2\right)\frac{1}{V^3}. & \text{L} \\ \frac{2}{\Delta_{\perp}^2 \Delta_{\parallel}}\left(L_{FS} + N_S L_{AS}\right) + \frac{L_{MS}}{\Delta_{\parallel}^3}. & \text{S} \end{cases} \qquad (1.12)$$

And the Coulomb logarithms "Zoo" has individual representative in each velocity region:

$$L_{MH} = \ln\frac{(\rho_{\max})_H}{2\langle\rho_{\perp}\rangle} \approx \ln\frac{V\omega_B}{2\Delta_{\perp}\omega_{pe}} \approx 2;$$

$$L_{FH} = \ln\frac{\rho_{FH}}{(\rho_{\min})_{\parallel}} \approx \ln\frac{mV^3}{ze^2\omega_B} \approx 10 - \ln z; \qquad \text{H}$$

$$L_{ML} = \ln\frac{(\rho_{\max})_L}{2\langle\rho_{\perp}\rangle} \approx \ln\left[\frac{\min\left\{\frac{V}{\omega_{pe}}, \left(\frac{3z}{n_e}\right)^{1/3}\right\}}{2\langle\rho_{\perp}\rangle}\right] \approx 3 \div 1;$$

$$L_{AL} = \ln\left[\frac{2\langle\rho_{\perp}\rangle}{\rho_{FL}}\right] \approx \ln\left[\frac{2\Delta_{\perp}}{(\Delta_{\perp} \to \Delta_{\parallel})}\right] \approx 0{,}7 \to 4; \qquad \text{L}$$

$$L_{FL} = \ln\left[\frac{\rho_{FL}}{(\rho_{\min})_L}\right] \approx \ln\left[\frac{mV^3}{ze^2\omega_B}\right] \approx 5 - \ln z;$$

$$L_{MS} = \ln\left[\frac{(\rho_{\max})_S}{\langle\rho_{\perp}\rangle}\right] \approx \ln\left[\frac{1}{2\langle\rho_{\perp}\rangle}\left(\frac{3z}{n_e}\right)^{1/3}\right] \approx 2;$$

$$L_{AS} = \ln\left[\frac{2\langle\rho_{\perp}\rangle}{\rho_{FS}}\right] \approx \ln\frac{2\Delta_{\perp}}{\Delta_{\parallel}} \approx 5; \qquad \text{S} \qquad (1.13)$$

$$L_{FS} = \ln\left[\frac{\rho_{FS}}{(\rho_{\min})_S}\right] \approx \ln\left[\frac{m\Delta_{\parallel}\Delta_{\perp}^2}{ze^2\omega_B}\right] \approx 5 - \ln z;$$

It is natural to incorporate this logic into a program for numerical computations. An example of such a calculation is shown in Fig.4. Here the approximate nature of Eqs. (1.12) and (1.13) - the nonmonotonicity of the friction force in the case of moderate magnetisation (curves 2) is especially clear. It should be noted that for an unmagnetized beam (curve 1) $F_{\perp}$ reaches a maximum for $V = \Delta_{\perp}$, while $F_{\parallel}$ reaches one for $V = \Delta_{\parallel}$.

The second fact worth noting is the increase of the friction force with decreasing particle velocities up to $V \approx \Delta_\parallel$ (Fig.4, curve 3). This effect distinguishes a magnetised beam from an unmagnetised one.

The friction force in a magnetised beam reaches a maximum for $V \approx \Delta_\parallel$. In this region the approximation of binary particle-electron collisions already works poorly. Nevertheless, from (1.12) we find

$$F_{\max} \sim -\frac{2\pi q}{\Delta_\parallel^2} L_{MS} \sim -2\pi z^2 e^2 n_e^{2/3} \quad . \tag{1.14}$$

The last expression is obtained if the value of $\Delta_\parallel$ from (1.1) is used and we take L_{MS}~1. On the other hand, $F_{\max}$ (1.14) is equal in order of magnitude to the force acting on a point charge ze by the same electron charge located at a distance $l_e = n_e^{-1/3}$ from it.

Finally, the third feature of the friction force is the great difference between its transverse and longitudinal components at low velocities. This is caused by the flattened distribution: the axial component of the field of a charged disk (Fig.1) close to its surface is much larger than the radial component. As a result, the longitudinal component of the particle velocity is cooled in the region $V < \Delta_\perp$ more quickly than the transverse component. The ratio $F_\parallel/F_\perp$ is of the order of $\Delta_\perp/V_\parallel$ for V~$\Delta_\perp$ and of order 3 for $\Delta_\perp >> V > \Delta_\parallel$.

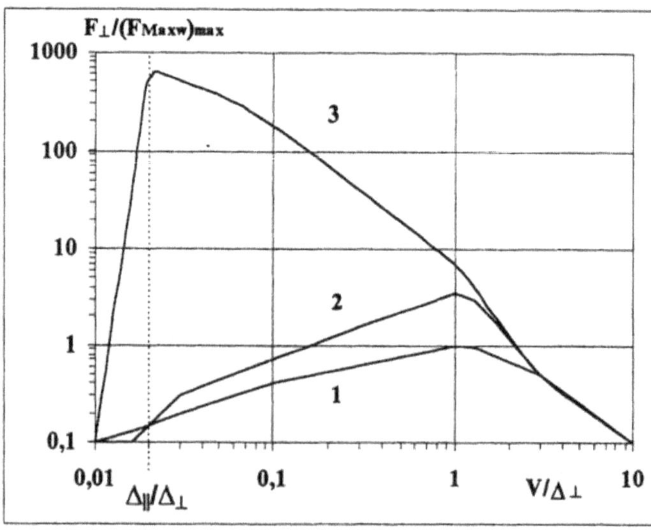

Fig.4. The dependence of the transversal friction force on particle velocity: 1 - the uniform Maxwellian plasma, 2 - the flattened distribution at B=0, 3 - the magnetised electron beam. $\Delta_\perp/\Delta_\parallel = 50$.

1.4. The equilibrium state

The lowest temperature of the particle beam, cooled down by an electron one, is dependent essentially of the electron velocity distribution. One can distinguish three outmost cases:

1)Uniform velocity distribution ($T_{\|}=T_{\perp}\equiv T_e$) and $B=0$. Then particles can be cooled down to the electron temperature Te:

$$T_p = T_e \quad , \quad \sqrt{\langle V^2 \rangle} = \sqrt{\frac{m}{M}} \cdot \Delta_e \tag{1.15}$$

2)Flattened distribution (Eq.1.1) $T_{e\|} << T_{e\perp}$, $B = 0$. The particle temperature limit has different magnitudes for transverse component and longitudinal one:

$$\begin{aligned} T_{p\perp} &= T_{e\perp} \quad , \quad \sqrt{\langle V_{\perp}^2 \rangle} = \sqrt{\frac{m}{M}} \cdot \Delta_{\perp} , \\ T_{p\|} &= \sqrt{T_{e\perp} T_{e\|}} \quad , \quad \sqrt{\langle V_{\|}^2 \rangle} = \sqrt{\frac{m}{M}} \cdot \sqrt{\Delta_{\perp} \Delta_{\|}} \end{aligned} \tag{1.16}$$

3)The magnetised electron beam with flattened distribution (Eq.1.1):

$$\begin{aligned} &T_p \sim T_{e\|} \quad , \quad \Delta_{p\perp} \sim \Delta_{p\|} \sim \sqrt{\frac{m}{M}} \cdot \Delta_{\|} \quad , \\ &F_{max} \sim -2\pi z^2 e^2 n_e^{2/3} \quad . \end{aligned} \tag{1.17}$$

The magnetisation criterion :

$$\rho_{\perp} \leq (\rho_{max})_S \quad \Rightarrow \quad B \geq \frac{m\Delta_{\perp} c}{e} \left(\frac{n_e}{3z} \right)^{1/3} \quad . \tag{1.18}$$

It corresponds, by the order of magnitude, to the condition

$$\rho_{\perp} \leq l_e \; . \tag{1.19}$$

For typical parameter values $n_e \sim 10^8 cm^{-3}, T_{\perp} \sim 0.2 eV$ ($\Delta_{\perp} \sim 2 \cdot 10^7 cm/s$) it gives B≥500 G.

1.5. The dielectric description and MD simulation

What is the difference of the approaches?

Binary collision approximation assumes the cut-off of Coulomb collisions by some impact parameter ρ_{max} and no collective phenomena in electron gas are taken into account. *Dielectric description* is an analysis given on the base of Fourier transform with involvement of dielectric permeability of the electron gas [4,5]. *Molecular dynamics* (*MD*) simulation is a numerical simulation of particle-electron Coulomb collisions at very low density of electron gas [6].

The results of the dielectric description analysis have rather contradictory character: the authors of the Ref.[4] claim for a disagreement between the "binary approximation" and the "dielectric" one. From other side, the analysis, presented in the Ref.[5], brought a good agreement of both approximations. One can admit, that the first statement (Ref.[4]) is initiated mostly by the use of unproper formulae for the friction force in the approximation of the binary collisions. This fact proves the comparison of the curves presented on Fig.5.

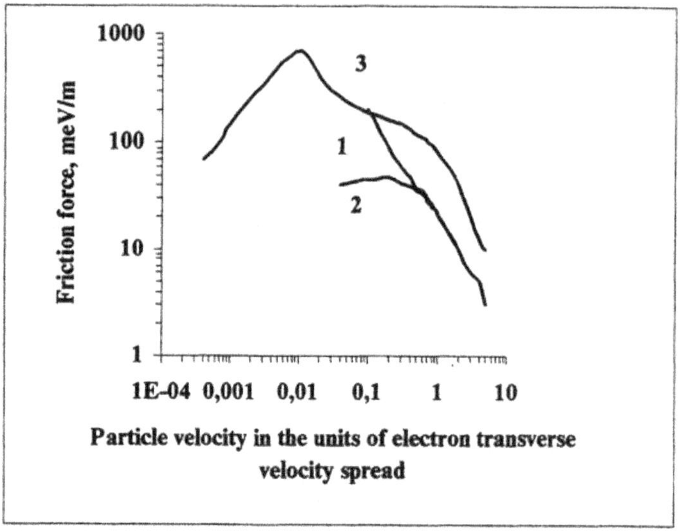

Fig.5.The dependence of longitudinal cooling force on the particle velocity $n_e = 3\times10^8$ cm^{-3}, $T_\perp = 1$ eV, $T_\parallel = 0$, $B = 700$ G, $z = A = 1$
1 - dielectric theory [4]
2 - binary collisions [4]
3 - the same dependence, calculated by Eqs. (1.12) and (1.13) ($T_\parallel=5\times10^{-5}$ eV)

MD simulation [6] are done still mainly for electron beam parameters, which are difficult for a comparison with two other approaches and experimental results. Nevertheless, these studies are in good progress and one can hope the necessary information will be available soon.

2. Review of Experimental Results

Three stages of experimental research can be distinguished in the development of the electron cooling method:

1974-1984: Experiments at proton storage rings at the Budker INP in Novosibirsk [7-10], at CERN [11], and Fermilab [12].

1985-1988: Experiments at the Model Solenoid setup at the INP in Novosibirsk [13,14].

1988-1994: The startup of operation of antiproton and ion storage rings and experimental setups can be found in the original studies and reviews [3,15-18].

2.1. Transverse component cooling

The measurements of the decrement of the damping of betatron oscillations carried out at the NAP-M accelerator [10] showed that for proton energies in the range 1.5-65 MeV the dependence of the decrement on the relative velocity of the particles and the electrons is approximated by the dependence (see Ref.3)

$$\lambda^* = \frac{A}{z^2}\frac{\lambda_{exp}}{n_e^*} = \frac{10^6}{\left(V_\perp^2\right)_{cm/sec}} \ cm^3/sec \ ,$$

$$10^6 \le V_\perp \le 10^7 \ cm/sec \ ,$$

$$n_e^* = 10^{-8}\eta n_e \ . \qquad (2.1)$$

Here $\Delta_\perp \approx 2\times10^7$ cm/sec and $\Delta_\parallel \approx 3\times10^5$ cm/sec, i.e. the data pertain to low velocities. Apparently, this fact is related to a strong influence of particle betatron oscillations and electron space charge (see Section 4). The latest TSR results [20] for $\lambda_\perp$ (Fig.12; C^{6+} and S^{16+} ions) are in good agreement with the NAP-M results.

2.2. The longitudinal friction force

The longitudinal component of the friction force is efficiently measured by using Schottky diagnostics (see [16] and [18]), which make it possible to accurately measure the frequency spectrum of the revolution in the cooled beam. Therefore, in the physics of the electron cooling of ions most of experimental data have been obtained for the longitudinal degree of freedom.

The properties of the longitudinal component of the friction force have been studied carefully at the Model Solenoid (MOSOL) setup at the Budker INP [14], where a beam of protons or negatively charged hydrogen ions accelerated in an electrostatic accelerator was cooled during the time of flight through an electron beam. The high time stability of the accelerator energy ensured that the spread of longitudinal velocities of the particle was small, and the use of an H^- source and a magnesium-vapour target which could be switched on remotely made it possible to change the sign of the particle charge while keeping the other parameters fixed. This made very accurate comparison of the values of the friction force for particles of both signs possible. The precision of the experiment was further ensured by the use of a special electrostatic spectrometer [13]. Another important feature of the experiment was the high degree of uniformity of the magnetic field.

The results of these experiments and also of most of the others which have been performed by now [13,21-24], show that at high and low velocities ($V>\Delta_\parallel$) there is good qualitative agreement with theory. In particular, the magnetisation effect is convincingly represented: the friction force continues to grow with decreasing V in the range $V<\Delta_e \approx 3\times10^7$ cm/sec (see Ref.3).

For superlow velocities, four different experiments (MOSOL, IUSF, ESR, and TSR) have revealed a linear dependence of the friction force on $V_\parallel$, as predicted by the theory. However, the absolute value of the force measured experimentally is almost an order of magnitude smaller than that calculated for the density $n_e=1\times10^8$ cm^{-3}. This difference is apparently related to insufficient stability of the electron energy. Actually, attainment of the maximum level of the friction force at superlow velocities requires the stability

$$\frac{\delta\mathcal{E}}{\mathcal{E}} << 2\frac{\Delta_\parallel}{\beta c} = 2\frac{1}{\beta}\sqrt{\frac{e^2 n_e^{1/3}}{mc^2}} \quad . \tag{2.2}$$

The cubic dependence of the force on $\Delta_\parallel$ requires a fairly strong inequality (if only a factor of 3) to eliminate the effect of $\delta\mathcal{E}$. For the MOSOL setup this estimate gives $\delta\mathcal{E}/\mathcal{E} < 2\times10^{-5}$, and for the IUCF and TSR setups it is still an order of magnitude smaller. The correctness of these explanations is also confirmed by the fact that the points of all three experiments lie well on a single curve, which implies the absence of a dependence of the reduced force on the density n_e [we recall that $\Delta_\parallel \propto n_e^{-1/6}$ in the absence of instability of the electron energy; see Eq. (1.1)]. Inhomogeneity of the magnetic field of the electron cooling system has an analogous effect. It is not impossible that the low level of the force in the early experiments at ICE, where also $F_\parallel \propto V_\parallel$ (Fig.7), was due to the same cause. If this is true, then the pulsations of the electron energy at ICE were about 10^{-3} ($\Delta_\parallel \sim 10^7$ cm/sec).

The cooling of multiply charged ions [19,24] merits special discussion. In the range $V_\parallel > \Delta_\parallel$ good agreement with the theory is observed. At the same time, the experimental date are quite contradictory at superlow velocities: the TSR points coincide well with the points for protons (Fig.7), but a weak (logarithmic) z dependence is observed. The ESR points [24] lie considerably lower, demonstrating a strong z dependence.

The question of the fine details of cooling in this region essentially remains open. Some of the specific features of the interaction at superlow velocities are discussed in Section 3.

2.3. *The friction force for negatively charged particles*

Negatively charged particles (antiprotons, negatively charged ions) repel electrons, while positively charged particles attract them. In collisions with parameters $\rho < \rho_{min}$ in the first case a magnetised electron is reflected and transfers a momentum $2mV$ to the particle, while in the second case it passes near the particle without transferring any appreciable momentum to it. Therefore, for a negatively charged particle the longitudinal component of the friction force is increased by

$$\Delta F_\parallel = -n_e V 2mV\pi\rho_{min}^2 \sim -\frac{2\pi q}{m} \times \begin{cases} 1/V^2, & V > \Delta \\ V^2/\Delta_\parallel^4, & V < \Delta \end{cases} \quad . \tag{2.3}$$

and the maximum value of the longitudinal component [see Eq.1.14] is

$$\left(F_\parallel^{(-)}\right)_{max} \sim -\frac{2\pi q}{m\Delta_\parallel^2}\left(1 + L_{MS}\right) \sim -2\pi e^2 n_e^{2/3}\left(1 + L_{MS}\right) \quad . \tag{2.4}$$

This effect has been convincingly demonstrated in experiments at the MOSOL setup [13,14](see also Ref.3, Section 5.1).

3. The cooled particle beam

3.1. The equilibrium momentum spread

The momentum spread of the particles in a cooled beam is to a large degree determined by intrabeam scattering. From this it follows (see Ref.3, Section 6.2) that the equilibrium momentum spread in an intense cooled beam must grow as $z\sqrt{N/A}\varepsilon^{3/4}$. The experimental results indicate that the physics of deep cooling of a beam is more complicated. For example, in the early experiments at NAP-M (Refs. 25 and 17) it was discovered that there is a suppression of the momentum spread to the level $\Delta p/p=1.2\times10^{-6}$ if the beam intensity does not exceed $N\sim3\times10^{7}$ particles (Fig.6). For $N>3\times10^{7}$ the spread increased sharply. We note that the minimum spread attained corresponds to a temperature of 10^{-4} eV in the particle frame, which for the parameters of the NAP-M experiment is in good agreement with the condition (1.17).

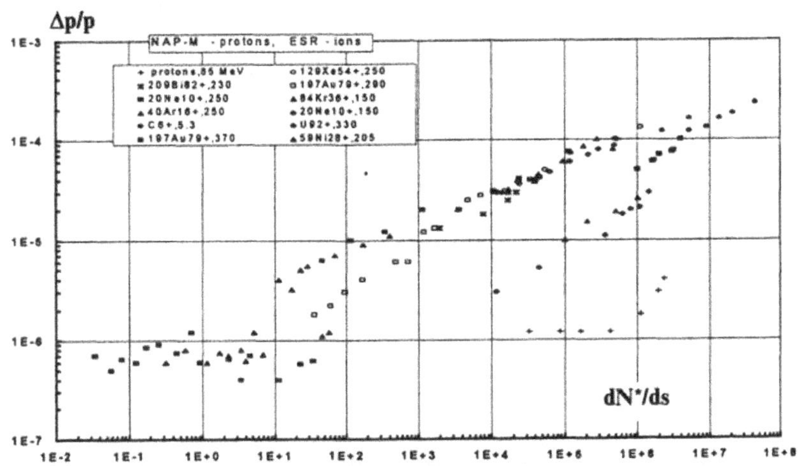

Fig.6.Dependence of the particle momentum spread on the reduced linear density of the cooled beam: $dN^*/ds = (z^2/A)(N/\eta_\infty\beta^2\gamma^2c)$.

In experiments involving heavy multiply charged ions at the ESR [24, 26, 27] and at the TSR [28] a different picture was observed (Fig.6): for a wide range of ions in the energy range 150-290 MeV/nucleon the momentum spread increases monotonically from the level 10^{-6}. (To be able to compare the experimental results, in Fig.6 we have used the reduced linear particle density; see Eq. (3.4) below.) However, the recent results obtained on ESR [29], demonstrate apparently the behaviour of ion momentum spread, similar to that in NAP-M (see Fig.8, Au^{79+} (370 MeV) and Ni^{28+}).

Before discussing these results, let us return to [25]. There the level of noise of a cooled proton beam was measured and used to find the particle momentum spread. The relation between these parameters was obtained in [30], where the physics of the noise of a particle beam in the presence of friction (cooling) was analysed. In particular, the formation of space-charge waves in a beam circulating in a storage ring was described. This results in a splitting of the Schottky noise spectrum of the beam into two peaks separated by an amount

$$\Delta\omega_n = n2\omega_p, \quad n = 1,2,... \tag{3.1}$$

Here ω_p is so called plasma frequency, equal to

$$\omega_p = \frac{c}{R_S}\sqrt{\frac{\eta_\omega r_p}{\pi\gamma}\frac{z^2N}{AR_S}\ln\frac{b}{a}} \geq \eta_\omega \frac{\beta c}{R_S}\frac{\Delta p}{p} \ ,$$

$$\eta_\omega = \frac{p}{\omega}\frac{d\omega}{dp} \ , \tag{3.2}$$

where $R_S=C/2\pi$ is the average radius of the orbit and a<<b are the radii of the beam and chamber cross sections, N - the numerical number; z, A - the ion charge and mass numbers. The sign ± indicates that in the particle frame there are two waves travelling parallel and antiparallel to the velocity v_0 with which they drift as a whole in the lab frame. This splitting exists if the following condition is satisfied in the cooled beam:

$$\omega_p = \Delta\omega_{\min} \geq \eta_\omega\omega_s\frac{\Delta p}{p} \ . \tag{3.3}$$

The stable state of the beam with split spectrum can, as a rule, be supported when there is sufficiently strong feedback which suppresses the growth of coherent oscillations of the particles: a so-called damper [31].

As a rule, the experimental values of $\Delta\omega_n$ are in good agreement with the calculated values (3.1). Taking into account (3.2), the condition (3.3) can be written as

$$N_{th} \geq \frac{A}{z^2}\frac{\eta_\omega\beta^2\gamma C}{2r_p\ln\frac{b}{a}}\left(\frac{\Delta p}{p}\right)^2 \ . \tag{3.4}$$

Excess over the threshold value is a necessary condition for splitting of the peaks. The sufficient condition is more complicated to formulate. There are essentially two: the beam must be "deeply" cooled, and the intensity must satisfy (3.4). What does deeply cooled mean?

Here the idea of the ordering of the particle locations in the cooled beam arises. This was first pointed out in [32]. The noise suppression for $N<3\times10^7$ in the NAP-M proton beam indicates the appearance of a significant correlation between the particle locations in the beam. The condition for such ordering to arise is known from solid-state physics (the so-called Wigner crystal):

$$\frac{z^2e^2}{l_pT} >> 1, \quad l_p = n_p^{-1/3} \ , \tag{3.5}$$

where T and n_p are the temperature and density of particles with charge ze, and l_p is the average distance between particles. In this case the potential energy of the interparticle interaction exceeds the kinetic energy of the thermal motion. In the particle beam in a storage ring there are two parameters characterising the distance between particles: the transverse dimension of the beam a, and the longitudinal "gap" between the particles in the beam $l_p=C/N$. The condition (3.5) is obviously satisfied if

$$a < l_\parallel << \frac{z^2e^2}{T} \quad . \tag{3.6}$$

It is interesting that the inequality (3.5) directly implies that the potential energy of a particle in a ordered beam chain is larger than the kinetic energy of the thermal motion:

$$\Gamma_\parallel \equiv \frac{z^2e^2}{l_\parallel} \bigg/ \frac{(\Delta p)^2}{2M_{eff}} > 1 \quad . \tag{3.7}$$

Here

$$M_{eff} = \frac{\gamma M}{\eta_\infty} \ln\frac{b}{a} \tag{3.8}$$

is some effective mass of the particle moving in a storage ring under the action of the wave field (the synchrotron mass multiplied by a logarithmic coefficient taking into account the mass of the wave field which carries the particles).

The two inequalities (3.6) also form the condition for "deep cooling". In early experiments involving heavy ions, the results of which are shown in Fig.8, this condition is not satisfied. As a result, the nature of the dependence of the momentum spread in the ESR and TSR ion beams and in the NAP-M proton beam differ fundamentally. It is interesting that in the NAP-M the noise suppression was observed up to $l_\parallel$~0.05 mm for a=0.1 mm. In the recent ESR experiments the points Au(370 MeV) and Ni fall outside the uniform dependence (the spread is roughly proportional to the square root of the linear particle density). For these points the dependence has a quasithreshold character with $(dN^*/ds)_{crit}$~10. At such a poor intensity of the cooled ion beam the parameter $l_\parallel$ becomes very large, of few tens mm (see Table 2.1) and the criterion (3.5) looses its meaning: the ions do not interact during much of the time. In this case one can introduce another criterion taking into account scattering of the colliding ions in a transverse direction. Then, to prevent an ion pass another one, the scattering angle should exceed $\pi/2$, or impact parameter (which is of order of betatron amplitude $a_\perp$) is to be small enough:

$$a_\perp \le \frac{(ze)^2}{T_\parallel} \quad . \tag{3.9}$$

This makes possible to formulate the second criterion of the ordering:

$$\Gamma^* = \frac{(ze)^2}{a_\perp T_\parallel} > 1 \quad . \tag{3.10}$$

The Table 2.1 presents the parameters of the NAP-M experiments and the ESR ones with Au and Ni ions. They demonstrate a significant difference in conditions of both experiments: in the case of NAP-M the criterion (3.5) is well satisfied, whereas the ESR experiment parameters are in a good agreement with criterion (3.10).

Table 3.1. Parameters of NAP-M and ESR experiments

Storage ring	NAP-M	ESR	
Point N (Fig.8)	1	2	3
Particles	protons	$^{197}Au^{79+}$	$^{59}Ni^{28+}$
$\mathcal{E}$, MeV/u	65	370	205
$\Delta p/p$	1.2×10^{-6}	4×10^{-7}	6×10^{-7}
$T_\parallel = \left(\beta\frac{\Delta p}{p}\right)^2 Mc^2$, MeV	0.17	14	6.5
ε, π mm mrad	2×10^{-3}	$\sim1\times10^{-3}$	
$a_\perp$, mm	0.1	~0.1	
N	2×10^{7}	1×10^{3}	700
$l_\parallel=C/\gamma N$, mm	2.3×10^{-3}	150	180
$\Gamma_\parallel = \frac{(ze)^2}{l_\parallel T_\parallel}$	3.7	4.3×10^{-3}	1×10^{-3}
$\Gamma^* = \frac{(ze)^2}{a_\perp T_\parallel}$	0.08	6.4	1.7
The ordering	Yes	Yes	Yes

All the experiments cited in the present study were carried out at an energy below the critical value ($\eta_\omega>0$). Another feature of the NAP-M was the fact its operating regime was close to the critical energy ($\eta_\omega \approx 0.08$), as a result of which the threshold N_{th} (3.4) was lowered significantly.

One can conclude, that for obtaining of an ordering effect the deep three dimensional cooling is necessary and cooled beam parameters are to be corresponding to criteria (3.7), (3.10).

One fact more deserves attention. When the beam is cooled properly in longitudinal dimension, but has rather large betatron amplitudes, the ion collisions can occur not so often as at low amplitudes due to the missing effect: one ion passes another one without significant scattering, if impact parameter $\rho \sim a_\perp$ is large. In this case the number of interacting ions is order of

$$\Delta N \sim \frac{\tau_{bet}}{\Delta t_{cool}} \,, \tag{3.11}$$

where τ_{bet} is period of betatron oscillations,

$$\Delta t_{coll} \sim \frac{1}{\Delta\omega_S R_S}\frac{C}{N} = \frac{\tau_{rev}}{N\eta_\omega \frac{\Delta p}{p}} \tag{3.12}$$

- the time duration between two ion collisions, τ_{rev} - the ion revolution period in the ring. Then, instead of well ordered ion chain we have an ion beam, consisting of particle groups ("clusters") with particle number in each cluster of order of ΔN. Nevertheless, such a beam ***clusterisation*** leads also to the noise reduction, as well as in an ordered beam.

From Formula (3.11), (3.12) we have the following result:

$$\frac{(\Delta N)_{cluster}}{N} \sim \frac{\eta_\omega}{Q_{bet}} \frac{\Delta p}{p} . \tag{3.13}$$

Q_{bet} is here the betatron number of the ring. The parameters of Table 3.1 give us $(\Delta N)cluster \sim 1$ in the NAP-M case and less than 10^{-4} in ESR one.

3.2. The coupling effect at superlow velocity

The particle beam and the cooling electrons meet on the cooling section at the exit from the toroidal section of an electron cooling device. In the particle frame the beam junction looks like a fast displacement of the electrons in the direction perpendicular to the axis of the beam of cooled particles:

$$\Delta z \approx \frac{(\beta c \Delta t)^2}{2R} , \tag{3.14}$$

where R is the radius of curvature of the electron trajectory in the toroidal solenoid. For typical parameters this estimate gives a displacement of order 1 mm over a cyclotron revolution period, i.e., the particle and the electron meet "instantaneously". If the particle are already cooled to the level $V \le \Delta_\parallel$ and have positive charge, the electron and the particle move together, and the electron osscilates along the direction of the magnetic field under the action of the particle field, while simultaneously rotating slowly with frequency ω_d (drifting) around the field line passing through the particle. This is the well known drift in crossed fields: the longitudinal magnetic field and the transverse electric field of the particle.

$$\omega_d \sim c\frac{ze}{\rho^3 B} . \tag{3.15}$$

The separation of the ion and the electron occurs in the second toroidal section and has the same "instant" character. Due to this fact the electron oscillation and rotation phase is practically random, which results in an additional diffusion (see details in Ref.3). Instead of the equilibrium temperature (1.17) we have

$$T_p^* = T_{e\parallel}(1+\delta) , \tag{3.16}$$

where

$$\delta \sim \begin{cases} \dfrac{1}{L_{MS}}\left(\dfrac{\omega_B^2 \tau}{\omega_{pe}}\right)^{1/3} & - \textit{protons;} \\ 0.04 z^{5/3} \dfrac{\omega_{pe}\tau}{L_{MS}} & - \textit{ions;} \end{cases} \tag{3.17}$$

$\tau = l_{cool}/\beta c$ - the time duration of the particle passing through the cooling device. This formula is correct, when

$$\omega_{pe}\tau << 1 \ , \qquad (3.18)$$

ω_{pe} - electron plasma frequency.

The numerical estimations (Table 3.2) show, that this effect can give a significant increase of the equilibrium temperature. The comparison with experimantal temperature values looks not too much contrary.

Table 3.2. Estimation of "the coupling effect"

Storage ring	NAP-M		ESR	
Particle	proton		$^{197}Au^{79+}$	$^{59}Ni^{28+}$
Point N (Fig.8)	-	1	2	3
$\mathcal{E}$, MeV/u	1.5	65	290	205
B, kG	0.3	1.0	1.5	
n_e, cm^{-3}	5×10^6	2×10^8	5×10^7	
$T_{e\parallel}$, meV	0.025	0.04	0.05	
δ	440	200	150	30
T_p^*, meV	11	8.4	8.0	1.5
$(\Delta p/p)_{exp}$	7×10^{-5} (θ_p)	1.2×10^{-6}	4×10^{-7}	6×10^{-7}
$T_{exp} = \left(\beta\frac{\Delta p}{p}\right)^2 Mc^2$, meV	7.3	0.17	14	6.5

4. Electtron Cooling Process in a Storage Ring

The calculatuion of an electron cooling process in a storage ring is the problem of a great practical interest. For many applications there is a need to analyse the beam parameters behaviour under action of a cooling electron beam. However, very intricate and multiparametric character of the friction force Eqs. (1.12), (1.13) makes the task of an analytical analysis highly complicated. And the problem becomes practically unresolvable analytically, when additional effects are to be taken into account, such as particle betatron and synhrotron oscilaltions in the ring, the influence of the space charge of the cooling electron beam, and so on. The analytical formulae for cooling time, obtained in different works, give rather rough estimation by the order of magnitude (see details in Ref.3)

The soluiton lies in use of the computer simulation of the process. Such an attempt was made and described in Ref.33. The computation program calculates the evolution of particle oscillation amplitude, using Eqs.(1.12), (1.13) for friction force. Computation is being produced "turn by turn" when the following effects are taken into account:

- betatron oscillations
- dispersion in cooling section
- electron beam space charge and its (possible) neutralisation
- phase oscillations if RF voltage is ON
- possible variation of electron energy ("sweeping").

For the demonstration in present report the "lead ion regime" of LEAR is used: the ions Pb^{54+} with the energy E_p=4.23 MeV/amu.

Few conclusions can be made from the computation results.

4.1. Particle dynamics.

The nonlinear character of the friction force as a function of particle velocity (see (1.12), (1.13)) influences significantly on the variation of the particle velocity (momentum) components in time. Particularly, the cooling rate increases drastically, when the particle velocity falls in the "S" region (see details in Ref.33).

4.2. Betatron oscillation and electron beam space charge.

The influence of betatron oscillations and electron beam space charge appears in some averaging process, which smoothes the "gaps" in the friction force function, described by Eqs. (1.12) and (1.13). On the other hand the betatron oscillations bring the particle motion across the electron beam, so it interacts with electrons, which have different velocity due to the beam space charge. Such an effect is taken into account in computation, which shows, that cooling time shortens very essentially, when the space charge is neutralized.

Also one should mention the next well-understandable result of the computation: the cooling process do not differ if either an angular or its equivalent coordinate amplitudes are chosen as initial parameter values.

The influence of electron space charge changes significantly the dependence of the friction force and cooling rate on electron current I_e: the cooling rate increases linearly with I_e, when η=1, but the increase becomes very modest at η<0.5.

The influence of the electron space charge can lead even to heating of the ion to be cooled. As the analysis shows, in the case of nonneutralised electron beam (the case of the well known parabolic dependence of the space charge potential on radial coordinate) the **heating** occurs, if initial magnitudes of the ion momentum components and dispersion of the ring satisfy to the following conditions:

$$\frac{\Delta p_{x,z}}{p} > \frac{1}{\alpha_{x,z}\gamma}, \quad \frac{\Delta p_S}{p} > \frac{1}{\alpha_x}, \quad D > \frac{R_S}{Q_x}\frac{\sqrt{\alpha_x}}{\gamma},$$

$$\alpha_{x,z} = \frac{eI_e}{\beta^3\gamma^3 mc^3}\left(\frac{R_S}{Q_{x,z}a_e}\right)^2 . \tag{4.1}$$

Here $\Delta p_i = p_i - \langle p_i \rangle$, R_s, $Q_{x,z}$ - the average radius and betatron numbers of the ring, a_e - electron beam radius.

This effect is significant for low energy coolers with an intense electron beam. The similar effect can limit also the equilibrium temperature of the cooled ion beam. If it is cooled down to the diameter $2a_\perp$, the space charge potential varies across this size through a range

$$\delta U_e = \frac{I_e}{\beta c}\left(\frac{a_\perp}{a_e}\right)^2 . \tag{4.2}$$

This gives the longitudinal temperature spread (betatron oscillations again!) equal to

$$\delta T_\parallel = \frac{(e\cdot\delta U_e)^2}{\beta^2\gamma^2 mc^2} . \tag{4.3}$$

4.3. The electron beam expansion

An adiabatic expansion of the cooling electron beam in decreasing magnetic field permits to reduce electron transverse temperature as

$$T_\perp(B) = T_\perp(B_0)\frac{B}{B_0} . \tag{4.4}$$

This effect was successfully demonstrated in the INP group experiments [34] and later applied to electron cooling of protons on CRYRING [35]. There was found an increase of the longitudinal cooling force, when the beam is expanded. However, this results does not agree with the theory: if the electron beam is magnetised, the cooling depends very stightly on electron transverse temperature. The same justifies the computation (Fig.7): the cooling rate does not depend practically on $\Delta_\perp$, when it is less than $\Delta_\perp \leq 4\times10^{-3}\beta c$.

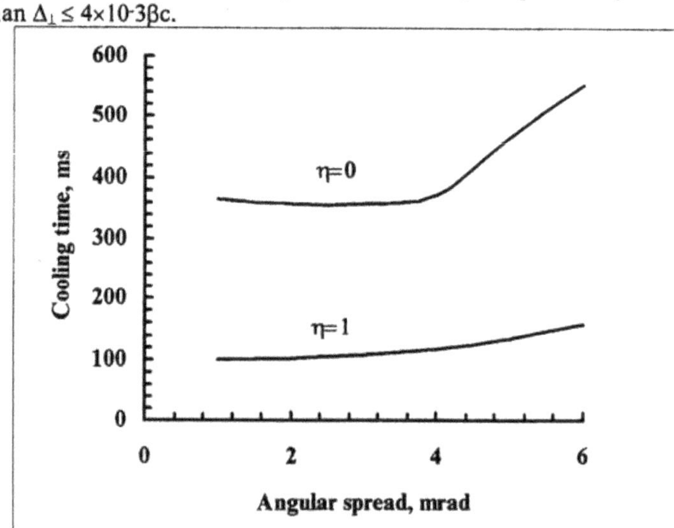

Fig.7. The dependence of cooling time on electron angular spread (Lead 54+, I_e=300 mA, B=600 G).

The effect, observed in the experiment [35] can be explained by the beam transfer, due to the expansion, from a nonmagnetised state to the magnetised one, which results in an increase of the friction force. Actually, when electron density and magnetic field are fixed in cooling section, a reduction of $\Delta_{\perp}$ leads to satisfaction of the conditions (1.18), (1.19). (This can be done by simalteneous increase of magnetic field on the gun cathode B_c and the beam current $I_e \propto B_c$). However, such an enhancement of the friction force has saturation, when magnetisation is reached.

4.4. "Sweeping" process

The problem of special practical interest is a possibility to cool down as fast as possible a particle beam with large momentum spread. The method, discussed since very beginning of electron cooling method development [2] but never tested, is so called sweeping - the variation of electron energy during cooling process. This method was simulated by computations, which gave very promising results [33].

4.5. Bunched beam

The cooling of a bunched particle beam has own peculiarities. One can say that the influence of the electron beam space charge and the ring dispersion is much more significant, then for a coasting beam (see Ref.33): even one can get regime, when the particle beam does not reach the cooled state at all.

References

1. G.Budker, A.Skrinsky. *Electron cooling.* Usp.Fiz.Nauk, **124** (1978) 561.
2. Ya.S.Derbenev, A.N.Skrinsky. *The Physics of Electron Cooling.* Physics Reviews, ser.Sov.Phys.Reviews, **3** (1981) 165.
3. I.Meshkov. *Electron cooling: status and perspective.* Part. & Nuclei, **25** (1994) 1487.
4. A.H.Sorensen & E.Bonderup. *Electron coolong,* Nucl.Instr.Meth., **215** (1983) 27-54.
5. V.A.Lebedev. In *The ultimate possibilities of electron cooling.* (Preprint of Budker INP 88-61, Novosibirsk, 1988, Appendix 1).
6. C.Toepfer et al. *Molecular Dynamics simulations* (see this Proceeding).
7. G.Budker, N.Dikansky, V.Kudelainen et al. *Particle Accelerators,* **7** (1976) 197.
8. G.Budker, A.Bulyshev, N.Dikansky et al. (Moscow: Nauka) **1** (1977) 236.
9. G.I.Derbenev, A.N.Skrinsky. *Sov.J. Fiz. Plasmy.* **4** (1978) 273.
10. N.Dikansky, V.Kononov, V.Kudelainen et al. In *Proc. of The VI-th All-Union Conf. on High Energy Part. Accel.* (JINR, Dubna) **1** (1979) 99.
11. M.Bell, J.Chaney, H.Herr et al. *Nucl.Instr.Meth.,* **190** (1981) 237.
12. T.Ellison, W.Kells, V.Kerner et al. *IEEE Trans. Nucl. Sci.,* NS-30 (1983) 2370.
13. N.Dikansky, V.Kudelainen, V.Lebedev. (Preprint INP 88-61, Novosibirsk, 1988).

14. N.Dikansky, V.Kokoulin, N.Kot et al. . In *Proc. of The XIII-th Intern. Conf. on High Energy Part. Accel.* (Novosibirsk: Nauka) **1** (1987) 330.
15. A.Skrinsky, V.Parkhomchuk. *Phys. Part. Nucl.*, **12** (1981) 557.
16. H.Poth. (Preprint CERN EP/90-04, Geneva, 1990).
17. V.Parkhomchuk, A.Skrynsky. *Rep. on Progr. in Phys.*, **54**, No.7 (1991) 919.
18. J.Bosser. *CERN Accelerator School*, (CERN 92-01, Geneva, 1992).
19. A.Wolf, C.Ellert, M.Grieser et al. *Proc. Workshop on Beam Cooling and Related Topics*. (Montreux, 3-8 Oct, 1993, CERN, Geneva, 1994), p.416.
20. M.Grieser, F.Albrecht, D.Habs et al. *Report on the 4-th Europ. Part. Accel. Conf.*, (London, June, 1994).
21. H.Danared. *Proc. Workshop on Beam Cooling and Related Topics*. (Montreux, 3-8 Oct, 1993, CERN, Geneva, 1994) p.322.
22. Ch.Ellert, D.Habs, E.Jaeschke et al. *Nucl. Instr. Meth.*, **A314** (1992) 399.
23. T.Ellison, M.Ball, B.Brown et al. *Scientific and Techn. Report of Indian Univ. Cyclotron Facility*. (Indian Univ., Bloomington, 1990) p.112.
24. M.Steck, K.Becker, F.Bosch et al. *Report on the 4-th Europ. Part Accel. Conf.*, (London, June, 1994).
25. E.Dementiev, N.Dikansky, A.Medvedko et al. (Preprint INP 79-70, Novosibirsk; Preprint CERN/PS/AA 79-41, Geneva, 1979).
26. M.Steck, N.Angert, K.Becker et al. *Proc. Workshop on Beam Cooling and Related Topics*. (Montreux, 3-8 Oct, 1993, CERN, Geneva, 1994) p.395.
27. G.Kalish, K.Becker, B.Franzke et al. *Proc. of the 3-th Europ. Part Accel. Conf.*, (Berlin) **1** (1992) p.780.
28. B.Hochadel, F.Albrecht, M.Grieser, D.Habs et al. *Proc. Workshop on Beam Cooling and Related Topics*. (Montreux, 3-8 Oct, 1993, CERN, Geneva, 1994) p.198.
29. M.Steck et al. *Suppression of intrabeam scattering in cooled ion beams. Hyperfine structures*. (To be published).
30. V.Parkhomchuk, D.Pestrikov. (Z. Tekh. Fiz., 1980,) **50**, No.5, 1411; (Sov. Phys. Tech. Phys.) **25** (1980) 875.
31. J.Bosser, M.Chanel, R.Ley, G.Tranquille. In: *Proc. Intern. Workshop "Physical Experiments and First Results on the Heavy Ion Storage and Cooler Rings"*. (Smolenice, June 1-5, 1992; JINR, Dubna, 1994).
32. V.Parkhomchuk. *Proc. of Workshop on Electron Cooling and Related Applications*. (Kernforschungszentrum Karlsruhe, 1984) p.71.
33. A.Lavrentev, I.Meshkov. (Preprint JINR E9-95-317, Dubna, 1995).
34. V.I.Kudelainen, V.A.Lebedev, I.N.Meshkov, V.V.Parkhomchuk et al. (Zh. Eksp. Teor. Fiz., 1982) **83**, 2056; (Sov. Phys. JETP) **56** (1982) 1191.
35. H.Danared et al. *In Proc. of the IV-th Europ. Part. Accel. Conf.*, (London) **1** (1994) 380.

SPECTROSCOPY OF TRAPPED IONS

H. WALTHER

Sektion Physik der Universität München and
Max-Planck-Institut für Quantenoptik
85748 Garching, Fed. Rep. of Germany

ABSTRACT

In this paper two groups of applications of ion traps will be discussed: those on high precision spectroscopy of single ions and on the investigation of phase transitions of a few ions in traps.

1. Introduction

Ion traps got an important tool for precision spectroscopy. After the pioneering work on single ion spectroscopy performed end of the seventies[1] it got clear that this system allows to observe an ion in an unperturbed environment with practically no influence of the trapping fields. Another advantage is that the interaction time with the laser beam is not limited by the motion of the particle, therefore no transit time broadening is present. The fact that the motion of the ions can be cooled by laser light allows to reach the so-called Dicke limit[2,3] which means that the residual motional amplitude of the ion is much smaller than a fraction of the light wavelength used to probe the resonance line. In this limit the Doppler broadening disappears and since the temperature is in the range of mK or below also the second-order Doppler effect is excluded. Thus the ultimate resolution is determined by the natural lifetime of the investigated transition.

High resolution experiments have also been performed in ion traps in the microwave region. In this case the Doppler effect does not show up due to the low frequency of the investigated transitions and the number of ions stored in the trap can be very large. There is also a lot of work performed on the development of frequency standards in the microwave region; this cannot be covered here (for details see a recent review[4,5]). It should be mentioned here that not all those experiments have to be performed using laser light for the state preparation and detection of the ions. For example one of the early experiments of this kind was the determination of the Zeeman splitting of $^4He^{+}$[6] and of the hyperfine splitting of the $^3He^+$ ground state by Schussler et al.[7]. The ions in the trap were polarized by spin exchange collisions with a polarized beam of Cs atoms. The microwave resonance destroyed the polarization of the $^3He^+$ leading to a smaller lifetime of the ions in the trap. Also optical pumping detection using the light of resonance lamps was performed in other cases.

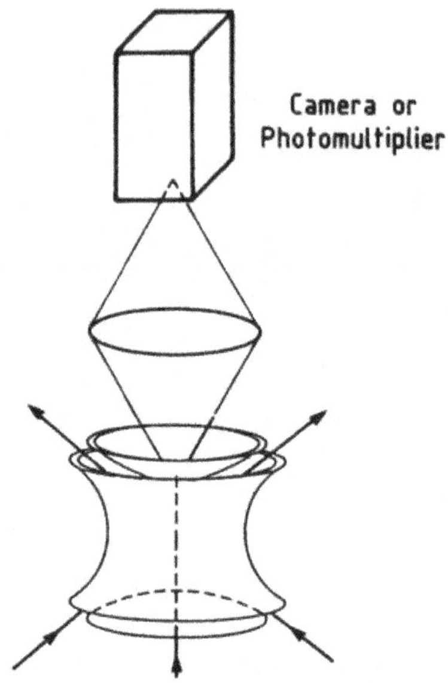

Fig. 1: Scheme of the Paul trap.

2. Spectroscopy of single ions

The Paul-trap[8,9] used for most of the experiments described here consists of a ring electrode and two end caps as shown in Fig. 1. Trapping can be achieved if time-varying electric fields are applied between ring and caps (the two caps are electrically connected). A dc voltage in addition changes the relation of the potential depth along the symmetry axis (vertical direction in Fig. 1) to that in a perpendicular direction. The equation of motion of an ion in such a situation is the Mathieu differential equation, well known in classical mechanics, which – depending on the voltages applied to the trap (dc and radio-frequency voltages) – allows stable and unstable solutions. Another way to achieve trapping is the use of a constant magnetic field aligned along the symmetry axis leading to the Penning trap.[10,11] In this case only a dc voltage has to be applied between ring and cap electrodes.

To produce the ions in the Paul trap mostly a neutral atomic beam is directed through the trap centre and ionized by electrons. The resulting trapped ions have a lot of kinetic energy rendering them useless for most applications, such as spectroscopy; therefore the ions have to be cooled. This is done by laser light. The laser frequency ν is tuned below the resonance frequency, so that the energy of the photon is not sufficient to excite the atom.[3] Crudely, the ion can extract the missing energy from its motion and thus reduce its kinetic energy. In other words, the atomic velocity Doppler

Table 1: Investigation of trapped ions in connection with the test of fundamental quantum phenomena. In the experiments on the linearity of quantum mechanics and on the Zeno effect ion clouds have been investigated.

Phenomenon	References
Antibunching and sub-Poissonian statistics in resonance fluorescence	Diedrich and Walther [13]
Quantum jumps	Nagourney *et al.* [14], Sauter *et al.* [15], Bergquist *et al.* [16]
Interference of the light of two ions	Eichmann et al. [17]
Linearity of quantum mechanics	Bollinger *et al.* [18]
Zeno-effect	Itano *et al.* [19]

Table 2: Potential ions for optical frequency standards.

Ion	Cooling transition	Clock transition	Lifetime clock transition	References
Ba^+	$6p\,^2P_{1/2}$-$6s\,^2S_{1/2}$ 493 nm	$5d\,^2D_{5/2}$-$6s\,^2S_{1/2}$/1.76 μm	32 s	Neuhauser *et al.* [1] Nagourney *et al.* [23]
Hg^+	$6p\,^2P_{1/2}$-$6s\,^2S_{1/2}$ 194 nm	$6d\,^2D_{5/2}$-$6s\,^2S_{1/2}$/282 nm	0.1 s	Bergquist *et al.* [21, 24] Diedrich *et al.* [22]
Yb^+	$6p\,^2P_{1/2}$-$6s\,^2S_{1/2}$	$6d\,^2D_{5/2}$-$6s\,^2S_{1/2}$ $6f\,^2F_{7/2}$-$6s\,^2S_{1/2}$	8d	Klein *et al.* [25] Bell *et al.* [26] Lehmitz *et al.* [27]
In^+	$5s5p\,^3P_1$-$5s^2\,^1S_0$ 230 nm	$5s5p\,^3P_0$-$5s^2\,^1S_0$/236 nm	0.13 s	Dehmelt [20] Peik *et al.* [28]
Sr^+	$5p\,^2P_{1/2}$-$5s\,^2S_{1/2}$ 422 nm	$4d\,^2D_{5/2}$-$5s\,^2S_{1/2}$/674 nm	0.372 s	Madej and Sankey [29]

shifts the atom into resonance to bridge the detuning gap Δ between laser and resonance frequency, and the atom absorbs the photon of momentum $\hbar k = h\nu/c$. After the absorption process the momentum of the atom is reduced, lowering its kinetic energy. The lowest temperature achievable is determined by the Doppler limit,[3,12] which is in the millikelvin region. The low temperatures can be obtained within a fraction of a second.

After the pioneering work of Dehmelt and Toschek on the spectroscopy of single Ba-ions[1] it got clear that single ion spectroscopy will be an important tool for high resolution spectroscopy and also for new frequency standards. A survey on the applications of single ion spectroscopy is given in tables 1 and 2. The applications reach from the investigation of fundamental quantum phenomena to high resolution spectroscopy with the aim of the development of new frequency standards.

The proposal to use a forbidden optical transition of a single laser-cooled ion stored in a Paul trap as the basis for an optical atomic frequency standard is more than ten years old.[20] Impressive progress has already been achieved along these lines and high resolution laser spectroscopy has been reported on Hg^+[21,22] and Ba^+ ions.[23] Other ions that are investigated in this context are ytterbium, indium and strontium (see table 2 for a survey). A severe restriction in the choice of ions for these experiments is the availability of suitable transitions: a strong dipole transition is needed for laser cooling and a long lived transition for the frequency standard and both lines have to be in the wavelength range accessible for continuous wave laser sources. The prospective clock transitions of all the ions mentioned above are electric quadrupole transitions like $^2S_{1/2} \rightarrow {}^2D_{5/2}$ or higher order multipole processes. The dependence of those transitions on external electric and magnetic fields may cause a problem for a frequency standard. Around zero magnetic field they show a linear Zeeman shift of the order MHz/G unless a Zeeman component $(m_F = 0) \rightarrow (m_F = 0)$ of an isotope with half integer nuclear spin is used.[21] In this case however the hyperfine splitting of the ground state leads to optical pumping which makes a continuous excitation for laser cooling more complicated. A small but difficult controllable shift arises from the interaction with the electric quadrupole field of the trap.[20]

To avoid these problems Dehmelt suggested the use of a transition between two states with vanishing angular momentum of the electrons such as $6s^2\,{}^1S_0 \rightarrow 6s6p\,{}^3P_0$ in Tl^+[20,30] or the corresponding lines of the other ions of the third main group of the periodic system: boron, aluminium, gallium and indium. The three lighter of these cannot be laser cooled presently because the intercombination line $^1S_0 \rightarrow {}^3P_1$ is too weak and the wavelength of the strong dipole transition $^1S_0 \rightarrow {}^1P_1$ lies in the VUV below 160 nm and is thus inaccessible for cw lasers. In the case of thallium the wavelengths for $^1S_0 \rightarrow {}^3P_0$ and $^1S_0 \rightarrow {}^3P_1$ are 202.2 nm and 190.8 nm respectively, and for indium they are 236.5 nm and 230.6 nm. The two indium wavelengths can be generated relatively easily using the β-bariumborate (BBO) crystal for the fequency doubling of blue lasers. Especially the clock transition at 236.5 nm is technically convenient, since it coincides with the fourth harmonic of the 946 nm transition $^4F_{3/2} \rightarrow {}^4I_{9/2}$ of the Nd:YAG laser. So this intrinsically frequency stable solid-state laser can be used as the oscillator to drive the clock transition. Indium is thus presently the most promising of the group III ions. The possibilities to use this ion for an optical frequency standard are presently investigated in our laboratory.[28] The results obtained so far will be reported in the following.

The decay $^3P_0 \rightarrow {}^1S_0$ between two levels with vanishing electronic angular momenta is forbidden in all orders of the multipole expansion of the radiation field. However for atoms with nonvanishing nuclear spin the hyperfine interaction induces a weak electric dipole transition between these states. This is due to small admixtures of the decaying states $^1P_1$ and $^3P_1$ to the otherwise stable $^3P_0$ level. This effect has been quantitatively investigated so far only in a few cases. In mercury the absorption coefficient for the forbidden transition was measured and the lifetime of $^3P_0$ was inferred from that.[31] The result was 1.7 s for ^{199}Hg and 2.6 s for ^{201}Hg. Recently lifetimes in the picosecond range have been measured for hyperfine quenched $^3P_0$ states of highly charged helium-like ions.[32] In the experiment reported here we use the perturbation free environment offered by an ion trap to measure the lifetime of the $^3P_0$ state in In^+. We observe the quantum jumps of a single ion into and out of this level using Dehmelt's electron shelving scheme.[20]

For the experiments with single laser-cooled ions a Paul-Straubel-trap with a 1 mm diameter ring electrode was used,[33] driven by a voltage of 1000 V at 10.7 MHz. Together with a small indium oven, an electron source for ionization and the detection optics for monitoring the fluorescence of the ions, the trap is mounted inside a stainless steel ultrahigh vacuum chamber.

The vacuum chamber is surrounded by three pairs of coils oriented along orthogonal axes, one of which coincides with the laser beam direction. These coils are used to compensate for the earth's magnetic field or to apply a controllable magnetic field. When the ions are optically pumped and the fluorescence intensity is observed as a function of the magnetic field, we get a narrow zero field resonance due to magnetic depolarisation of the ground state, analogous to the Hanle effect of an excited state.[34] This signal can be used to cancel magnetic stray fields inside the trap with a precision of a few mG.

The used Paul-Straubel trap produces a steep potential for a good localization of the stored ion and simultaneously the laser straylight is very small owing to the open electrode structure.[33] Without damping through a buffergas the trap still can store several ions but at a typical well depth of 4 eV these can be evaporated from the trap by heating them with the laser detuned to the blue side of the resonance. After this procedure usually a single ion remains which is then trapped with high stability; it can be kept for hours even when it is not permanently laser cooled. We also observed Coulomb crystals[35] of two laser cooled indium ions separated by $3\mu m$.

In order to laser cool the ion on the narrow $^1S_0 \rightarrow {}^3P_1$ transition its hyperfine component $F = \frac{9}{2} \rightarrow \frac{11}{2}$ was excited with circular polarized light in zero magnetic field. This effectively prepares a two level system $(F = m_F = \frac{9}{2}) \rightarrow (F = m_F = \frac{11}{2})$ by means of optical pumping. We observed photon count rates up to 1.2 kHz from a single ion in saturation, indicating that our overall detection efficiency is about 0.1 %. Usually we worked below saturation to get a better ratio of fluorescence signal to laser straylight, where we achieved a value of 12. By scanning the laser we recorded excitation spectra of the cooling transition with a linewidth of about 20 MHz. This is limited partly by laser frequency jitter and also to a larger extent by frequency modulation caused by the micromotion of the ion at the trap frequency of 10.7 MHz. A linewidth of 20 MHz corresponds to a modulation index of 2 and a residual micromotion amplitude of 70 nm that might be caused by a slight displacement ($\approx 1\mu m$) of the ion from the trap center. Very precise control of stray DC fields is required to minimize this effect. The presence of micromotion sidebands leads to a strong variation in the temperature as the laser is scanned across the line profile so that we can presently only give an upper limit of 100 mK for the temperature. If the

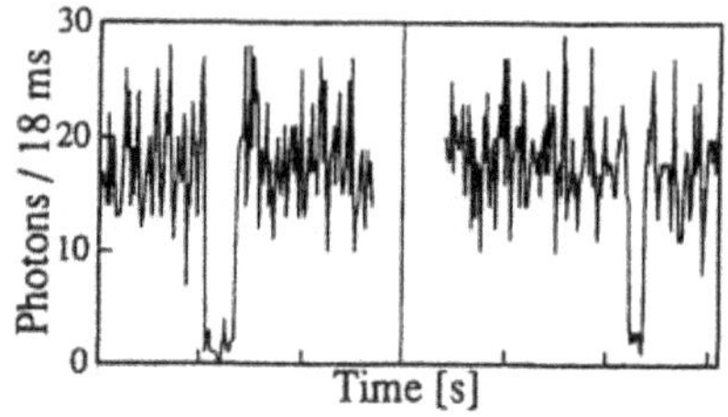

Fig. 2: Fluorescence from a single indium ion, showing dark periods due to quantum jumps to the metastable $^3P_0$ state.

laser frequency is stabilized to a linewidth below the natural linewidth of the cooling transition corresponding to 360 kHz and the micromotion further reduced then much lower temperatures will be feasible.

If we look at the temporal behaviour of the resonance fluorescence of a single indium ion on the transition $^1S_0 \rightarrow {}^3P_1$ we observe dark periods whenever the metastable $^3P_0$ level is populated. This can happen either by direct laser excitation from the ground state, as it will be done for the frequency standard, or through a magnetic dipole decay from the $^3P_1$ level. The possibility of observing quantum jumps in such a three level system with only one driven transition has been discussed theoretically (see e.g. Ref. 36) and has been realized for the Hg^+ ion[37] (see also table 1 for more references).

In a simple rate equation model let A_0 be the decay rate from $^3P_0 \rightarrow {}^1S_0$, A_1 that from $^3P_1 \rightarrow {}^1S_0$ and A_2 that from $^3P_1 \rightarrow {}^3P_0$. Since $A_1 \gg A_0, A_2$ the fluorescence intensity switches from "on" $\sim A_1\rho_{11}$ (ρ_{11} : population of $^3P_1$) to "off" at a rate $A_2\rho_{11}$ and back "on" at the rate A_0. The magnetic dipole decay rate A_2 for the transition $^3P_1 \rightarrow {}^3P_0$ at $\lambda = 9.32\mu m$ wavelength can be estimated as:[38]

$$A_2 = \frac{2\pi\sigma^2 h}{9\epsilon_0 m^2 c^2 \lambda^3} = 0.0223s^{-1}$$

The branching ratio A_1/A_2 is about 10^8 and with beginning saturation of the cooling transition ($\rho_{11} = 0.25$) we expect one $^3P_1 \rightarrow {}^3P_0$ decay every 3 minutes.

To observe these dark periods in the experiment we counted the photomultiplier pulses in time intervals of 18 ms and stored these numbers in the computer. The laser power was measured synchronously and checked to be constant. Samples of these data exhibiting quantum jumps are shown in Fig. 2. We see an abrupt decrease of the count rate to the stray light level and the reappearance of the fluorescence after some 100 ms. These events occur at a rate that is a little lower but still consistent with the estimate given above. The duration of the dark periods can be used to evaluate the lifetime of the $^3P_0$ level. We observed about 150 decays and calculated from those data the function $d(t)$ as the number of dark periods with a duration shorter than t. This function was fitted to $d(t) = a(1 - e^{-(t-t_o)/\tau})$ to get the lifetime τ. Events with a duration shorter than $t_o = 70ms$ were not considered. Fig. 3 shows a logarithmic If the

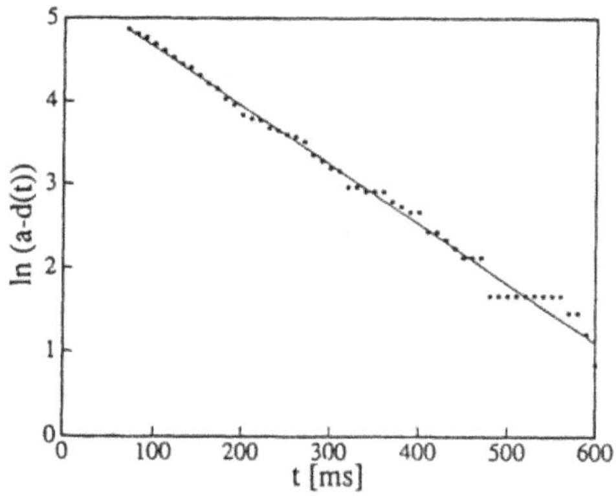

Fig. 3: Distribution of the durations of observed dark periods (see text for the definition of $a - d(t)$). The fitted decay constant is the inverse lifetime of the ${}^3P_0$ level.

plot of $a - d(t) = ae^{-(t-t_o)/\tau}$ together with the fit, indicating a simple exponential behaviour. The result for the lifetime of ${}^3P_0$ is

$$\tau\,({}^3P_0) = 0.14(2)s.$$

The uncertainty results from the number of observed decays.

The experiments show that the indium ion is a possible candidate for an optical frequency standard. Although the cooling transition in indium is a relatively slow intercombination line with only 360 kHz natural linewidth we were able to cool a single ion using this transition and to obtain a rate of fluorescence photons that is high enough to detect the dark periods due to electron shelving in the ${}^3P_0$ level. Besides these technical difficulties the small linewidth of the cooling transition enabled us to reach the strong binding regime using moderate trapping field strength. This opens the possibility of cooling the ion into low quantum levels or even the ground state of the single ion oscillator.

A frequency standard will benefit from the insensitivity of the indium ion's ${}^1S_0 \rightarrow {}^3P_0$ transition frequency against external perturbations as well as from the availability of a frequency stable oscillator like a diode laser pumped Nd:YAG laser. A suitable source has been developed in our laboratory for this purpose. The accuracy to which the wavelength of this forbidden transition is known has recently been improved significantly by the measurement of Larkins and Hannaford,[39] so that direct laser excitation shall be possible in the near future.

3. Observation of phase transitions in a Paul trap - Order versus chaos: crystal versus cloud

In the next part of the paper we would like to describe experiments on phase

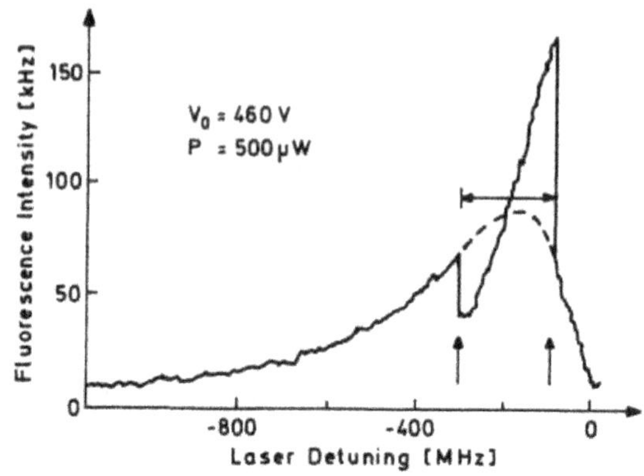

Fig. 4: Fluorescence intensity, that is, photon counts per second, from five ions as a function of the laser detuning Δ. The vertical arrows indicate the detunings where phase transitions occur. The horizontal arrow shows the range of detunings in which a stable five-ion crystal is observed. The spectrum was scanned from left to right.

transitions of a few ions in a Paul trap. The results discussed were obtained using a Paul trap with a ring diameter of 5 mm and an end-cap separation of 3.54 mm.[13,40] This trap is larger than most of the other ion traps used in laser experiments.[1,22] The radio frequency of the field used for dynamic trapping is 11 MHz. The trap is mounted inside a stainless steel, ultrahigh vacuum chamber. The ions are loaded into the trap by means of a thermal beam of neutral atoms (magnesium atoms in our case), which are then ionized close to the center of the trap through a small hole in the lower end-cap (see Fig. 1). In order not to distort the trap potential, the hole is covered by a fine molybdenum mesh. The neutral Mg beam and the laser beam pass through the gaps between the end-cap and the ring electrodes. The laser frequency is shifted by an amount Δ below the $3S_{1/2} \rightarrow 3P_{3/2}$ resonance-transition of $^{24}Mg^+$ at 280 nm to extract kinetic energy from the ions by radiation pressure as discussed previously. In this way, a single ion can be cooled to a temperature below 10 mK. The fluorescence from the ions is observed through a hole in the upper end-cap (see Fig. 1), again covered by a molybdenum mesh. The large size of the trap allows a large solid angle for detecting the fluorescence radiation, either with a photomultiplier, or by means of a photon-counting imaging system. To observe the ions, the cathode of the imaging system is placed in the image plane of the microscope objective attached to the trap; in this way images of the ions could be obtained.[41,42]

The existence of phase transitions in a Paul trap manifests itself by significant jumps in the fluorescence intensity of the ions as a function of the detuning Δ between the laser frequency and the atomic transition frequency. These discontinuities are indicated in Fig. 4 by vertical arrows, and occur between two types of spectra: a broad and a narrow one, analogous to the fluorescence spectrum of a single, cooled ion. We interpret the broad spectrum as a fingerprint of an ion cloud and the narrow

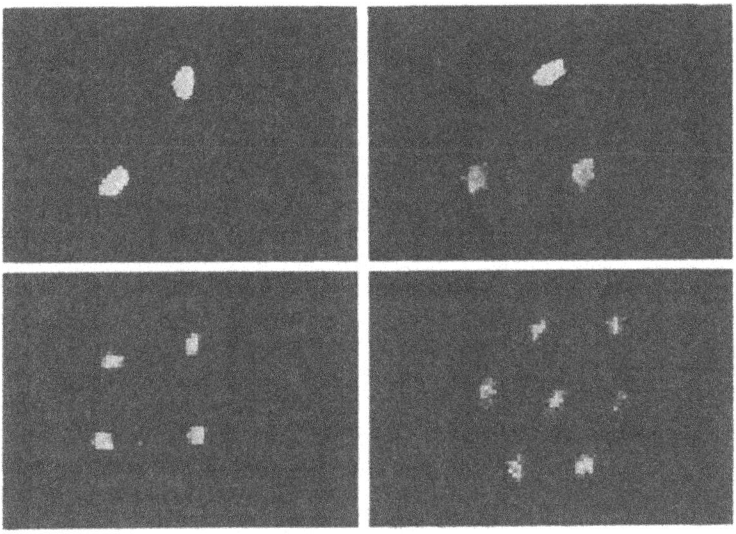

Fig. 5: Two, three, four and seven ions confined by the dynamical potential of a Paul trap and crystallized into an ordered structure in a plane perpendicular to the symmetry axis of the trap. The average separation of the ions is 20 μm.

spectrum as being characteristic for an ordered many-ion situation with a "single-ion signature". Thus the jumps clearly indicate a transition from a state of erratic motion within a cloud to a situation where the ions arrange themselves in regular structures. In such a crystalline ion structure the mutual Coulomb repulsion is compensated by the external, dynamic trap potential. The regime of detunings in which such crystals exist is depicted in Fig. 4 by the horizontal arrow. The existence of the two phases - crystal and cloud - can be verified experimentally by direct observation with the help of a highly sensitive imaging system and theoretically by analyzing ion trajectories via Monte Carlo computer simulations.[42]

The excitation spectrum of Fig. 4 and the jumps in it were recorded by altering the detuning Δ from large negative values to zero. When we scan the spectrum in the opposite direction the jumps occur at different values of Δ, which means hysteresis is associated with these phase transitions.[42] Such hysteresis behavior can be expected with laser-cooled ions because the cooling power of the laser strongly depends upon the details of the velocity distribution of the ions. The behavior of the ions in the trap is governed by the trap voltage, the laser detuning and the laser power. Hysteresis loops appear whenever one of these parameters is changed up or down while the others are kept constant.

Figure 5 shows the ion structures as measured with the imaging system. For the measurements only a radio-frequency voltage was applied to the trap electrodes:

in this case the potential is a factor of two deeper along the symmetry axis than perpendicular to it, and therefore plane ion structures are observed being perpendicular to the symmetry axis.[41,42]

After we have accepted the existence of the ion crystals and the corresponding phase transitions, how do they actually occur? Would the cooling laser not force any cloud immediately to crystallize? A heating mechanism balancing the cooling effect of the laser must be the answer to this puzzle, but what heating mechanism? Since the early days of Paul traps this so-called radio-frequency heating has repeatedly been cited.[3,10] A deeper understanding, however, was missing and was provided only recently by a detailed study of the dependence of the cloud → crystal and crystal → cloud phase transitions on the relevant parameters.[41–43]

The ions are subjected to essentially four different forces: the first one arising from the trapping field, then the Coulomb interaction between the ions, the laser cooling force, and finally a random force arising from the spontaneously emitted photons. Using these forces computer simulations of the motion of the ions can be performed.[42] Depending on the external parameters such as the laser power, the laser detuning, and the radio-frequency voltage, the experimentally observed phenomena could be reproduced.[42,43] Some of the results of the simulations are summarized in Fig. 6. Plotted is the radio-frequency heating parameter κ of five ions versus their mean separation.[43] For zero laser power and large r, we did not observe any net heating of the ions. This is confirmed by our experiments in which, even in the absence of a cooling laser, large clouds of ions can be stored in a Paul trap over several hours without being heated out of the trap. When the ions are far apart, the Coulomb force is small, and on short time scales the ions behave essentially like independent single stored ions. For this reason, we call this part of the heating diagram the *Mathieu regime*.[41,43] Turning on a small laser, the rms radius r reduces drastically, but comes to a halt at about 14 μm. At this distance the nonlinear Coulomb force between the ions plays an important role and the motion of the ions gets chaotic. In this situation the power spectrum of the ions becomes a continuum leading to a radio-frequency heating process.

Increasing the laser power further results in an even smaller cloud. The smaller cloud produces more chaotic radio-frequency heating, as seen clearly by the negative slope of the heating curve in the range $8<r<14$ μm. Finally, in the range $4<r<8$ μm there is still chaotic heating but the slope of the heating curve is positive. As a consequence of the resulting triangular shape of the heating curve at a laser power of about $P = 150$ μW, corresponding to $r \approx 8$ μm, the chaotic heating power can no longer balance the cooling power of the laser light and the cloud collapses into the crystalline state located at $r \approx 3.8$ μm. At this point the amplitude of the ion motion is so small that the nonlinear part of the repulsive Coulomb force is negligible again, so that chaotic heating disappears and the phase transition occurs.

Due to this collapse of the cloud state, the behavior of the heating rate in the range $3.8<r<8$ μm cannot be studied by balancing laser cooling and radio-frequency heating. In this case we start out from the crystal state and slightly displace the ions to explore the vicinity of the crystal. We observe no heating for $3.8<r<4$ μm, but

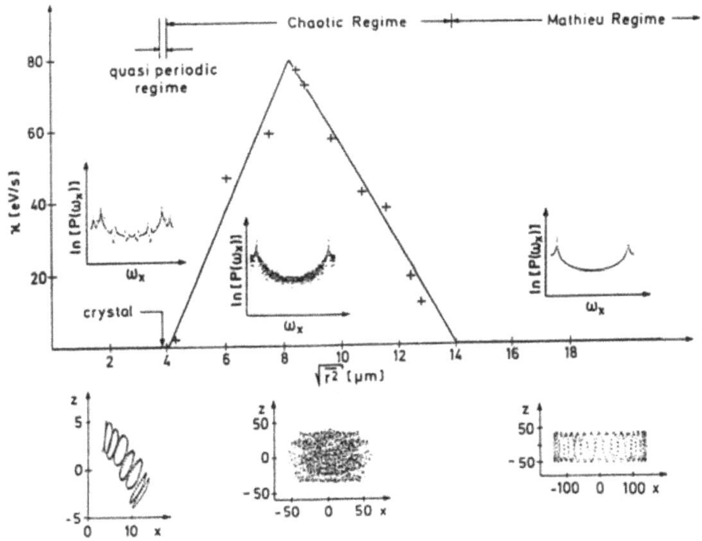

Fig. 6: Average heating rate κ (increase of kinetic energy per unit time), of five ions in a Paul trap versus mean ion separation. The insets show the power spectrum and the corresponding stroboscopic Poincaré sections (plane perpendicular to the symmetry axis) of relative separation of the two ions in three characteristic domains: The crystal state, the chaotic regime, and the Mathieu-regime. The units on the axes are in μm. In order to calculate the power spectrum of the "crystal" shown on the left hand side, the distance of the two ions was displaced by 1 μm from the equilibrium position. The Mathieu-regime shown on the right is dominated by the secular motion.

quasiperiodic motion, and thus dub this the *quasiperiodic regime*. We call the upper edge of the quasiperiodic regime ($r \approx 4$ μm) the *chaos threshold*. An initial condition beyond the chaos threshold, i.e., satisfying $r \approx 4$ μm, leads to heating, and expansion of the ion configuration.

For the quasiperiodic, the chaotic, and the Mathieu regimes, respectively, we display the corresponding type of power spectrum as the insets above the abscissa of Fig. 6. The data were actually taken for the case of two ions, but would not look much different in the five-ion case. We obtain a discrete spectrum in the quasiperiodic regime and a complicated noisy spectrum in the chaotic regime. The spectrum in the Mathieu regime is again quite simple and dominated by the secular motion frequency. We also show stroboscopic pictures of the locations of the ions in the plane perpendicular to the symmetry axis of the trap characterizing the three regions (inserts below the abscissa in Fig. 6).

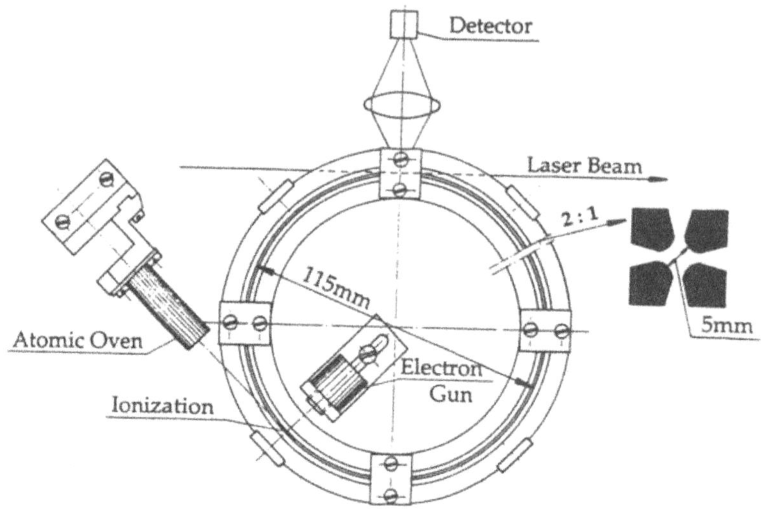

Fig. 7: Quadrupole storage ring. The cross section of the electrode configuration is shown on the insert on the right hand side of the figure.

4. The ion storage ring

A completely new era of accelerator physics could begin if it were possible to produce, store and accelerate Coulomb crystals in particle accelerators and storage rings. To work with crystals instead of the usual dilute, weakly coupled particle clouds has at least one advantage: the luminosity of accelerators (storage rings) could be greatly enhanced and (nuclear) reactions whose cross sections are too small to be investigated in currently existing accelerators would become amenable to experimentation.

In the following, we would like to discuss very briefly our experiments using a miniature quadrupole storage ring. The storage ring is similar to the one described by Drees and Paul[44] or by Church.[45] We can observe phase transitions of the stored ions and the observed ordered-ion structures are quite similar to the ones expected in relativistic storage rings, however, much easier to achieve. The motivation for building this small storage ring came from the fact that micromotion perturbs the ion structures in a Paul trap and only a single trapped ion is free of micromotion.[3] The ring trap used consists of a quadrupole field, leading to harmonic binding of the ions in a plane transverse to the electrodes of the quadrupole and no confinement along the axis (see Fig. 7). Confinement along the axis is achieved, however, by the Coulomb interaction between the ions when the ring is filled; then the total number of ions in the ring determines the average distance between them.

A scheme of the ring trap used for our experiment is shown in Fig. 7.[46,47] It consists of four electrode rings shown in the insert on the right. The hyperbolic

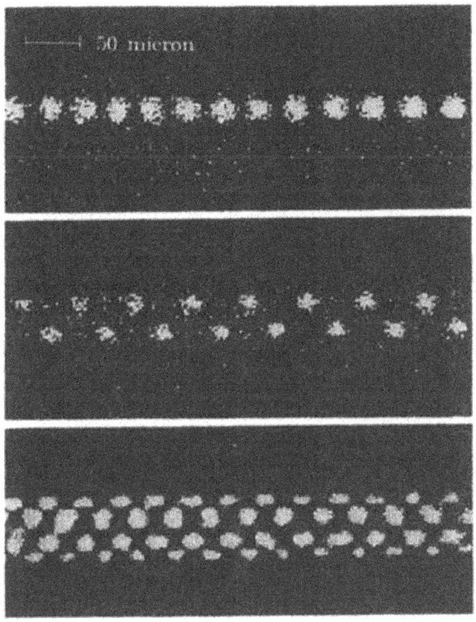

Fig. 8: Crystalline structures of laser-cooled $^{24}Mg^+$ ions in the quadrupole storage ring. At a low ion density ($\lambda = 0.29$) the ions form a string along the field axis (a, upper). Increasing the ion density transforms the configuration to a zigzag (b, middle; $\lambda = 0.92$). At still higher ion densities, the ions form ordered helical structures on the surface of a cylinder, e.g., three interwoven helices at $\lambda = 2.6$ (c, lower). As the fluorescent light is projected onto the plane of observation in this case the inner spots are each created by two ions seated on opposite sides of the cylindrical surface, resulting in a single, bright spot.

cross section of the electrodes required for an ideal quadrupole field was approximated by a circular one. The experiments were also performed with Mg^+ ions. The ions were produced between the ring electrodes by ionizing the atoms of a weak atomic beam produced in an oven that injected the atoms tangentially into the trap region. The electrons used for the ionization came from an electron gun, the electron beam of which was perpendicular to the direction of the atomic beam. A shutter in front of the atomic beam oven allowed the interruption of the atomic flux. The ultrahigh vacuum chamber was pumped by an ion getter pump. After baking the chamber a vacuum of 10^{-10} mbar could be reached. Under these conditions the number of ions stored in the trap stayed practically constant for several hours.

When laser cooling of the ions is started a sudden change in the fluorescence intensity is observed, resembling very much that seen with stored ions in a Paul trap (Fig. 4), which indicates a phase transition and the formation of an ordered structure of ions. The ion structure can also be observed using an ultrasensitive imaging system. Pictures of typical ion structures are shown on Fig. 8. The ions are excited by a frequency-tunable laser beam that enters the storage ring tangentially.

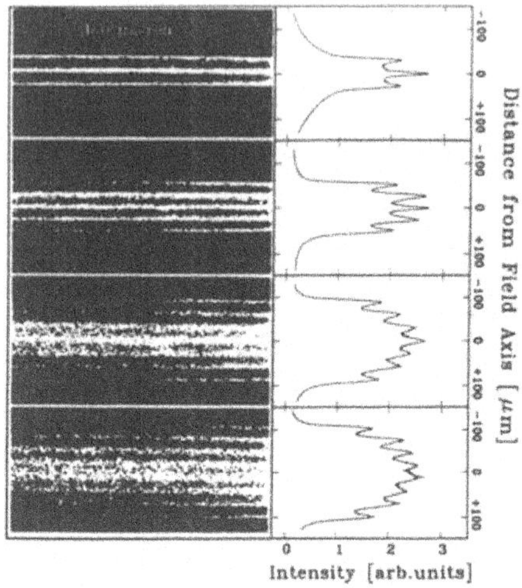

Fig. 9: Images and intensity profiles of (from the top): one shell plus string (a; $\lambda \approx 4.3$), two shells plus string (b: $\lambda \approx 12.2$), three shells plus string (c; $\lambda \approx 26$), and four shells (d; $\lambda \approx 34$). There are up to $\approx 8 \times 10^5$ ions stored in the ring for the four-shell structure.

In the linear configuration the ions are all sitting in the center of the quadrupole field, therefore they do not show micromotion and it is possible to cool them further to temperatures in the microkelvin region. The new cooling methods proposed by Dalibard et al.[12] can be applied to the Mg^+ ions so that the single-photon recoil limit can be achieved for the cooling process. At this limit the kinetic energy of the ions is smaller than the energy resulting from the "zero-energy" motion of the harmonically bound ions; the ion structure then reaches its vibrational ground state, i.e. a Mössbauer situation is generated. That means that the recoil limit for laser cooling of the ions does not exist anymore, the ion structure then corresponds to a Wigner crystal.

The observed configurations (Figs. 8 and 9) in the quadrupole ring trap are described in a recent paper by Birkl et al.[48] We will review the major results reported in that paper in the next section as well as the comparison to the theory (Rahman and Schiffer;[49] Hasse and Schiffer[50]).

In the molecular dynamics calculations,[50] a cylindrically symmetric, static harmonic potential is assumed to describe the confining field. Each particle is subjected to the Coulomb forces of all other particles and to the confining field. The classical equations of motion are integrated for a system of several thousand particles, starting with random positions and velocities, to give the time evolution of the system. Cooling the stored particles is simulated by scaling down the resulting velocities of the stored particles at defined instants of the integration process. After sufficient cooling, ordered structures such as strings, zigzags, shells and multiple shells should arise, owing to the confining field's harmonic potential. Our experiments are well suited to checking these predictions. To compare the experimental results with theory, we can use the normalized "linear particle density" λ, which is given by the ion density multiplied by the ratio of Coloumb repulsion and confining force of the trap.[50] Low λ values correspond to a deep potential well or a small number of ions, resulting in an equilibrium structure closely confined to the field axis making up a string of ions (Fig. 8a). This is the micromotion-free configuration discussed previously, and the analogue of the single stored ion in a Paul trap, as in both cases the ions sit in the potential minimum and show no micromotion. For higher values of λ, the structures extend more and more into the off-axis region, giving rise to (in the order of increasing λ) a plane zigzag structure (Fig. 8b) and cylindrical structures with the ions forming helices on cylindrical surfaces. The structure in Fig. 8c consists of three interwoven helices with six ions per pitch. The string and the zigzag have also been observed with laser-cooled Hg^+ ions in a linear trap.[51]

Increasing further the number of ions leads to structures with smaller spacings between the ions where we cannot optically resolve individual ions any more. Images of these structures are presented in Fig. 9. The radial intensity profiles displayed on the right-hand side of the figure provide information about the structures as they give a measure of the radial distribution of the ions. For increasing λ it becomes energetically more favourable to create a string inside the first ion shell (Fig. 9a) to provide space for more particles. This results in a structure that is a three-dimensional analogue of the plane seven-ion crystal for a Paul trap (Fig. 5). The inner string turns into a second shell at higher densities: a string then develops inside this second shell (Fig. 9b), and so on. Figures 9c and 9d show structures consisting of three shells plus string, and four shells, respectively. We have been able to observe all possible structures, from the string up to four shells plus string. The formation of multiple-shell crystalline structures in the quadrupole storage ring contrasts with the observation of shell structures in Penning traps, where the ions do not occupy fixed positions inside the shells.[52]

Figure 10a gives a summary of experimental data for all recorded images in which the ions were individually resolved. The depth of ψ_o of the potential well and the ion density per unit length are the experimental parameters. The theoretical boundaries between the different shell structures, predicted by Hasse and Schiffer[50] in terms of the functional dependence of λ on ψ_o and the ion density are given by the straight lines with constant λ. String structures are expected for $\lambda < 0.709$, zigzag structures in the range $0.709 < \lambda < 0.964$ and single shells in the range $0.964 < \lambda < 3.10$. Many different structures that are degenerate in energy are expected within the single-shell regime. We obtained stable configurations near $\lambda = 1.3$ and $\lambda = 2.0$ (two interwoven helices) and near $\lambda = 3.0$ (three interwoven helices - Fig. 8c). The observed structures agree with the predicted scheme for a large range of experimental parameters, thus confirming the theoretical results.

A summary of the experimental observations for ordered-shell structures with up to four shells plus string and without resolution of individual ions is presented in Fig. 10b. The depth ψ_o of the confining potential well is again one of the experimental

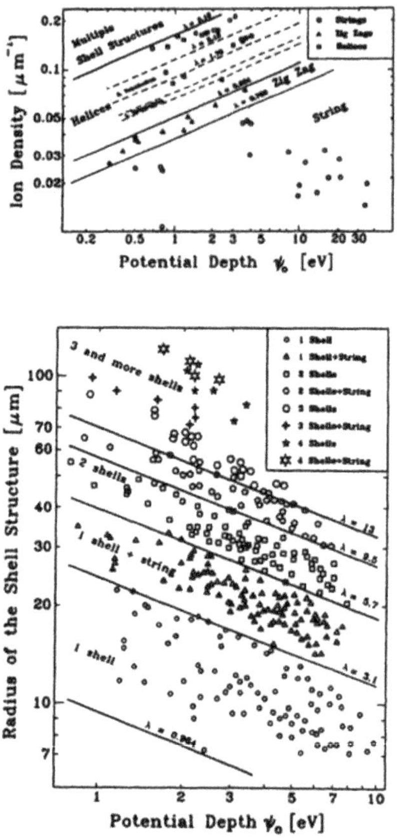

Fig. 10: Summary of the experimental results. Part a: Individual ions resolved, where the observed structures are characterised by the ion density per unit length and the depth of the potential well ψ_o. These two parameters can be combined to give the normalised linear particle density λ which fully determines the ion configuration. The straight lines show critical λ values separating the regions corresponding to the various theoretically expected structures. The observed configurations are labelled with different symbols for each structure. Part b: Individual ions unresolved, where the observed shell structures with up to four shells plus string are characterised by their radius ρ and the potential depth ψ_o. The various observed structures are again separated by lines of theoretically determined critical λ (for details see Ref. 48).

parameters. As the ion density cannot be determined directly from the images, the radius ρ of the structures is used instead as the second parameter. The theoretically predicted boundaries between the different shell structures are again given as straight lines of constant λ following Hasse and Schiffer.[50] The observed ion configurations are seen once more to be determined by λ for a wide range of potential depths and ion densities.

Our results have important implications for two very different fields. Consider first the physics of low-energy particles: an ion string in a quadrupole ring, being free of micromotion, can be cooled to its vibrational ground state in the confining potential using recently proposed laser cooling techniques.[12] This would place the string in the Lamb-Dicke regime with a vanishing first-order Doppler effect because the spatial amplitude of the motion is smaller than the wavelength of the atomic transition. Furthermore, the second-order Doppler effect, which can only be reduced by further cooling, also disappears making the stored ions very interesting for frequency standards. The large number of ultracold ions available in the ring will lead to a high signal-to-noise ratio. Finally, cooled ions in the ring represent a quantum object of macroscopic dimensions: a Wigner crystal.

With the experimental confirmation that the ordered structures expected in high-energy ion storage rings can indeed be formed, a completely new era of accelerator physics would emerge. The enhanced luminosity of the corresponding beams would allow studies of ionic reactions whose cross sections are too small for investigations in existing accelerators.

A long-term objective for our experiments is to cool an ion crystal to still lower temperatures. Due to the motion driven by the trapping field (micromotion) this is not possible for more than a single ion in a three-dimensional quadrupole trap (Paul trap), as discussed already above. On the other hand, this can be achieved in two-dimensional trapping fields like the quadrupole storage ring or linear ion traps where the free motion in the third dimension is limited by other means. The weak binding of ions in the quadrupole storage ring allows us to apply the wide variety of cooling techniques developed for free atoms. One of the most efficient methods is polarization gradient cooling (PGC).[53–55] In the following, we present the first application of this cooling technique to trapped ions. For details see also Ref. 56).

As the other experiments described here, the laser cooling was performed upon $^{24}Mg^+$ ions using the $J_g = 1/2 \rightarrow J_e = 3/2$ transition at a wavelength of 280 nm (natural linewidth $\Gamma = 2\pi \cdot 43MHz$). PGC on this type of transition has been studied extensively in the theoretical literature although the corresponding experimental system has not previously been investigated. The trap of this experiment is the same as the one described above.

However, only a small fraction of the whole storage volume is investigated this time. A weak, electrostatic potential well along the ring circumference with a minimum close to the centre of the observation region enables us to study ensembles of few or even single trapped ions (Fig. 11). This essentially one-dimensional (1D) potential is created by a pair of electrodes separated by about 50 mm, one on either side of the observation region in close proximity to the ring electrodes ($\sim 100\mu m$). Its shape has been mapped by three separate experimental techniques. They are in agreement on the fact that the potential is harmonic with an oscillation frequency ω_z for the ion of $\omega_z = 2\pi \cdot 11.9kHz$ with an error of less than 5 %. The cooling laser field is directed along the ring tangent through the observation region and coincides with

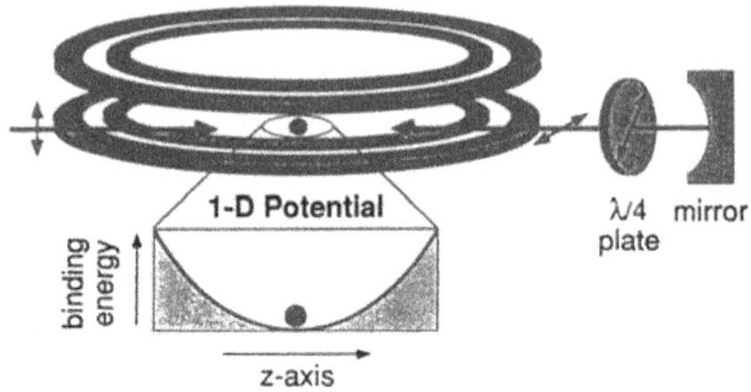

Fig. 11: Experimental schematic with a single ion confined in a quadrupole storage ring. The radial confinement in the secular potential leads to an oscillation frequency of $2\pi \cdot 1MHz$. The ion is only weakly bound along the circumference (see insert). The polarization gradient light field is created by retroreflection of the incoming laser beam with a subsequent double pass through a quarter wave plate. The laser axis is defined as the z-axis.

the z-axis of the 1D potential (Fig. 11). The fluorescence light is detected radially. Despite the curvature of the storage ring, the laser beam deviates less than 5 mrad from the curved minimum of the secular potential over the 500 μm field of view. On the other hand, the coupling of the various degrees of motion in the trapping field is strong enough to cool the radial ion motion to a temperature of several tens of millikelvin for a single trapped ion. Cooling an ensemble of ions reduces the kinetic energy in this component even further due to Coulomb coupling.

Figure 11 displays the scheme for the PGC of the trapped ions. The light field consisting of two counterpropagating orthogonal linearly polarized laser beams (lin $\perp$ lin-configuration). The detuning from the atomic resonance frequency is known to within 2 MHz.

The localization of the ions in a shallow potential well along the laser cooling axis gives us an accurate measure of the temperature. In our experiments, the fluorescence light scattered by the ions is spatially resolved using a photon-counting imaging system. The integration time is five seconds. Consider for the moment a single trapped ion. Determining the spread of the ion's spatial distribution around the equilibrium position allows us to calculate the mean potential energy of the ion. According to the Virial theorem, in a harmonic potential well, this is identical to the mean kinetic energy. Figure 12a shows the position distribution of a single trapped ion in the 1D potential when PGC is applied. As shown, the distribution is well fit by a Gaussian function. Therefore, the kinetic energy of the ion displays the characteristics of a thermal distribution allowing us to assign a temperature T to a specific value of the mean kinetic energy. The following definition has been used. $(1/2)k_BT = \langle E_{kin}\rangle = \langle E_{pot}\rangle = (1/2)M\omega_z^2\langle z^2\rangle$, where k_B is Boltzmann's constant, M is the ion mass, and $\langle z^2\rangle$ the variance of the position spread. For the large spatial distribution associated with a shallow potential, this method is quite accurate. However, the addition of more ions to the trap effectively produces a steeper potential due

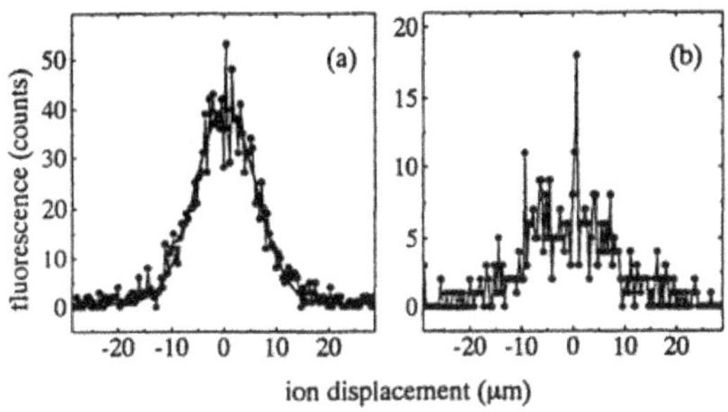

Fig. 12: Measured position distributions for single ions. a) This typical distribution is obtained for $I = 3.3 \cdot 10^{-2} I_{sat}$ and $\delta = -0.5\Gamma$. Its width gives a temperature of $(503 \pm 50)\mu K$. b) A non-Gaussian distribution is found for $I = 3.3 \cdot 10^{-2} I_{sat}$ and $\delta = -1.6\Gamma$. This distribution shows a narrow central structure, however assigning a temperature to such a distribution is not straightforward (see text).

to the mutual Coulomb repulsion and in turn increases the experimental error. For this reason, we focus upon the experiments addressing the cooling of a single trapped ion. In addition, results for a two-ion crystal are presented also.

The minimum observable temperature is defined by the diffraction limit of $0.6\mu m$ of the detection system. Using the known curvature of the potential well, such a diffraction spot leads to a temperature of $1\mu K$. The error in the temperature is due to the finite pixel size of $0.4\mu m$. This error increases with T leading to $\Delta T = 50\mu K$ for a typical $T = 400\mu K$.

Gaussian distributions such as shown in Fig. 12a are found throughout most of the parameter space. However, in the regime where the variance of the ion distribution is minimized, the distributions are not Gaussian. They are characterized by a narrow structure atop a broad Gaussian background (see Fig. 12b). The low kinetic energies associated with these non-Gaussian distributions will be discussed later. For the moment, only the results in the Gaussian regime will be considered.

One of the basic theoretical[57] and experimental[55] results for PGC of neutral atoms is that for large detunings the equilibrium temperature is solely dependent on the amplitude U_o of the light shift, which the atoms experience in a polarization gradient field. The inset of Fig. 13 shows the periodically shifted energies $U_\pm$ of the $m_F = \pm 1/2$ ground states together with U_o. The cooling stems from spontaneous Raman scattering between the two light-shifted ground states. The most probable transition is from the higher-energy ground state to the lower-energy ground state and it reduces the kinetic energy of the atom by an amount close to U_o.

Figure 13 gives a summary of the data presented as a function of $I/\mid\delta\mid$. For large detuning ($\mid\delta\mid \geq \Gamma$) this is proportional to U_o. For small values of $I/\mid\delta\mid$, the two curves follow the same functional dependence. Here, the final temperature is solely defined by $I/\mid\delta\mid$ and so by U_o. As expected, the two curves deviate

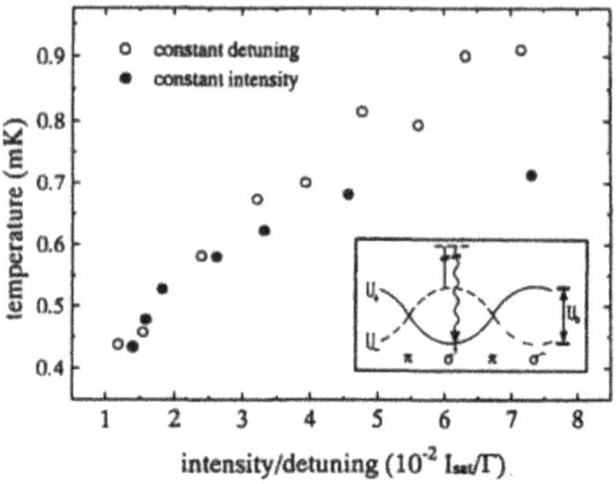

Fig. 13: Ion temperature vs. intensity divided by detuning (o constant detuning, • constant intensity). Qualitatively, the results agree with the predictions of the free-atom theory. On the other hand, the quantitative values leading to the lowest temperatures are two orders of magnitude smaller than for unbound ions. The inset shows the periodic energy shift of the atomic ground states and the fundamental cooling cycle. The miminum temperature we have observed in such a measurement is $T = (300 \pm 50)\mu K$. This is a factor of three below the Doppler limit. This figure is taken from Ref. 56.

for large values of $I/\mid\delta\mid$ since along the constant intensity curve the detuning becomes too small for this simple single-parameter dependence to hold. These results are in qualitative agreement with the free-atom theory. However, the quantitative values of $I/\mid\delta\mid$ leading to the lowest temperatures for the trapped ion are approximately 100 times smaller than for a free atom. In fact, for these smaller values of $I/\mid\delta\mid$, polarization gradient cooling is ineffective for free atoms. We find that for trapped ions the minimum temperature occurs when the depth of the optical potential, U_o, approaches $\hbar\omega_z$. The minimum values displayed in this plot correspond to $U_o = 5\hbar\omega_z$.

This suggests a simple physical picture for the polarization gradient cooling of trapped ions. The trapped ion is truly a quantum-mechanical harmonic oscillator with discrete energy levels. Therefore, the kinetic energy of the ion can best be extracted in units of $\hbar\omega_z$, i.e. the energy quanta of the harmonic oscillator. For that reason, the cooling process is most effective when the energy removed in a PGC cycle, approximately U_o, is equal to a quantum of vibrational energy $\hbar\omega_z$ or a low multiple of this value. Successive spontaneous anti-Stokes Raman-scattering events lead to a ladder of downward transitions through the vibrational states of the harmonic oscillator. In contrast to the free-atom case the final temperature is determined by U_o and ω_z. This reduces to the sole dependence on U_o when ω_z is fixed as in our experiment.

Recent theoretical studies[58-60] of polarization gradient cooling for trapped ions in the weak binding limit ($\omega_z \ll \Gamma$) confirm this physical picture, i.e. the minimum temperature occurs for $U_o = \hbar\omega_z$. However, this theory makes the additional assumption that the ions are confined to a spatial extent small compared to the wavelength of the cooling transition (Lamb-Dicke limit). The latter condition is not fulfilled in our experiment. This theoretical requirement is not a necessary condition for the functioning of the cooling mechanism, but rather is necessary for the analytical solution of the problem.

For the moment, let us consider one of the more fascinating predictions of the current PGC theories that the minimum kinetic energy attainable is on the order of the energy of the first excited state of the trapping potential. For the case of a weakly bound ion, this can be much lower than the minimum temperature for a free atom. Our observation of non-Gaussian ion distributions supports this prediction. In Fig. 12b, a sharp spike rises at the centre of the trapping potential. This distribution was found for a relatively large detuning and U_o of eleven times $\hbar\omega_z$. The width of this spike is less than 3 picture elements of the imaging system of $1.2\mu m$. If we treat this narrow structure as above, such a localization corresponds to an ion that on average has a temperature of $4\mu K$ and is in the sixth excited state of our 1D trapping potential. Our measurements indicate that the ion maintains this sharp localization for approximately a tenth of our observation time of five seconds. The instability in the localization can be attributed to frequency and intensity fluctuations or perhaps more importantly to the coupling of this 1D motion to the much hotter radial motion. Of course, for such low energies, the single-photon recoil energy plays a significant role in defining the cooling limit. In fact, both the kinetic energy of the ion and the optical trap depth are on the order of the recoil energy. This raises the distinct possibility that the ion is actually trapped in an optical potential well. For our parameters, these wells are quite shallow and at most only the ground state and a single excited state (or band) are bound. The energy of an ion in this first excited state corresponds to a temperature of $7\mu K$. At present, our experimental results are unable to distinguish between these physical descriptions of the sharp localization of the ion. In either case, cooling to a low vibrational state was observed.

We extended the study of PGC of trapped ions to the case of low-density ion crystals (ion strings). Only in a two-dimensional quadrupole trapping field it is possible to cool more than a single ion down to a temperature which is similar to the one obtained for a single ion. For both ions, the position distribution around their mean separation can be fitted by a Gaussian function. In the course of these studies, we found the same quantitative functional dependence of the ion temperature as observed for a single ion. The minimum temperature for a two-ion crystal was found to be $T = (280 \pm 75)\mu K$. This is the coldest 1D ion crystal reported to date. These results clearly show that in a quadrupole storage ring with PGC or other cooling techniques, ensembles of ions can also be cooled far below the Doppler limit.

5. Conclusion

In this paper single ion spectroscopy and the generation of crystalline ion structures in Paul traps have been reviewed. In the last part of the paper experiments using polarization gradient cooling of ions have been described. In those experiments it was demonstrated that a one-dimensional ion crystal could be cooled far below the Doppler limit.

References

1. W. Neuhauser, M. Hohenstatt, P. Toschek and H. Dehmelt, *Phys. Rev. Lett.* **41** (1978) 233 and *Phys. Rev.* **A22** (1980) 1137.
2. R. H. Dicke, *Phys. Rev.* **89** (1953) 472.
3. D.J. Wineland, W.I. Itano and R.S. Van Dyck, in *Advances in Atomic and Molecular Physics 19,* eds. D. Bates and B. Bederson, p. 136 (Academic Press, Boston 1983).
4. R. Blatt, P. Gill and R.C. Thompson, *Journal of Modern Optics* **39** (1992) 193.
5. R.C. Thompson, in *Advances in Atomic, Molecular and Optical Physics 30*, eds. D. Bates and B. Bederson (Academic Press Inc., Boston 1993), in print.
6. H.G. Dehmelt and F.G. Major, *Phys. Rev. Lett.* **8** (1962) 213.
7. H.A. Schüssler, F.N. Fortson and H. G. Dehmelt, *Phys. Rev.* **187** (1969) 5.
8. W. Paul, O. Osberghaus and E. Fischer, *Forschungsberichte des Wirtschafts- und Verkehrsministeriums Nordrhein-Westfalen*, **415** (1958) 1.
9. E. Fischer, *Z. Phys.* **156**, (1959) 1.
10. D.J. Wineland, W.M. Itano, J.C. Bergquist, J.J. Bollinger and J.D. Prestage, in *Atomic Physics 9*, eds. R.S. van Dyck, Jr. and E.N. Fortson, p. 3 (World Scientific Publishing, Singapore, 1984).
11. H.G. Dehmelt, in *Adv. Atom. Molec. Phys. 3*, eds. D.R. Bates and I. Estermann, p. 53 (Academic, New York, 1967).
12. J. Dalibard, C. Salomon, A. Aspect, E. Arimondo, R. Kaiser, N. Vansteenkiste and C. Cohen-Tannoudji, in *Atomic Physics 11*, eds. S. Haroche, J.C. Gay and G. Grynberg, p. 199 (World Scientific Publishing, Singapore, 1988).
13. F. Diedrich and H. Walther, *Phys. Rev. Lett.* **58** (1987) 203.
14. W. Nagourney, J. Sandberg and H. Dehmelt, *Phys. Rev. Lett.* **56** (1986) 2767.
15. Th. Sauter, W. Neuhauser, R. Blatt and P.E. Toschek, *Phys. Rev. Lett.* **57** (1986) 1696.
16. J.C. Bergquist, R.G. Hulet, W.M. Itano and D.J. Wineland, *Phys. Rev. Lett.* **57** (1986) 1699.
17. U. Eichmann, J.C. Bergquist, J.J. Bollinger, J.M. Gilligan, W.M. Itano, D.J. Wineland and M.G. Raizen, *Phys. Rev. Lett.* **70** (1993) 2359.
18. J.J. Bollinger, D.J. Heinzen, W.M. Itano, S.L. Gilbert and D.J.Wineland, *Phys. Rev. Lett.* **63** (1989) 1031.
19. W.M. Itano, D.J. Heinzen, J.J. Bollinger and D.J. Wineland, *Phys. Rev.* **A41** (1990) 2295.

20. H. Dehmelt, *IEEE Trans. Instrum. Meas.* **31** (1982) 83.

21. J.C. Bergquist, F. Diedrich, W.M. Itano and D.J. Wineland, in *Laser Spectroscopy IX*, eds. M.S. Feld, J.E. Thomas and A. Mooradian (Academic Press, Boston, 1989) p. 274.

22. F. Diedrich, J.C. Bergquist, W.M. Itano and D.J. Wineland, D.J., *Phys. Rev. Lett.* **62** (1989) 403.

23. W. Nagourney, N. Yu and H. Dehmelt, *Opt. Comm.* **79** (1990) 176.

24. J.C. Bergquist, W.M. Itano and D.J. Wineland, *Phys. Rev.* **A36** (1987) 428.

25. H.A. Klein, A.S. Bell, G.P. Barwood and P. Gill, *Appl. Phys.* **B50** (1990) 13.

26. A.S. Bell, P. Gill, H.A. Klein, A.P. Levick, Chr. Tamm and D. Schnier, *Phys. Rev.* **A44** (1991) R20.

27. H. Lehmitz, J. Hattendorf-Ledwoch, R. Blatt and R. Harde, *Phys. Rev. Lett* **62** (1989) 2108.

28. E. Peik, G. Hollemann and H. Walther, *Phys. Rev. A.*, in print.

29. A.A. Madej and J.D. Sankey, *Opt. Lett.* **15** (1990) 634.

30. H. Dehmelt, *Bull. Am. Phys. Soc.* **18** (1973) 1521, ibid. **20** (1975) 60, D. Wineland and H. Dehmelt, *Bull. Am. Phys. Soc.* **20** (1975) 637.

31. M.C. Bigeon, *J. Physique* **28** (1967) 51.

32. R.W. Dunford et al., *Phys. Rev.* **A44** (1991) 764; P. Indelicato et al., *Phys. Rev. Lett.* **68** (1992) 1307, B.B. Birkett et al., *Phys. Rev.* **A47** (1993) R2454.

33. C.A. Schrama, E. Peik, W.W. Smith and H. Walther, *Opt. Comm.* **101** (1993) 32.

34. J.C. Lehmann and C. Cohen-Tannoudji *C. R. Acad. Sc. Paris* **258** (1964) 4463.

35. F. Diedrich, E. Peik, J.M. Chen, W. Quint and H. Walther, *Phys. Rev. Lett.* **59** (1987) 2931.

36. J. Javanainen, *Phys. Rev.* **A33** (1986) 2121; A. Schenzle, R.G. DeVoe and R.G. Brewer, ibid. **33** (1986) 2127; M. Merz, A. Schenzle, *Appl. Phys.* **B50** (1990) 115.

37. J.C. Bergquist, R.G. Hulet, W.M. Itano and D.J. Wineland, *Phys. Rev. Lett.* **57** (1987) 1699.

38. I.I. Sobel'man, in *Atomic Spectra and Radiative Transitions 2* (Springer, Berlin, 1992).

39. P.L. Larkin and P. Hannaford, *Z. Physik D* (in print).

40. F. Diedrich, E. Peik, J.M. Chen, W. Quint and H. Walther, *Phys. Rev. Lett.* **59** (1987) 2931.

41. R. Blümel, J.M. Chen, F. Diedrich, E. Peik, W. Quint, W. Schleich, Y.R. Shen and H. Walther, in *Atomic Physics 11*, eds. S. Haroche, J.C. Gay and G. Grynberg, p. 243 (World Scientific Publishing, Singapore, 1988).

42. R. Blümel, J.M. Chen, W. Quint, W. Schleich, Y.R. Shen and H. Walther, *Nature* **334** (1988) 309.

43. R. Blümel, C. Kappler, W. Quint and H. Walther, *Phys. Rev.* **A40** (1989) 808.

44. J. Drees and W. Paul, *Z. Phys.* **180** (1964) 340.

45. D.A. Church, *J. Appl. Phys.* **40** (1969) 3127.

46. H. Walther, in *Proc. of the Workshop on Light Induced Kinetic Effects on Atoms, Ions and Molecules*, eds. L. Moi, S. Gozzini, C. Gabbanini, E. Arimondo and F. Strumia) (ETS Editrice, Pisa, 1991).

47. I. Waki, S. Kassner, G. Birkl and H. Walther, *Phys. Rev. Lett.* **68** (1992) 2007.

48. G. Birkl, S. Kassner and H. Walther, *Nature* **357** (1992) 310.

49. A. Rahman and J.P. Schiffer, Phys. Rev. Lett. **57** (1986) 1133.

50. R.W. Hasse and J.P. Schiffer, *Ann. Phys.* **203** (1990) 419.

51. M.G. Raizen, J.M. Gilligan, J.C. Berquist, W.M. Itano, and D.J. Wineland, *Journ. Mod. Opt.* **39** (1992) 233.

52. S.L. Gilbert, J.J. Bollinger and D.J. Wineland, *Phys. Rev. Lett.* **60** (1988) 2022.

53. J. Dalibard and C. Cohen-Tannoudji, *J. Opt. Soc. Am. B* **6** 2023 (1989).

54. P.D. Lett, R.N. Watts, C.I. Westbrook, W.D. Philips, P.L. Gould and H.J. Metcalf, *Phys. Rev. Lett.* **61** (1988) 169.

55. C. Salomon, J. Dalibard, W.D. Philips, A. Clairon and S. Guellati, *Europhys. Lett.* **12** (1990) 683.

56. G. Birkl, J.A. Yeazell, R. Rückerl, H. Walther, *Europhys. Lett.* **27** (1994) 197.

57. Y. Castin and J. Dalibard, *Europhys. Lett.* **14** (1991) 761.

58. Yoo Sung Mi and J. Javanainen, *Phys. Rev. A* **48** (1993) R30; J. Javanainen and Yoo Sung Mi, *Phys. Rev. A* **48** (1993) 3776.

59. J.I. Cirac, R. Blatt, A.S. Parkins and P. Zoller, *Phys. Rev. A* **48** (1993) 1434.

60. Li Yongqing and K. Mølmer, *Laser Physics* **4** (1994) 829.

Three-Dimensional Laser Cooling by Synchrobetatron Coupling

Hiromi Okamoto
Accelerator Laboratory, Institute for Chemical Research, Kyoto University
Gokanoshou, Uji, Kyoto 611, Japan

Shinji Machida
National Laboratory for High Energy Physics (KEK)
Oho, Tsukuba, Ibaraki 305, Japan

Andrew M. Sessler
Lawrence Berkeley Laboratory
1 Cyclotron Road, Berkeley, California 94720, U.S.A.

ABSTRACT

We describe a possible method to achieve three-dimensional laser cooling of fast circulating beams in a storage ring. Two different schemes are developed in order to apply the proposed idea to real machines. We confirm the practical feasibility of the present schemes with detailed simulations, in particular, taking into account the lattice parameters of the storage ring ASTRID in Denmark.

1. Introduction

The laser cooling technique[1] has been widely accepted, due to recent experimental success[2], as a powerful tool to manipulate the longitudinal phase space of stored and circulating ion beams. The beam temperature reachable with this new technique is now, at least, three orders of magnitude lower than that with the traditional techniques like electron cooling[3] and stochastic cooling[4]. In fact, the longitudinal temperature of 1 mK has been obtained in the ASTRID ring with 100 keV $^7Li^+$ beams[5]. This is the main reason why laser cooling is often linked to beam crystallization[6,7] for which the beam temperature must be reduced to an extremely low level.

Contrary to this promising potential, laser cooling, if applied to beams in a storage ring, is essentially one-dimensional; namely, it operates only upon the longitudinal motion. Although, not surprisingly, we have experimentally observed sympathetic transverse cooling due to Coulomb interactions, the resulting transverse temperature is still much higher than the longitudinal temperature[8]. Further, the recent results of molecular dynamics simulations by J.Wei *et al.* indicate that, as the temperature is decreased, sympathetic cooling becomes less and less effective so that longitudinal cooling alone is not adequate to reach a crystalline state[9].

In the recent papers, we proposed a novel method to overcome this difficulty in laser cooling; namely, an artificial coupling between the transverse and longitudinal degree of freedom is introduced to indirectly cool transverse motions. In principle, this method enables us to reduce temperatures of all three degrees of freedom simultaneously and effectively. According to the theoretical predictions, transverse cooling rates can be enhanced, under resonance conditions, to the same level as the longitudinal cooling rate. Further, since the transverse cooling effect basically comes from a simple single-particle dynamics, the fundamental beam quantities such as beam current are left as free parameters. Therefore, the proposed 3D cooling technique is quite suitable for the future systematic study of the various crystalline structures.

In the present paper, we first review the essence of the 3D laser cooling method, introducing two different coupling sources, *i.e.* the so-called *coupling cavity* operated in TM_{210} mode[10] and dispersive coupling induced in an ordinary RF cavity[11], both produce a linear synchrobetatron coupling in the lowest-order approximation. Tracking results are then given to demonstrate the validity of theoretical predictions. Finally, proof-of-principle numerical experiments are performed, employing the lattice parameters of the storage ring ASTRID in Aarhus University, Denmark[12], one of the two rings in the world having a laser cooling system[13].

2. The Sources of Synchrobetatron Coupling

2.1. *The Dispersion-Coupling Scheme*

Let us first consider a usual storage-ring system involving an *ordinary* RF cavity. Starting from the exact Hamiltonian of the system and applying several canonical transformations and scalings of the variables, we eventually obtain the approximate Hamiltonian

$$H = \frac{1}{2}(p_x^2 + \nu_x^2 x^2) + \frac{1}{2}(p_y^2 + \nu_y^2 y^2) - \frac{\xi_0 W^2}{2} + \Gamma_q y\left(x - \frac{\eta}{R}W\right)\delta_p(\theta - \theta_q)$$
$$- \frac{h_b q V_b}{p_0 \beta_0 c}\cos\left[\psi - \frac{1}{R}\left(\eta p_x - \frac{d\eta}{d\theta}x\right)\right]\delta_p(\theta - \theta_b), \qquad (1)$$

where the transverse motions have been smoothed, and a pure skew quadrupole magnet has been added. The notation used here is:

ν_x=horizontal tune, ν_y=vertical tune, ξ_0=phase slip factor,
R=average ring radius, η=dispersion function,
h_b=harmonic number of the cavity, V_b=maximum RF voltage,
θ_q=position of the skew quadrupole magnet, θ_b=position of the RF cavity
q=charge state of stored ions, β_0=relativistic parameter,
$\delta_p(\theta)$=periodic delta function with the periodicity of 2π,
p_0=momentum of the reference particle, and c=speed of light.

Note that the independent variable θ has been defined by θ=s/R with s being the distance measured along the reference particle orbit. Further, the coupling strength of the skew quadrupole, Γ_q, can be related to the field gradient as

$$\frac{L}{B\rho}\left(\frac{\partial B}{\partial y}\right) = \frac{\Gamma_q}{R}, \qquad (2)$$

where L is the length of the magnet.

It is evident from Eq.(1) that the longitudinal motion can be coupled with the horizontal motion, provided that either the dispersion η or the derivative of dispersion $d\eta/d\theta$ is non-zero at $\theta=\theta_b$. The fourth term in the right hand side of Eq.(1) corresponds to the skew quadrupole potential which generates a linear coupling between the horizontal and vertical motion. The vertical degree of freedom is thus indirectly connected to the longitudinal degree of freedom.

Specifically, the significance of the dispersion-coupling scheme is its simplicity. In fact, in the ASTRID case, we need nothing new but it is only necessary to install a skew quadrupole magnet at an open location, as an RF cavity is already sitting at a dispersive position. However, in order to enhance the coupling, the ring should be re-tuned such that the following two conditions are roughly satisfied:

$$\nu_x - \nu_L = \text{integer}, \quad \nu_x - \nu_y = \text{integer}, \tag{3}$$

where ν_L is the longitudinal tune. These conditions imply that the non-integer parts of the three tunes must be identical. Hence, it is required to set the longitudinal tune as large as possible to avoid the excitation of severe integer resonances in the transverse motions.

The easiest way to increase ν_L is, indeed, the use of a higher RF voltage but it is clearly dangerous since the longitudinal energy spread of laser-cooled beams eventually

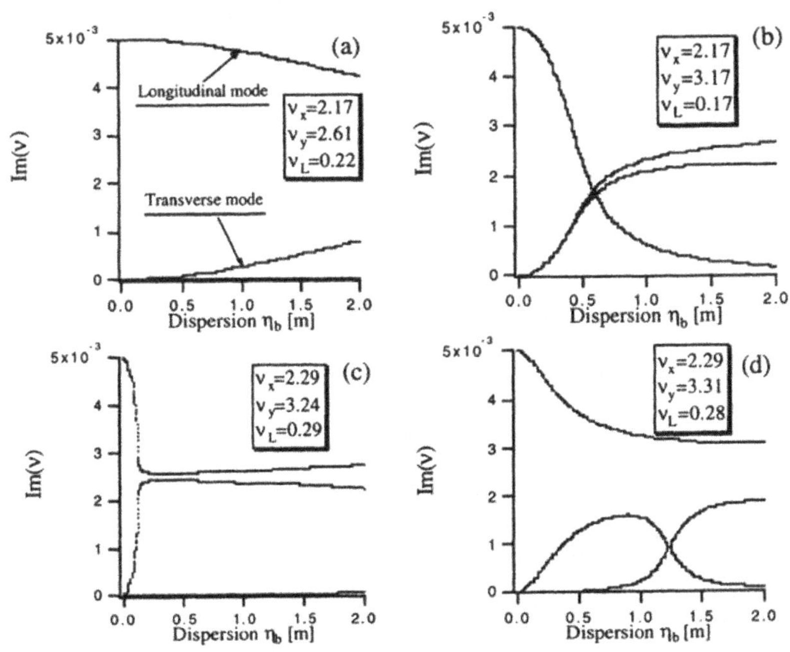

Fig.1 Cooling rates vs. the size of dispersion at the cavity location; The RF cavity sits at $\theta_b=\pi$, the position opposite to the cooling section, while the skew quadrupole location is assumed to be $\theta_q=3\pi/2$. The coupling constant is set at $\Gamma_q=0.1$.

reaches even less than 0.1 eV. Further, recalling that the beam is initially continuous, we should adiabatically capture it first, ramping the RF voltage. This adiabatic capturing process causes the initial blowup of the longitudinal emittance where a larger bucket height results in a larger initial $\delta p/p$ to be scanned. A lower RF voltage is thus preferable for shortening the cooling time as well. In ASTRID, the RF voltages of only 70 volts have actually been used for laser cooling experiments of bunched beams[14].

In order to see how much the coupling is strengthened under the resonance conditions, let us take a look at the cooling rates evaluated from a linear analysis. As shown in Ref. 11, it is straightforward to derive dispersion relation after introducing a simple linear friction model to represent the laser-cooling effect. The cooling-rate curves of four typical cases are given in Fig. 1 where $\eta_b=\eta(\theta=\theta_b)$. The case (b) is clearly the best one as we have the optimum point where the cooling rates of the three directions becomes approximately the same. In other cases, one or two of the three cooling rates look quite small at any value of dispersion.

Whenever the tunes satisfy the conditions in Eq.(3), the cooling rates vary, with respect to η_b, in such a way as seen in Fig. 1-(b); namely, as η_b increases, the longitudinal rate goes down while both transverse rates simultaneously grow from zero, and then the three curves eventually intersect at roughly a single point. Of practical importance is the fact that the point of the intersection can be moved horizontally by changing the strength of the skew quadrupole gradient. This means that, even if the dispersion at the cavity position involves some error due to, *e.g.*, the misalignments of magnetic elements, we can easily search for the optimum operating point, compensating for the error. All we have to do for this purpose is simply to adjust the skew field strength.

2.2. *The Coupling-Cavity Scheme*

An alternate way of generating synchrobetatron coupling is the use of the so-called *coupling cavity* operating in a TM_{210} mode. In this scheme, we need two RF cavities; namely, a coupling cavity and an ordinary RF cavity which we call a *bunching cavity* from now on. The role of the bunching cavity is to make ions execute synchrotron oscillations so that the first condition in Eq.(3) is satisfied.

Assuming, for simplicity, a rectangular cavity which has a width of 2a and the height of 2b, the vector potential can be written as

$$\mathbf{A}_c = \left(0,\ \ 0,\ \ \frac{V_c}{\omega_c}\sin\left(\frac{\pi x}{a}\right)\cos\left(\frac{\pi y}{b}\right)\sin(\omega_c t + \phi_c)\right) \tag{4}$$

where V_c is the maximum voltage of the coupling cavity, ϕ_c is constant, and $(\pi/a)^2+(\pi/2b)^2=(\omega_c/c)^2$. The Hamiltonian for the coupling-cavity scheme is then found, neglecting the effect of dispersion, to be

$$H = \frac{1}{2}(p_x^2 + \nu_x^2 x^2) + \frac{1}{2}(p_y^2 + \nu_y^2 y^2) - \frac{\xi_0 W^2}{2} - \frac{h_b q V_b}{p_0 \beta_0 c}\cos\psi\,\delta_p(\theta-\theta_b)$$
$$+\Gamma_q xy\,\delta_p(\theta-\theta_q) - \frac{2\pi h_b \Gamma_c}{h_c} x\cdot\sin\left(\frac{h_c}{h_b}\psi\right)\delta_p(\theta-\theta_c), \tag{5}$$

where h_c is the harmonic number of the coupling cavity set at the position $\theta=\theta_c$, and the paraxial approximation has been applied to the coupling-cavity potential. In general, it is difficult to obtain the analytic form of the coupling constant Γ_c which strongly depends on the cavity design, but, for a rectangular cavity, we have $\Gamma_c=qV_cR/2\beta_0cp_0a$.

We should here notice that it is impracticable to provide the potential in Eq.(4) by means of a simple rectangular resonator because of its relatively low operating frequency. Since, in our applications, the frequency ω_c should mostly be of the order of 1~100 MHz, the structure dimension becomes quite large if an ordinary RF cavity is used. It is thus desirable to design a special type of resonator. A possible solution is the use of a lumped circuit, which will be discussed elsewhere.

Mathematically speaking, the coupling-cavity scheme is almost equivalent to the dispersion-coupling scheme discussed in the last subsection. Both schemes, therefore, work in a similar way. However, the coupling-cavity scheme has the advantage in controlling the coupling property, since we have at our disposal several additional parameters such as the operating frequency and RF voltage of the coupling cavity. Moreover, as pointed out later, the transverse cooling rates could be sufficiently large even when the resonance conditions is not satisfied. It is thus worthy to pursue this scheme although it seems a bit more complicated than the dispersion-coupling scheme.

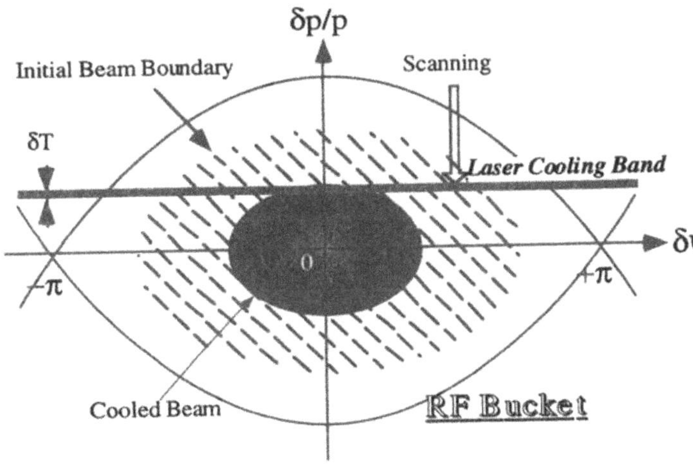

Fig.2 Schematic drawing of the laser cooling process

3. Tracking Results

In the numerical simulations given in this section, the particle motions governed by the Hamiltonians (1) and (5) are tracked in a symplectic manner, apart from the laser cooling section. Needless to say, it is possible to take into account the effect of dispersion neglected in Eq.(5). The longitudinal dissipative effect due to the cooling laser is incorporated in a way as shown in Fig.2; namely, a narrow laser-cooling band with the width of δT is scanned from the top of the RF bucket down to the $\delta\psi$-axis. Simulation particles which come into the laser-cooling band while passing through the cooling section are kicked downward by the amount of δT corresponding to the limiting temperature. The speed of laser scanning is assumed to be constant in the present simulations.

As an example, let us observe how efficiently the dispersion coupling transfers the cooling effect into transverse directions, using the parameters summarized in Table.1. Fig.3 demonstrates the time evolutions of the horizontal, vertical, and longitudinal RMS emittance when the three tunes are exactly on resonance, *i.e.* $(\nu_x,\nu_y,\nu_L)=(3.1,3.1,0.1)$. The corresponding phase-space configurations are illustrated in Fig.4. The stored beam assumed here is 100keV $^{24}Mg^+$ which has been employed in recent laser-cooling experiments in ASTRID. The skew coupling strength has been taken to be $\Gamma_q=0.35$, based on the theoretical prediction for the optimum operation. Since the input beam is continuous, we have used the first 250 turns to adiabatically capture it. The laser-cooling band is then scanned during the following 20000 turns. It is seen, from Fig.3, that the 3D laser-cooling scheme proposed works excellently. We have done a lot of tracking simulations, all confirming the validity of theoretical predictions and the promising possibility of the present cooling schemes.

Finally, we give another example in order to approximately reproduce the current situation of ASTRID where no effective transverse cooling has been observed although a cavity is located in a dispersive region. The transverse tunes, which have usually been used in ASTRID, are far from resonance and, in addition, the skew quadrupole is not installed. The value of the applied RF voltage adopted here is also close to that employed for bunched-beam cooling experiments, but the harmonic number has been taken ten times larger. Fig.5 shows the result. As expected, almost nothing takes place in the transverse phase spaces.

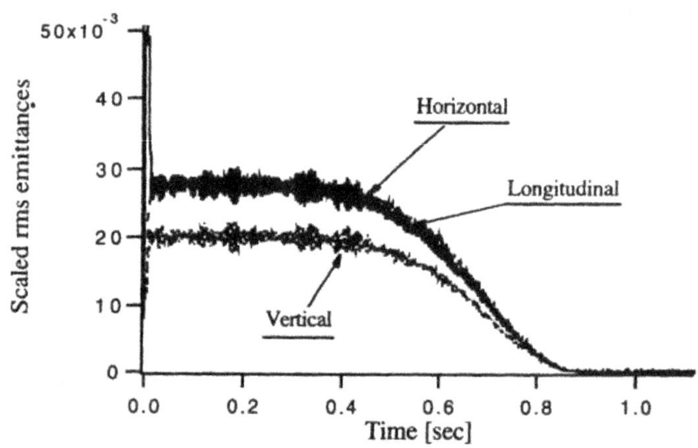

Fig.3 Tracking result; The parameters adopted in this simulation are consistent, except for the position of the skew quadrupole magnet, with those employed in Fig.6 in which the exact lattice of the ASTRID ring has been considered. The skew quadrupole has been set at $\theta_q=3\pi/5$ measured from the center of the cooling section, while $\theta_q=\pi$ in Fig.6. But the change of θ_q gives rise to no essential difference in tracking results.

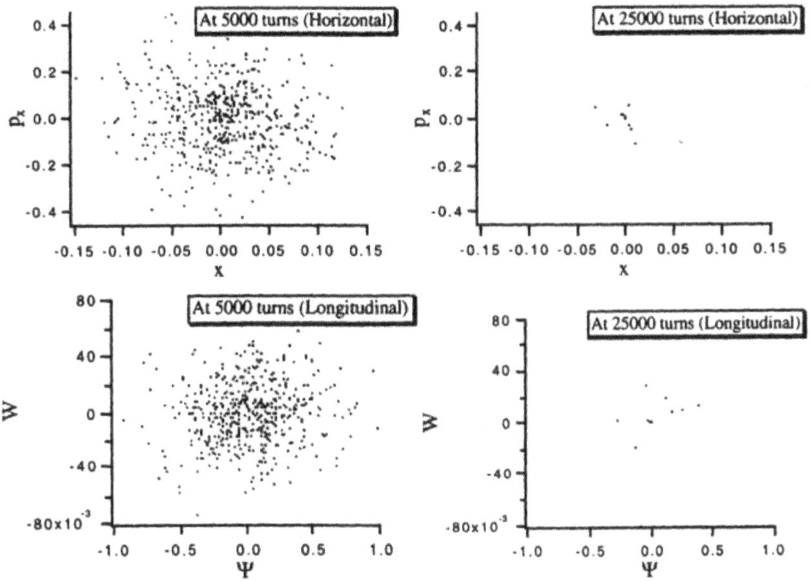

Fig.4 Phase-space configurations corresponding to the simulation in Fig.3. The vertical phase-space not shown is roughly the same as the horizontal one.

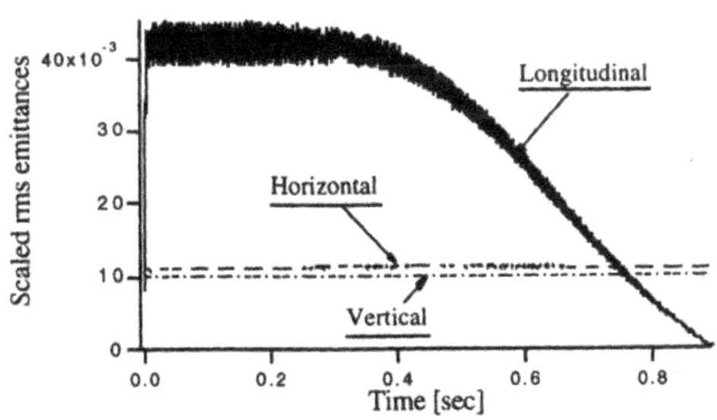

Fig.5 Tracking result roughly corresponding to the current ASTRID situation. The fundamental parameters adopted are as follows: (ν_x, ν_y, ν_L)=(2.29, 2.73, 0.038), $\theta_b=\pi/2$, $h_b=260$, $\eta(\theta=\theta_b)=2.7$ m and $\xi_{x0}=-1$. After the adiabatic capture of the beam, the RF voltage reaches about 7 volts. The skew quadrupole magnet has been turned off.

4. Feasibility of 3D Laser Cooling in the ASTRID ring

We now perform a more realistic numerical experiment, exactly considering the lattice parameters of an existing storage ring. For this purpose, we here employ the tracking code SIMPSONS[15] originally written for the study of space-charge effects in synchrotrons. It is basically a thin-element code but enables us to do realistic investigation of the 3D cooling scheme with a specific lattice design. Furthermore, since time is taken as the independent variable, it is easy to incorporate the laser cooling process as well as to change lattice parameters in a manner similar to that done in real machines.

In the following, we take the lattice of the storage ring ASTRID, assuming again 100 keV $^{24}Mg^+$ beams initially continuous. While two independent schemes have been proposed, only the dispersive coupling scheme is explored here since it requires no new devices and is, therefore, easier to test in a real storage ring with little modification of the lattice. But, it is clear that, if this scheme worked, the coupling-cavity scheme should also work because the mathematical features of these schemes are almost equivalent.

In the normal operation mode of ASTRID, ν_x=2.29 and ν_y=2.73, and the RF cavity has usually been switched off, which means ν_L=0. Even when the cavity was turned on for laser cooling experiments of bunched beams, the corresponding longitudinal tune was too small for the resonance condition to be satisfied. In such a situation, there is no reason to anticipate the effective decrease of transverse temperatures as already demonstrated in Fig.5, even if the dispersion at the cavity position is finite. We must thus start from modifying the strength of the focusing magnetic fields to enhance the transverse cooling rates; namely, the transverse tunes should be adjusted to be ν_x=3.1 and ν_y=3.1, for example. In this 3D cooling mode, the phase slip factor ξ_0 becomes over one in the ASTRID case, *i.e.* ~1.38, which is beneficial to reducing the RF voltage. Further, to excite the linear synchrobetatron resonance, ν_L must be 0.1. Although this value of ν_L seems too large to achieve with a very low RF voltage, the small total energy of the $^{24}Mg^+$ beams allows us to have a large ν_L with no difficulty. In fact, assuming the harmonic number h_b=260 corresponding to 5.8 MHz, we need only 34 volts to achieve ν_L=0.1.

We put a pure skew quadrupole at the opposite side of the laser cooling section. The strength of the skew field is of essential importance since the cooling rates of the three directions are altered depending on it. The optimum value of the parameter Γ_q is about 0.35 under the setup assumed here. Noting that R~6.366 m in ASTRID, the skew gradient required is $(L/B\rho)\cdot(\partial B/\partial y) \sim 0.055$ [m^{-1}].

The input continuous beam of 100 keV $^{24}Mg^+$ ions has been assumed to have a Gaussian shape with respect to the momentum spread $\delta p/p$. The initial particle distributions in the horizontal and vertical phase space are also taken to be Gaussian, and the RMS emittances are both set at about 2.5×10^{-3} πmm·mrad roughly representing the actual injection beams in ASTRID. To adiabatically capture the beam into the RF bucket, the RF voltage is linearly increased from 0 to 34 volts in the first 10 mili-seconds. During this process, the longitudinal RMS emittance becomes several times larger than the initial value. Then, the cooling laser is turned on and is scanned from $\delta p/p \sim 7.5\times10^{-4}$ down to $\delta p/p \sim 0$ within 0.99 second at a constant speed. The width of the laser cooling band, δT, is fixed at about 1.5×10^{-5} in the present example. Fig.6 demonstrates the time evolution of the three emittances perfectly consistent with the predictions drawn from the theory.

Table. 1 The fundamental parameters used in the tracking simulations given in Figs.3 and 6.

Stored ion	$^{24}Mg^{+}$
Total beam energy	100 keV (β~0.003)
Storage ring circumference	R=40 m
Betatron tunes	$\nu_x=\nu_y=3.1$
Phase slip factor	$\xi_0 \equiv 1/\gamma_T^2-1/\gamma^2=-1.38$
Dispersion	$\eta(\theta=\theta_h)=\eta(\theta=\theta_a)=3.2$ m
RF frequency	5.8 MHz (h=260)
Maximum RF voltage	34 volts

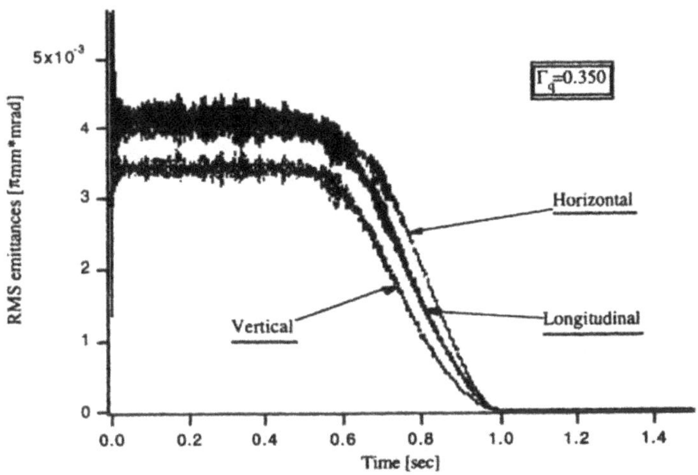

Fig.6 The result of the numerical experiment of 3D laser cooling in ASTRID

5. Off-Resonant Regime with the Coupling Cavity

The resonance conditions in Eq.(3) impose a severe restriction upon the choice of the longitudinal tune ν_L. Since, as already mentioned, a smaller value of ν_L is preferable due to some practical reasons, it is beneficial if transverse cooling rates can be sufficiently enhanced even with off-resonant value of ν_L. In this section, we show how far from the resonance we are allowed to go.

Let us again set the transverse tunes to be 3.1 where the corresponding best choice of ν_L is 0.1. According to the theory, in the dispersion-coupling scheme, the transverse cooling rates fall quickly as ν_L becomes smaller than 0.1. Contrarily, in the region $\nu_L>0.1$, the transverse rates are not damped so fast, which suggests that we could obtain off-resonant 3D cooling even when ν_L is larger than the resonant value. This off-resonant operation is, however, clearly impractical since we generally want to employ as low an RF voltage as possible.

Interestingly, in the coupling-cavity scheme, the situation is reversed; namely, it can be shown that the transverse cooling rates are larger in the region $\nu_L<0.1$ rather than the region $\nu_L>0.1$. According to the cooling-rate curves corresponding to the present case, we have small, but still sufficient, transverse cooling rates for ν_L even less than 0.04. Therefore, provided that cooling laser is scanned at a reasonably slow speed, it would be no problem to see effective transverse cooling, as shown in Fig.7, where the longitudinal tune have been set at 0.038.

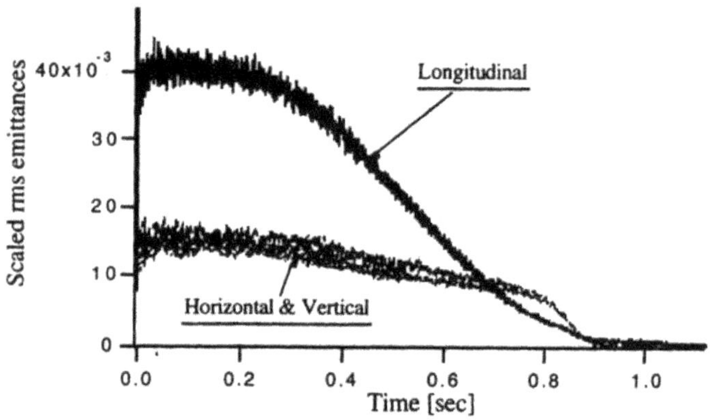

Fig.7 Tracking result in the coupling-cavity scheme; The fundamental parameters adopted are as follows: $(\nu_x,\nu_y,\nu_L)=(3.1,\ 3.1,\ 0.038)$, $\theta_b=\pi/2$, $\theta_c=\pi$, $\theta_q=3\pi/5$, $h_b=h_c=260$, $\Gamma_q=0.2$, $\Gamma_c=0.01$, and $\xi_{s0}=-1.38$. Since the effect of dispersion has been neglected for simplicity, the synchrobetatron coupling is generated only by the fields of the coupling cavity.

6. Summary

As we have seen in this paper, there seems to be practically no difficulty in applying the proposed 3D laser cooling schemes to real machines. It has been proven both theoretically and numerically that the transverse cooling rates can be considerably enhanced by driving the lattice operating point close to the coupling resonances. Provided that the conditions in Eq.(3) are precisely satisfied, we can even anticipate transverse cooling rates as high as the longitudinal cooling rate.

The only concern about the present schemes is the effect of the Coulomb interactions neglected here, which will become increasingly dominant as the beam gets to be more dense. In fact, noting that the magnitude of space-charge-induced tune shift could be quite large in a very cold beam, it would no longer be possible to maintain the resonance conditions throughout the cooling process. However, unless the tune shift is too large, the transverse cooling rates is still expected to be sufficiently high. Even after getting into a space-charge-dominated state, the transverse cooling should be much more effective than in the case where we only have weak sympathetic coupling. Further, as discussed in Section 5, the use of the coupling-cavity scheme will probably lighten this difficulty because of its effectiveness in the off-resonance region. We thus believe that, in order to approach closer to, or even achieve, a crystalline state, the present 3D cooling schemes should be tried.

Another interesting possibility of the present schemes is its application to stochastic cooling. The theory in Ref.10 and 11 would also be applicable to stochastic cooling system because a general frictional force is taken to represent the cooling effect, in other words, no specific cooling process is assumed. Therefore, if we have a longitudinal (or transverse) stochastic cooling device, it might be possible to indirectly cool the transverse (or longitudinal) motion through the coupling resonances.

Finally, it is worthwhile to notice that the coupling cavity is substantially the same as the so-called *conditioner cavity* which has been proposed to improve free-electron-laser performance[16]. In this sense, the coupling-cavity might be of practical importance and will be explored in more detail in future work.

7. References

1. T. Hänsch and Schawlow, Opt. Commun. **13**, 68 (1975).
2. S. Schröder *et al.*, Phys. Rev. Lett. **64**, 2901 (1990).
3. G. I. Budker, At. Energ. **22**, 345 (1967).
4. S. van der Meer, CERN Report No. CERN/ISR-PO/72-31 (1972) (unpublished).
5. J. S. Hangst *et al.*, Phys. Rev. Lett. **67**, 1238 (1991).
6. J. P. Schiffer and P. Kienle, Z. Phys. A **321**, 181 (1985).
7. J. Wei, X.-P. Li, and A. M. Sessler, Phys. Rev. Lett. **73**, 3089 (1994).
8. Miesner *et al.*, in this proceedings.
9. J. Wei, A. Draeseke, A. M. Sessler, and X.-P. Li, in this proceedings.
10. H. Okamoto, A. M. Sessler, and D. Möhl, Phys. Rev. Lett. **72**, 3977 (1994).
11. H. Okamoto, Phys. Rev. E **50** (6),4982 (1994); LBL Report No.35594 (1994).
12. J. S. Hangst *et al.*, Nucl. Instrum. Meth. B68, 17 (1992).
13. Another is the TSR ring in Heidelberg, Germany; See, for example, W. Petrich *et al.*, Phys. Rev. A **48**, 2127 (1993).
14. J. S. Hangst *et al.*, Phys. Rev. Lett. **74**, 4432 (1995); Note that the application of RF field is not essential to the laser cooling process itself but is only necessary in the 3D cooling scheme in order to *resonantly* couple the betatron motions with the synchrotron motion. However, provided that ions are executing synchrotron oscillations, the scanning range of laser frequency can be halved.
15. S. Machida, AIP Conf. Proc. **297**, Computational Accelerator Physics, 459 (1993).
16. A. M. Sessler, D. H. Whittum, and L.-H. Yu, Phys. Rev. Lett. **68**, 309 (1992).

IONS IN ELECTRON PLASMAS AT STRONG COUPLING: STOPPING POWER AND ELECTRON–ION RECOMBINATIONS

G. ZWICKNAGEL
Laboratoire de Physique des Gaz et des Plasmas, Bâtiment 212, Université Paris XI, F-91405 Orsay, France

Q. SPREITER, M. MILLER, C. TOEPFFER, P. –G. REINHARD
Institut für Theoretische Physik, Universität Erlangen, Staudtstr. 7, D-91058 Erlangen, Germany

ABSTRACT

We employ molecular dynamics (MD) computer simulations to investigate ions in electron plasmas at strong coupling. We consider the stopping power on heavy ions for strong, non–linear target-ion coupling in non-ideal plasmas. In this regime, the energy loss of the ions scales at low velocities with Z as approximatively $Z^{1.5}$ and thus clearly deviates from the $Z^2 \ln(\text{const.}/Z)$ scaling of the conventional weak coupling theories. This is connected with an enhanced local density of electrons around highly charged, slow ions. We report in particular on such enhancements caused by the trapping of electrons in high Rydberg states due to multi–particle collisions at strong coupling . These loosely bound electrons are likely to be responsible for enhanced recombination rates observed in recent experiments on highly charged ions in storage rings.

1. Introduction

Merging the ion beam with the cold electron beam in the electron cooler of a storage ring offers the possibility to reduce the phase–space volume of the stored heavy ion beam. This electron cooling is achieved by extracting thermal energy out of the ion beam by Coulomb interaction between the ions and the cold electron plasma in which the beam is embedded [1][2][3]. Furthermore the electron cooler delivers an ideal environment for the study of electron–ion recombination processes at low relative energy. These recombinations of ions with electrons can be detected by separating the lower–charge–state ions in one of the storage ring magnets and counting them. Besides spontaneous radiative recombinations also laser–induced resonant transitions can be investigated and allow spectroscopic studies of the populated Rydberg levels[4]. Due to the low electron temperatures in the longitudinal component and the possible high ion charge states, one deals with strongly coupled Coulomb systems. Thus the description of the processes in the electron cooler requires a treatment beyond the usual weak coupling approaches. To investigate some aspects of this physics of highly charged ions in electron plasmas at strong coupling we perform molecular dynamics (MD) computer simulations, i.e. the numerical integration of the classical equations of motion for point particles interacting by the Coulomb force. These simulations are an appropriate tool to study the dynamical properties of strongly coupled non–equilibrium Coulomb systems, since strong correlations and non–linear collective effects are automatically taken into account[5]–[9]. Our main interest lies in the new features occurring at strong coupling due to the non–linearity of the ion–target coupling and the non–ideality of the target, i.e. the electron plasma. The new features we observed in our simulations are: (1) In the strong ion-target coupling low velocity regime the stopping power of the ions scales with Z approximatively as $Z^{1.5}$ and thus clearly deviates from the $Z^2 \ln(\text{const.}/Z)$ scaling of the conventional weak coupling theories. (2) This behavior of the stopping is accompanied by electrons trapped by the ion into high Rydberg states through recombinations by multi-particle collisions. (3) Due to these recombinations the electron density around the ion can be $10^1 - 10^2$ times higher for strong coupling as predicted by the linear, weak coupling approaches.

In such electron–ion recombinations the electrons are trapped in high Rydberg states of the ion by exchanging energy and momentum with the ion and the surrounding electrons in multi–particle collisions. This multi–particle recombination can only occur for strong coupling, where at any time many particles interact with each other. It is thus quite different from the usual three–body–recombination in weakly coupled electron plasmas, where it is very unlikely for the ion and two electrons to come very close together to interact sufficiently strong. These enhanced electron densities may be important to understand the large discrepances between the at GSI[11], at CERN[12], at the TSR[4] and at the CRYRING[13] experimentally found enhanced radiative/laser-induced recombination rates for highly charged ions and the available theoretical results. These experiments show, besides strong enhancements which must have their origin in the atomic structure (see e.g.[12]), general enhancements of recombination rates by factors of the order of five. As the radiative/laser–induced recombination rates are proportional to the electron density, or more precisely, to the density of electrons with a given energy, an enhancement in the local electron density around the ion directly results in an enhancement of these recombination rates. The further radiative and atomic processes themselves are not considered in our present simulations.

In order to work out the different, fundamentally new features when going from the weak to the strong coupling regime, most of our activities are devoted to heavy ions in a non–magnetized, isotropic electron plasma. Such investigations are also of particular interest in a different application of strongly coupled Coulomb systems, which is the plasma–beam interaction in a heavy ion driven inertial confinement fusion scenario[6][9] The results and conclusions for the isotropic, non–magnetized plasmas presented below should mainly demonstrate the new effects to be expected for strong coupling and provide a qualitative description of the processes in the electron cooler rather than a quantitative one. For the relevant quantitative statements we present some first, preliminary results for the more complex situation in the electron cooler with anisotropic electron temperatures and an external, longitudinal magnetic field, guiding the electron beam.

In the following we consider systems consisting of a heavy ion (charge Ze, mass M and a non-relativistic velocity $\mathbf{v}$) in a plasma of classical electrons (density n and mass m). In the case of an isotropic plasma with temperature T the electrons have an isotropic Maxwell velocity distribution corresponding to T. The electron–electron coupling strength within the plasma is given by the plasma parameter $\Gamma = e^2(4\pi n/3)^{\frac{1}{3}}/k_BT$, representing the ratio of mean potential to thermal energy, whereas the coupling between the electron plasma and the ion is expressed by the coupling parameter $Z\,\Gamma^{3/2}$, which represents the ratio of the ion charge to the charge of the electrons in a Debye–sphere ($Z\,\Gamma^{3/2} \propto Z/n\lambda_D^3$ with the Debye–length $\lambda_D = (k_BT/4\pi e^2 n)^{1/2}$). These systems (isotropic, no magnetic field) and the corresponding simulations are completely characterized by Γ, Z and v, where the plasma parameter Γ is a measure for the ideality of the electron plasma and the coupling parameter $Z\,\Gamma^{3/2}$ quantifies the linearity of the ion–target coupling. In the case of an anisotropic plasma characterized by two temperatures for the transversal and the longitudinal degrees of freedom the velocity distribution of the electrons is given by two separate Maxwell distributions for the transversal and the longitudinal direction. Now one has two quantities for the electron–electron coupling strength as well as for the ion–target coupling defined by the two existing temperatures. In the case of a magnetized plasma a homogeneous magnetic field will be present in the simulation and acting on the particle movement in addition to the Coulomb forces.

After some discussion of the different interesting coupling regimes in section 2, we present in section 3 our results on the stopping power (i.e. the cooling force) and demonstrate the new features for strong coupling . Section 4 is devoted to our observations concerning the enhanced electron densities around an ion, before a summary is given in section 5.

2. The Different Coupling Regimes

With respect to the plasma and coupling parameters described above one has to distinguish different coupling regimes of interest. We restrict this discussion to the isotropic case without magnetic field. The additional effects on the stopping and the screening of ions due to an anisotropic electron distribution and a magnetic field have been investigated in[1]–[3][14] for the case of weak coupling.

2.1. Weak Coupling

In the weak coupling regime the electron plasma is ideal (negligible electron–electron collisions, $\Gamma \ll 1$) and the coupling is linear ($Z\,\Gamma^{3/2} \ll 1$). In this regime there are two well-known approaches to describe the stopping power of charged particles in a plasma. In the binary collision approximation the stopping is determined by regarding the momentum or energy transfer in isolated binary collisions of the ion with the electrons and averaging over all possible relative velocities and impact parameters. This yields the stopping power $-dE/ds$ for an ion as an integral over the electron velocity distribution and the transport cross section for Rutherford scattering, depending on the ion–electron relative velocity v_r[15][16][10]. The screening effects are accounted for by a cut-off b_m which excludes the soft collisions at large impact parameters and provides a finite cross section. Usually b_m is given by interpolating between the static and the adiabatic screening lengths, i.e. $b_m(v) = [\lambda_D^2 + (v/\omega_p)^2]^{1/2}$, where $\omega_p = (e^2 n/m)^{1/2}$ is the plasma frequency. Alternatively the stopping can be obtained using the dielectric linear response formalism. There the stopping is given by the force between the ion and the polarization cloud caused in the plasma by the ion itself, where the corresponding polarization potential is calculated self-consistently from a linearized Vlasov–Poisson equation for the electron distribution. This automatically includes the screening effects and yields the stopping power in terms of the well-known dielectric response function $\varepsilon(k,\omega)$ of a classical electron plasma [17][18][10]. Here it is necessary to exclude hard collisions with small impact parameters. The corresponding cutoff parameter k_m (see [18][10]) is chosen to exclude close Coulomb collisions with a scattering angle larger than 90°, i.e. $k_m = mv_r^2/Ze^2$.

In the ideal weak coupling regime with $\Gamma \ll 1, Z\Gamma^{3/2} \ll 1$ or in the regime of high ion velocities ($v \gg v_{th} = (kT/m)^{1/2}$) both theories yield essentially the same results. In either approximation only unimportant scattering events are excluded. Scaling the velocities in units of $v_{th} = (kT_e/m_e)^{1/2}$, the lengths in λ_D and dE/ds in units of $3^{1/2}\Gamma^{-3/2}k_B T/\lambda_D$ the stopping power is in both above approaches a function only of the coupling parameter $Z\Gamma^{3/2}$ and the velocity v . In the high velocity range ($v \gg 1$) both descriptions yield the stopping power[18][10]

$$\frac{dE}{ds} \sim -\frac{Z^2\,\Gamma^3}{v^2}\ln\left(\frac{v^3}{\sqrt{3}|Z|\Gamma^{3/2}}\right) . \tag{1}$$

In the limit of low velocities ($v \ll 1$) the stopping power behaves like

$$\frac{dE}{ds} = -R_1 v - R_3 v^3 + O(v^5) \quad , \tag{2}$$

where the friction coefficient R_1 in the binary collision approximation is given by[10]

$$R_1 = Z^2\,\Gamma^3\frac{1}{6}\sqrt{\frac{2}{\pi}}\int_0^\infty dx\, e^{-x}\ln\left(1+\frac{4\,b_m^2 x^2}{3\,Z^2\,\Gamma^3}\right) ; \qquad b_m = 1 \tag{3}$$

and in the dielectric linear response formalism by[18][10]

$$R_1 = \frac{Z^2\Gamma^3}{3\sqrt{2\pi}}\left[\ln(k_m^2+1) - \frac{k_m^2}{k_m^2+1}\right] ; \qquad k_m = \frac{2}{\sqrt{3}\,|Z|\,\Gamma^{3/2}} \, . \tag{4}$$

In the limit $Z\Gamma^{3/2} \ll 1$ Eqs.(3),(4) reduce to the common result

$$R_1 \approx Z^2\Gamma^3\frac{1}{3}\sqrt{\frac{2}{\pi}}\,\ln\left(\frac{2}{\sqrt{3}|Z|\Gamma^{3/2}}\right) \, . \tag{5}$$

The screening of an ion (at rest) in a weakly coupled plasma is described by the well-known Debye–Hückel result, i.e. the static screening is characterized by an exponential decrease on the scale of λ_D for the electric potential as well as for the induced electron density. In the weak coupling regime, this length for the screening of an ion due to the surrounding electrons is independent of the ion charge. The induced electron density is $n_{in} = n\, Z\Gamma^{3/2}\exp(-r)/r$ (with r in λ_D) and thus the total electron density around an ion of charge Z is

$$n_e(r) = n\left(1 + Z\Gamma^{3/2}\,\frac{\exp(-r)}{r}\right) . \tag{6}$$

Here n denotes the homogeneous density of the electrons when no ion is present as well as the ideal gas limit ($\Gamma^{3/2} \to 0$) of $n_e(r)$.

2.2. Ideal Targets and Non-linear Coupling

In the regime of an ideal target plasma (collisionless $\Gamma \ll 1$) and a non-linear coupling ($Z\,\Gamma^{3/2} \gtrsim 1$), the stopping power as well as the screening and the induced densities are functions only of the coupling parameter $Z\,\Gamma^{3/2}$ and the velocity v, but no analytical expressions are available. Only for the regime of the validity of an expansion in the coupling parameter $Z\,\Gamma^{3/2}$, i.e. ($Z\,\Gamma^{3/2} \lesssim 1$), some analytical work on the stopping power was done by Peter[18]. An important point of this analysis, however, is the fact, that for high ion velocities v the linear coupling condition is $Z\,\Gamma^{3/2}/v^3 \ll 1$ instead of $Z\,\Gamma^{3/2} \ll 1$ for low velocities. In the full non-linear coupling regime the required solutions of the corresponding non-linear Vlasov–Poisson equations have to be determined numerically. In this direction there is some recent, first work of direct numerical solution[19][20] or particle-in-cell (PIC) simulations[21][20]. To compare our MD-results, which account for electron–electron collisions and all kinds of multi-particle collisions, to the limit of an ideal target where these collisions are absent, we have also developed a PIC-code (see section 2.3).

In contrast to the linear behavior the screening-length for non-linear coupling depends on the coupling parameter $Z\,\Gamma^{3/2}$. A numerical solution of the corresponding Poisson-equation shows that the static screening-length (due to only free electrons) increases with increasing coupling, i.e. with the ion charge Z when the plasma parameter Γ is fixed.

2.3. Non–ideal Targets and Non–linear Coupling

For non–ideal plasmas $\Gamma \gtrsim 0.1$ and non–linear coupling $Z\,\Gamma^{3/2} \gtrsim 1$ a full kinetic description including collisions and higher correlations is required. In this regime the stopping power as well as the screening–length depend on the plasma parameter Γ and the ion charge Z separately. To investigate the interaction of ions in plasmas for such strongly coupled situations we perform MD computer simulations. In the MD–simulations the equations of motion for point particles interacting by the Coulomb force are integrated. Each particle interacts with all other particles and the numerical effort scales with the square of the particle number. An infinite plasma is approximated by an elementary cubic simulation volume which contains typically 500 particles and which is continued periodically in all directions. Despite the relative small number of particles this approximation turns out to be less severe for strong coupling, where the Debye length is small compared to the size of the simulation volume. In order to separate effects due to non–linearity from those due to non–ideality we perform in addition to this full MD simulations particle-in-cell (PIC) simulations with test–particles. There each real electron is split off in many (typically 350) test–particles with correspondingly smaller mass and charge. The simulation cube with its periodic continuations remains unchanged. While the collisions of the test–particles with the ion are treated exactly, the interaction between the test–particles is now due to a mean–field, where the source of this mean–field is given through the smoothed charge density of the test–particles themselves. The calculation of the mean–field is provided by solving the Poisson–equation on a 3-D-grid. The numerical effort is here linear in the test–particle number. The mesh–size of this grid now determines a smallest length scale for phenomena, which can be described by such a procedure. Thus all hard collisions with a closest approaches of the particles of order or below this scale cannot be accounted for. Hence one deals with an (almost) ideal, collisionless plasma and describes only (collective) phenomena with wave–lengths larger than the mesh–size. In the PIC–simulations, the electrons represented by the test–particles interact like extended charge distributions instead of point–particles as in the MD–simulations, where of course now smallest length exists.

2.4. The Situation in the Electron Cooler

In the electron cooler, the motion of the electrons in the transversal direction is hindered by the magnetic field. For this reason we assume that the interesting 'effective' temperature is close to the smaller longitudinal temperature. Taking the longitudinal temperature $T_{\|}$ and the further parameters at the TSR in Heidelberg[22] ($n_e \approx 3 \cdot 10^7 \mathrm{cm}^{-3}, k_B T_{\|} \approx 1 \ldots 5 \cdot 10^{-4} \mathrm{eV}, Z = 1 \ldots 16$) and the ESR in Darmstadt[24]

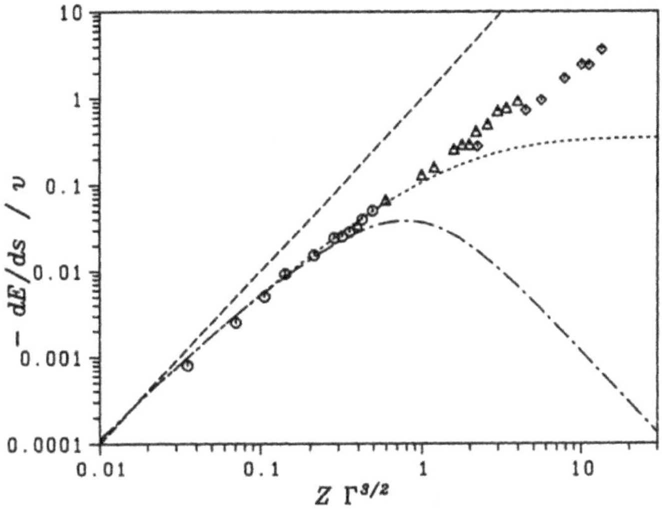

Figure 1: Friction coefficient, i.e. $(-dE/ds)/v$ at low ion velocities, in units of $3^{1/2}\Gamma^{-3/2}k_BT/\lambda_D v_{th}$ as a function of $Z\Gamma^{3/2}$. MD–results for different Z and $\Gamma = 0.11$ (circles), $\Gamma = 0.34$ (triangles) and $\Gamma = 1.08$ (diamonds) compared to the binary collision approximation (dotted curve, Eq.(3)) and the linear dielectric response treatment (dash-dotted curve, Eq.(4)). The dashed line is a pure $Z^2\Gamma^3$-law.

($n_e \approx 10^6\mathrm{cm}^{-3}, k_BT_{\|} \approx 5\cdot 10^{-5}\mathrm{eV}, Z = 6\ldots 92$) one obtains $\Gamma\lesssim 0.7$, $Z\,\Gamma^{3/2}\lesssim 9$ and $\Gamma\lesssim 0.4$, $Z\,\Gamma^{3/2}\lesssim 23$, respectively. This is clearly beyond the linear regime and near the transition from an ideal to a non-ideal plasma.

3. Stopping Power of Heavy Ions

3.1. Isotropic, Non-magnetized Electron Plasmas

From a large number of performed MD–Simulations, with various ion velocities v, plasma parameters Γ and ion charge states Z we obtained the stopping power of heavy ions as a function of the velocity, the strength of coupling and the ion charge. For high ion velocities ($v \gg 1$) the stopping power found in the simulations agrees for all investigated cases with the weak coupling result (Eq.1). This can be understood immediately regarding the much more relaxed condition for the linear regime $Z\Gamma^{3/2}/v^3 \ll 1$ at high velocities (see section 2.2). Concentrating on the non-ideal strong coupling–low velocity regime, where the stopping power dE/ds and hence the cooling force is linear in the velocity, we observe deviations from the weak coupling

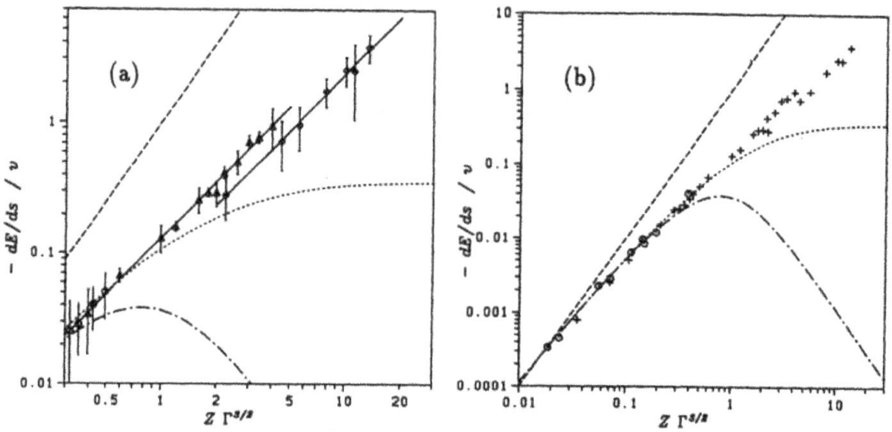

Figure 2: Friction coefficient scaled as in Fig.1. (a) Same as Fig.1, but now only for $Z\Gamma^{3/2} > 0.3$ and the MD–results with error–bars. The solid lines are fits to the MD–data going with $Z^{1.43}$ (for fixed Γ). (b) Here, in contrast to Fig.1, the MD–results (crosses) are not further distinguished with respect to Γ. The circles are the experimental results from A.Wolf[22] assuming an effective temperature of $T = 3.1T_{||}$ (see text). All other curves are the same as in Fig.1.

behavior which increase with the coupling strength. In Fig. 1 we show the friction coefficients R_1, i.e. the slope of the stopping power with respect to the velocity at low velocities ($dE/ds/v$ see Eq.(2)), as a function of the coupling parameter $Z\Gamma^{3/2}$ determined from our MD simulations in comparison to the conventional theories (R_1 from Eqs.(3)(4)). For reference purposes the extreme weak-coupling $Z^2\Gamma^3$-behavior of the friction coefficient has been drawn as dashed line. The double-logarithmic scale should be noted. In Fig. 2a the region of large coupling parameters is plotted in more detail. A fit (solid lines) to the MD results yields the same scaling of $R_1 \propto Z^{1.43}$ for both plasma parameters $\Gamma = 0.34$ and 1.08. The friction coefficient is in this regime a function depending separately on Z and Γ. Applying the PIC treatment to this regime one finds in contrast a pure $Z\Gamma^{3/2}$ dependence and the PIC–results differ at strong coupling by factors of 1–5 from the MD–results (see [9]). In Fig. 2b we show a comparison of our MD–results (crosses) with the experimental results from A.Wolf et al.[22] (circles), where we adapted their data for the cooling forces at low velocities measured in the TSR electron cooler to our situation of an isotropic electron plasma by assuming an effective Temperature of $T = 3.1T_{||}$, where $T_{||}$ is the longitudinal temperature present in the experiments. In Fig. 3 the stopping power is given as a function of the velocity v for $\Gamma = 1.08$, $Z = 10$. Both the MD-simulations (diamonds) and the PIC-treatment (circles) show that the stopping power remains proportional

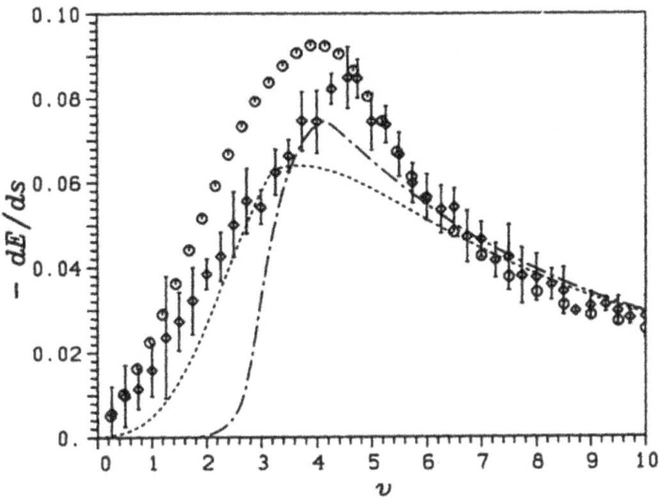

Figure 3: Ions with $Z = 10$ in an electron plasma with $\Gamma = 1.08$. Stopping power dE/ds in units of $3^{1/2}\Gamma^{-3/2}k_BT/\lambda_D$ as a function of v in $(k_BT/m)^{1/2}$. The diamonds with error-bars (representing runs with different initial configurations) are MD results, the circles are PIC results. The binary collision approximation yields the dotted curve, the dielectric linear response treatment the dash-dotted curves.

to the velocity for $v \lesssim 3$. The slopes $-dE/ds/v$ (friction coefficients) are marginally smaller in the MD-simulation than in the PIC-treatment which shows less fluctuations because of the large number of test particles.

While there is good agreement with the simulation results for weak coupling (see e.g. small $Z\Gamma^{3/2}$ in Fig. 1) and high velocities (see Fig. 3), we always find that the stopping power is underestimated by the conventional treatments in the non-ideal strong coupling-low velocity regime (see curves in Fig. 1 and Fig. 3). There the used cut-off procedures become unrealistic by excluding too many collisions. For $\Gamma \gtrsim 0.1$ the continuum treatment and for $Z\Gamma^{3/2} \gtrsim 1$ the linearization of the dielectric response treatment become doubtful and it becomes also questionable to use the screening length from the linearized Debye theory as upper cut-off in the binary collision approximation. The binary collision result in Fig. 1 (Eq.(3)) which saturates at large coupling due to the dominance of close collisions with maximal kinematical momentum transfer could be improved upon by employing a larger effective screening length in the cut-off.

The friction coefficient (Fig. 1) obviously increases slower than Z^2 for strong coupling. In agreement with the experiments by Steck et al.[23], Winkler et al.[24]

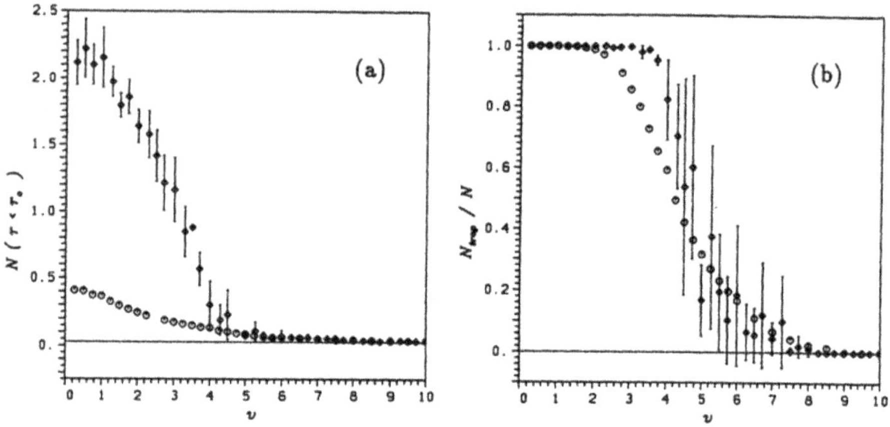

Figure 4: Ions with $Z = 10$ in an electron plasma with $\Gamma = 1.08$. The diamonds with error-bars are MD results, the circles are PIC results. (a) Number of electrons N within a sphere of radius $r_c = 0.3a$ ($a = (4\pi n/3)^{-1/3}$) around an ion as function of the ion velocity v in v_{th}. The dotted line indicates the ideal gas value $N_{id} = (0.3)^3$. (b) Fraction of trapped electrons N_{trap}/N in this sphere as a function of the ion velocity v in v_{th}.

and Wolf et al.[22] and other PIC simulations by Hofmann[21] we find an approximate $Z^{1.5}$ dependence. This change in the Z–scaling is accompanied by an enhanced local density of electrons around a highly charged, slow ion. In Fig. 4a we show the mean number of electrons around an ion within a sphere of radius $r_c = 0.3a$ ($a = (4\pi n/3)^{-1/3}$) and in Fig. 4b the fraction N_{trap}/N of bound electrons in this sphere, i.e. electrons with negative energy concerning the motion relative to the ion, as a function of the velocity v for $\Gamma = 1.08$, $Z = 10$. The number of electrons in a sphere is used instead of the density for reasons of better statistics. The MD–results show an enhancement of the number of electrons at low velocities – for $v < 1$ this corresponds to a factor of about 80 compared to the ideal gas value – and a large fraction of these electrons are trapped in the ion–potential. The enhancement vanishes for large ion velocities, i.e. large relative energies.

The density or the number of electrons around an ion are physical quantities which are sensitive to the non–ideality of the plasma, i.e. to the collisions between the electrons and further typical many–body effects, in particular three-body and multi-particle collisions involving the ion and two or more electrons. While these collisions are correctly treated in the MD simulations they are suppressed in the PIC treatment which is essentially mean field, as it is clearly visible in Fig. 4a. Thus a study of the electron density around an ion always requires a full MD–simulation. The enhance-

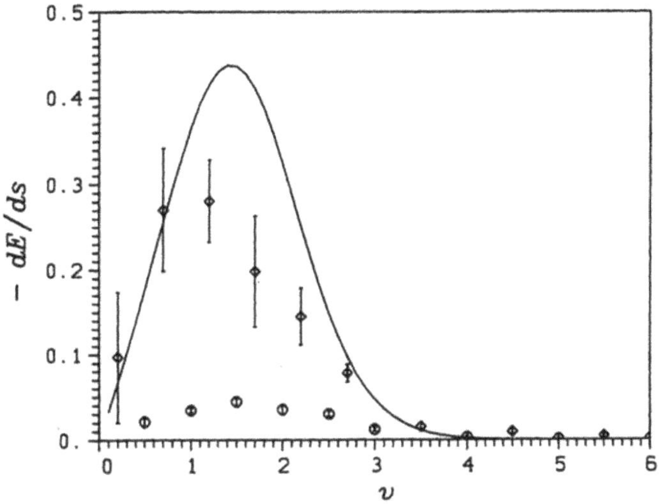

Figure 5: Infinitely strong magnetized electron plasma with $\Gamma = 0.11$. Stopping power dE/ds in $3^{1/2}\Gamma^{-3/2}k_BT/\lambda_D$ as a function of v in $(k_BT/m)^{1/2}$ for an ion moving in the direction of the magnetic field. MD–results $Z = 8$ (circles) and -8 (diamonds) together with the result from the dielectric response formalism (solid curve).

ments of the local electron density around an ion are studied and described in more detail in section 3.

3.2. Infinite magnetic field strength

The usual theoretical investigations on the influence of a magnetic field on the cooling force are based on an extension of the weak coupling approaches for the stopping power described above to include the magnetic field. It typically requires very sophisticated cut–off procedures (e.g. [2][3][20]). This becomes very doubtful for increasing coupling and a satisfactory answer on the effects of the magnetic field for the stopping power is still under discussion. As a contribution to this discussion we consider here the limiting case of an infinite strong magnetic field, where the motion of the electrons is assumed to become one–dimensional along the direction of the magnetic field. The stopping power on an ion moving in this direction observed in MD–simulations[25] is shown in Fig. 5 for charge states $Z = 8$ (circles) and -8 (diamonds) and a plasma parameter $\Gamma = 0.11$ ($Z\Gamma^{3/2} = 0.31$) corresponding to the Heidelberg TSR conditions $n = 3 \cdot 10^7 \mathrm{cm}^{-3}$ and a longitudinal temperature of 5.8 K. The solid curve represents the linear response treatment, using a dielectric function

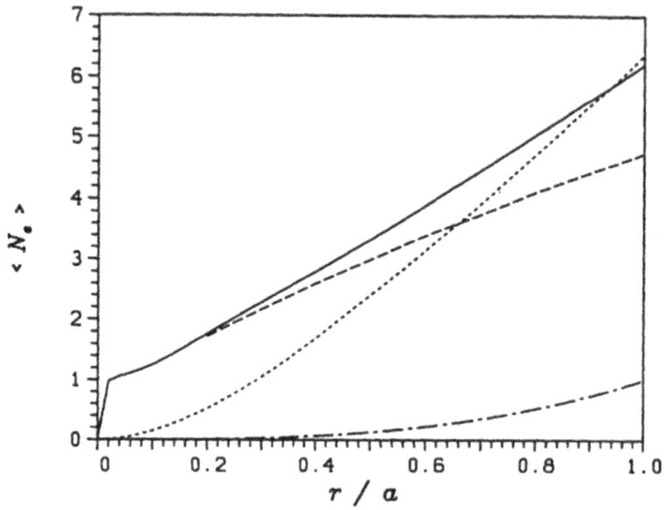

Figure 6: Time averaged number of electrons N_e in a sphere of radius r around an ion of charge $Z = 10$ for $\Gamma = 1.08$ in the stationary state ($t > 3.5\tau_p$) as a function of r in $a = (4\pi n/3)^{-1/3}$. MD-results for all (full curve) and the trapped (dashed) electrons together with the weak coupling prediction (dotted) Eq.(6) and the ideal gas value $N_e(r) = (r/a)^3$ (dash-dotted).

derived for this case of an infinite magnetic field (1-dim electron movement)[17][2] and the same cut-off k_m as for the isotropic case (see section 2.1). The large difference between the stopping power for positive and (with $Z = -8$ hypothetical) negative ions clearly indicates that a mean-field description (which is independent on the sign of the charge) cannot succeed under these circumstances. Furthermore, the stopping power for positive ions is much more smaller in the MD-simulations than predicted by the linear response treatment. Quantitative very similar results were also obtain for this infinitely strong magnetized case for a little bit different coupling parameters by Boine and D'Avanzo[20], using both a direct numerical solution of the Vlasov-Poisson equations and PIC-simulations. These authors show further, that the stopping power remains considerably below the values predicted by the dielectric treatment even for a much more weaker coupling ($Z\Gamma^{3/2} = 4.6 \cdot 10^{-3}$). There, an agreement can however be achieved by fitting the cut-off k_m. Of course, this just demonstrates the difficulties and problems connected with the cut-off procedures in the conventional descriptions. Taking all these results together and applying it to the situation in the electron cooler, we find that a very strong magnetic field (Larmor radius $r_L \ll a$) will significantly reduce the stopping power (for all velocities) compared to the case of an anisotropic,

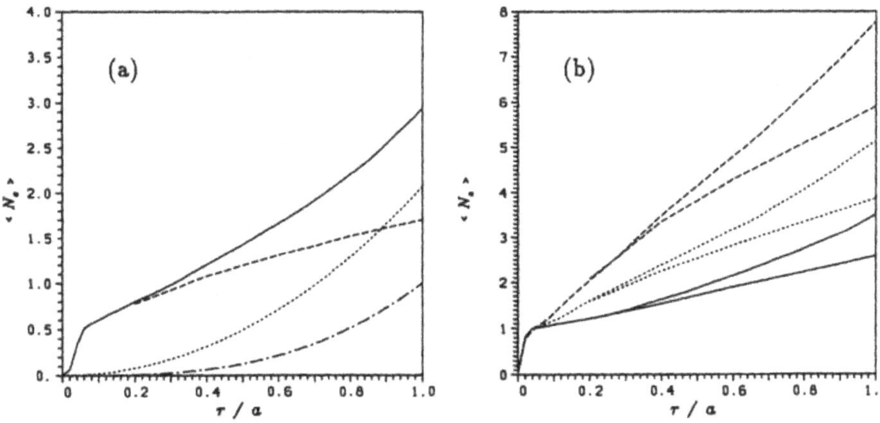

Figure 7: Time averaged number of electrons N_e around an ion in the stationary state (here $t > 4\tau_p$) as a function of r in $a = (4\pi n/3)^{-1/3}$. (a) MD–results, weak coupling prediction and ideal gas value as in Fig. 6 but now for $Z = 4$ and $\Gamma = 0.34$. (b) MD–results for $\Gamma = 1.08$ and ion charge $Z = 4$ (solid curves), 8 (dotted) and 16 (dashed) for the total number of electrons (upper curve of each pair) and the number of trapped electrons (lower curve of each pair).

non–magnetized electron plasma (see [25]). This yields however no statement for the case of moderate magnetic fields $r_L \approx a$.

4. Electron Densities around an Ion

To investigate the enhancement of the local electron density around an ion at very low relative energies mentioned above we performed MD–simulations where the ion is at rest in the electron plasma ($v = 0$). In these simulations the Coulomb interaction between an ion of charge Z and an electron plasma with Γ is switched on at time $t = 0$, which can be considered as the time when the ion beam gets in contact with the cold electron beam at the entrance of the electron cooler. After switching on the interaction the ion attracts electrons and the density of electrons around the ion grows and reaches on a time scale given by the plasma period τ_p ($= 2 \cdot 10^{-8}$s for an electron density $n = 3 \cdot 10^7 \text{cm}^{-3}$) an almost stationary state. In the case of strong coupling a lot of electrons are trapped in the field of the ion by multi–particle collisions, i.e. the energy of these trapped electrons, consisting of their kinetic energy and the potential energy in the field of the ion and the other electrons, is negative.

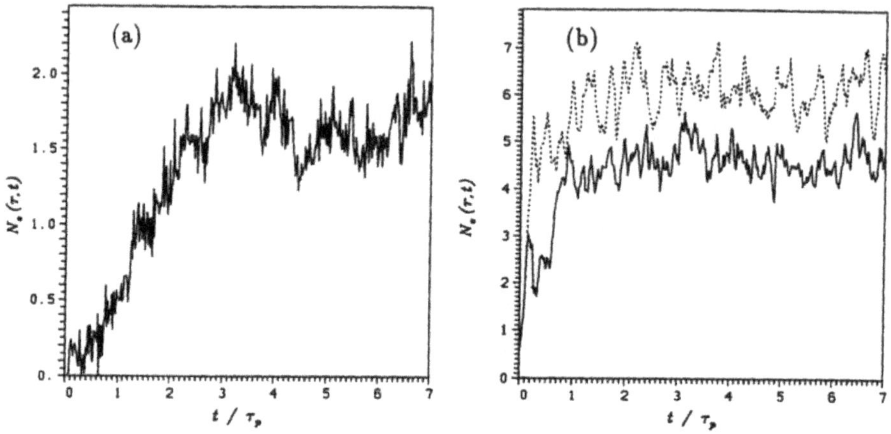

Figure 8: Time dependence of the number of electrons $N_e(r,t)$ around an ion of charge $Z = 10$ and for $\Gamma = 1.08$. (a) For a radius $r/a = 0.2$; all electrons are trapped ones. (b) For a radius $r/a = 1.0$; total electron number (dotted curve) and the number of trapped electrons (solid).

4.1. Isotropic, Non-magnetized Electron Plasmas

Typical results of these simulations for the case of an isotropic electron plasma are presented in Figs.6–9. In Fig.6 we show the MD-results for the stationary mean values (in this case for $t > 3.5\tau_p$) of the number of electrons $N_e(r)$ in a sphere of radius r around an ion with $Z = 10$ for $\Gamma = 1.08$ ($Z\Gamma^{3/2} = 11.2$) together with the number of trapped electrons. For comparison the prediction for the weak coupling regime, i.e. the density $n_e(r)$ of Eq.6 integrated over a sphere of radius r, and the ideal gas limit $N_e(r) = (r/a)^3$, $a = (4\pi n/3)^{-1/3}$, are added. In Fig.7a the same is given for $Z = 4$ and $\Gamma = 0.34$ ($Z\Gamma^{3/2} = 0.8$), whereas Fig.7b presents a comparison of the total electron number and the number of trapped electrons for different $Z = 4, 8, 16$ and $\Gamma = 1.08$. At strong coupling, the observed number of electrons and thus the local electron density is strongly enhanced within small spheres compared to the weak coupling (Debye–Hückel) value, which itself is larger than the ideal gas value. For $r < 0.1 - 0.3$ the number of electrons found in the MD-simulations can become $10^2 - 10^3$ times larger than the ideal gas value. This remains also valid for more moderate coupling (see Fig. 7a compared to Fig. 6). The electrons very close to the ion are all trapped in the ion field. For a density of $n = 3 \cdot 10^7 \mathrm{cm}^{-3}$ and $Z = 10$ the trapped electrons in the case of Fig.6 have energies $E = -12.5\mathrm{meV} \ldots - 0.25\mathrm{meV}$ for the plotted distances $r/a = 0.02 \ldots 1.0$ corresponding to Rydberg–states of principal quantum numbers $\approx 330 \ldots 2350$. From the charge–dependence of the number of

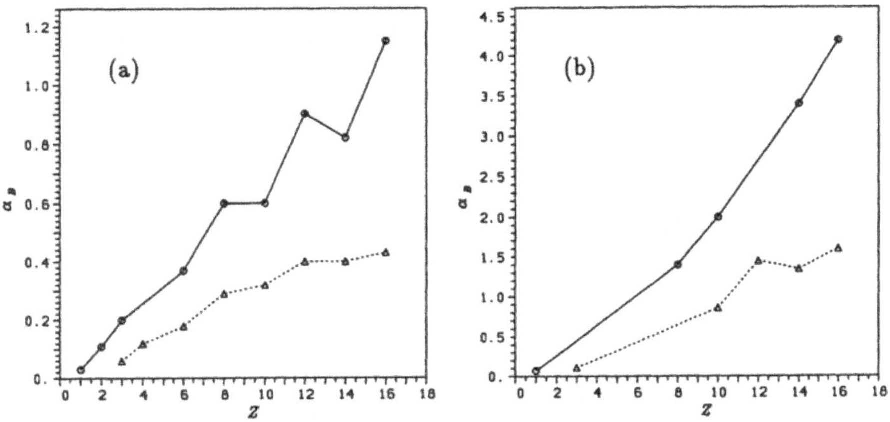

Figure 9: Growth rates α_B for the number of trapped electrons around an ion for $\Gamma = 0.34$ (dotted curves and triangles) and 1.08 (solid and circles) in units of inverse plasma periods $1/\tau_p = \omega_p/2\pi = (e^2n/m)^{1/2}/2\pi$ as a function of the ion charge state Z. (a) For a sphere with $r/a = 0.2$. (b) For a sphere with $r/a = 0.6$.

electrons in Fig. 7b it is obvious that the induced electron density and number of electrons is *not* linear in Z in the strong coupling regime in contrast to the weak coupling behavior Eq.(6). Thus the electron cloud around the ion becomes larger with increasing Z causing a growing enlargement of the screening length above λ_D (see [8]).

So far we have only considered the enhanced density in the stationary state. In Fig.8 the time evolution of the number of electrons $N_e(r,t)$ is shown for $Z = 10$, $\Gamma = 1.08$ and spheres with $r/a = 0.2$ and 1.0. The increase to its (quasi-) stationary value is faster for a larger radius r than for a smaller one, since the trapped electrons approach the ion by transferring energy to the surrounding electrons. Here, the transient behavior of N_e takes, depending on r, typically $1-3$ plasma periods. Regarding the time evolution of the electron number roughly as a linear increase to the stationary mean value and as a constant thereafter, one can simply define a corresponding growth rate $dN_e(r,t)/dt$ as the slope in the first part of the time evolution. These growth rates $\alpha_B(r)$ (here for the number of trapped electrons around an ion $dN_{trap}(r,t)/dt$) extracted from the MD–simulations are shown in Fig. 9 as a function of the ion charge state Z and for spheres with $r/a = 0.2$ and 0.6 and for $\Gamma = 0.34$ and 1.08. These growth rates are more or less linear in the ion charge Z. This clearly contradicts the Z^3–charge dependence of the usual three–body recombination and indicates the very different behavior in a strongly coupled Coulomb system compared to the weak coupling features. As mentioned above, the multi–particle collisions are

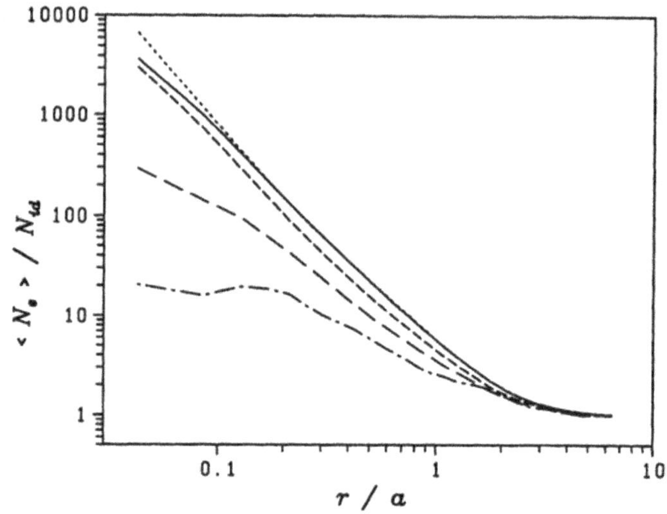

Figure 10: An ion of charge $Z = 10$ in an isotropic electron plasma with $\Gamma = 0.89$. MD–results for the time averaged ($t = 0 \ldots 3\tau_p$) number of electrons N_e around the ion divided by the ideal gas value N_{id} as a function of r in $a = (4\pi n/3)^{-1/3}$ and for different strength of an applied magnetic field with $r_L = mv_{th}/eB$ and $r_L/a = \infty$ (solid curve), 10 (dotted), 1 (short–dashed) 0.3 (long–dashed) and 0.1 (dash-dotted).

responsible for the trapping in the high Rydberg–states. Such collisions are of course extremely unlikely in the weak coupling regime.

4.2. Isotropic, Magnetized Electron Plasmas

To study the effects of a magnetic field in an isotropic electron plasma, we extended our simulations by including a homogeneous magnetic field. The magnetic field strength B is measured by the ratio r_L/a, where r_L is the Larmor radius defined as $r_L = mv_{th}/eB$ (for the conditions in the Heidelberg TSR $r_L/a \approx 1$). The MD–results for the number of electrons as a function of the sphere radius r around an ion of charge $Z = 10$ in an isotropic plasma of $\Gamma = 0.89$ are divided by the corresponding ideal gas value and presented in Fig. 10 for different magnetic field strengths. In this strongly coupled situation, a reduction of the enhancement of the electron number/density for close distances is found when the magnetic field increases. Here, the values shown for the electron numbers are time averages over $\approx 3\tau_p$ after switching on the interaction between ion and electrons. This time corresponds to the time of flight of the ions through the cooler section at the TSR.

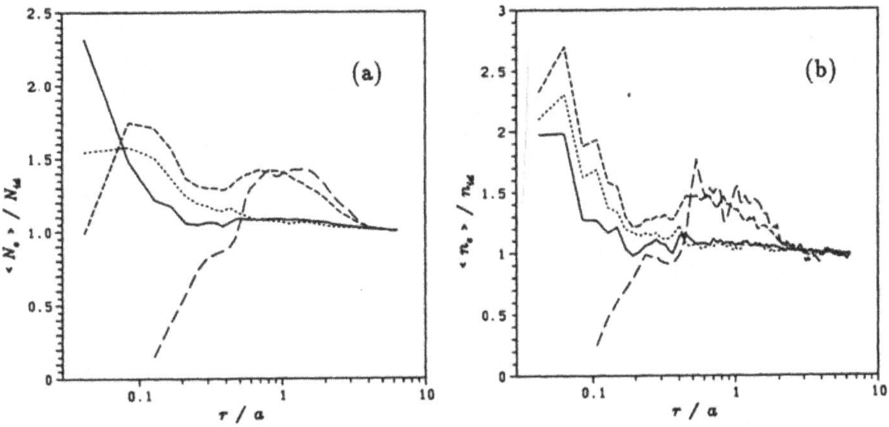

Figure 11: An ion of charge $Z = 10$ in an anisotropic electron plasma with temperatures of 1000 K and 5 K in the transversal und longitudinal direction, respectively. MD-results for (a) the time averaged ($t = 0 \ldots 3\tau_p$) number of electrons N_e in a sphere of radius r around the ion and (b) the time averaged density at a distance r, both quantities divided by the corresponding ideal gas value. Here r is in units of $a = (4\pi n/3)^{-1/3}$ and the strength of the magnetic field is $r_L/a = mv_{th}/eBa = \infty$ (solid curve), 10 (dotted), 1 (short-dashed) and 0.1 (long-dashed). v_{th} is defined through the transversal temperature of 1000 K.

4.3. Anisotropic, Magnetized Electron Plasmas

When, in addition, an anisotropic electron distribution is used, the situation changes significantly. We now regard an ion of $Z = 10$ in an electron plasma of $n = 3 \cdot 10^7 \mathrm{cm}^{-3}$ with a transversal temperature of 1000 K and a longitudinal one of 5 K, corresponding to the situation in the Heidelberg TSR. Now, the Larmor radius r_L is defined by taking the thermal velocity corresponding to the transversal temperature. In Fig. 11 the time averaged number and density of electrons around the ion relative to their ideal gas values are shown. Without magnetic field (solid curves) there is only for very close distances (note the logarithmic scaling of r) a rather moderate enhancement of $2-3$. Since the effective temperature in this situation is close to the very large transversal temperature, one has in fact a very weakly coupled plasma with $\Gamma \approx 10^{-3}$. For increasing magnetic field an additional moderate enhancement of the number of electrons of $1.5 - 2$ develops at larger distances, while for strong magnetic fields a strong reduction of the electron number takes place at small distance accompanied by an enhancement for larger r. A similar behavior is observed for the densities, which in addition show also an enhancement for close distances and moderate magnetic fields. The growing hindrance of the transversal motion of the electrons with increasing magnetic field yields obviously not only an effective temperature approaching more and more the cold longitudinal one, but also suppresses the multi-particle collisions responsible for the enhancement of the electron numbers and densities. From these preliminary results, it seems that both these competitive processes yield at small distances and for medium magnetic field strengths some maximum of the enhancement . When comparing the above data with the results shown for the isotropic, non-magnetized plasma one should take into account that we have not necessarily reached here the stationary state of the time evolution of the electron number/density. So far we have not considered this aspect in detail. Although the enhancements for the above parameters are rather moderate they could explain, at least in part, the experimentally observed enhanced recombination rates. Investigations on this will be the subject of further work.

5. Summary

5.1. New Features at Strong Coupling

The new features in the behavior of ions in electron plasmas at strong coupling discussed in detail above are : (1) For strong ion-target coupling the energy loss of the ions at low velocities scales with Z approximately as $Z^{1.5}$ in agreement with recent experimental and other theoretical results, but in disagreement with the $Z^2 \ln(\text{const.}/Z)$ scaling of the conventional weak coupling theories. (2) This is accompanied by the appearance of a cloud of electrons trapped by the ion into high Rydberg states through multi-particle collisions. This trapping causes an enhanced electron density around the ion . (3) These electron–ion recombinations show growth rates almost linear in the ion charge Z. (4) The size of the electron clouds increases with Z and Γ and decreases with v. This results in a different screening, with a (static) screening length above the Debye-Hückel result of the linear theory.

5.2. Influence of the Magnetic Field

One effect of the presence of a magnetic field is clearly the hindrance of the motion perpendicular to the field resulting in a reduction of energy– and momentum transfer between the charged particles. Thus in a strongly magnetized plasma the stopping of ions as well as the multi–particle collisions responsible for the enhancement of the local electron densities around an ion become less efficient. For moderate magnetic field strengths and an anisotropic plasma this may, up to a certain extend, be balanced by an decrease of the effective temperature and thus an increase in the coupling strength.

5.3. Remarks

Future work on this subject will improve the simulation result for the situation of an anisotropic electron distribution together with an external magnetic field, addressing also the question of an anisotropy of the local electron densities (up to now we always regarded averages over spheres). A further point is a necessary detailed investigation of the dependence of radiative recombination rates on an energy density of electrons, which is space dependent in the surrounding of the ion. Also influences of further electrical fields on the trapping of electrons should be considered. Such fields are created in the cooling section by the guiding magnetic field, which is not perfect homogeneous, as well as at the entrance (and the end) of the cooler, when the ion beam and the electron beam are merged (separated).

6. Acknowledgements

This work was supported by the Bundesministerium für Bildung und Forschung (BMBF), the Gesellschaft für Schwerionenforschung Darmstadt (GSI), the European Community and the Deutsche Forschungsgemeinschaft (DFG).

7. References

1. H. Poth, Electron cooling: theory. experiment, application, Phys.Rep., 196 (1994) 135.
2. A.H. Sørensen and E. Bonderup, Electron cooling, Nucl.Instr.Meth., 215 (1983) 27.
3. I.N. Meshkov, Electron cooling: Status and perspectives, Phys.Part.Nucl. 25 (1994) 631.
4. U. Schramm et al., Laser-induced recombination studies with the adiabatically expanded electron beam of the Heidelberg TSR, in: Proceedings of the 1st Euroconference on Atomic Physics with Stored Highly Charged Ions, Heidelberg, 1995, to be published in Hyperfine Interactions.
5. G. Zwicknagel, C. Toepffer and P.-G. Reinhard, Stopping power of heavy-ion beams in strongly coupled plasmas, Il Nuovo Cimento, 106A (1993) 1857.
6. G. Zwicknagel, C. Toepffer and P.-G. Reinhard, Molecular dynamics simulation of electron cooling, in *Beam Cooling and Related Topics*, J.Bosser ed., CERN, Genf, 1994, pp. 245–248.
7. G. Zwicknagel, C. Toepffer and P.-G. Reinhard, Stopping power of heavy ions in strongly coupled plasmas, Laser and Particle Beams, 13 (1995) 311.
8. G.Zwicknagel, C.Toepffer and P.-G.Reinhard, Molecular dynamic simulation of ions in electron plasmas at strong coupling, in: Proceedings of the 1st Euroconference on Atomic Physics with Stored Highly Charged Ions, Heidelberg, 1995, to be published in Hyperfine Interactions.

9. G.Zwicknagel, C.Toepffer and P.-G.Reinhard, Energy loss of heavy ions in strongly coupled plasma, in: Proceedings of the International Symposium on Heavy Ion Fusion, Princeton, 1995, to be published in Fusion, Engineering and Design.
10. G. Zwicknagel, *Energieverlust schwerer Ionen in stark gekoppelten Plasmen*, Ph.D. Thesis, Universität Erlangen, 1994.
11. O. Uwira et al., Recombination of electrons with U^{28+} ions in a dense, cold electron target, GSI scientific report 1994, GSI Darmstadt, Darmstadt, 1995, p. 137
12. S.Baird et al, Measurement of the lifetime of Pb^{52+}, Pb^{53+} and Pb^{54+} beams at 4.2 MeV per nucleon subject to electron cooling, to be published in Physics Letters B
13. H. Gao et al., Observation of enhanced electron–ion recombination at very low energies, Phys.Rev.Lett 75 (1995) 4381.
14. Y.S. Derbenev and A.N. Skrinsky, The kinetics of electron cooling of beams in heavy particle storage rings, Part.Acc., 8 (1977) 1.
15. P. Sigmund, Kinetic theory of particle stopping in a medium with internal motion, Phys.Rev. A, 26 (1982) 2497.
16. L. De Ferrariis and N.R. Arista, Classical and quantum–mechanical treatments of the energy loss of charged particles in dilute plasmas, Phys. Rev. A, 29 (1984) 2145.
17. S. Ichimaru, *Basic Principles of Plasma Physics*, Benjamin, Reading, Massachusetts, 1973.
18. Th. Peter and J. Meyer-ter-Vehn, Energy loss of heavy ions in dense plasma, Phys.Rev. A, 43 (1991) 1998.
19. O. Boine–Frankenheim, Nonlinear stopping power of ions in plasmas, to be published in Laser and Particle Beams.
20. O. Boine–Frankenheim and J. D'Avanzo, Stopping power of heavy ions in a strongly magnetized Plasma, to be published in Physics of Plasmas.
21. I. Hofmann, What is electron cooling for high Z ?, in *Beam Cooling and Related Topics*, J.Bosser ed., CERN, Genf, 1994, pp. 330–335.
22. A. Wolf et al., Charge dependence of the electron cooling force for heavy ions, in *Beam Cooling and Related Topics*, J.Bosser ed., CERN, Genf, 1994, pp. 416–421.
23. M. Steck et al., Electron cooling of highly charged ions at the ESR, in *Beam Cooling and Related Topics*, J.Bosser ed., CERN, Genf, 1994, pp. 395–399.
24. Th. Winkler et al., Charge dependence of the electron cooling force, GSI scientific report 1994, GSI Darmstadt, Darmstadt, 1995, p. 245, and Electron cooling force measurements for highly charged ions at the ESR, in: Proceedings of the 1st Euroconference on Atomic Physics with Stored Highly Charged Ions, Heidelberg, 1995, to be published in Hyperfine Interactions
25. M. Miller, *Einfluß eines Magnetfeldes auf die Bremskraft schwerer Ionen*, Diploma Thesis, Universität Erlangen, 1994.

Electron Cooler Impedances

A. V. Burov*
Laboratori Nazionali di Legnaro, I-35020, Legnaro, Italy

November 14, 1995

Abstract

An electron beam of a cooler in an ion storage ring can be considered as a medium which responds to fields generated by an ion beam. Electron density perturbations awaken by the ion beam act back on it, which can give rise to instabilities. This reaction can be described in terms of an impedance introduced in the ring by the electron beam. Longitudinal and transverse impedances of an electron cooler are derived here, increments and thresholds of corresponding instabilities are estimated.

1 Introduction

The interaction of an ion beam with different elements of the vacuum chamber can give rise to coherent instabilities of the beam. If the beam temperature is sufficiently high, its collective modes are stabilized due to the Landau damping. Under cooling, the temperature is going down and the Landau damping is switching off; but the cooling brings about its own decrement in the collective motion of the beam. However, if the decrement of some mode is smaller than its increment caused by the interaction with the environment, the beam is either stopped at the instability threshold or lost. So, these increments determine a minimum cooling rate needed to achieve a beam temperature below the Landau damping threshold.

The beam - environment interaction is usually described in terms of a vacuum chamber impedance [1]. The impedance concept is developed mainly for an ultra-relativistic case, where a difference between the beam velocity and the velocity of light is insignificant [2]. This condition is not satisfied at existing ion storage rings equipped with coolers. In a relativistic case, a significant contribution in the beam - surrounding interaction is given by a wall-structure impedance. Its broad-band component is exponentially damped , as $\sim \exp(-4.8/(\beta\gamma))$ for $\gamma - 1 \leq 1$, a narrow-band component at the eigenfrequency ω_r is damped as $\sim \exp\left(-2\omega_r b/(\beta\gamma c)\right)$, where b is an aperture radius [3]. The influence of the wall resistivity is too small to play any role in the cooling process. Therefore, the electron cooler itself can be the main reason of instabilities; this problem was discussed in Ref.[6, 7].

The longitudinal and transverse impedances of the cooler are calculated here and shown to be normally orders of magnitude higher then the resistive wall ones.

*On leave from the Budker Institute of Nuclear Physics, 630090, Novosibirsk, Russia

That is why increment rates of the corresponding instabilities can be higher than typical cooling rates, even for rather low ion currents (e. g. ~ few μA).

2 Main Equations

The dynamics of the magnetized electron medium of a cooler excited by fluctuations of the cooled ion beam can be described in a reference frame by the following set of equations [9]:

$$\begin{aligned} \frac{\partial \tilde{n}_e}{\partial t} + n_e \frac{\partial \tilde{v}_e}{\partial z} &= 0 \\ \frac{\partial \tilde{v}_e}{\partial t} - \frac{e}{m}\frac{\partial \Phi}{\partial z} &= 0 \\ \frac{1}{r}\frac{\partial}{\partial r}\left(r\frac{\partial \Phi}{\partial r}\right) + \frac{\partial^2 \Phi}{\partial z^2} + \frac{1}{r^2}\frac{\partial^2 \Phi}{\partial \phi^2} &= 4\pi e(\tilde{n}_e - \tilde{n}_i) \end{aligned} \tag{1}$$

Here n_e is an electron density, $\tilde{n}_e, \tilde{n}_i$ are electron and ion density perturbations, $\tilde{v}_e$ is a perturbation of electron velocity, Φ is an electrostatic potential. In the impedance calculations ions are assumed to be protons, the impedances are linear responce functions and do not depend on the charge of exciting particles. Unperturbed velocities of the beams are supposed to be equal, the solenoidal magnetic field assumed to be directed along the longitudinal axis z everywhere. The former assumption is justified if the velocities coincide with a sufficiently good accuracy, the criterium is discussed in the next chapter. The later assumption is only warranted for long-wave perturbations,

$$ka \ll 1, \tag{2}$$

where k is a longitudinal wavenumber, a is a radius of the electron beam. As it is demonstrated in the following, the cooler impedances reach their maxima somewhere in the intermediate region, $ka \simeq 1$. In this wave band the results obtained from Eqs. (1) can serve only as estimations with accuracy about $\simeq$ 100%. More accurate results in the intermediate band could be found with the electron beam curvature in the ions entrance to be taken into account, this rather complicated problem is not considered here.

Thermal velocities of electrons and ions $\Delta v_{e,i}$ are neglected in this impedance calculation, which is justified if $\Delta v_{e,i}\tau < 1/k$, where $\tau = g/v$ is a time of the electron-ion interaction when the particles pass with a velocity v through the cooler with a length g. Taking into account (2) , the validity condition of this hydrodynamic approximation can be expressed as $\Delta v_{e,i}/v < a/g$, which normally is well satisfied.

The equations of motion (1) have to be supplemented by zero boundary conditions imposed on the electron perturbations at the entrance of the ions in the electron beam; in the reference frame:

$$\tilde{n}_e|_{z=-vt} = 0, \quad \tilde{v}_e|_{z=-vt} = 0 \tag{3}$$

A particular solution of Eqs. (1) is obvious:

$$\tilde{n}_e = \tilde{n}_i = \tilde{n}_{il} \cos(l\phi) \exp ikz, \quad \tilde{v}_e = 0, \quad \Phi = 0. \tag{4}$$

The solution of the problem with the boundary conditions Eq. (3) can be presented as the particular solution Eqs. (4) of the unhomogeneous set of equations (1) plus a general solution of the homogeneous set (Eqs. (1) with $\tilde{n}_i = 0$). The arbitrary constants of the general solution are found from the boundary conditions.

Looking for the general solution as $\propto \exp(iqz - i\omega t)\cos(l\phi)$, the result is found as:

$$\begin{aligned} \Phi(r) = & \; AJ_l(\kappa r/a), \quad r \leq a \\ \Phi(r) = & \; BK_l(qr) + CI_l(qr), \quad r \geq a \\ \tilde{v}_e = - & \; \frac{e}{m}\frac{q}{\omega}\Phi, \quad \tilde{n}_e = -\frac{\bar{n}_e e}{m}\frac{q^2}{\omega^2}\Phi \\ & \kappa = qa\omega_e/\omega, \end{aligned} \tag{5}$$

where $\omega_e = \sqrt{4\pi\bar{n}_e e^2/m}$ is the electron plasma frequency, J_l denotes the Bessel function, K_l, I_l are modified Bessel functions, $l \geq 0$. The constants B, C are connected with the constant A by means of conditions for the potential and its radial derivative to be continuous at $r = a$:

$$\Phi|_{r=a-0} = \Phi|_{r=a+0}, \quad \frac{\partial\Phi}{\partial r}|_{r=a-0} = \frac{\partial\Phi}{\partial r}|_{r=a+0} \tag{6}$$

The requirement for the potential to be zero on the vacuum chamber surface

$$\Phi(b) = 0 \tag{7}$$

gives a sequence of eigennumbers $\kappa = \kappa_{l\mu}$ with $\mu = 1, 2, 3, ...$ as a radial mode counter.

The eigenfrequencies are described by sound-like dispersion equations

$$\omega_{l\mu\pm} = \pm q u_{l\mu}, \quad u_{l\mu} = \omega_e a/\kappa_{l\mu} \tag{8}$$

A time dependence of the electron perturbation $\propto \exp(iqz - i\omega t)$ at the ions entrance, where $z = -vt$, has to be the same as a time dependence of the exciting perturbation of the ion beam, assumed to be $\propto \exp(ikz)|_{z=-vt} = \exp(-ikvt)$. This requirement determines the (Doppler) connection between the electron and ion longitudinal wave numbers:

$$q = q_{l\mu\pm} = k/(1 \pm \alpha_{l\mu}), \quad \alpha_{l\mu} = u_{l\mu}/v. \tag{9}$$

The solution of the problem is simplified in two limiting cases. The first corresponds to the vacuum chamber to be radially remote from the electron beam: $b \gg a, q^{-1}$; the second - almost to touch it: $b - a \ll b$. In the first case, when the vacuum chamber is wide, the eigennumbers $\kappa_{l\mu}$ are zeroes of the Bessel function J_{l-1}:

$$J_{l-1}(\kappa_{l\mu}) = 0, \quad qb \gg 1, \quad (l,\mu) \neq (0,1)$$
$$\kappa_{01} = \sqrt{2/\ln(1/(qa))}, \tag{10}$$

as it follows from the boundary conditions.

If the metal surface of the vacuum chamber adjoins the electron beam, the eigennumbers are zeroes of the Bessel function J_l:

$$J_l(\kappa_{l\mu}) = 0, \quad b - a \ll b. \tag{11}$$

Orthogonality relations follow

$$\int_0^a J_l(\kappa_{l\mu} r/a) J_l(\kappa_{l\nu} r/a) r dr = \delta_{\mu\nu} \frac{a^2}{2} F_{l\mu}^2,$$
$$F_{l\mu} = \begin{cases} J_l(\kappa_{l\mu}) & \text{if } qb \gg 1 \\ J_l'(\kappa_{l\mu}) & \text{if } b - a \ll b \end{cases} \tag{12}$$

Summing up, the solution of the problem Eqs. (1) can be presented in the following form:

$$\Phi = \sum_{\mu\pm} A_{l\mu\pm} J_l(\kappa_{l\mu} r/a) \cos(l\phi) \exp(iq_{l\mu\pm} z \mp i\omega_{l\mu} t)$$

$$\tilde{v}_e = -\frac{e}{m} \sum_{\mu\pm} \pm \frac{A_{l\mu\pm}}{u_{l\mu}} J_l(\kappa_{l\mu} r/a) \cos(l\phi) \exp(iq_{l\mu\pm} z \mp i\omega_{l\mu} t)$$

$$\tilde{n}_e = \tilde{n}_{il} \cos(l\phi) \exp(ikz) - \frac{\omega_e^2}{4\pi} \sum_{\mu\pm} \frac{A_{l\mu\pm}}{u_{l\mu}^2} J_l(\kappa_{l\mu} r/a) \cos(l\phi) \exp(iq_{l\mu\pm} z \mp i\omega_{l\mu} t) \tag{13}$$

As it follows from the boundary condition imposed on $\tilde{v}_e$ at the entrance Eqs. (3) the waves travelling along and against the beams motion (+ and −) have the same amplitudes:

$$A_{\mu+} = A_{\mu-}. \tag{14}$$

The values of the amplitudes have to be found from the zero boundary condition imposed on the density perturbation $\tilde{n}_e$.

3 Longitudinal Impedance

In this chapter longitudinal perturbations are considered, i. e. this is the case $l = 0$. Making use of the orthogonality relations Eq. (12) and assuming the radius of the ion beam much smaller then the radius of the electron beam, the amplitudes of excited modes can be found:

$$A_\mu = \frac{2\tilde{\rho}_{i0}}{\kappa_\mu^2 F_\mu^2}, \quad \tilde{\rho}_{i0} = 2\pi \int \tilde{n}_{i0}(r) r dr \tag{15}$$

To derive the back influence of the electron beam on ions, an average of the electric field over a time of the flight through the cooler τ

$$< E_z >= -\frac{1}{\tau}\int_0^\tau \frac{\partial\Phi}{\partial z}|_{z=0}\, dt \tag{16}$$

have to be calculated. The connection between this value and the exciting perturbation $\tilde{\rho_{i0}}$ can be described in terms of the longitudinal impedance $Z^{\parallel}$ [1]:

$$< E_z > \tau = -e\tilde{\rho_{i0}}Z^{\parallel} \tag{17}$$

An imaginary part of the impedance is responsible for a coherent frequency shift of the ions at the wavenumber k. This shift is small due to the small portion of the ring occupied by the cooler. A real part of the impedance is responsible for an instability , thus, it is of the main interest. Making use of the formulae above it can be presented as follows:

$$\begin{aligned}
\mathrm{Re}Z^{\parallel} = & \; 2\sum_\mu \frac{(1-\cos(\omega_{\mu+}\tau)) - (1-\cos(\omega_{\mu-}\tau))}{\kappa_\mu^2 F_\mu^2 u_\mu} = \\
= & -4\sum_\mu \frac{\sin\left(\frac{\omega_{\mu-}\tau+\omega_{\mu+}\tau}{2}\right)\sin\left(\frac{\omega_{\mu-}\tau-\omega_{\mu+}\tau}{2}\right)}{\kappa_\mu^2 F_\mu^2 u_\mu}
\end{aligned} \tag{18}$$

$$\omega_{\mu\pm} = kv\frac{\alpha_\mu}{1\pm\alpha_\mu}, \quad \alpha_\mu = \frac{u_\mu}{v}.$$

The contributions in the real part of the impedance (18) from the positive (+) and negative (−) waves can be seen to have opposite signs: an emission of the positive wave takes away a longitudinal momentum of ions, i. e. decelerates them; on the contrary, radiation of the negative wave with a negative momentum accelerates the ion beam.

The condition for the electron current to be small in the comparison with its limit value [9] can be expressed as

$$\alpha_\mu \ll 1.$$

Under this condition phase advances of the positive and negative waves $\omega_{\mu\pm}\tau$ can be rather close even for $\mu = 1$:

$$(\omega_{1-} - \omega_{1+})\tau/2 \ll 1. \tag{19}$$

In this case the (+) and (−) contributions in the impedance almost cancel each other. If so, the series over radial modes in Eq. (18) converges at least as $1/\mu^2$, and the high radial modes ($\mu \geq 2$) can be neglected in the summation. The impedance is proportional in this case to the difference of phase advances of (1−) and (1+) modes:

$$\mathrm{Re}Z^{\parallel} = -4\omega\tau\frac{\sin(\omega\tau)}{\kappa^2F^2v} = -\frac{Z_0\omega\tau}{\pi\beta\kappa^2F^2}\sin(\omega\tau), \tag{20}$$

where the counter $\mu = 1$ is omitted, $Z_0 = 4\pi/c = 377\Omega$, $\beta = v/c$ and $\omega = \omega_e ka/\kappa$ is the frequency of the first radial mode.

Under the assumed condition of the small difference between $\pm$ phase advances (18) the real part of the impedance Eq.(18) linearly increases with the longitudinal wave number k up to $k \simeq 1/a$.

In the short wave band ($ka > 1$), the part of the electron beam with the transverse size $\simeq 1/k$ only effectively interacts with ions, the influence of the remote parts of the electron beam drops with the distance r as $\simeq \exp(-2kr)$. Hence, the impedance in the short wave band can be estimated by a substitution in Eq. (18) or Eq.(20) the effective transverse size $1/k$ instead of the real size a and considering the electron beam as adjoining a metal surface with a radius $\simeq 1/k$. It follows, that the impedance achieves the maximum at $ka \simeq 1$, the later can be estimated from Eq. (20) with $\kappa = 2.4$, $F = 0.5$,:

$$|\mathrm{Re}Z^{\parallel}|_{\max} \simeq 0.1 Z_0 \frac{\omega_e \tau}{\beta}, \tag{21}$$

where $\omega = \omega_e/\kappa$ was substituted (which corresponds to $ka = 1$), $\omega\tau \gg 1$ was assumed.

Substituting, for example, $n_e = 3 \cdot 10^7\ \mathrm{cm}^{-3}$, $g = 1\mathrm{m}$, $\beta = v/c = 0.06$ (Li_7^{+1} in the CRYSTAL ring [8]), it gives the cooler impedance $\mathrm{Re}Z^{\parallel} = 10K\Omega$, which is three orders of magnitude more than typical values of the resistive-wall impedance.

Velocities of the electron and ion beams are assumed to be zero in this calculations. It is warranted, if an additional difference between phase advances of the (+) and (−) modes introduced by a discrepancy of the velocities δv is smaller then this difference for equal velocities: $k\delta v < \alpha^2 kv$, or

$$\delta v/v < \alpha^2. \tag{22}$$

The real part of the impedance is responsible for a coherent instability with an increment

$$\begin{aligned} \Lambda_k &= \left| k u_i \frac{\mathrm{Re}Z^{\parallel}}{2Z_{sc}} \right|, \quad Z_{sc} = i\frac{Z_0 L_i}{\beta} kR, \\ L_i &= \ln(1/(ka_i)) + 1/2, \quad u_i = c\sqrt{2\bar{\rho}_i r_0 \left(\frac{1}{\gamma^2} - \frac{1}{\gamma_t^2}\right)}, \end{aligned} \tag{23}$$

where r_0 is a classical radius of the ion, R is a radius of the storage ring, a_i is a radius of the ion beam, $\bar{\rho}_i = 2\pi \int \bar{n}_i r dr$ is its linear density. The increment rate for the coasting Li beam with $2 \cdot 10^7$ particles, $L_i = 4$, $\mathrm{Re}Z^{\parallel} = 10\,K\Omega$, is found to be: $\Lambda = 5\mathrm{s}^{-1}$.

To damp the instability and to continue a cooling process, the cooling rate $\lambda_{\parallel}$ needs to be twice more than the increment [6, 7], $\lambda_{\parallel} > 10\mathrm{s}^{-1}$ in the last example.

For a 1mA beam of C_{12}^{+6} with $\beta = 0.041$ at TSR, cooled by a 350 mA, $a = 5$ cm electron beam, the increment Λ_k (23) occurs to be equal to $5 \cdot 10^2\,\mathrm{s}^{-1}$, which is far above the cooling rate. In such a case a steady temperature of the cooled ion beam is determined by Landau damping. Assuming the Gaussian distribution, the Landau-damping threshold can be expressed as

$$T^{\parallel}(K) \geq T^{\parallel}_{th}(K) = 0.02 f_Z \frac{I(\mu\mathrm{A}) Z_i L_i}{\beta(1-\gamma^2/\gamma_t^2)}, \tag{24}$$

where the factor f_Z logarithmically depends on the impedance: at $\mathrm{Re}Z^{\parallel}/|Z_{sc}| = 3 \cdot 10^{-5}$ $f_Z \approx 1$, at $\mathrm{Re}Z^{\parallel}/|Z_{sc}| = 1 \cdot 10^{-2}$ $f_Z \approx 2$.

An application of this formula to the mentioned C_{12}^{+6} beam gives the threshold temperature $T^{\parallel}_{th} = 2 \cdot 10^4$K, which is rather close to the observed value $3 \cdot 10^4$K

4 Transverse Impedance

An excitation of transverse dipole modes ($l = 1$) of the electron beam and their back action on ions can be described in terms of a transverse impedance $Z^{\perp}$ [2]:

$$\Delta p_x = < E_x > \tau = i\bar{\rho}_i x_i Z^{\perp}, \tag{25}$$

where Δp_x is a transverse momentum change of a test particle with a unit charge, $< E_x >$ is an average of the electric field,

$$< E_x > = -\frac{1}{\tau} \int_0^{\tau} \frac{\partial \Phi}{\partial x}|_{x=0} dt, \tag{26}$$

$\bar{\rho}_i$ is a linear density of the ion beam, x_i is an amplitude of its deviation along the $x-$ direction. The potential Φ is given by Eq.(13) with $l = 1$, wave amplitudes A_μ are found from the zero boundary conditions at the entrance Eqs.(3):

$$A_\mu = -\frac{2\bar{\rho}_i x_i}{\kappa_\mu a F_\mu^2}, \quad \bar{\rho}_i = 2\pi \int \bar{n}_i(r) r dr. \tag{27}$$

Asymptotically, at $\kappa_\mu \gg 1$, $F_\mu = (\pi\kappa_\mu/2)^{-1/2}$ and A_μ tends to be a constant:

$$A_\mu = A = -\pi\bar{\rho}_i x_i / a, \quad \text{for } \kappa_\mu \simeq \pi\mu \gg 1 \tag{28}$$

From here, the real part of the transverse impedance, which is responsible for a transverse instability [2], is found:

$$\mathrm{Re}Z^{\perp} = -\frac{Z_0}{4\beta a} \sum_\mu \frac{\kappa_\mu v}{a} \int_0^{\tau} dt \cos\left(\frac{\omega_{\mu-} + \omega_{\mu+}}{2} t\right) \sin\left(\frac{\omega_{\mu-} - \omega_{\mu+}}{2} t\right). \tag{29}$$

Contrary to the longitudinal case, this series is diverging. It means a large number of the contributing modes and approves an application of the asymptotic equation (28).

An upper limit of the summation $\mu_{max} = m$ is determined by a finite transverse size of the ion beam a_i, assumed to be zero in the previous calculation of the wave amplitudes A_μ; it gives $m \simeq a/a_i$. If the difference of $\pm$ phase advances at $\mu \sim m$ is small, $(\omega_- \tau - \omega_+ \tau)/2 < 1$, the expression (29) can be simplified:

$$\mathrm{Re}Z^{\perp} = -\frac{Z_0 \omega_e \tau}{4\beta a} S, \quad S = \sum_{\mu=1}^{m} \begin{cases} \sin(\omega_\mu \tau) & \text{if } \omega_\mu \tau \gg 1 \\ \omega_\mu \tau / 2 & \text{if } \omega_\mu \tau \leq 1 \end{cases} \tag{30}$$

The result of the summation depends on the phase advance ψ and its difference between neighbor modes $\Delta\psi$ at the upper limit of the summation $m \simeq a/a_i$, where

$$\psi = \psi_m = \omega_m \tau \simeq \frac{\omega_e \tau}{\pi m} ka \simeq \frac{\omega_e \tau}{\pi} ka_i$$

$$\Delta\psi = \psi_{m-1} - \psi_m \approx \psi/m.$$

The first case corresponds to a high value of the phase advance difference: $\Delta\psi > 1$. The phase difference between neighbor modes is so high in this case, that an interference between modes can be neglected, the phases are effectively random and the result is found by a summation of the r.m.s. values:

$$|S| \simeq \sqrt{m/2}, \quad \text{if } m < m_r = \sqrt{\omega_e \tau ka/\pi} \tag{31}$$

The second possibility is a high value of the phase advance but a small phase difference: $\psi > 1$, $\Delta\psi < 1$. It takes place for $m_r < m < m_r^2$. In this case the first m_r modes summarizes as in the previous case, giving a contribution S_r in the total sum: $|S_r| \simeq \sqrt{m_r/2}$. The components of the series with the small phase difference, $m_r < \mu < m$ can be calculated by a change of the summation on the integration: $\sum_\mu \approx \int d\mu$, which gives $|S_c| \simeq m^2/m_r^2$ and

$$|S| \simeq |S_r| + |S_c| = \sqrt{m_r/2} + m^2/m_r^2, \quad \text{if } m_r < m < m_r^2. \tag{32}$$

The third case corresponds to the phases themselves to be small near the cut-off, $\psi < 1$. Here the series S consists of the random S_r, coherent S_c and small phase S_s parts. The first two of them, with the numbers $\mu < m_r^2$, are estimated as in the previous cases, the last diverges logarithmically, with the cut-off at $\mu = m$:

$$\begin{aligned} |S_r| &\simeq \sqrt{m_r/2}, \quad |S_c| \simeq \Delta m = m_r^2, \quad S_s \simeq -(m_r^2/2)\ln(m/m_r^2) \\ |S| &\simeq |S_r| + |S_c| + |S_s| \approx |S_s|, \quad \text{if } m > m_r^2. \end{aligned} \tag{33}$$

Finally, the result of the summation can be presented as follows:

$$|S| \simeq \begin{cases} \sqrt{m/2}, & \text{if } m < m_r \\ \sqrt{m_r/2} + m^2/m_r^2, & \text{if } m_r < m < m_r^2 \\ \sqrt{m_r/2} + m_r^2 + \frac{m_r^2}{2}\ln(m/m_r^2) & \text{if } m > m_r^2 \end{cases} \tag{34}$$

$$m \simeq a/a_i, \quad m_r \simeq \sqrt{\omega_e \tau ka/\pi}.$$

Turning back to Eq. (30) and assuming, for instance, the electron density $\bar{n}_e = 3 \cdot 10^7\ \text{cm}^{-3}$, the cooler length $g = 1$ m, the velocity $v = 2 \cdot 10^9$ cm/s, the electron beam radius $a = 1$cm, taking the sum factor $S = 10$, the transverse impedance is found:

$$|\text{Re}Z^\perp| \simeq 30 M\Omega/m. \tag{35}$$

Generally, a real part of an impedance is responsible for an instability, an increment for the coasting beam is [2]:

$$\Lambda^{\perp} = \frac{\bar{\rho}_i r_0 c^2}{\gamma Q_b} \frac{|\mathrm{Re} Z^{\perp}|}{Z_0}, \tag{36}$$

where Q_b is a betatron tune, $r_0 = Z_i e^2/(M_i c^2)$ is the classical radius of an ion of the beam, γ is a relativistic factor.

Taking as an example the Li_7^{+1} beam with $\bar{\rho}_i = 3 \cdot 10^5 \mathrm{m}^{-1}$, $\beta = 0.06$, $Q_b = 2$ [8], assuming the transverse impedance of 30 MΩ/m, the increment is calculated: $\Lambda^{\perp} = 3\ \mathrm{s}^{-1}$.

To suppress the instability at low temperatures, where the Landau damping does not exist, a transverse cooling rate of the electron cooler $\lambda_e^{\perp}$ has to be twice higher than the increment [6]:

$$\lambda_e^{\perp}/2 > \Lambda^{\perp}.$$

For an example above it means that $\lambda_e^{\perp} > 6\ \mathrm{s}^{-1}$ has to be achieved.

5 Conclusions

Longitudinal and transverse impedances introduced in a storage ring by the electron beam of an electron cooler have been found here; both of them occurs to be orders of magnitude higher than resistive-wall ones. The corresponding values of coherent increments can be comparable with the cooling decrements even for such small ion currents, which are usually suggested for crystallization.

6 Acknowledgments

The report has been done in the framework of the CRYSTAL project. The author is thankful to V. V. Parkhomchuk, L. Tecchio, G. Bisoffi, M. Grieser, G. Lamanna and S. Gustafsson for discussions and information.

References

[1] A. Sessler and V. Vaccaro, CERN Report ISR-RF/67-2 (1967)

[2] A. W. Chao, *Physics of Collective Beam Instabilities in High Energy Accelerators,* J. Wiley & Sons, Inc., New York, 1993.

[3] A. V. Burov, *Wall Impedance for Low and Moderate Beam Energies,* CSR Note, Dec. 1994.

[4] N. S. Dikansky, D. V. Pestrikov, *The Physics of Intense Beams and Storage Rings,* AIP Press, New York, 1994.

[5] V. V. Parkhomchuk, private communication, 1992.

[6] V. V. Parkhomchuk, D. V. Pestrikov, in *Proc. of the Workshop on Beam Cooling and Related Topics,* p. 327, Montreux, 1993.

[7] A. V. Burov, in *Proc. of the Workshop on Beam Cooling and Related Topics,* p. 230, Montreux, 1993.

[8] G. I. Budker et al. In Proc. of 10 Int. Conf. on High Energy Accel., v. 1, p. 498, Serpukhov, 1977.

[9] M. Grieser, private communication.

[10] D. Reistad, private communication.

[11] L. S. Bogdankevich, A. A. Rukhadze, Sov. Uspekhi Phys. Nauk, **103**, 4, p. 609 (1971).

[12] P. L. Morton, V. K. Neil and A. M. Sessler, J. Appl. Phys. **37**, 3875 (1966).

[13] L. Tecchio, G. Bisoffi et al., *CRYSTAL. A Storage Ring for Crystalline Beams. Feasibility Study.* LNL-INFN Int. Rep., 1994.

III. CRYSTALLINE BEAMS

THE PHYSICS OF CRYSTALLINE BEAMS

JOHN P. SCHIFFER
Physics Division, Argonne National Laboratory, 9700 S. Cass Avenue, Argonne, IL 60439 and The University of Chicago, 5720 S. Ellis Avenue, Chicago, IL 60637

The possibility of seeing ordering phenomena in particle beams that are cooled to very low temperatures has been discussed for some time[1-5]. The types of ordering depend on the linear particle density: as the density increases, the expected structures change from one dimensional strings to two dimensional zig-zag patterns, then to three-dimensional arrangements on the surface of a cylinder, growing into a multiple shell structure of concentric cylinders[4,5]. All of these shapes are seen in simulations where the "temperature" of the system is allowed to reach a low value; they are illustrated in figure 1.

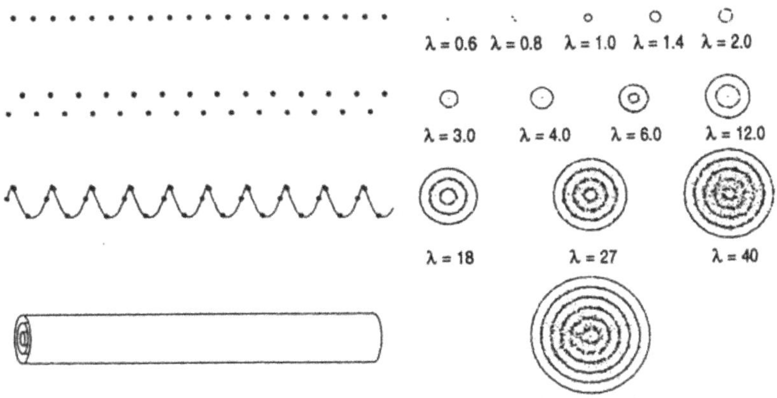

Figure 1. Configurations of ions in a cold beam for various values of the linear particle density λ. On the left are schematic pictures for the one-dimensional "string", two-dimensional "zig-zag", and three-dimensional simple helix and multi-shell configurations. On the right the output of Molecular Dynamics simulations for these various configurations is shown.

Fundamentals

What quantities matter in considering a cold beam? For cylindrically symmetric focusing, the shell structure is determined by a single parameter, the dimensionless linear particle density[4] λ: $\lambda \equiv (N/C)\, a_{SW}$, where N is the number of ions, C the circumference of the ring, and a_{SW} the Seitz-Wigner radius $a_{SW} = [3mq^2/\omega_\beta{}^2]^{3/2}$, with m the ion's mass, q its charge, and ω_β the betatron frequency. Note that a_{SW} also enters in the plasma coupling parameter: $\Gamma \equiv (q^2/a_{SW})/kT$. A system with $\lambda < 0.9$ will be a "string" for $0.9 < \lambda < 1.3$ it is a two-dimensional "zig-zag", for $1.3 < \lambda < 2.4$ it forms a single shell, and for $\lambda = 27$, for instance there would be four shells -- the number of shells increases as the $\sqrt{\lambda}$. Asymmetric focusing introduces an additional parameter that influences the shell structure. Finite length beam bunches become elongated spheroids, and an example is shown in figure 2.

Figure 2. Simulations of a bunched beam, of 20,000 ions with length on the order of 20 cm and the maximum radius on the order of 100 microns.

Intrabeam Scattering -- Good or Bad?

In the normal thinking about cooling beams in a storage ring, intrabeam scattering is seen as a limit to low temperatures. Such scattering arises from the Coulomb field between particles, and the Rutherford scattering between particles on different betatron trajectories is the perturbing problem. However, the ordering between particles in a very cold beam, just discussed, exists only because of the same Coulomb interaction between the particles in the beam, an interaction that requires that in the cold limit the ions in the beam be as nearly equally spaced as possible. The factors limiting beam cooling (as, for instance, mentioned in the announced purpose of this Workshop) are usually considered to be in intrabeam scattering. This thinking must be reconsidered in this cold limit. This subject is the main purpose of this talk. The extent to which the periodic transverse oscillations from the focusing elements and the shear from the bending elements in the accelerator lattice[2,6,7] couple into "heating" a beam (in the sense of moving it away from an ordered configuration) is perhaps the best measure of intrabeam scattering (in the damaging sense) in a cold beam -- it is not quite the same as the usual intrabeam scattering and will depend sensitively on many details.

The concept of beam "temperature" poses no problems for a beam that is moving in a straight line and is focused continuously. The problem is that this is not what one is dealing with in real beams, and the definition of temperature that is needed to approach ordering is not the same as the state that is imposed on a beam circulating in a storage ring. The subject is further complicated by the fact that neither the bending nor the focusing of the beam is achieved continuously, but by discrete focusing elements.

I would therefore like to discuss in a little more detail what is needed for a cold beam, and what should be meant by temperature. Much of this has been stated before[2,3], but I feel that clarification of these issues is at the heart of what is keeping us from attaining an ordered state in a circulating beam.

Bending with Cooling to *Fixed Linear Velocity*

First consider the case of a beam, uniformly bent and with uniform focusing. When the beam is hot, the particles undergo independent betatron oscillation in the focusing field and their interactions (intrabeam scattering) are small. But in the other limit, when a beam is at very low temperatures at the space-charge limit, the betatron oscillations have been frozen out. In this cold equilibrium state the particles are pushed together by the focusing force to the point that the focusing force is exactly canceled by the space-charge repulsion. If the linear density of particles is sufficiently low for the beam to be a one-dimensional string, or a zig-zag pattern in the vertical plane, there is no problem. But when the number of particles is sufficiently high to condense into one or more cylinders, then the particles that are on the inside of the curving beam *must* travel more slowly than their neighbors, and the particles on the outside faster. This is illustrated schematically in figure 3. Since current cooling techniques force particles to travel with the *same linear velocity* the "cooling" forces in a storage ring enforce a velocity distribution that is different from the one required for an ordered state.

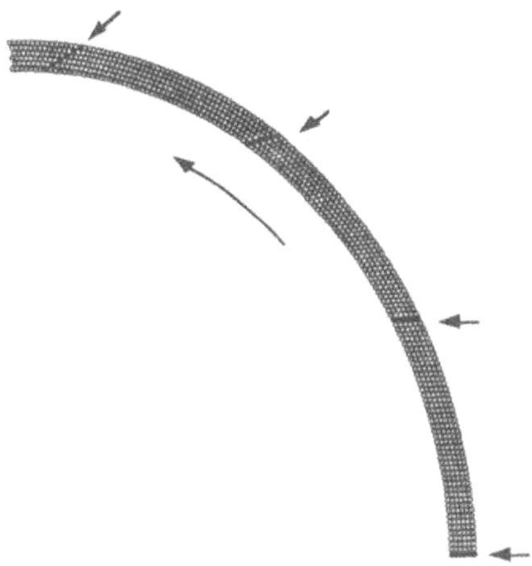

Figure 3. Schematic illustrating the shear that is experienced by a beam where particles are moving with a constant linear velocity. The curved arrow indicates the direction of motion, the short arrows and the darkened points indicate the relative positions of ions that start out, at the lowest arrow, at the same z-coordinate.

One may easily estimate the relative kinetic energies between particles in the perfect ordered beam: non-relativistically they are $\Delta E = E\,(\Delta r/R)^2$, where E is the kinetic energy of the circulating beam, R is the bend radius of the storage ring and Δr is the radial separation between particles, typically few x 10^{-3} cm in the ordered state. Clearly, what would be desirable, is some mechanism for introducing *graded cooling* which enforces the requisite velocity gradient for a constant *angular velocity* on the beam. At present there is no obvious technique for achieving this.

How should we think of a relevant "temperature" in the non-inertial frame that is co-rotating with the beam? It is the deviation from the velocity profile that corresponds to constant angular velocity. So if we had a beam with this profile, perfectly ordered, what would its "temperature" look like with the available measurement techniques, for instance with a laser scan of the beam velocity spread? We can estimate the apparent "temperature" that would appear in such a scan as

$$T_{apparent} \approx \left(\frac{r_{beam}}{R_{ring}}\right)^2 E_{kin} = \frac{(2\pi)^2}{9}\lambda \frac{q^2}{a_{sw}} \frac{1}{\nu_\beta^2} \tag{1}$$

where r_{beam} is the radius of the beam, E_{kin} is its kinetic energy in the laboratory frame, q the charge of the ions, and ν_β the betatron tune of the storage ring without space-charge effects. (With $r_{beam} = 0.6$ mm, $R_{ring} = 5$ m, $E_{kin} = 10$ MeV, $T_{apparent} \approx 1000°$ K. Note that q^2/a_{sw} is the energy numerator in the definition of the plasma parameter Γ.)

Conversely, one must remember that a beam that is at the space-charge limit is a strongly-coupled plasma and therefore the longitudinal and transverse degrees of freedom are strongly coupled -- with equilibration times on the order of $1/\omega_{plasma}$ where the plasma frequency is equal to the betatron frequency for a cold cylindrical beam and only slightly different when the beam is not cylindrically symmetric. When a constant linear velocity is forced onto the beam the "apparent temperature" defined above will be mixed into transverse (and random) velocities when the beam passes through bending magnets, thereby the above estimate for the apparent temperature, mixes into "heat" through a process, which is the form "intrabeam scattering" takes for cold beams in the space-charge limit. This heating can be avoided for very thin beams that become one- or two-dimensional, as long as they have no finite extent in the bend plane.

The parameter relevant to this heating is the rate at which the beam is sheared compared to the radius of the beam. The former is the cyclotron frequency, the latter is determined by the strength of the focusing whose measure is the betatron frequency. The ratio of the two frequencies is the betatron tune ν_β. Equation (1) may then be rewritten in terms of the value of the plasma coupling parameter that one gets assuming the temperature is determined only by this source of heating as

$$\Gamma_{shear} \approx 0.7 \, \nu_\beta^2 / \lambda \tag{1a}$$

where the estimate of the apparent longitudinal temperature of (1) was divided by 3. This expression is an estimate of a limiting value of temperature for a multi-shell beam, say for $\lambda >> 4$ or so, with constant linear-velocity "cooling" forced on the beam.

There are two other, not very realistic, options to conceivably deal with the shear problem. One of these, for a very stiff, strongly focused beam, will be discussed below. The other has been seen in simulations[3] but may be very difficult to realize in practice. If the beam cannot resist the imposed bending shear and the longitudinal cooling is applied continuously (easy in simulations not so easy in practice) then the particles segregate into a set of strings that slip by each other. The cooling needs to be strong because any transverse disorder would cause the particles to scatter from each other and as soon as that happens the system will heat up as discussed above. The strings themselves form an ordered pattern within the beam envelope, a lattice of equilateral triangles -- a pattern that has also been seen in sheared colloids[8]. An example is shown in figure 4. But the stability of this type of order is very delicate against any type of change (e.g. the capture of an ion from the beam causing a lower particle density in one string and a migration of an ion from one string to the next would be a source of very strong heating until the system can settle down again. Simulations have not been done to test such instability.

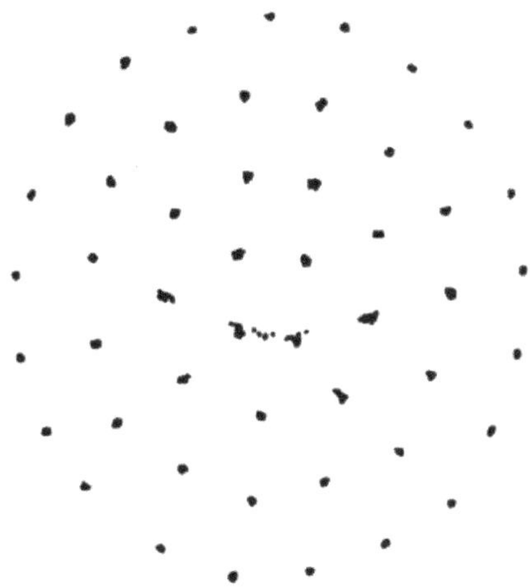

Figure 4. Separation of particles into sliding strings shown in a simulations for a constant focusing field in which the particles are subjected to continuing shear. The simulation contains 2000 particles and is for a linear particle density $\lambda = 27$.

Periodic Focusing

The above considerations apply to beams for which both the focusing and bending is uniformly distributed. In a real storage ring both are periodic and the effects of this need to be taken into account. First, however, we need to consider, in general terms, what it implies about the properties of the beam when the focusing lattice consists of discrete elements. Forgetting for the moment about the alternating gradient (the fact that the beam is alternately focused and defocused) consider first what happens if the focusing of the beam is achieved only through periodic focusing lenses. This will mean that the beam particles are pushed together suddenly in focusing element and then are pushed apart gradually, because of their mutual space charge repulsion. In the steady focusing case the system was in equilibrium -- now it will undergo oscillations. Since deviations from equilibrium can be approximated with harmonic motion, the amplitude of the envelope oscillations of the beam between focusing elements will increase quadratically with the interval between focusing elements, and the amplitude of these oscillations determines the kinetic energy contained in them. This is illustrated schematically in figure 5. A perfectly

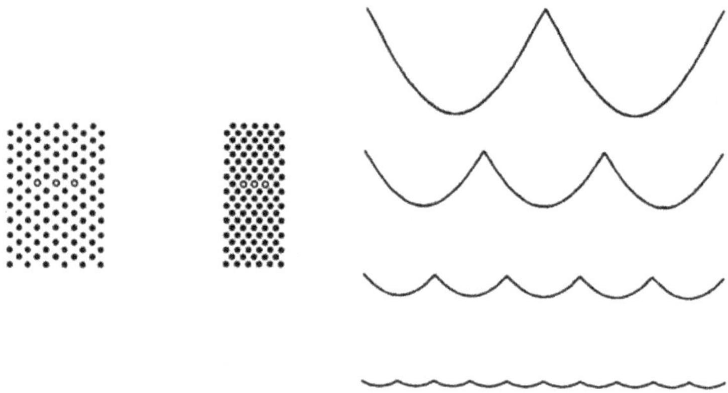

Figure 5. Schematic representation of a section of beam in a periodic focusing lattice with the beam traveling in the vertical direction. On the right are shown the excursions of the beam envelope in a simple lattice where the interval between focusing elements changes by a factor of two for each plot and the amplitude of oscillations changes quadratically with this interval.

ordered beam in the focusing lattice will undergo these oscillations and if we had a way of determining the spread in the transverse beam velocities there would be again an apparent temperature which may be written as

$$T_{apparent} \approx \left(\frac{\Delta r_{beam}}{r_{beam}}\right)^2 \left(\frac{r_{beam}}{R_{ring}}\right)^2 \frac{v_\beta^2 E_{kin}}{N_{f\beta}^2} = \frac{\pi^2}{6} \lambda \frac{q^2}{a_{sw}} \frac{1}{N_{f\beta}^2} \tag{2}$$

where Δr_{beam} is the change in the beam radius between focusing elements, and $N_{f\beta}$ is the number of focusing cells per betatron period. The important point is that these oscillations imposed by a periodic focusing field are not, by themselves, as destructive to ordered structure as the shear effect considered above, the ordered system can survive these oscillations as a rather plastic one and the restoring forces balance the perturbation so that an ordered array with superimposed envelope oscillations are obtained with no (or very little) coupling into random, "thermal" excitations[2,4,9]. The only caution is that the beam has normal modes[3,6] illustrated in figure 6 that should be avoided: for the pure radial volume oscillation the eigen-frequency of the beam is at the betatron frequency, while for the quadrupole shape oscillation it is $\sqrt{2}$ times that. So the period of focusing cells should avoid harmonics of these modes. Still, the kinetic energy associated with these oscillations is best minimized, and it is desirable to have as many focusing elements per betatron period as practical.

It should also be noted that any transverse cooling, should it become practical, would freeze these transverse oscillations at that point in the lattice and the location for transverse cooling in the lattice would need to be considered with caution.

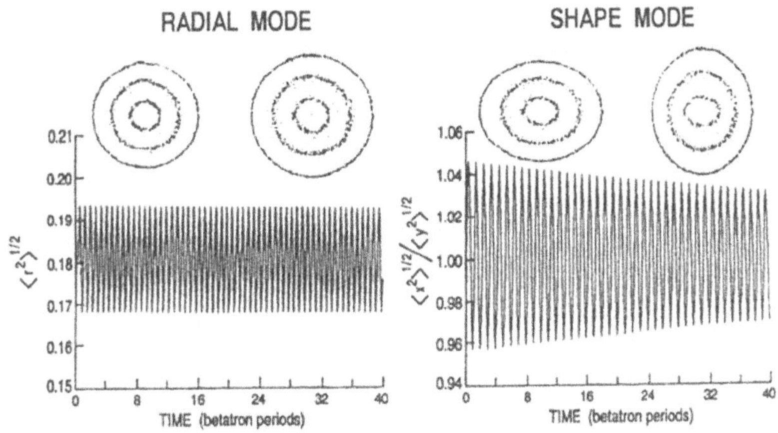

Figure 6. Normal modes of a beam in a constant focusing field are illustrated. On the left a radial "monopole" volume oscillation was excited; this seems to be a true eigen-mode of the system and does not couple into other degrees of freedom. On the right a quadrupole shape oscillation appears to be damped into more complicated modes. The simulation had 1000 ions.

Periodic Bending with Cooling to *Constant Angular Velocity* in the "Soft Beam" Limit

A technique that would cool the beam to a velocity profile, with a gradient appropriate to the bend radius, such that the particles all follow their trajectories isochronously, would be ideal for a lattice where bending is continuous -- all longitudinal shear effects would disappear. However, in a real lattice, that consists of discrete bending elements and straight sections, even such cooling can present problems. In the limit when the bending elements are short compared to the straight sections they will impose a shear on the particles; they will be sheared in one direction in the bending elements and gradually catch up in the straight sections. Figure 7 illustrates this. For instance, a 90° bend is pretty violent: for two particles separated radially by a distance d, such a bend would introduce a longitudinal displacement $\pi/2$ times d, in other words about 1.6 times the interparticle spacing. It is difficult to see how an ordered system could displace such violent effects. Of course, the ideal state would be one where the particles are displaced half that amount before the bend in one direction ($\pm\,\pi/4$), and then by the same amount in the other direction after the bend, but the displacement is till large. Clearly the more the bending elements are distributed the better -- a 60° bend would imply that the total displacement would be $\pi/3$ and from before to after it could be $\pm\,\pi/6$, so the excursions may become survivable. Simulations also confirm this in various ways. But any slight disorder in the beam will cause particles to rub against each other and become a source of heat.

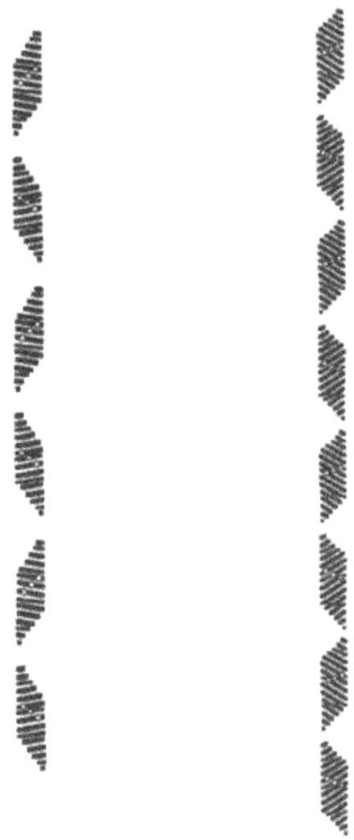

Figure 7. Schematic picture, in the same sense as figure 2, of a section of beam that is cooled to a constant angular velocity, and subjected to discrete bending elements. Each of the successive pairs represents a segment of beam before and after a bend (90° on the left and 60° on the right), with the change between segments taking place in the straight sections. The dark lines appear because of merging of the dots used to represent particles; three "adjacent" particles in the center of each segment are represented by open circles.

Bending of Beams in the "Stiff Beam" Limit

There is a further way out: *if* one can form the beam in an ordered state, then the crystalline structure produces a certain resistance to the shearing that is imposed on the beam by bends[2,3] -- in other words the particles *can* be forced to move faster or slower in the bends by the interparticle forces. The elastic shear modulus of these structures has been studied in simulations, though the results are somewhat contradictory. The results[3] seem to indicate that there is a limit to the resistance to shear that an ion beam, cooled to a constant linear velocity can sustain. This is shown in figure 8.

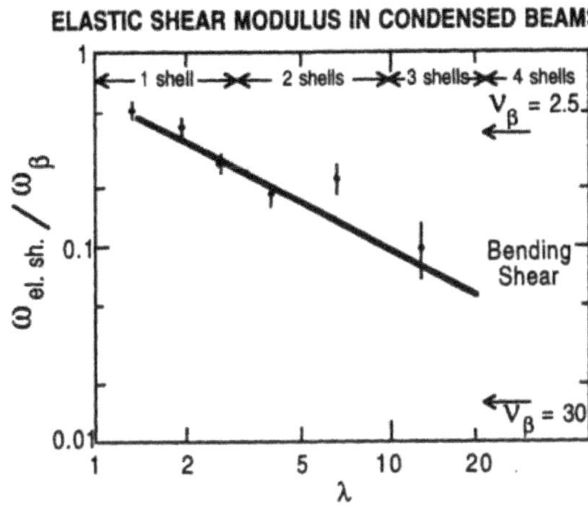

Figure 8. Values of the elastic shear modulus (relative to the focusing strength, which is represented by the betatron frequency) of a cylindrical beam, extracted empirically from Molecular Dynamics simulations. When the imposed shear from the circular orbit, represented by the reciprocal of the betatron tune ν_β is larger than the elastic shear modulus. Two values representing "small" and "large" storage rings are shown on the right. The elastic shear modulus is a function of the linear particle density λ, plotted on the bottom scale and the corresponding number of shells is shown on top.

As with any solid, the crystallized beam has a natural elastic resistance to shear. The elastic shear modulus may be characterized by a limiting natural shear frequency of the beam, and the requirement is that this natural frequency be larger than the driving frequency: $\omega_{shear} \geq \omega_{cyclotron}$. Simulations suggest that the values for this is $\omega_{shear} \approx \omega_\beta/\lambda$, in other words the condition for a beam to survive bending shear is that

$$\nu_\beta \geq \lambda \qquad (3)$$

With such a beam, cooling to constant linear velocity would be sufficient, for instance. Such lattices do exist in large storage rings, such as LEP and RHIC, or could have existed in the SSC, with the rate of bending relatively slow compared to the strength of the focusing -- and beams with 4-10 shells could survive in such rings. The value of ν_β for smaller rings is typically around 2.5, corresponding to the single-shell regime.

Options

a) Minimizing Heating from Bending Shear

1. Find a method of cooling to <u>*constant angular velocity*</u>. (even a rough gradient will ease the problem), and/or

2. Find a method to *compensate for bends* by changing the velocity profile in special inserts (rf or electrostatic solutions suggested -- none seem viable at present).

3. Use *high ν_β* and *high multiplicity* to increase the elastic shear modulus of the beam and minimize shear heating effects (appropriate in any case).

4. Put up with bending shear and settle for "*sliding strings*" (but this needs extremely strong cooling, which does not seem practical at present).

5. Stick to *sparse beams* (one-dimensional "strings" or single shells). (This may seem like giving up on a proper crystal -- also, it is not clear whether longitudinal cooling can be effective in this regime, since the transverse degrees of freedom will be very weakly coupled to the longitudinal ones.) If achievable, this could be a start.

b) Minimizing Heating from Discrete Focusing Elements

1. *Maximize* number of *focusing elements* in lattice per betatron period. (amplitude of oscillation decreases quadratically with number)

2. Focusing lattice and all aspects of ring should be as *evenly spaced* (and with as short a repeating cycle) as possible.

3. *Avoid harmonics* of the betatron frequency f_β —also harmonics of $\sqrt{2}f_\beta$ for equal vertical and horizontal frequencies (these are the frequencies of the normal modes and one should avoid driving them.) N.B. These values need to be modified for asymmetric focusing .

4. Any transverse cooling (should a practical scheme develop) will be most effective at a waist or maximum in the cold beam envelope.

Stability of the Ordered State and a Point about Simulations

I wish to make a point about simulations that may be important. There have been simulations in which configurations were found, with an accelerator lattice that included both periodic bending and "realistic" alternating-gradient focusing, similar but not identical to the solutions found with uniform focusing and bending. Now in such simulations the nature of the ordering, the cooling, and the stability of the configuration against perturbations is crucial. The principal point I wish to make is about the size of the sample that is followed in simulations.

In Molecular Dynamics simulations usually a section of the beam is treated, but then the question is what to do at the ends of this section. The usual procedure is to make the system self-enclosed -- that is let the particles at one end of the system interact with the forces from the other end. Thus the system must be periodic, at least with the length of the cell. In figure 9 there are examples for a simple case, showing that the configuration that comes from a simulation with Molecular Dynamics is strongly influenced by the choice of the length of the repeating cell, whether the number of ions in it is divisible by 2, by 3, or by neither, even when the cell length is many times the repeating length for the lowest state as may be seen in figure 9. This can be particularly important for beams with many more ions, where there are multiple shells. A short repeating cell will have a symmetry forced

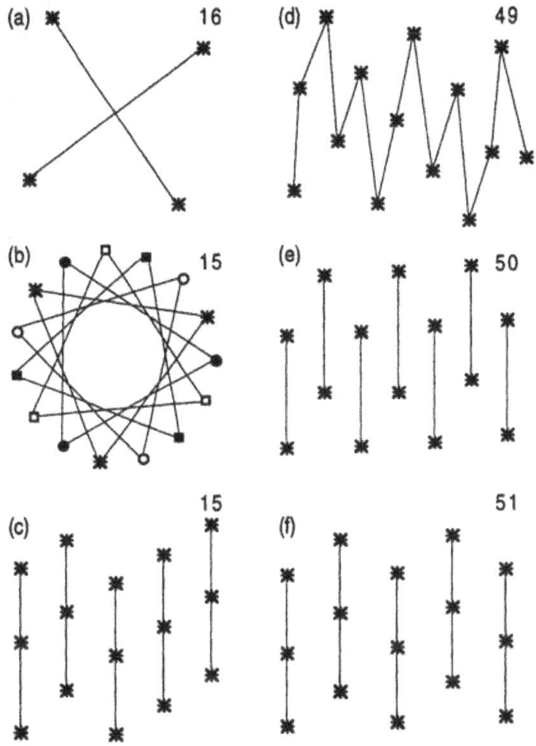

Figure 9. The effect of choice of repeating cell is shown on simulations of a beam with a fixed linear particle density $\lambda = 3.0$. The patterns seen are determined by the length of the cell. (a) shows the head on view for 16 ions in the cell -- they are lined up in pairs, along two perpendicular direction, (b) and (c) show results for 15 ions in the x-y plane and the x-z plane respectively -- the pattern is different with sets of three particles at a given z-coordinate along the beam forming triangles and the triangles rotating in orientation by 144 degrees; (d), (e), and (f) show a portion of the x-z plane for 49, 50, and 51 ions respectively.

on it that may not be the actual lowest state for a long beam, and the extent to which a particular configuration is stable under the repeating bending shear and focusing may very much depend on the symmetry of the configuration. It is therefore important to do simulations with as long a repeating unit cell as possible.

Summary

It seems that the time has come in the pursuit of lower and lower beam temperatures to start focusing more detailed attention to the reality of storage rings -- conventional cooling techniques and measures of temperature are generally not the appropriate ones at the lowest temperatures. Finding solutions to these serious problems does not appear to be impossible, but these considerations must be kept in mind in designing new storage rings with the aim to approach the regime of ordered three-dimensional beams. In particular, such rings will have to

a. Use calculations of the lattice *with the full effects of space charge included*. (N.B. averaged over time, space charge exactly cancels the focusing fields for a cold beam and therefore must be explicitly included.)

b. Find technical solutions and incorporate several of:

 i. cooling to introduce a longitudinal *velocity gradient* and favor *constant angular velocity;*
 ii. *high multiplicity* in bending and focusing elements;
 iii. stronger focusing *(high betatron tune)*;
 iv. *high symmetry* in the ring design.

c. Finally, simulations should try to incorporate as much realism as possible, with larger repeating cells and more detailed descriptions of the lattice.

This research was supported by the U.S. Department of Energy, Nuclear Physics Division, under Contract W-31-109-Eng-38.

References

1. J. P. Schiffer and P. Kienle, *Z. Phys.* **A 321** (1985) 181; J. P. Schiffer and O. Poulsen, *Europhys. Lett.* **1** (1986) 55; A. Rahman and J. P. Schiffer, *Phys. Rev. Lett.* **57** (1986) 1133.
2. J. P. Schiffer and A. Rahman, *Z. Phys.* **A 331** (1988) 71; A. Rahman and J. P. Schiffer, *Physica Scripta* **T22** (1988) 133.
3. J. P. Schiffer, Proceedings of Workshop on Crystalline Ion Beams, GSI-89-10 Report, April 1989, ISSN 0171-4546, p. 2.
4. R. W. Hasse and J. P. Schiffer, *Annals of Physics* **203** (1990) 441.
5. D. Habs, MPIH-1987-V10 (1987).
6. J. P. Schiffer, Proceedings of the Workshop on "Beam Cooling and Related Topics", Montreux, Switzerland, October 4-8, 1993, ed. J. Bosser, CERN Report 94-03, pg. 455-457 (1994).
7. J. P. Schiffer, Proceedings of the 1995 Particle Accelerator Conference and International Conference on High-Energy Accelerators, Dallas, TX, May 1-5, 1995.
8. W. Xue and G. S. Grest, *Phys. Rev. Lett.* **64** (1990) 419.
9. J. Wei, X.-P. Li, and A. M. Sessler, *Phys. Rev. Lett.* **73** (1994) 3089.

DIVERSE TOPICS IN CRYSTALLINE BEAMS*

JIE WEI

Brookhaven National Laboratory, Upton, New York 11973

ANDREW DRAESEKE and ANDREW M. SESSLER

Lawrence Berkeley Laboratory, 1 Cyclotron Road, Berkeley, California 94720

XIAO-PING LI

BIOSYM Technologies Inc., 9685 Scranton Rd., San Diego, CA 92121

ABSTRACT

Equations of motion are presented, appropriate to interacting charged particles of diverse charge and mass, subject to the external forces produced by various kinds of magnetic fields and radio-frequency (rf) electric fields in storage rings. These equations are employed in the molecular dynamics simulations to study the properties of crystalline beams. The two necessary conditions for the formation and maintenance of crystalline beams are summarized. The transition from 1D to 2D, and from 2D to 3D is explored, and the scaling behavior of the heating rates is discussed especially in the high temperature limit. The effectiveness of various cooling techniques in achieving crystalline states has been investigated. Crystalline beams made of two different species of ions via sympathetic cooling are presented, as well as circulating "crystal balls" bunched in all directions by magnetic focusing and rf field. By numerically reconstructing the original experimental conditions of the NAP-M ring, it is found that only at extremely low beam intensities, outside of the range of the original measurement, proton particles can form occasionally-passing disks. The proposed New ASTRID ring is shown to be suitable for the formation and maintenance of crystalline beams of all dimensions.

1. Introduction

In previous work, the nature of crystalline beams has been studied by many different investigators.[1]–[15] In particular, we have made a systematic study in a series of papers[16]–[22] which has consisted of deriving the equations of motion for charged particles undergoing Coulomb interaction in a real storage ring, and then using them to study many properties of crystalline beams. We have shown that the storage ring

*Work performed under the auspice of the U.S. Department of Energy, supported by NSF Grant DMR-91-15342, and by the DOE, Office of Energy Research, Office of High Energy and Nuclear Physics, under Contract No.DE-AC03-76SF00098.

must be alternating-gradient (AG) focusing, that operation must be below the transition energy of the ring, and that the storage ring lattice periodicity must be much greater than twice the maximum betatron (transverse) tune.

In this paper, we carry on our studies with investigations of a number of disparate crystalline beam subjects. First, we present in Section 2 the generalized equations of motion appropriate to the study. Previously, we have presented a rather lengthy derivation[16] which considered one species of ions with no longitudinal external force; since the generalization is straight-forward, we only present the results. Using these equations, we then discuss in Section 3 some results of the molecular dynamics (MD) simulations including the method of determining the ground state, the evaluation of the heating rates, and the scaling laws in comparison with that obtained from the intra-beam scattering (IBS) theory.

In Section 4, we re-state the conditions for crystallization, emphasizing the second condition pertaining to the machine lattice periodicity and the maximum transverse tune. The phonon spectrum, which is the key to this condition, is investigated analytically for one-dimensional (1D), and numerically for 2D and 3D crystalline structures.

Previously, we achieved crystalline states in the MD simulation by imposing once per lattice period a condition of periodicity while correcting the longitudinal slippage.[16] To investigate the effectiveness of conventional cooling methods, we employ in Section 5 more realistic models of conventional stochastic, electron, and laser cooling, as well as the proposed longitudinally tapered cooling.

In Section 6, we study the development of crystalline beams consisting of two different ion species of similar rigidity, while only one species of which is subject to cooling. We show that sympathetic cooling due to intra-particle interaction cools the other species, resulting in crystalline structures of two types of ions. These structures possess more complicated properties than single-species structures.

To extend our knowledge into bunched beams, we further investigate in Section 7 the effect of a radio-frequency (rf) electric field upon the formation of 3D bunched "crystalline balls" circulating in the storage ring. Such "blobs of matter" should have interesting properties and applications different from that of usual crystalline beams which are finite only in the transverse directions.

In the late 1970's, anomalous behavior in the Schottky signals was observed on the NAP-M ring, leading to the suggestion of crystalline beams and precipitating a great deal of theoretical and experimental work. It is thus interesting to use the tools that we have developed to reconstruct the original conditions of the NAP-M ring, and to see whether, theoretically, a crystalline state could have been realized. The results are summarized in Section 8.

In Section 9, we apply the methodology to a ring which has been proposed for the careful study of crystalline beams; namely the New ASTRID ring. We determine the conditions under which 1D, 2D, and 3D crystalline structures can be formed, and determine the cooling rate needed for crystallization.

Finally, the conclusions are presented in Section 10.

2. Equations of Motion

In order to adopt the molecular dynamics (MD) methods, we employ[16] a rotating frame (x, y, z, t) of a reference particle in which the orientation of the axes is rotating so that the axes are constantly aligned to the radial (x), vertical (y), and tangential (z) direction of motion. The equations previously derived for the motion of the particles in the presence of bending dipole and focusing quadrupole magnetic fields, can be generalized for multi-species of ions under various external dipole, quadrupole, and sextupole magnetic fields and longitudinal electric fields.

It is convenient to define a reference particle with charge state Z_0 and atomic mass M_0. We scale dimensions in terms of the characteristic distance ξ, with $\xi^3 = r_0\rho^2/\beta^2\gamma^2$, where $r_0 = Z_0^2e^2/M_0c^2$, the velocity of the reference particle is βc, its energy is γM_0c^2, and it moves on an orbit with bending radius ρ in magnetic field B_0. We measure time in units of $\rho/\beta\gamma c$ and energy in units of $\beta^2\gamma^2Z_0^2e^2/\xi$. For the ith species of ions with electric charge state Z_i and mass M_i, we define the relative charge and mass with respect to the reference values

$$\bar{Z}_i \equiv Z_i/Z_0, \quad \text{and} \quad m_i \equiv M_i/M_0. \tag{1}$$

In a bending region with bending radius ρ, the equations of motion for particles of the ith species can be derived as

$$\begin{cases} \ddot{x} - \left(2 - \dfrac{\bar{Z}_i}{m_i}\right)\gamma\dot{z} + \left[1 - \left(2 - \dfrac{\bar{Z}_i}{m_i}\right)\gamma^2\right]x + \left(\dfrac{\bar{Z}_i}{m_i} - 1\right)\dfrac{\rho}{\xi} & = & -\dfrac{\partial V_{Ci}}{\partial x}, \\ \ddot{y} & = & -\dfrac{\partial V_{Ci}}{\partial y}, \\ \ddot{z} + \left(2 - \dfrac{\bar{Z}_i}{m_i}\right)\gamma\dot{x} + \left(\dfrac{\bar{Z}_i}{m_i} - 1\right)\gamma^2 z & = & -\dfrac{\partial V_{Ci}}{\partial z}, \end{cases} \tag{2}$$

where the dots denote differentiations with respect to the time t measured in the above units, the Coulomb potential is

$$V_{Ci}(x, y, z) = \sum_j \frac{\bar{Z}_i\bar{Z}_j}{m_i\sqrt{(x_j - x)^2 + (y_j - y)^2 + (z_j - z)^2}}, \tag{3}$$

and the summation, j, is over all the other particles. The Hamiltonian that corresponds to Eqs. 2 and 3 is

$$\begin{aligned} H_i(x, P_x, y, P_y, z, P_z; t) &= \frac{1}{2}\left(P_x^2 + P_y^2 + P_z^2\right) + \frac{1}{2}\left[x^2 + \left(\frac{\bar{Z}_i}{m_i} - 1\right)\gamma^2 z^2\right] \\ &\quad - \left(2 - \frac{\bar{Z}_i}{m_i}\right)\gamma x P_z + \left(\frac{\bar{Z}_i}{m_i} - 1\right)\frac{\rho}{\xi}\, x + V_{Ci}(x, y, z). \end{aligned} \tag{4}$$

In a straight section, where there is no bending of particles, there often are quadrupole magnets for focusing, sextupole magnets for chromatic correction, and electric field for acceleration or bunching. If the quadrupole and sextupole strengths are represented by

$$n_1 = -\frac{\rho}{B_0}\frac{\partial B_y}{\partial x} \quad \text{and} \quad n_2 = -\frac{\rho}{B_0}\frac{\partial^2 B_y}{\partial x^2}, \tag{5}$$

respectively, the equations of motion can be derived as

$$\begin{cases} \ddot{x} - \dfrac{\bar{Z}_i}{m_i}\left[n_1 x + \dfrac{n_2\xi}{2}(x^2 - y^2)\right] & = & -\dfrac{\partial V_{Ci}}{\partial x}, \\ \ddot{y} + \dfrac{\bar{Z}_i}{m_i}(n_1 y + n_2\xi xy) & = & -\dfrac{\partial V_{Ci}}{\partial y}, \\ \ddot{z} & = & -\dfrac{\partial V_{Ci}}{\partial z} + F_s, \end{cases} \tag{6}$$

where the electrical force F_s in the reduced units can be expressed in terms of electric field E_s measured in the laboratory frame,

$$F_s \equiv -\frac{\partial U_s}{\partial z} = \frac{\bar{Z}_i}{m_i}\frac{Z_0 e E_s \xi}{M_0 c^2}\left(\frac{\rho}{\xi\beta\gamma}\right)^2. \tag{7}$$

The corresponding Hamiltonian is

$$\begin{aligned} H_i(x, P_x, y, P_y, z, P_z; t) &= \frac{1}{2}\left(P_x^2 + P_y^2 + P_z^2\right) - \frac{n_1}{2}\frac{\bar{Z}_i}{m_i}(x^2 - y^2) \\ & \quad -\frac{n_2\xi}{6}\frac{\bar{Z}_i}{m_i}(x^3 - 3xy^2) + V_{Ci}(x, y, z) + U_s. \end{aligned} \tag{8}$$

In the reduced units, the revolution period of the reference particle in the storage ring is C_0/ρ, where $C_0 = 2\pi R$ is the circumference of the ring.

Obviously, in the case that only one species of ions is studied, $\bar{Z}_i = m_i = 1$, Eqs. 2 - 8 become much simplified.

3. Molecular Dynamics Results

In previous papers, we have presented many computational results which the reader may wish to consult. In this section, we present results from several different MD studies of single-species un-bunched beam, emphasizing the method of determining the ground states, the scaling behavior of the heating rates, and the comparison with the IBS theory.

3.1 A Study of 1D to 2D, and 2D to 3D Transitions

It has been well established[7] that at very low line density (beam intensity), a crystalline beam is a 1D string. As the density is increased beyond critical values, the ground state first converts from 1D into 2D extending in the plane of weaker focusing, then from 2D into 3D. As the density is increased even further, the structure expands in transverse dimension, while the intra-particle distance remains approximately the same.

In order to determine the ground state at a given density and to study the transition from 1D to 2D, and from 2D to 3D, we have undertaken extensive numerical computations. For simplicity but without losing generality, we present an example using a constant gradient (CG) harmonic oscillator model, instead of an AG focusing ring, with the transverse focusing strengths $K_x = 2.8$ and $K_y = 3.2$. Fig. 1 shows

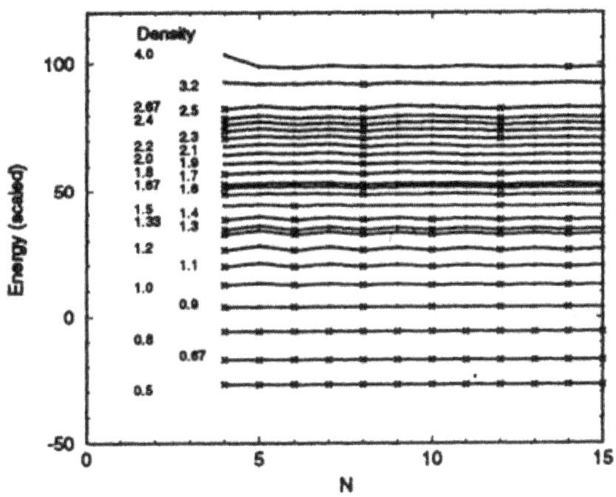

Figure 1: Lowest energies achievable with a finite number of particles N per MD cell of length L in the MD calculations at various densities N/L. The crosses show the lowest energy achievable at the particular density.

the lowest energies achievable with a finite number of particles N per MD cell of length L in the MD calculations at various densities N/L (The vertical axis shows scaled energy obtained by the expression $15N/L + 100E$, where E is the sum of the average kinetic and potential energy of the particles, and L is in units of ξ.). When $N/L \leq 0.8$, the ground state is a 1D chain of equally spaced particles, which can be obtained with any number of test particles per MD cell, as shown by the crosses in Fig. 1. Within the range $0.9 \leq N/L \leq 1.4$, the ground state is a 2D zig-zag, which can be obtained only with even number of test particles per MD cell, again as shown by the crosses in Fig. 1. When $N/L \geq 1.5$, the ground state becomes 3D. The minimum number of test particles needed to achieve the lowest-energy state increases. For example, the lowest energy state for the $N/L = 2$ case can be obtained only when the number of test particles is an integral multiple of 8. Fig. 2 shows a 3D view of

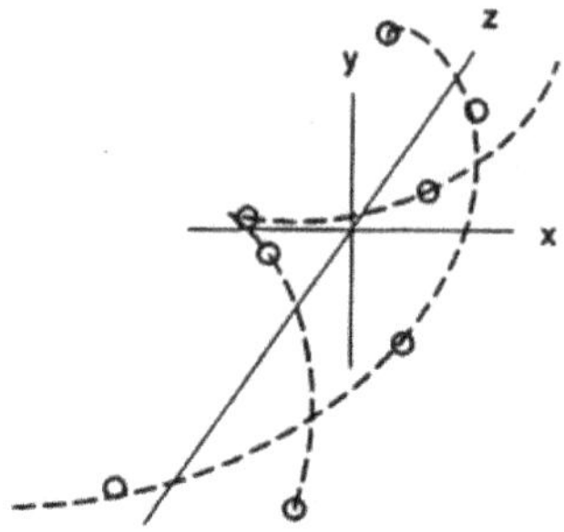

Figure 2: A 3D view of a 3D double-helix structure at $N/L = 2$ in a constant-gradient potential. In this study, the transverse focusing strengths are $K_x = 2.8$ and $K_y = 3.2$.

such a double-helix structure.

3.2 Temperatures and Heating Rates

For the time dependent Hamiltonian system discussed in Section 2, it is advantageous to define the normalized beam temperature in terms of the deviation of P_x, P_y and P_z from their ground-state values,

$$T_x = \langle(\Delta P_x)^2\rangle, \quad T_y = \langle(\Delta P_y)^2\rangle, \quad \text{and} \quad T_z = \langle(\Delta P_z)^2\rangle, \tag{9}$$

squared and averaged both over different particles and over a time period (typically 20 lattice periods) that is long compared with the focusing period but short compared with the total time of observation. When the system temperature is sufficiently higher than the break-up temperature (transition between crystalline and "running" state[20]) of the crystalline state,[20] the normalized temperature T can be directly related to the conventionally used beam temperature T_B, un-normalized rms beam emittance $\epsilon_{x,y}$, and rms momentum spread $\Delta p/p$ by the relations

$$\frac{\xi^2}{\rho^2}[T_x, T_y, T_z] \approx \frac{2k_B}{\beta^2\gamma^2 M_0 c^2}[T_{Bx}, T_{By}, T_{Bz}] = \left[\frac{\epsilon_x}{\bar{\beta}_x}, \frac{\epsilon_y}{\bar{\beta}_y}, \frac{1}{\gamma^2}\left(\frac{\Delta p}{p}\right)^2\right], \tag{10}$$

where k_B is the Boltzmann constant, and $\bar{\beta}_{x,y}$ are the average amplitude functions of the machine lattice. When the system temperature in one or more direction is lower than the break-up temperature, the conventionally used quantities T_B, $\epsilon_{x,y}$, and $\Delta p/p$ can not properly characterize the ordered state.

In a typical storage ring, the particles under Coulomb interaction absorb energy from the AG lattice via a multi-phonon process.[22] In the absence of cooling, Fig. 3 shows the rate of temperature increase (heating rate) as a function of temperature at

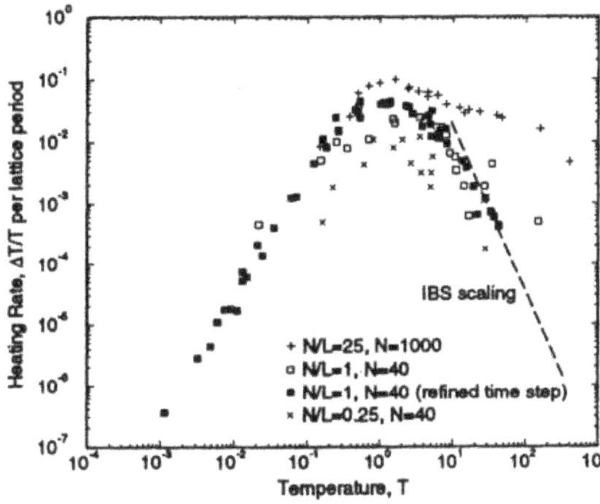

Figure 3: Heating rates as functions of temperature at various beam densities N/L. Here, N is the number of particles per MD cell of length L in units of ξ.

various beam densities that correspond to 1D ($N/L = 0.25$), single-shell 3D ($N/L = 1$), and multi-shell 3D ($N/L = 25$) ground states. The machine consists of 10 FODO cells with 25% bending with $\nu_x = 2.8$, $\nu_y = 2.1$, and the beam energy is $\gamma = 1.4$. It is observed that the break-up temperature $T \approx 1$ corresponds to the peak heating rate, independent of the beam density. On the other hand, the peak heating rate increases with the beam density. Typically, strong correlation in particle position appears in all directions when the temperature is below $T \approx 0.05$.

At low temperature, the heating rate is low. Numerically, a large number of iteration steps is required to accumulate a noticeable amount of temperature increase. On the other hand, at high temperature the particle velocity is high. In both low and high temperature range, we often need refined iteration step size and increased number N of particles per MD cell to achieve high numerical accuracy. The comparison between different step sizes is shown by the filled and empty squares in Fig. 3.

3.3 High Temperature Limit

At temperatures high compared with that for the formation of crystalline structure, intra-beam Columb scattering (IBS) theory can be readily used to describe the beam heating. We expresses the heating rate in terms of the growth rate of the beam temperature T_B as

$$\begin{bmatrix} \dfrac{1}{T_{Bz}}\dfrac{dT_{Bz}}{dt} \\ \dfrac{1}{T_{Bx}}\dfrac{dT_{Bx}}{dt} \\ \dfrac{1}{T_{By}}\dfrac{dT_{By}}{dt} \end{bmatrix} = \frac{Z^4 N_0 r_0^2 L_c \beta^2 c}{2^{3/2} A^2 \bar{\beta}_x \bar{\beta}_y C_0 T_{Bx} T_{By} T_{Bz}^{1/2}} \left(\frac{M_0 c^2}{2k_B}\right)^{5/2} F\left(\frac{a^2+b^2}{2}\right) \begin{bmatrix} 2(1-d^2) \\ -\dfrac{a^2}{2}+d^2 \\ -\dfrac{b^2}{2} \end{bmatrix}, \tag{11}$$

where

$$d^2 \approx \frac{\gamma^2 T_{Bz}}{\gamma_T^2 T_{Bx} + \gamma^2 T_{Bz}},\ a^2 \approx \frac{\gamma_T^2}{\gamma^2} d^2,\ b \approx \frac{T_{Bx}}{T_{By}} d^2, \tag{12}$$

$F(\,)$ is a "form factor" (Ref. [23]), L_c is the Coulomb logarithm, N_0 is the total number of particles in the storage ring of circumference C_0, γ_T is the relativistic factor of the transition energy, and A is the atomic mass number. Ignoring the dependence on the detailed beam configuration, the heating rate can be rewritten in terms of the normalized temperature T as

$$\frac{1}{T}\frac{dT}{dt} \sim \frac{N}{L} \frac{\beta c L_c \rho}{2\pi^{1/2} \bar{\beta}_x \bar{\beta}_y T_x T_y T_z^{1/2}}, \tag{13}$$

where $N/L \equiv N_0 \xi / C_0 \gamma$ is the number of particles per unit length in units of ξ. The heating rate at high temperature is shown to be linearly proportional to the number of particles N_0 (or linear density N/L) of the beam. When the temperature increases, the rate decreases proportionally with $T^{5/2}$. Furthermore, the rate is linearly proportional to the beam velocity βc, similar to the way that conventional cooling rate behaves.

The results from the MD simulation can be readily compared with the IBS theory. In this temperature regime, we choose the MD time step so that the particle displacement in one time step is much smaller than the typical intra-particle distance. The heating rate is evaluated as a function of the beam temperature for density $N/L = 1$. As shown in Fig. 3, the temperature dependence of the heating rate obtained from the MD simulation (filled squares) agrees well with that obtained from the IBS theory (dashed line). On the other hand, the MD results completely disagree with the IBS extrapolation at the low temperature regime.

4. Conditions for Crystallization

In previous works, we have derived two conditions which are necessary to form and maintain the crystalline beams. Since the present storage rings do not satisfy these conditions, there is considerable interest in designing and building rings suitable for the study of crystalline beams. In view of this fact, we present once again these conditions with more detailed discussion.

4.1 First Condition

The first condition requires that the ring is alternating-gradient focusing, as contrasted to constant-gradient focusing. In addition, it is required that the energy of the particles is less than the transition energy, i.e.

$$\gamma < \gamma_T. \tag{14}$$

In a typical ring, γ_T is approximately equal to the horizontal tune ν_x (number of betatron oscillation wavelengths in a circumference).

4.2 Second Condition

The second condition arises from the requirement that there is no linear resonance between the phonon modes of the crystalline structure and the machine lattice periodicity. Note that this is a condition for the maintenance of a crystalline beam, in contrast to the first condition which is for the formation of a crystalline beam.

In 1D case, the phonon spectrum can be analytically derived in the smooth approximation.[22] Typically, the highest phonon frequency is close to the maximum betatron tune. In order for the beam to avoid the linear resonance produced by the periodic modulation of the machine AG lattice, the lattice periodicity ω_l should be at least twice as high as the maximum betatron tune $\max(\nu_x, \nu_y)$.

For 3D crystalline structures having many shells, the phonon spectrum may be found approximately by ignoring the detailed structure and considering the properties of a uniformly charged rod. The frequency of the monopole mode, which is likely to be the highest frequency of the beam, has been previously derived[24] to be $\sqrt{2}\nu_{x,y}$. Thus, in the high density limit, the second condition becomes

$$2\sqrt{2}\ \max(\nu_x, \nu_y) < \omega_l. \tag{15}$$

At intermediate densities, the phonon modes can no longer be found analytically. In this case, we first employ the molecular dynamics to determine the ground-state structures. Then, using a newly written computer program, we calculate the phonon spectrum in smooth approximation (The detailed description of the method and results will be published elsewhere.). It has been verified that the highest phonon frequency is always less than $\sqrt{2}\max(\nu_x, \nu_y)$.

Therefore, in order to maintain crystalline beams at any beam density, the storage ring should be designed such that the lattice periodicity is at least $2\sqrt{2}$ as high as the maximum betatron frequency tune (Eq. 15).

5. Study of Cooling Methods

In finding the ground state in most of our early numerical simulations, we "cool" once per lattice period by simply imposing a condition of periodicity (by averaging initial and final coordinates and momenta) while correcting P_z according to the amount of slippage in z for each particle. Imposing the periodicity condition is a fast and effective way of finding a periodic or ground state. However, it is experimentally unrealistic at least at the current stage of technology. Therefore, we made numerical models that describe the actual experimental cooling. Our work, quite clearly, sheds light upon the subject of what experimental cooling methods are effective for the crystalline formation.

5.1 Model for Stochastic, Electron, and Laser Cooling

We simulate the actual cooling by reducing once per lattice period the momentum component P_i of each particle by a certain fraction,

$$\Delta P_i = -f_i P_i, \quad i = x, y, z. \tag{16}$$

By observing the reduction of temperature as a function of time, we determine the effectiveness of cooling when f_x, f_y, and f_z are varied.

When the beam density is low such that the ground state is a 1D chain, the ground state can be achieved by cooling in three directions with a rate higher than the maximum heating rate (Numerically, the cooling time is often several orders of magnitude slower than by imposing the periodicity condition). If cooling is only applied to the longitudinal direction (similar to regular laser cooling), the longitudinal temperature can be greatly reduced. Not surprisingly, the transverse temperature often can not be effectively reduced in this case for lack of sufficient IBS interaction.

When the beam density is high so that the ground state is 2D or 3D extending in the radial direction, all particles must have the same average angular velocity and, therefore, have different average linear velocity. Typically, there is no place in the ring lattice where all the particles have the same linear velocity in the ground state. Therefore, cooling in only the longitudinal direction ($f_x = f_y = 0$) often fails to produce a periodic crystalline state. On the other hand, cooling in the transverse direction is very effective for attaining a periodic crystalline state.

As an example, we study the case of a single-shell ground state ($N/L = 1$) in Fig. 3 with the machine consisting of 10 FODO cells with 25% bending with $\nu_x = 2.8$, $\nu_y = 2.1$, and $\gamma = 1.4$. When cooling is applied only in the longitudinal direction ($f_x = f_y = 0$), temperatures in all directions can be greatly reduced (often to $T \leq 0.1$) through IBS interaction reaching an intermediate non-gaseous state, as shown in Fig. 4 along the $f_x = f_y = 0$ axis with $f_z \geq 0.03$. However, since further reduction in temperature can not be achieved (except for the $f_z = 0.03$ case), ordering in the beam can not be realized.

When cooling is applied to the transverse direction ($f_z = 0$), temperature in all directions can be greatly reduced. Ordering in the beam can be achieved when the cooling strength is in the range $0.04 < f_{x,y} < 0.45$. After the ordering is achieved, the same ground state as given by imposing the periodicity condition can be achieved when the actual transverse cooling is adiabatically turned off. On the other hand, rapid actual cooling in the longitudinal direction ($f_z > 0.06$) almost always destroys ordering.

The implication of these studies is that the present cooling techniques suffice in making 1D crystals. In order to make higher density structures, it is necessary to

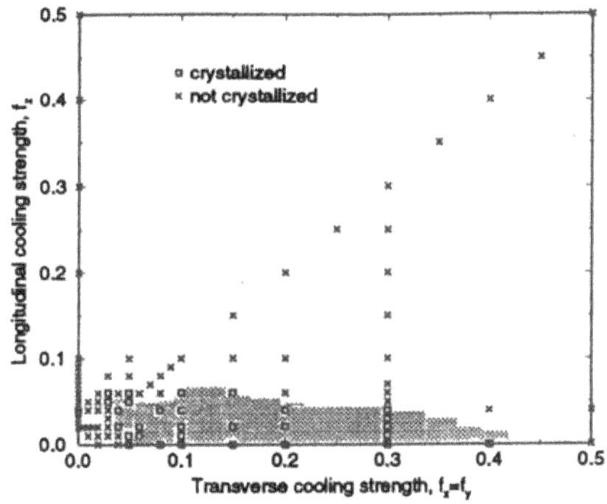

Figure 4: Beam crystallization by reducing the transverse (with $f_x = f_y$) and longitudinal (linear) momentum deviation. The shaded area and the squares indicate the region of cooling strengths with which the beam is successfully crystallized.

develop cooling methods that are effective in the transverse as well as the longitudinal direction.[25]

5.2 Model for Tapered Cooling

At the crystalline ground state, the radial displacement of the particle is linearly proportional to the particle momentum P_z. Therefore, an alternative method of beam cooling, as previously discussed widely, is to impose a condition of longitudinal momentum (or velocity) variation with respect to the radial displacement (tapered cooling). We simulate this method by varying once per lattice period the longitudinal momentum P_z of each particle according to

$$\Delta P_z = -f_z (P_i - C_{xz} x), \tag{17}$$

where, at the final stage of cooling when the temperature is extremely low ($T \leq 0.1$), we often assume $f_z = 1$. In the example of single-shell ground state (Section 5.1), the ordering can be achieved when the coefficient C_{xz} is within about $\pm 5\%$ of the ideal ground-state value ($C_{xz} \approx 0.285$).

6. Two-Species Crystals

Different species of charged ion beams travelling together with the same velocity will exchange energy via Coulomb interaction. The beam with lower temperature will absorb thermal energy from the beam with higher temperature. If cooling is applied on one ion species, the other ion species, with its temperature reduced, will be cooled "sympathetically".

In this section, we consider two species of ion beams with similar charge-to-mass ratio (Z/M) and similar rigidity, stored in the same storage ring. As we cool one species of ions, the total temperature of the two-species system decreases and, eventually, a crystalline state can be reached. As an illustrative example, we study the system of equal density Ar^+ ($Z = 1$, $A = 39.9624$) and Se^{2+} ($Z = 2$, $A = 79.9225$) ions. The machine consists of 10 FODO cells with 25% bending with $\nu_x = 2.8$, $\nu_y = 2.1$. The bending radius is $\rho = 1$ m, and the average machine radius is $R = 4$ m. The beam energy is $\gamma = 1.00001$. The characteristic distance is $\xi = 1.2 \times 10^{-5}$ m.

Only the Ar^+ beam is externally cooled by imposing the periodicity condition in three directions (Section 5). Figs. 5 and 6 show the beam structure during the

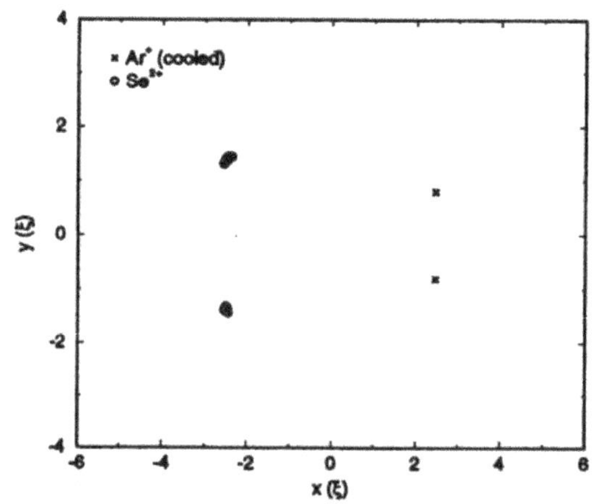

Figure 5: A two-species crystalline structure formed by Ar^+ ($A = 39.9624$) and Se^{2+} ($A = 79.9225$) ion beams of equal density ($N(Ar^+)/L = N(Se^{2+})/L = 0.5$, $L = 40\xi$) projected into the $x-y$ plane. Only Ar^+ beam is externally cooled. The temperatures are $T(Ar^+) = 5 \times 10^{-6}$, and $T(Se^{2+}) = 3 \times 10^{-4}$, respectively.

cooling process (after 100 revolutions of cooling on Ar^+) when the temperature of the Ar^+ beam is reduced to $T = 5 \times 10^{-6}$, and the temperature of the Se^{2+} beam is reduced sympathetically to $T = 3 \times 10^{-4}$. The crystalline state is stable when the external cooling is removed.

Due to the different charge-to-mass ratio Z/M between the different ion species, the average radial positions of the two beams are different. From Eq. 2, it can be

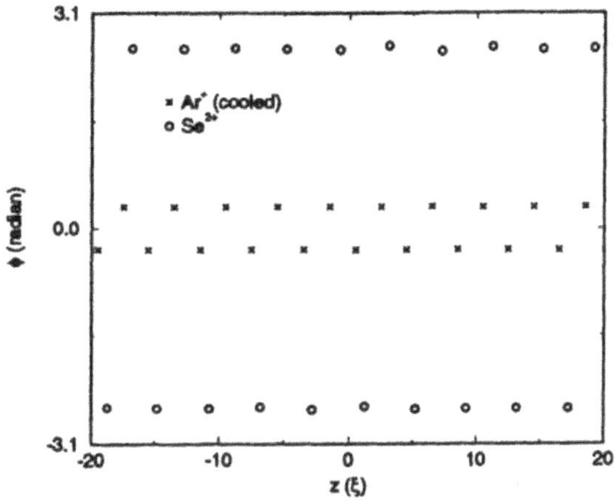

Figure 6: The same crystalline structure as shown in Fig. 5 projected into the $z - \phi$ plane, where ϕ is the polar angle.

estimated that the distance between the average radial beam centers x_c is

$$\Delta(x_c) \sim -\Delta\left(\frac{\bar{Z}}{m}\right)\frac{\rho}{\xi}, \tag{18}$$

i.e., it is linearly proportional to the difference in charge-to-mass ratio, and to the ratio ρ/ξ between the bending radius of the dipole magnets and the characteristic distance ξ. In Fig. 5, the positions of the Ar^+ ions are radially more outward from the center of the machine because of their relatively smaller Z/M.

7. Longitudinally Bunched "Crystal Balls"

We longitudinally bunch the beam of charged particles using rf cavities, located at one or more positions in the azimuth, with the rf electric voltage varying sinusoidally with time,

$$V(t) = V_0 \sin\left(h\omega_0 t\right), \tag{19}$$

where V_0 is the peak voltage, $\omega_0 \equiv \beta c/R$ is the revolution frequency of the reference particle, and h is the cavity harmonic number. Relative to the reference particle ($x = y = z = 0$), a particle of positive (or negative) z arrives at the cavity at a later (or earlier) phase and, therefore, experiences an energy gain per cavity:

$$eV_s = e\int E_s ds = eV_0 \sin\left(\frac{h\xi}{\gamma R}\,z\right). \tag{20}$$

In order to simulate the effect of a rf cavity, we employ Eq. 6 with $F_s = 0$ except in the region of the rf cavity. Because the cavity is a small fraction of the circumference, its effect can be approximated in a single numerical step using Eq. 20.

We find from the MD numerical studies that an rf field is no impediment to the formation of crystalline beams. Clearly, the rf field produces longitudinal bunching. Hence, the crystalline structures are finite in extent in all three directions. Among the many cases (1D, 2D, and 3D structures) that we have explored, we give one example in the series of figures from Fig. 7 to 11. The machine, having a periodicity

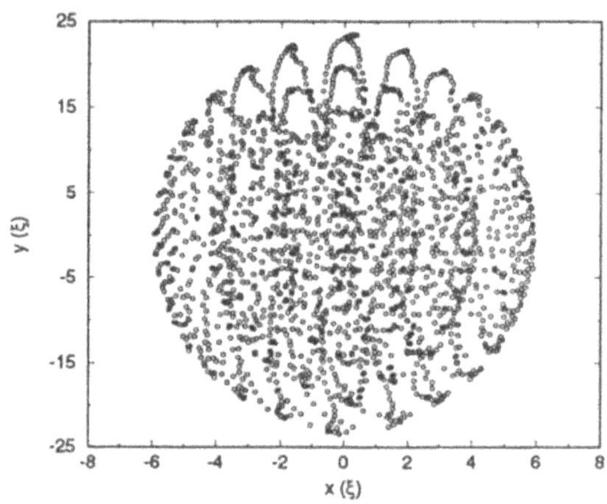

Figure 7: A multi-shell "crystal ball" structure with particle positions projected into the $x - y$ plane. Note that the structure is bunched in all three directions. The normalized temperature is $T \approx 2 \times 10^{-10}$.

of 10, consists of 10 FODO cells with 25% bending with $\nu_x = 3.1$, $\nu_y = 1.6$, and $\gamma_T = 2.9$. The beam energy corresponds to $\gamma = 1.4$. The rf voltage per lattice period is $V_0 = 200$ V. The machine parameters are chosen such that $\pi\gamma R/h\xi = 1000$, i.e., one rf wavelength corresponds to a distance from -1000ξ to 1000ξ (e.g. for $^{24}Mg^{2+}$ beam, $\xi = 6.4 \times 10^{-7}$; if $\rho = 1$ m, $R = 4$ m, then $h = 2.7 \times 10^4$.).

Fig. 7 shows a projection of the multi-shell structure into the transverse plane. In this example, $N = 2000$ particles are used for simulation. The structure is obtained by cooling (imposing the periodicity condition) the beam for 2,000 lattice periods (40,000 numerical steps). Figs. 8 and 9 show projections of the same structure into the $x - z$ and $y - z$ planes, respectively. Since one rf wavelength corresponds to a distance from -1000ξ to 1000ξ, this crystalline structure extends over about 1/10 of the rf period. The tilt in the $x - z$ plane in Fig. 8 is a result of the particular azimuthal location relative to the rf cavity at which the projection is made. Of course,

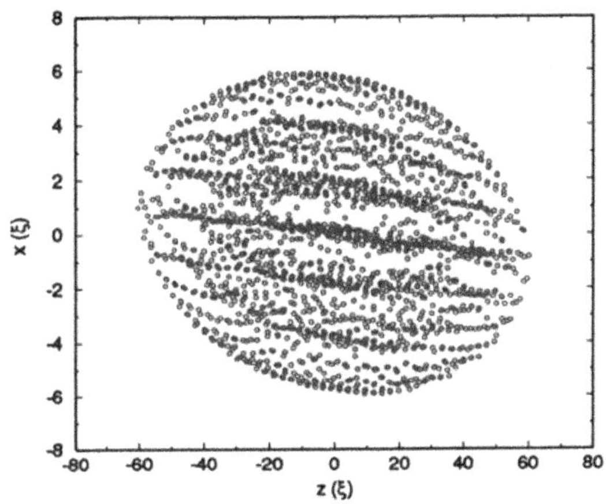

Figure 8: The same multi-shell structure as shown in Fig. 7 with particle positions projected into the $x - z$ plane.

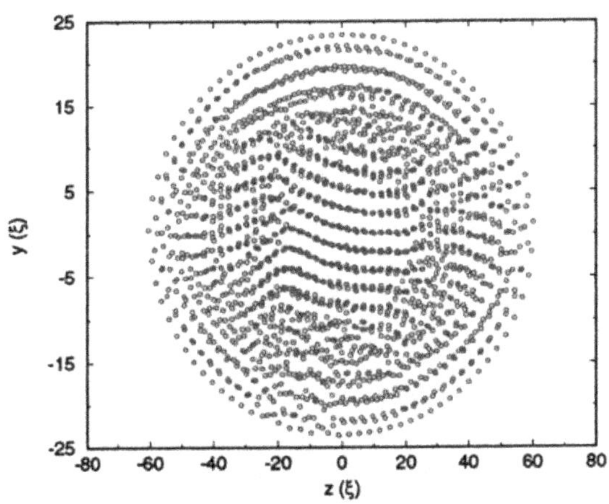

Figure 9: The same multi-shell structure as shown in Fig. 7 with particle positions projected into the $y - z$ plane.

the crystalline beam breathes and shears as it moves along the FODO lattice. The trajectories of positions and momenta are displayed for one out of the 2000 particles in Figs. 10 and 11, respectively. The rapid change in P_z in Fig. 11 is caused by the

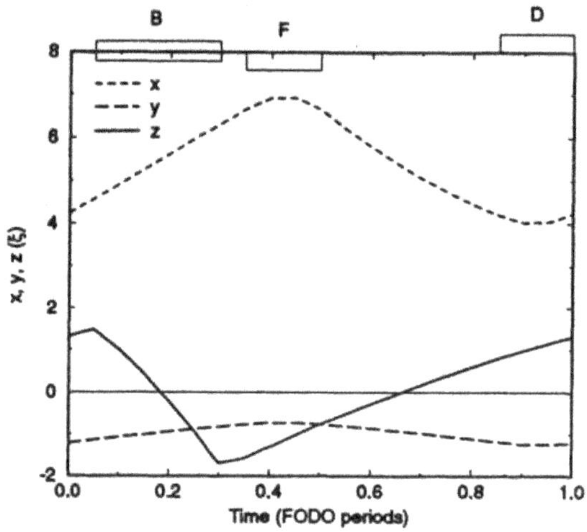

Figure 10: Typical particle trajectory of a bunched crystalline beam. Lattice components are displayed on the figure: B is a bending section, F and D are focusing and de-focusing sections.

energy gain at the rf cavity.

The structure shown in Fig. 7 corresponds to a low temperature state of $T \approx 2\times10^{-10}$. After the cooling is turned off, the structure essentially remains unchanged. Further studies indicate that the heating rate for this bunched crystalline beam is similar to that for the un-bunched beam (Fig. 3).

With given beam and machine parameters and cavity h, the peak voltage V_0 determines the strength of the longitudinal focusing and, therefore, the longitudinal size of the crystal. A higher (or lower) voltage results in a shorter (or longer) crystal length and higher (or lower) density in the longitudinal direction. Since the transverse intra-particle distance is mainly determine by the strength of the transverse focusing and the Coulomb force, this higher (or lower) longitudinal density results in a larger (or smaller) transverse crystal size. Obviously, if the voltage is sufficiently small, the crystal becomes de-bunched.

With given machine and rf cavity parameters, the nature of the crystalline ground state depends upon the number of particles confined in the distance of one rf wavelength (i.e. beam density). When the number of particles is small, the ground state is a 1D chain of finite length with smaller intra-particle distance for particles near the chain center. This chain tilts along x direction in the $x-z$ plane due to the energy

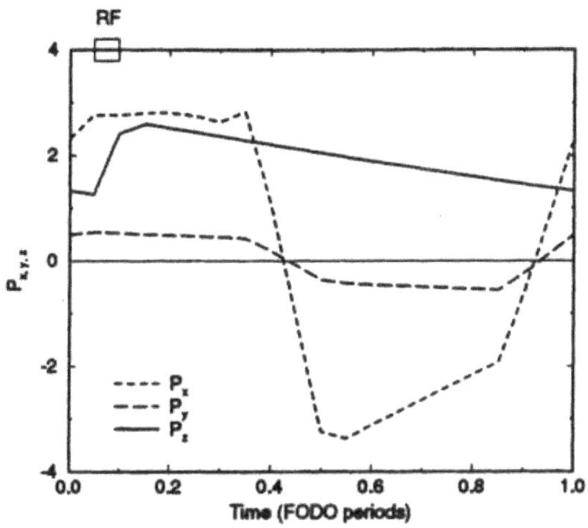

Figure 11: Typical particle momentum trajectory of a bunched crystalline beam. The location of the rf cavity is displayed on the figure.

increment caused by the rf cavity. As the beam density increases, the ground state becomes 2D extending in the plane of weaker transverse focusing. As the density increases further, the ground state becomes single and then multi-shell 3D, as shown in Figs. 7 - 9. The intra-particle distance, which is mainly determined by the strength of focusing and Coulomb force, is independent of the beam density. The conditions for the formation and maintenance of crystalline beams as discussed in Section 4 also apply to the bunched beam.

8. NAP-M Simulations

Having developed the formalism which allows us to study both highly developed and incipient crystalline beam structures, we are in a position to study, theoretically, the experimental situation on the NAP-M ring when it was used to study crystal formation in the late 1970's.[26]-[29] Actually in the experiment, the physicists were simply using electron cooling to reach the lowest possible temperatures. Compared with the un-cooled beam, the Schottky noise power for the cooled beam is reduced by approximately two orders of magnitude. Typically, the Schottky power decreases with the decreasing beam intensity. However, when the beam intensity is below $N_0 = 2\times10^7$, the Schottky power becomes independent of the intensity, corresponding to a longitudinal beam temperature of $T_{Bz} = 1$ K. It has later[29] been estimated that a crystalline state could have been observed at NAP-M at low intensity of $N_0 < 10^6$, outside of the range of the original measurement. At that time, no detailed theoretical analysis was performed; now, we are in a position to do just that.

We obtained the description of the NAP-M lattice and beam parameters primarily by private communication with Nicoli Dikanskiĭ. The machine circumference was 47.25 m, the lattice periodicity was 4, the transverse tunes were $\nu_x = 1.35$ and $\nu_y = 1.26$, the transition energy corresponded to $\gamma_T = 1.18$, and the beam energy corresponded to $\gamma = 1.07$. Electron cooling was employed to reduce the beam temperature. The lowest achieved temperatures were transverse T_{Bx}, $T_{By} = 50 \sim 100$ K, and longitudinal $T_{Bz} = 1$ K.

Using these parameters, we perform MD simulations in which electron cooling is simulated by reducing the momenta in three directions. Numerically starting from a high-temperature state, we turn off the cooling when the temperature falls into the desired range. Within the range of the experimentally measured beam intensity (from $N_0 = 2 \times 10^6$ to 10^8), our MD simulation indicates no crystalline structure.

In order to verify Dikanskiĭ's speculation[29] that a crystalline state could have been observed at NAP-M at low intensity of $N_0 < 10^6$, we perform detailed study at the intensity of $N_0 = 1.5 \times 10^5$, outside of the range of the original experimental measurement. Fig. 12 shows the density of the beam as a function of radial distance ρ,

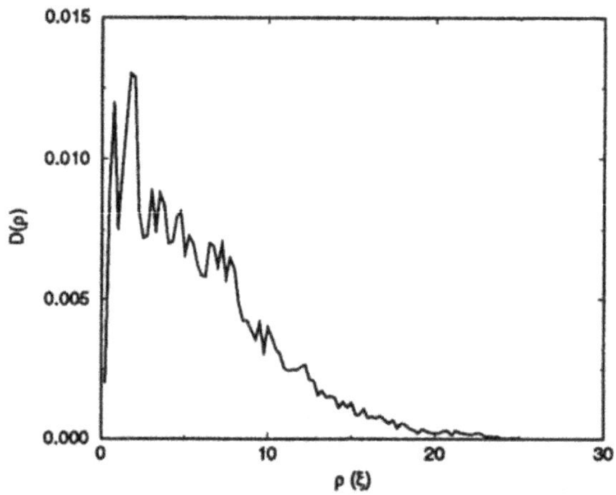

Figure 12: Density as a function of the radial distance ρ for the NAP-M simulation with $T_{Bx} \approx T_{By} \approx 50$ K, and $T_{Bz} = 1$ K, averaged over 22 μs and assuming $N_0 = 1.5 \times 10^5$. The radius ρ is measured in units of $\xi = 4.6\mu$m.

obtained using 40 particles per MD cell averaging over 50 revolutions (or a time period of 22 μs). The beam extends transversely to about 20ξ, or 0.11 mm. Fig. 13 shows the density of the beam as a function of longitudinal distance z, again averaging over 50 revolutions. Due to the relatively low beam intensity and relatively high transverse

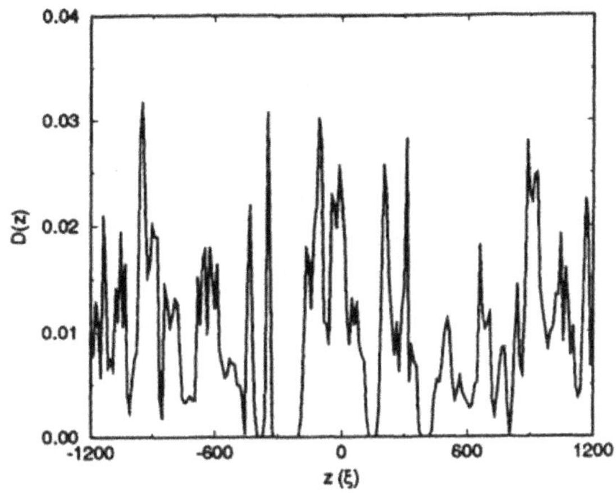

Figure 13: Density as a function of the longitudinal distance z for the NAP-M simulation with $T_{Bx} \approx T_{By} \approx 50$ K, and $T_{Bz} = 1$ K, averaged over 22 μs and assuming $N_0 = 1.5 \times 10^5$. The distance z is measured in units of $\xi = 4.6\mu$m.

temperature, the beam behaves like a 1D chain of "disks" formed by transversely oscillating particles. Because of the non-zero longitudinal momentum spread (P_z) or temperature (T_{Bz}), the disks drift in the longitudinal direction, occasionally passing each other for lack of IBS interaction at low particle density.

In order to quantitatively determine the amount of longitudinal drift, we define the one-body correlation function $G_z(\tau)$,

$$G_z(\tau) = \langle (z_i(t+\tau) - z_i(t))^2 \rangle_{i,t}, \tag{21}$$

where $\langle\ \rangle$ denotes the average over all the particles i and over time t. Fig. 14 shows $G_z(\tau)$ as a function of time interval τ at three different longitudinal temperatures $T_{Bz} = 10$ K, 1 K, and 0.1 K, respectively, while the transverse temperatures are kept constant at $T_{Bx} \approx T_{By} \approx 50$ K (note the logarithmic scale). The zig-zag along each curve, with their amplitude proportional to the transverse temperature and their frequency the same as the horizontal betatron tune ν_x, is caused by the coupling between longitudinal (z) and horizontal (x) motion. At low temperature $T_{Bz} < 0.1$ K, G_z is approximately a constant in τ, and the disks are "frozen" in their longitudinal positions during the time of observation. At intermediate temperature $T_{Bz} \sim 1$ K, G_z increases linearly with τ, and disks drift and occasionally pass each other. In both cases, the longitudinal motion of the particles is much restricted, and the longitudinal Schottky signal will be greatly suppressed as observed experimentally. At high temperature, i.e. $T_{Bz} > 10$ K, the Coulomb interaction becomes negligible, and G_z is

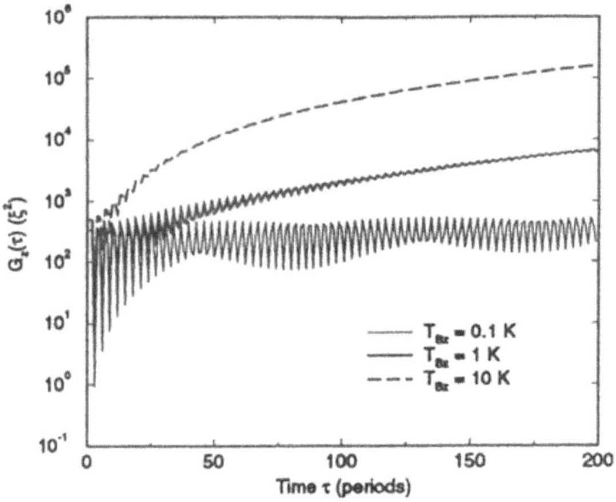

Figure 14: Longitudinal correlation functions as functions of time interval τ for the NAP-M simulations assuming $N_0 = 1.5 \times 10^5$. In the three cases shown, $T_{Bx} \approx T_{By} \approx 50$ K, and $T_{Bz} = 10$ K, 1 K, and 0.1 K, respectively.

proportional to τ^2 in this gaseous state.

As long as the particle density is low so that the amplitude of the coupled z oscillation (which is of the same order as the transverse amplitude when T_{Bz} is much smaller than $T_{Bx,y}$) is much smaller than the longitudinal intra-particle distance, the rate of longitudinal drift is insensitive to the transverse temperature and beam intensity. When the particle density is high so that the amplitude in z is larger than the intra-particle distance, the disk structure no longer exists due to the large Coulomb interaction. In this case, the beam behavior can be described by the conventional IBS theory as discussed in Section 3.3.

9. New ASTRID Simulations

Detailed study of crystalline beams in the present ASTRID ring was presented in Ref. [21]. There, we have shown that only 1D structures were stable, and that 1D structures can only be achieved when the number of particles in the ring is less than 3.4×10^5. Furthermore, the maximum heating rate for the 100 keV $^7Li^+$ beam was shown to be $\Delta T/T \approx 4 \times 10^{-2}$ per lattice period, which corresponds to the minimum cooling rate needed for the crystal formation of 8×10^3 (K/sec) at the temperature of 1.5 K.

Even without developing a 1D crystal due to the high transverse temperature, one can develop an incipient crystal, analogous to what could have been observed at NAP-

M at low beam intensity. Based on the machine parameters given in Ref. [21], we have performed numerical simulations similar to those for NAP-M. Assume that the transverse temperature is about 10^3 K. In order to form longitudinally well separated "disks", the number of particles in the ring should be less than about 3×10^4. On the other hand, the currently achievable longitudinal temperature of 1 mK is sufficiently low to keep the disks "frozen" in their longitudinal positions.

In the present ASTRID ring, 2D and 3D crystalline structures can not be maintained because the second condition (Section 4.2) is not satisfied. As a consequence, consideration is given to a proposed New ASTRID ring which satisfies both necessary conditions for the formation and maintenance of crystalline beams. According to the machine lattice obtained from Jeffrey Hangst, this machine has a periodicity of 4, and a circumference of 48.6 m. The transverse tunes are $\nu_x = 1.38$ and $\nu_y = 1.32$, and $\gamma_T = 1.29$. The beam energy is $\gamma = 1.000016$. Not surprisingly, ground states of all dimensions can be readily achieved. An example of a 3D state is given in Fig. 15. The heating rates obtained for 1D, 2D, and 3D crystalline states are similar to those

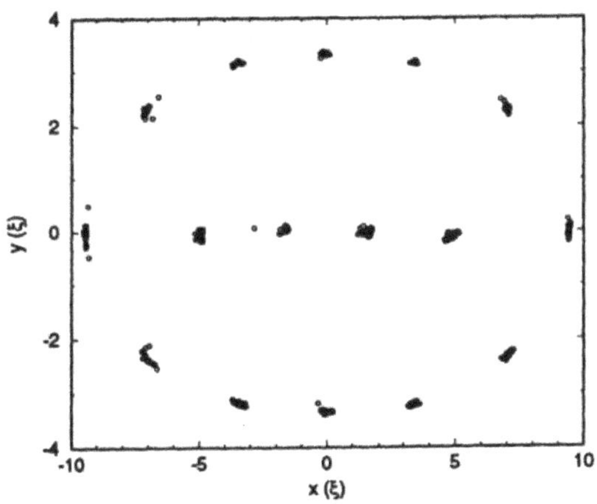

Figure 15: A 3D structure with particle positions projected into the $x-y$ plane. The New ASTRID ring has a periodicity of 4 with $\nu_x = 1.38$, $\nu_y = 1.32$. The beam energy is $\gamma = 1.000016$. The calculation is done with $N = 400$, and $L = 100\xi$.

shown in Fig. 3.

Numerical studies were made also for the case of $\nu_x = 1.67$ and $\nu_y = 1.68$. Since the second condition was in this case violated, it was again not surprising to find that only 1D crystals could be formed and maintained, as shown in Fig. 16.

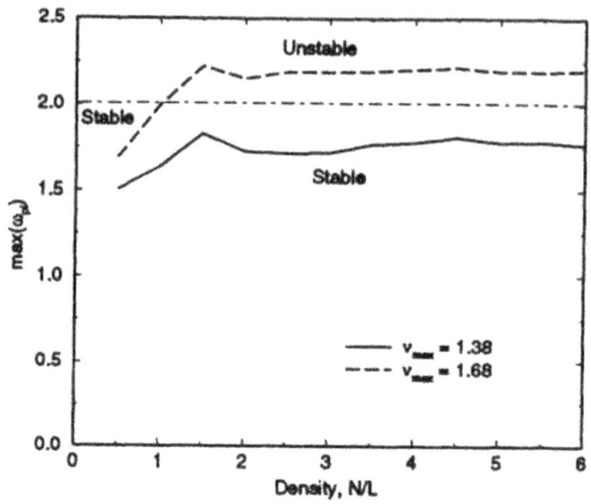

Figure 16: The maximum phonon frequencies as functions of beam density at various design transverse tunes. The crystalline structure is unstable when the maximum phonon frequency is higher than half of the lattice frequency. Here, the lattice periodicity is 4.

10. Conclusions

We have presented the equations of motion appropriate to interacting charged particles of diverse charge and mass, subject to the external forces produced by bending dipole, AG focusing quadrupole, and sextupole magnets, and by bunching rf cavities in real storage rings. Employing these equations in the molecular dynamics simulations, we obtained heating rates of the crystalline beam as functions of beam density and temperature. In the high temperature limit, it has been shown that the MD results agree well with the intra-beam scattering theory.

As presented in our previous papers, there are two necessary conditions for the formation and maintenance of the crystalline beam. The first condition requires that the storage ring is AG focusing operating below the transition energy. The second condition requires that the lattice periodicity of the ring is at least $2\sqrt{2}$ as high as the maximum betatron frequency tune.

Various cooling methods have been studied using the MD simulation. Provided that the cooling rate is adequate, the present cooling techniques suffice in making 1D crystals. In order to achieve higher-density ordered states, it is necessary to develop cooling methods that are more effective in the transverse direction.

The formalism has been used to study the formation of crystalline beams of two different ion species of similar charge-to-mass ratio. It is shown that cooling only one of the two species can result in a crystalline state of both species; i.e., that the mutual interaction of the particles leads to sympathetic cooling.

The formalism has also been used to study the effect of rf fields upon a bunch of charged particles. It is shown that "crystal balls" with finite extent in all three directions can be formed and maintained.

We have studied the NAP-M ring where anomalous Schottky behavior was observed. Using the tools that we have developed, we reconstruct the conditions of the NAP-M experiment. It is found that no crystalline state can be reached at the original experimental range of beam intensity and temperature. On the other hand, at extremely low beam intensities, an incipient crystalline structure could be observed at the NAP-M installation with the particles forming into occasionally-passing disks.

Finally, we have studied the New ASTRID ring which has been proposed for the formation and study of crystalline beams. Note that the existing rings in Denmark and Germany do not satisfy the second necessary condition and, consequently, would only allow the formation of 1D structures. It is shown that the New ASTRID ring would be suitable for the formation of crystalline beams of all dimensions.

In summary, we have shown that the formalism and numerical methods that we have developed allow one to study a diversity of topics in crystalline beams. Such subject seems rich, indeed, and one eagerly awaits experimental realization of crystalline beams.

Acknowledgements

The authors would like to thank J. Hangst, D. Habs, N.S. Dikanskiĭ, and I. Hofmann for useful discussions, and D. Reistad for pointing out an error in our assumption of the NAP-M experimental condition.

References

[1] N.S. Dikanskiĭ and D.V. Pestrikov, Proc. of the Workshop on Electron Cooling and Related Applications, KfK 3846, ed. H. Poth, (1984).

[2] J.P. Schiffer and P. Kienle, Z. Phys. A **321**, 181 (1985).

[3] J.P. Schiffer and O. Poulsen, Europhys. Lett. **1**, 55 (1986).

[4] A. Rahman and J.P. Schiffer, Phys. Rev. Lett. **57**, 1133 (1986).

[5] D. Habs, MPI Heidelberg preprint MPIH-1987-V10 (1987).

[6] J.P. Schiffer, Phys. Rev. Lett. **61**, 1843 (1988).

[7] R.W. Hasse and J.P. Schiffer, Ann. Phys. **203**, 419 (1990).

[8] R.W. Hasse, GSI preprint GSI-90-23 (1990).

[9] R.W. Hasse and V.V. Avilov, Phys. Rev. A **44**, 4506 (1991).

[10] J.P. Schiffer, Phys. Rev. Lett. **70**, 818 (1993).

[11] J.P. Schiffer and A. Rahman, Z. Phys.A – Atomic Nuclei 331, 71–74 (1988).

[12] J.P. Schiffer, Proc. of the Workshop on Crystalline Ion Beams, Werheim, Germany, 2 (1988), ed. R.W. Hasse, I. Hofmann, D. Liesen.

[13] J.P. Schiffer and A. Rahman, Z. Phys.A-Atomic Nuclei **331**, 71 (1988).

[14] J.P. Schiffer, Proc. of the Workshop on Crystalline Ion Beams, Werheim, Germany, editors R.W. Hasse, I. Hofmann, and D. Liesen, p.2 (1988).

[15] J. Beebe-Wang, N. Elander and R. Schuch, Nucl. Instrum. Methods **B79**, 806 (1993).

[16] J. Wei, X-P Li, and A.M. Sessler, "Crystalline Beam Ground State", Brookhaven National Laboratory Report BNL-52381, Upton, New York (1993).

[17] J. Wei, X-P Li, and A.M. Sessler, "Crystalline Beam Ground State", Proc. of The 1993 Particle Accelerator Conference,Washington, May 1993, p. 3527.

[18] J. Wei, X-P Li, and A.M. Sessler, "Critical Temperatures for Crystalline Beams", Proc. of the Workshop on Beam Cooling and Related Topics, Montreux, October 1993, p. 366.

[19] J. Wei, X-P Li, and A.M. Sessler, "The Low-Energy States of Circulating Stored Ion Beams: Crystalline Beams", Physical Review Letters **73**, 3089 (1994).

[20] J. Wei, X-P Li, and A.M. Sessler, "Crystalline Beams", Advanced Accelerator Concepts, Fontana, June 1994, AIP Conference Proceedings **335**, ed. P. Schoessow, p.224, (1995).

[21] J. Wei, X-P Li, and A.M. Sessler, "Crystalline Beam Properties as Predicted for the Storage Rings ASTRID and TSR", Proc. of the 1995 Particle Accelerator Conference, Dallas, 1995 (to be published). Note that the 2D ground state that we found for the TSR ring rotates once per lattice period along its longitudinal axis. Further study of the structure has shown that the rotation can be either clockwise or counter-clockwise, i.e., the two rotation states are degenerate.

[22] X-P Li, A.M. Sessler, and J. Wei, "Crystalline Beams in a Storage Ring: How Long Can It Last", Proc. of the 1994 European Particle Accelerator Conference, London, 1379 (1994).

[23] J. Wei, "Evolution of Hadron Beams under Intrabeam Scattering", Proc. 1993 Part. Accel. Conf. Washington D.C., p.3651 (1993).

[24] J. S. Hangst, *Laser Cooling of a Stored Ion Beam - A First Step Towards Crystalline Beams*, Ph.D Thesis, 1992, Argonne National Laboratory Report ANL/PHY-93/1.

[25] H. Okamoto, A.M. Sessler, and D. Möhl, Phys. Rev. Lett., bf 72, 3977 (1994).

[26] V.V. Parkhomchuk and N.S. Dikanskiĭ, Sov. Journ. of Tech. Phys., bf 50, 1411 (1980).

[27] E.E. Dement'ev, N.S. Dikanskiĭ, A.S. Medvedko, V.V. Parkhomchuk and D.V. Pestrikov, Zh. Tekh. Fiz **50**, 1717 (1980); English transl. Sov. Phys. Tech. Phys. **50**, 1001 (1980).

[28] V.V. Parkhomchuk and A.H. Skrinsky, Report on Progress in Physics, bf 54, 919 (1991).

[29] N.S. Dikanskiĭ and D.V. Pestrikov, *The Physics of Intense Beams and Storage Rings*, AIP Press (1994).

Analytical Studies of Crystalline Beams

A.F. Haffmans, D.M. Maletić, A.G. Ruggiero
Brookhaven National Laboratory
Upton, New York

We present an analytical analysis of the confinement and stability of "Crystalline Beams" in storage rings, in terms of such fundamental accelerator concepts as tune shift and stopband. We consider a "Crystalline Beam" consisting of sub-strings, arranged symmetrically around the reference trajectory, and examine the motion of a slightly perturbed test particle in the beam. We find that a "Crystalline Beam" has an upper and lower limit of the spacing between the ions. The upper limit is determined by condition that the "Crystalline Beam" has the same periodicity as the storage ring. The lower limit is set by the stability of the motion of the test particle around the equilibrium.

1 Introduction

The crystallization of laser cooled ions in a trap is by now a well established physical phenomenon [1,2]. It has been conjectured, that a similar experiment on ions moving in a storage ring will result in a "Crystalline Beam" [3,4]. When cooled sufficiently, the ions in the beam will form crystalline structures, similar to ones observed in ion traps. Because the ions are stacked closely together, the luminosity of such a "Crystalline Beam" would be very high, and high luminosity beams have important engineering applications, e.g. in nuclear waste management, or nuclear fusion [5] [22]. No experimentally conclusive observation of a "Crystalline Beam" has been made yet, except for one tantalizing observation in Novosibirsk. Indeed, the strongest support for this conjecture has come from molecular dynamics (MD) simulations, the first of which was performed almost a decade ago [4], and its subsequent improvements [6,7]. The crystalline structures predicted by the MD simulations have been observed in ion traps [1,2] and dusty plasmas [8].
Several laboratories are trying to either observe "Crystalline Beams" using existing facilities, or to design storage rings in which this is possible [9]. Therefore, we believe that there is a need for a relatively fast design tool to determine whether a particular lattice can store crystalline structure. This study is also motivated by the desire to generate fusion in an accelerator using "Crystalline Beams" [22]. Our goal is to obtain analytical estimates for the general properties of the "Crystalline Beam" and the storage ring containing it using as few á priori asumptions as possible.

Because "Crystalline Beams" have such a regular structure, they lend themselves particularly well for a simple analytical analysis. The cooling of the beam is not included in the analysis. That is, we start with a beam already at zero temperature. In addition, the linear theory we derive below ignores the effect of non-linearities storage rings may contain, either by design (e.g. sextupoles), or accident (e.g. field errors in linear elements) .

2 General outline

A typical "Crystalline Beam" consists of one or more substrings of equally spaced ions, arranged symmetrically around the refernce trajectory of the storage ring. The subtrings are shifted relative to each other. In Fig. 1 we show a horizontal zigzag as an example of a fairly simple "Crystalline Beam" which has been observed in ion traps [1,2] and MD simulations [6]. The spacing between the ion is denoted by λ, and the shift between the two substrings is denoted by τ.

We investigate the proporties of the "Crystalline Beam" in the storage ring by looking at the motion of a slightly perturbed test particle in the beam. In order to contain a "Crystalline Beam" , the storage ring needs to satisfy two criteria. First, the transversal focussing of the ring needs to be strong enough to balance the repulsion of the ions in beam. Confinement of the beam in the logitudinal direction is assured by the ion repulsion. Secondly, the equilibrium needs to be stable, and the stability of the equilibrium floows from the motion of the test particle around it.

To determine whether these requirements are met, the equations of motion of the test particle in the "Crystalline Beam" are linearized, and integrated approximately, either using the "smooth approximation", or by deriving a transfer matrix by integrating over a linear beam element. The beam's confinement and stability follow from an examination of the eigenvalues of the product of the transfer matrices representing the different elements in the storage ring.

But which "Crystalline Beam" should be studied ? We don't consider the cooling process, so starting with a "hot" beam, cool it down, and studying the resulting "Crystalline Beam" is out of the question.. Alternatively, we could examine all possible combinations of ions strings. But this is time consuming and not à priori.

Rather we use what we have dubbed the "Bifurcation Method". We start with a single string of ions along the reference trajectory, and increase the linear ion density in the string. The reference orbit will undergo a Pitchfork Bifurcation [14,15]. In a Pitchfork Bifurcation the original orbit becomes unstable, and generates two stable orbits, with the same period as the original one [14,15].

This is identical to a linear resonance or stopband: one set of eigenvalues of the transfer matrix become real [13]. So, we have the following succession of "Crystalline Beam" structures:

$$\underbrace{\text{String} \to \text{zigzag};}_{\text{increase } N \to 2N}\quad \underbrace{\text{zigzag} \to \text{helix};}_{\text{increase } 2N \to 4N}\quad \underbrace{\text{helix} \to \text{shell};}_{\text{increase } 4N \to 8N}\quad \underbrace{\text{shell} \to \ldots}_{\text{increase } 8N \to 16N}$$

where N is the number of ions in the accelerator (see also Fig. 2). In this

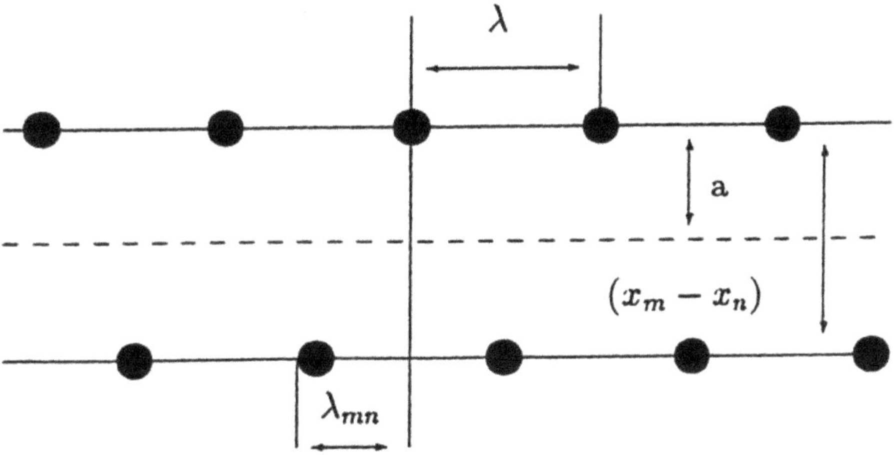

Figure 1: The horizontal zigzag. λ is the ion lattice constant; λ_{mn} is the relative spatial shift, and a is the separation from the (dilute beam) reference trajectory.

picture, the number of ions doubles at each transition, and the ion seperation increases roughly by a factor of two. There is an upper (formation) and lower (stability) limit for the ion seperation λ, where the upper limit is roughly twice the lower limit of the previous structure. We repeat this process until the next structure is no longer stable.

3 Stability and Confinement Conditions

In this section we derive the basic equations, following the method outlined in the previous paragraph.

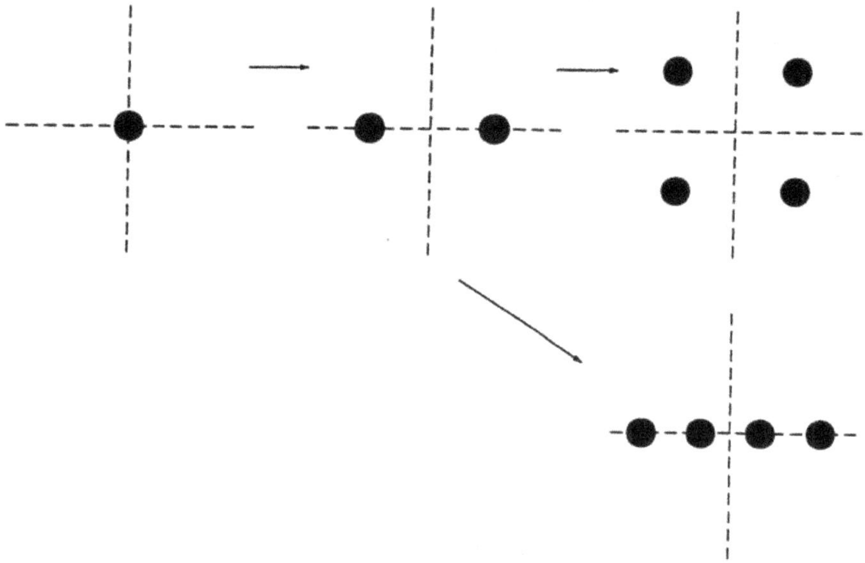

Figure 2: The bifurcation picture: At the critical linear ion density the string along the reference trajectory splits into a horizontal sigzag, which in turn might generate a "double" sigzag in the horizontal plane, or a helix, depending on the focusing properties of the storage ring. Structures involving more substrings are generated accordingly.

3.1 Equations of Motion

The derivation of the equations of motion of the test particle is discussed in detail in [11,12]. We use a curvilinear orthogonal coordinate system [13], with the path length s as the independent variable. x and y are in the radial (horizontal) and vertical direction, respectively. In addition, we introduce the path length difference σ and its canonical conjugate, the momentum error δ. The equations of motion are:

$$y'' + K_v(s)y + \frac{Qr_0}{eA\beta^2\gamma^3}\frac{\partial\,\Phi_{sc}}{\partial\,y} \;=\; 0; \tag{1}$$

$$x'' + K_h(s)x + \frac{Q r_0}{e A \beta^2 \gamma^3} \frac{\partial \Phi_{sc}}{\partial x} = h(s)\delta; \tag{2}$$

$$\sigma' = \frac{\delta}{\gamma^2} + h\,x; \tag{3}$$

$$\delta' = \frac{Q r_0}{e A \beta^2 \gamma^3} \frac{\partial \Phi_{sc}}{\partial \sigma}. \tag{4}$$

r_0 is the classical proton radius. The substring with the test particle is labeled 1. Its coordinates are $(x_1, y_1, \sigma_1 = 0)$.
λ_{1k} is the relative shift of substring k with respect to substring 1; $\lambda_{11} = 0$.
n_s substrings contribute to the space charge potential, viz.

$$\Phi_{sc} = \sum_{k=1}^{k=n_s} \Phi_k \tag{5}$$

$$\Phi_k = 2Qe\gamma \sum_i \frac{1}{[(x_1 - x_k)^2 + (y_1 - y_k)^2 + \gamma^2(\lambda\, i - \lambda_{1k})^2]^{\frac{1}{2}}} \tag{6}$$

3.2 Scaled Variables

We introduce the following dimensionless scaled coordinates:

$$\alpha_{hk} = \frac{x_1 - x_k}{\gamma\lambda} \quad \text{Horizontal aspect ratio.} \tag{7}$$

$$\alpha_{vk} = \frac{y_1 - y_k}{\gamma\lambda} \quad \text{Vertical aspect ratio.} \tag{8}$$

$$\tau_k = \frac{\lambda_{1k}}{\lambda} \quad \text{Fractional relative shift.} \tag{9}$$

The (derivatives) of the space charge potential are then proportional to:

$$f_n(\alpha_{hk}, \alpha_{vk}, \tau_k) = \sum_i \frac{1}{[\alpha_{hk}^2 + \alpha_{vk}^2 + (i - \tau_k)^2]^{\frac{1}{2}}} \tag{10}$$

The constant of proportionality contains all the "specifics" of the ion.

3.3 Expansion around the equilibrium

The space charge potential Φ_{sc} is expanded around the equilibrium position of the test particle:

$$x_1 \rightarrow x_t = x_1 + x \tag{11}$$

$$y_1 \rightarrow y_t = y_1 + y \tag{12}$$

$$\sigma_1 \rightarrow \sigma_t = \sigma_c + \sigma \tag{13}$$

with x, y, σ_c, σ small. The expansion in the tranverse plane follows from expanding the space charge term to first order in x,y, viz.:

$$\frac{\partial\Phi_{sc}}{\partial y} \to \frac{\partial\Phi_{sc}}{\partial y}|_{eq} + \frac{\partial^2\Phi_{sc}^2}{\partial y}|_{eq}\, y \tag{14}$$

Recall that there is no net longitudinal force. The longitudinal force due solely to the ions, and therefore the first order expansion in the longitudinal direction in σ is zero. This condition determines the shift between the different substrings:

$$\sum_{k=1}^{n_s} \sum_{i} f_3(\alpha_{hk}, \alpha_{vk}, \tau_k)(i - \tau_k) = 0 \tag{15}$$

τ_k is choosen in such a way, so that this condition is satisfied.

3.4 *Longitudinal Force and Shear*

The trajectory in the ring is curved. The previous condition does not take this into account. Because of the curvature, the particles must be allowed to "shear" a little bit with respect to each other. The confinement and stability under shear is determined by the second order expansion in the longitudinal coordinate:

$$\frac{\partial\,\Phi_{sc}}{\partial\,\sigma} \to \frac{\partial^2\Phi_{sc}}{\partial\sigma^2}|_{eq}\,\sigma_c + \frac{\partial^2\Phi_{sc}}{\partial\sigma^2}|_{eq}\sigma \tag{16}$$

3.5 *Envelope and Stability Equations*

Inserting these expansions into the equations of motion yields two sets of equations: Envelope equations, which determine the equilibrium, and Stability Equations, which determine its stability. The non-linear envelope equations are:

$$y_1'' + [K_v(s) - K_{sc}F_v]y_1 = 0 \tag{17}$$

$$x_1'' + [K_h(s) - K_{sc}F_h]x_1 = h(s)\delta \tag{18}$$

$$\sigma_c' = h(s)x_1 - \frac{\delta_1}{\gamma^2} \tag{19}$$

$$\delta_1' = 2K_{sc}F_l\sigma. \tag{20}$$

The space charge factor K_{sc} is given by:

$$K_{sc} = 2\frac{Q^2 r_0 g_0}{A\beta^2\gamma^5\lambda^3} \tag{21}$$

The form factors F_h, F_v and F_l depend only on the scaled coordinates. They follow directly from the structure of the cb, and are proportional to $f_{3k}(\alpha_{hk}, \alpha_{vk}, \tau_k)$.

A closed solution to this equation has to meet the isochronism condition: the "Crystalline Beam" has the same periodicity as the storage ring. In vector form:

$$\vec{\xi} = M\,\vec{\xi} \tag{22}$$

The solution is an eigenvector with eigenvalue 1,and a set of eigenvalues on the real positive axis. This corresponds to a (horizontal or vertical) stopband, and consequently no "Crystalline Beam" is possible at a coupling resonance. The linear stablity equations are:

$$\begin{aligned} y'' + [K_v(s) - K_{sc}\eta_v]y &= 0 && (23)\\ x'' + [K_h(s) - K_{sc}\eta_h]x &= h(s)\delta && (24)\\ \sigma' &= h(s)x - \frac{\delta}{\gamma^2} && (25)\\ \delta' &= 2K_{sc}\eta_l\sigma. && (26) \end{aligned}$$

The stability form factors are of the form:

$$\begin{aligned} \eta_v &= 1 + \frac{1}{g_0}\sum_{k=2}^{n_s} f_{3k} - 3\alpha_{vk} f_{5k} && (27)\\ \eta_h &= 1 + \frac{1}{g_0}\sum_{k=2}^{n_s} f_{3k} - 3\alpha_{hk} f_{5k} && (28)\\ \eta_l &= \frac{1}{2}(\eta_h + \eta_v) && (29) \end{aligned}$$

The first term (i.e. the '1') is due to the reference string; the second term due to the other strings. $\eta_{h,v,l}$ are constants, evaluated at the equilibrium, which follows from the envelope equations.

4 The string

The string consists of equally spaced ions along the reference trajectory, i.e. $x_1 = y_1 = \delta_1 = 0$. Therefore, the envelope equations are always satisfied. There are no other substrings, and the stability form factors $\eta_{h,v,l}$ are all 1.

4.1 *Determining the Stability using Transfer Matrices*

In Appendix A we derive transfer matrices for the drift, dipole, and quadrupole. For a detailed derivation we refer to [18]. The transfer matrix over one super period is:

$$M = \ldots \circ \; M_n \; \circ M_{n+1} \; \circ \ldots \tag{30}$$

The eigenvalues of M are determined for increasing values of λ. For λ small, the dilute (hot) beam values are found. This is done for a variety of different storage rings, and the results are shown in table 1.

4.2 *Stability of the String in the smooth approximation.*

In Appendix B, we give the derivation of the limits of stability in the verical, and radial direction using the smooth appriximation [18]. The string is stable in the vertical plane as long as the the ion separation λ is larger than the critical value, viz.:

$$\lambda > \lambda_c \, c_v, \tag{31}$$

$$c_v = (\nu_v \delta_v)^{-\frac{1}{3}}, \tag{32}$$

where λ_c is the critical lattice spacing:

$$\lambda_c \;\; = \;\; \left(\frac{Q^2 r_0 g_0 R^2}{A\beta^2\gamma^5}\right)^{\frac{1}{3}} . \tag{33}$$

Instability in the radial plane occurs when a radial or a longitudinal stopband is crossed or when a coupling resonance is encountered. The dilute beam tune $\nu_{v,0}$ lies between mN_p and $(m+1)N_p$. The following can be distinguished [18]:

- $\mathbf{mN_p < \nu_h < (m + \sqrt{m^2 + \frac{1}{2}})N_p/2}$
  The instability is caused by approaching a radial stopband. The stability limit is given by:

$$\lambda \;\; > \;\; \lambda_c c_h \tag{34}$$

$$c_h \;\; = \;\; (\nu_h \delta \nu_h)^{-\frac{1}{3}} \tag{35}$$

- $\mathbf{(m + \sqrt{m^2 + \frac{1}{2}})N_p/2 < \nu_h < (m + \frac{1}{2})N_p}$
  The instabilty is caused by the longitudinal stopband, and the stability

limit is given by:

$$\lambda > \lambda_c c_s \quad (36)$$

$$c_s = \left(\frac{4}{N_p}\right)^{\frac{2}{3}}. \quad (37)$$

- $(m+\frac{1}{2})N_P < \nu_h < (\frac{(6m+1)+4\sqrt{4(m+1)^2-2m}}{14})N_P$
  The instability is again caused by the radial stopband.

- $(\frac{(6m+1)+4\sqrt{4(m+1)^2-2m}}{14})N_P < \nu_h < (m+1)N_P$
  The instability is lost due to coupling of the modes, and the limit is set by:

$$\lambda > \lambda_c c_s \quad (38)$$

$$c_c = (\nu_h q)^{-\frac{2}{3}} \quad (39)$$

and q is the smallest positive value of

$$q \cdot = 1 \pm \sqrt{1 + \frac{(\nu_h - m_+ N_P)}{\nu_h}}. \quad (40)$$

4.3 Application

We have applied our method to five lattices for the case of partially stripped gold ions, viz. : $Q = 51, A = 197$ and $E_0 = 0.93126$ GeV /u. E_0 is the rest energy per nucleon. All cases correspond to the same magnetic rigidity $B\rho = 1.2$ Tm [18]. All rings consist of dipoles, FODO cells, and drift regions. The integer in the name refers to the superperiod of the particular ring. A detailed description of the different lattice geometries may be found in [20]. The results are listed in tables 1 and 2. In table 1 we list the limits of stability obtained using transfer matrices. The first column shows the name of the ring. The second column shows the dilute beam tune. The third column shows the value of the space charge factor K_{sc}, given by (21), and the fourth column the value of λ at which the string breaks up. The sixth column shows the type of instability at the breakup point: P denotes an unstable eigenvalue in the plane, with L the longitudinal, and R the radial (x) direction; C is a coupling resonance. V denotes instabilty in the vertical direction. Note, that some lattices have several regions of stability. E.g. the first instability in the FODO4 lattice, with an initial tune of 5.8 is caused by a longitudinal stopband.

However, if the number of ions in the string is increased further , λ decreases, and the string becomes stable again until a coupling resonance is encountered, when $\lambda = 52.8\mu m$. In table 2 we compare the transfer matrix results with those obtained with the smooth apprixmation. The results are similar, with the largets descrepencies in the case of coupling resonances.

name	total tune	K_{sc} $[m^{-2}]$	λ μm	N 10^6	unstab. eigenval.	$2\pi R$ [m]
FODO16	4.8	0.39	23.2	1.8	P,R	40.60
	4.0	0.28	25.7	1.6	P,R	
	3.2	0.19	29.2	1.4	P,R	
	2.4	0.11	35.7	1.1	P,R	
FODO4	5.8	0.02	62.6	0.98	P,L	60.91
		0.03	52.8	1.2	P,C	
		0.10	36.2	1.7	P,R	
		0.18	29.7	2.1	P,L	
		0.27	26.3	2.3	P,C	
	4.8	0.02	60.9	1.0	P,L	
		0.12	34.6	1.8	P,C	
		0.17	30.4	2.0	P,L	
		0.20	28.8	2.1	P,R	
	3.6	0.007	195.3	0.3	P,C	
		0.02	59.7	1.0	P,L	
		0.07	41.4	1.5	P,R	
FODO8	7.2	0.003	122.4	0.50	P,C	60.91
		0.09	38.1	1.6	P,L	
		0.25	26.8	2.3	P,R	
		0.33	24.4	2.5	P,L	
	6.0	0.02	60.4	1.0	P,C	
		0.09	37.7	1.6	P,L	
		0.28	25.9	2.4	P,C	
FODO4R	7.30	0.003	119.6	0.51	P,C	60.91
		0.02	60.5	1.0	P,L	
		0.05	47.3	1.3	P,R	
CRYSTAL8		0.01	76.1	0.85	V	64.82

Table 1: Overview of the calculations for five different lattices, for partially stripped gold ions. For CRYSTAL8 $\nu_{h,0} = 1.91$ and $\nu_{v,0} = 1.09$; for other lattices $\nu_{h,0} \approx \nu_{v,0}$. The dilute beam tune , the highest K_{sc} obtainable, the inter-ion spacing λ, and the number of particles stored are shown. P denotes an unstable eigenvalue in the plane, with L the longitudinal, and R the radial (x) direction; C is a coupling resonance. V denotes instabilty in the vertical direction.

5 Horizontal Zigzag

The horizontal zigzag consists of two substrings in the horizontal plane, at equal distance from the reference trajectory, with a spatial shift $\tau = 0.5$ [16] (see Fig. 1). The envelope equation is now:

$$x_1'''' + K_h(s)a - K_{sc}\frac{2f_3}{g_0}a = h(s)\,\delta, \tag{41}$$

$$\sigma_1' = h(s)x_1 - \frac{\delta_1}{\gamma^2}, \tag{42}$$

$$\delta_1' = 2K_{sc}\gamma^2(2\eta_e)\sigma_1 \tag{43}$$

We need to use two sets of transfer matrices: one for confinement, the other for stability. The derivation of these transfer matrices is similar to the ones for the string. Reference [16] can be consulted for a detailed derivation. The horizontal zigzag introduces the complication of dealing with the shear force between the two substrings when the trajectory curves. We deal with this below in section 5.1

5.1 Confinement of the horizontal zigzag in the smooth approximation.

The confinement in the radial plane is detremined by the isochronism condition. mentioned earlier: the "Crystalline Beam" has the same periodicity as the storage ring. In that case there is no net longitudinal displacement, i.e $\sigma' = 0$. We use this to integrate the equtions of motion approximately:

$$\int \sigma' ds = \int hads - \frac{1}{\gamma^2}\int \delta ds = 0, \tag{44}$$

which relates the momentum spread δ to the separation between the two substrings. On average,

$$\delta \approx \gamma^2 \frac{x_1}{R}. \tag{45}$$

Insert this into the equation of motion for x_1:

$$x_1'' + K_h(s)x_1 \underbrace{- \gamma^2\frac{h}{R}x_1 - K_{sc}\frac{2f_3}{g_0}x_1}_{\text{tune depression}} = 0. \tag{46}$$

The requirement that the tune depression is equal to the nearest half integral stopband sets an upper limit:

$$\lambda_u = \lambda_c \overline{c_h} F_{h,i}^{\frac{1}{3}} \tag{47}$$

$$\overline{c_h} = (\nu_h \overline{\delta\nu_h}) \tag{48}$$
$$\overline{\delta\nu_h} = \delta\nu_h - \frac{\gamma^2}{2\nu_h} \tag{49}$$

From (49) follows that a horizontally extended beam requires operation of the storage ring below the transition energy [7], viz:

$$\gamma < \sqrt{2\nu_h\delta\nu_h} \tag{50}$$

The upperlimit for a horizontal zigzag is given by:

$$\lambda_{u,horz.zigzag} = 2.4\lambda_{l,string} \tag{51}$$

That is, the ion seperation when the horizontal zigzag is first formed, is roughly double that of the string, when it became unstable. This is in accordance with the "Bifurcation Picture" outlined in section 2.

5.2 *Stability of the horizontal zigzag in the smooth approximation.*

The derivation of the limits of stability is pretty much the same as for the string, except that the stability form factors $\eta_{v,h,e}$ are now functions of the horizontal aspect ratio α_h. [18]. The horizontal zigzag becomes unstable when the tune shift due to the space charge hits a stopband. This determines a lower limit on the ion seperation:

$$\lambda_l = \lambda_c c_{v,h,s,c} \eta_{v,h,s,c}^{\frac{1}{3}} \tag{52}$$

where λ_c, and the coefficients $c_{v,h,s,c}$ are given by Eqs. [33], [32], [35], [37] and [39], respectively.
In fact, equations of the form [47] and [52] govern the confinement and stability of all "Crystalline Beams" , and can be used to obtain quick estimates for the properties of the storage ring.

5.3 *Horizontal Zigzag: Application*

We obtain the limits of stability using the smooth approximation and transfer matrices for the same ring configurations, and ions as for the string. Again [20] can be consulted for detailed descriptions of the lattice. The results are shown in table 3. The results for the smooth approximation and the transfer matrix method agree fairly well. Note, that the initial value of the ion spacing λ is roughly twice that of the string at instability (see table 1). Not all rings are able to support a horizontal zigzag. Specifically, the FODO8 and FODO4R lattices cannot maintain a horizontal zigzag, according the both the transfer matrix and smooth apprixımation estimates. For the FODO4 lattice we see differences between the two estimates. For $\nu_h = 4.8$ we find the lattice supports a horizontal zigzag in the smooth approximation only. For $\nu_h = 5.8$ it is the other way around. The explanation for this discrepancy is that the condition for isochronism really assumes a smooth dispersion function. This is the case in all the FODO16 lattices, and the FODO4 lattice with $\nu_h = 5.8$, but not in the other ones.

6 Smooth Approximation: from String to Shell

In this section we follow one complete bifurcation sequence, starting with a string, and ending with a shell. Using the smooth approximation, together with the familiar incoherent tune shift estimate [13] to impose this condition, we obtain the horizontal and vertical confinement conditions [10]:

$$\frac{\lambda^3}{F_{h,v}} = (\lambda_c \overline{c_{h,v}})^3, \tag{53}$$

with:

$$\begin{aligned} \overline{c_{h,v}} &= (\nu_{h,v}\overline{\delta_{h,v}})^{-\frac{1}{3}}, && (54) \\ \overline{\delta\nu_h} &= \frac{1}{2}(\nu_h - m\,N_p) - \frac{\gamma^2}{2\nu_h} && (55) \\ \delta\nu_v &= \frac{1}{2}(\nu_v - m\,N_p), && (56) \end{aligned}$$

λ is the same for all substrings, hence:

$$W = \frac{\mathcal{F}_h}{\mathcal{F}_v} = \frac{\overline{c}_v^3}{\overline{c}_h^3} = 1 - \Delta \propto \frac{K_h}{K_v}, \tag{57}$$

where Δ is the detuning from complete symmetrical tuning. For the structure to be stable, the ion spacing should exceed the lower limit given by (52).
We solved (57) for gold ions (A=157, Q=57) in a ring with radius 6.5 m, magnetic rigidity of 1.2 Tm and a horizontal tune $\nu_h = 4.8$. The vertical tune changed from $\nu_h = 4.8$ ($\Delta = 0.001$) to $\nu_h = 5.5$ ($\Delta = 0.25$), to $\nu_h = 6.2$ ($\Delta = 0.40$), and finally to $\nu_h = 6.7$ ($\Delta = 0.49$).
The results are shown in Fig. 3. The figure shows a cross section of the beam. We start with a string in O. As λ decreases, the ion density in string increases. When the critical value, given by (52) is reached, the string becomes unstable, undergoes a pitchfork bifurcation, and splits into two substrings, forming

name	total tune	λ_{tm} μm	λ_{sm} μm	unstab. eigenval.
FODO16	4.8	23	21	P,R
	4.0	26	23	P,R
	3.2	29	27	P,R
	2.4	36	33	P,R
FODO4	5.8	62	61	P,L
		36	35	P,R
	4.8	61	61	P,L
	3.6	195	174	P,C
		60	60	P,R
		41	43	P,L
FODO8	7.2	122	111	P,C
		38	38	P,L
		27	27	P,R
	6.0	60	61	P,C
		38	38	P,L
FODO4R	7.30	120	122	P,C
		61	61	P,L
		47	36	P,R
CRYSTAL8		76	75	V

Table 2: We compare the limits of stability determined with transfer matrices (λ_{tm}) with their smooth approximation estimates (λ_{sm}).

analytical results						
name	ν_v	λ_u μm	λ_l μm	α	δ (10^{-6})	
FODO16	4.8	50.1	25.7	0.86	2.94	Vertical
	4.0	57.4	29.7	0.84	3.03	Vertical
	3.2	66.9	32.3	0.92	3.42	Vertical
	2.4	83.5	39.2	1.01	4.08	Vertical
FODO4	5.8	87.1	–	–	–	
	4.8	130.0	60.8	1.09	3.47	Longitudinal
	3.6	80.9	–	–	–	
FODO8	7.2	49.7	–	–	–	
	6.0	56.2	–	–	–	
FODO4R	7.3	64.6	–	–	–	
numerical results						
name	ν_v	λ_u μm	λ_l μm	α	δ (10^{-6})	
FODO16	4.8	56	28.2	1.33	2.14	Vertical
	4.0	61	31.9	1.21	2.29	Vertical
	3.2	70	33.8	1.29	2.70	Vertical
	2.4	86	38.5	1.35	3.41	Vertical
FODO4	5.8	92	62.2	5.3	139	Coupling
	4.8	–	–	–	–	
	3.6	–	–	–	–	
FODO8	7.2	–	–	–	–	
	6.0	–	–	–	–	
FODO4R	7.3	–	–	–	–	

Table 3: Comparison of the smooth approximation ("analytical results") and transfer matrix ("numerical") results. The aspect ratio α and the momentum separation δ correspond to the limit of stability at $\lambda = \lambda_l$. The direction of the instability is shown in the last columns.

a horizontal zigzag. The initial ion separation in the horizontal zigzag is given by (47), and is twice the ion seperation of the string that gave rise to it.
The two substrings move away from the center, as λ decreases. When points A and B are reached, the substrings become unstable, undergo a pitchfork bifurcation, split in two, and form a helix. Again the initial ion separation in the helix follows from (47), which is roughly twice the value of the ion separation in the substrings of the unstable horizontal zigzag, given by (52).
As λ decreases, the substrings move away from A and B. They shift slightly with respect to each other, such that each consequative substring, as one goes clockwise from one substring to the next, is shifted by $\tau = 0.5$. When points C.D,E and F are reached, the conditions for the formation of a shell are satisfied, according to (47). The point of instability of the helix lies a little bit beyond this point. So, the substrings of the helix will undergo a pitchfork bifurcation, forming a shell with eight substrings.
The evolution of these substrings, as λ decreases, is similar to that of the helix. However, when they undergo a pitchfork bifurcation, the next structure is no longer stable. That is, a hollow shell, consisting of 16 substrings does not exist. We verified, that putting a substring along the refernce trajectory, will stabilize this shell structure.
As the focussing in the vertical direction increases the beam flattens, but the evalution of the substrings remains the same. When Δ is larger than 0.5,the horizontal zigzag no longer goes over into a helix, and a different set of structures is formed.
For this system the critical latice spacing is $\lambda_c = 47\mu m$. Instability occurs when $\lambda \approx 20\mu m$ and for each structure λ varies between $50\mu m$ and $20\mu m$.

7 Summary and Conclusions

Our goal from the outset was to obtain estimates for the storage ring, and "Crystalline Beam" properties, with as few a priori assumptions as possible. We did this by introducing the bifurcation picture: as the number of ions in the subtrings of the "Crystalline Beam" reach a critical limit, they undergo a pitchfork bifurcation. This implies, that the ion density in each "Crystalline Beam" has a upper (formation) and lower (stability) limit, and that it varies between these two values. Looking at the motion of a slightly perturbed test particle in the "Crystalline Beam" we derived the conditions for equilibrium and stability. The limits of formation and stability obtained using the smooth approximation, as well as transfer matrices were in good agreement.

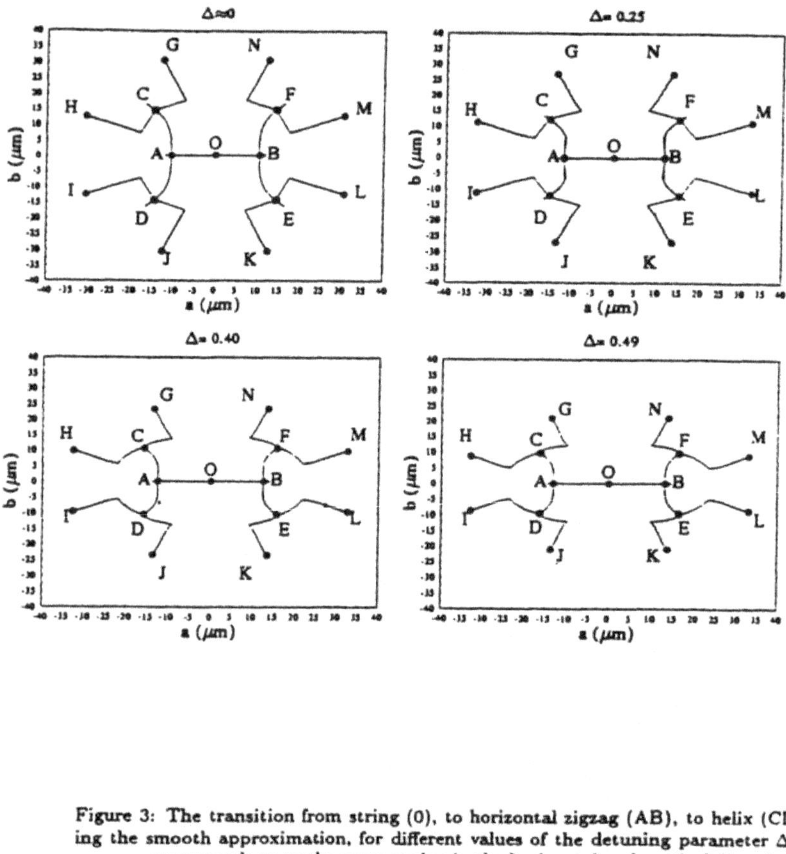

Figure 3: The transition from string (0), to horizontal zigzag (AB), to helix (CDEF), using the smooth approximation, for different values of the detuning parameter Δ. Δ = 0 corresponds to to the same tuning in the horizontal and vertical direction.

Acknowledgments

This work was conducted at Brookhaven Laboratory, in close cooperation with A.G. Ruggiero and D. Maletić. This could not have been done without their helpful suggestions.

Appendix A: Derivation of transfer matrices for the string.

String in the Drift

In a drift $K_h = K_v = h = 0$. Let l_Δ be the length of the drift and call $\zeta = l_\Delta h_{sc}$. The equations of motion are decoupled:

$$y'' - h_{sc}^2 y = 0, \tag{58}$$
$$x'' - h_{sc}^2 x = 0, \tag{59}$$
$$\sigma'' + 2h_{sc}^2 \sigma = 0. \tag{60}$$

The motion is defocused in the transverse directions, but is oscillatory in the longitudinal direction. The transfer matrix has the general form:

$$M_D = \begin{pmatrix} M_H & 0 & 0 \\ 0 & M_V & 0 \\ 0 & 0 & M_E \end{pmatrix}. \tag{61}$$

Each of the entries is a 2×2 matrix.

$$M_H = M_V = \begin{pmatrix} \cosh\zeta & \rho_{sc}\sinh\zeta \\ h_{sc}\sinh\zeta & \cosh\zeta \end{pmatrix}, \tag{62}$$
$$M_E = \begin{pmatrix} \cos\sqrt{2}\zeta & -\frac{\rho_{sc}}{\sqrt{2}\gamma^2}\sin\sqrt{2}\zeta \\ \gamma^2\sqrt{2}h_{sc}\sin\sqrt{2}\zeta & \cos\sqrt{2}\zeta \end{pmatrix}. \tag{63}$$

String in the Quadrupole

In a quadrupole of length l_q, $K_h = -K_v \equiv K$ and $h = 0$.

$$y'' - (K + K_{sc})y = 0, \tag{64}$$
$$x'' + (K - K_{sc})x = 0. \tag{65}$$

To obtain a convenient, compact notation, we introduce the variable $\bar{K} = |K \mp K_{sc}|$, the upper (lower) sign holding for the horizontal (vertical) motion.

The general form of the transfer matrix is once again:

$$M_Q = \begin{pmatrix} M_H & 0 & 0 \\ 0 & M_V & 0 \\ 0 & 0 & M_E \end{pmatrix}. \tag{66}$$

M_E is the same as the one already given for the drift. When $\pm K > K_{sc}$,the other two matrices are:

$$M_{V,H} = \begin{pmatrix} \cos\zeta & \frac{1}{\sqrt{K}}\sin\zeta \\ -\sqrt{K}\sin\zeta & \cos\zeta \end{pmatrix}, \tag{67}$$

where $\zeta = l_q\sqrt{K}$. When $\pm K < K_{sc}$ we obtain:

$$M_{V,H} = \begin{pmatrix} \cosh\zeta & \frac{1}{\sqrt{K}}\sinh\zeta \\ \sqrt{K}\sinh\zeta & \cosh\zeta \end{pmatrix}. \tag{68}$$

String in the Dipole

The equations of motion in a dipole with field index n and curvature $h = 1/\rho$ are:

$$y'' + (nh^2 - h_{sc}^2)y = 0, \tag{69}$$

$$x'' + \left[(1-n)h^2 - h_{sc}^2\right]x = h\delta, \tag{70}$$

$$\sigma' = hx - \frac{\delta}{\gamma^2}, \tag{71}$$

$$\delta' = 2\gamma^2 h_{sc}^2 \sigma. \tag{72}$$

The motion in the radial and longitudinal directions is coupled due to the curvature. The general transfer matrix is:

$$M_B = \begin{pmatrix} M_H & 0 & M_a \\ 0 & M_V & 0 \\ M_b & 0 & M_E \end{pmatrix}. \tag{73}$$

If $n = 0$, M_V is the same as in a drift; otherwise it is the same as in the quadrupole, with $K = -nh^2$.

To deal with the coupled radial and longitudinal motion we introduce the 4-dimensional state vector $\mathrm{U} = (x, x_1', \sigma, \delta)$, and the 4×4 matrix

$$\mathcal{M}_B = \begin{pmatrix} M_H & M_a \\ M_b & M_E \end{pmatrix} \tag{74}$$

as the transfer matrix which applies to U. For convenience, we define the ratio of the bending radius due to the dipole magnet and the space charge:

$$\xi = \frac{h_{sc}}{h}. \tag{75}$$

The equations of motion of the components of U are:

$$\begin{aligned} x_1' &= p, &(76)\\ p' + h^2(1-n-\xi^2)x &= h\delta, &(77)\\ \sigma' &= hx - \frac{\delta}{\gamma^2}, &(78)\\ \delta' &= 2\gamma^2\xi^2h^2\sigma. &(79) \end{aligned}$$

Solving this system of differential equations is straightforward.

We obtain a general solution of the homogenous equations by inserting $e^{\alpha s}$ and solving the resultant eigenvalue problem.

The complete transfer matrix reads:

$$\mathcal{M}_B = \tag{80}$$

$$\begin{pmatrix} F_3 + BCF_1 & F_2 + BCF_0 & hCF_0 & hF_1 \\ -AF_2 - (ABC - h^2C)F_0 & F_3 + BCF_1 & hCF_1 & hF_2 \\ hF_2 & hF_1 & F_3 + AF_1 & -BF_2 - (AB - h^2)F_0 \\ hCF_1 & hCF_0 & CF_2 + ACF_0 & F_3 + AF_1 \end{pmatrix}$$

where

$$\begin{aligned} A &= h^2(1-n-\xi^2), &(81)\\ B &= \frac{1}{\gamma^2}, &(82)\\ C &= 2\gamma^2\xi^2h^2. &(83) \end{aligned}$$

$$\begin{aligned} F_0 &= \frac{\sinh(\alpha_r l)}{D_r} + \frac{\sin(\alpha_i l)}{D_i}, &(84)\\ F_1 &= \alpha_r \frac{\cosh(\alpha_r l)}{D_r} + \alpha_i \frac{\cos(\alpha_i l)}{D_i}, &(85)\\ F_2 &= \alpha_r^2 \frac{\sinh(\alpha_r l)}{D_r} - \alpha_i^2 \frac{\sin(\alpha_i l)}{D_i}, &(86) \end{aligned}$$

$$
\begin{aligned}
F_3 &= \alpha_r^3 \frac{\cosh(\alpha_r l)}{D_r} - \alpha_i^3 \frac{\cos(\alpha_i l)}{D_i}, && (87)\\
\Gamma^2 &= h^2(1 - n + \xi^2), && (88)\\
D_r &= \alpha_r(2\alpha_r^2 + \Gamma^2), && (89)\\
D_i &= \alpha_i(-2\alpha_i^2 + \Gamma^2) && (90)
\end{aligned}
$$

with

$$
\begin{aligned}
\alpha_r &= \frac{1}{\sqrt{2}}\left(-\Gamma^2 + \sqrt{\Gamma^4 + 4\Upsilon^2}\right)^{\frac{1}{2}}, && (91)\\
\alpha_i &= \frac{1}{\sqrt{2}}\left(\Gamma^2 + \sqrt{\Gamma^4 + 4\Upsilon^2}\right)^{\frac{1}{2}}, && (92)\\
& && (93)
\end{aligned}
$$

Appendix B: Derivation of the limits of stability for the string using the smooth approximation.

The space charge force increases, as λ decreases, and the resulting tune shift due to the space charge can be estimated as:

$$
\Delta\nu_v = \frac{1}{4\pi} \oint_C \beta_v K_{sc}\, ds, \qquad (94)
$$

$$
\beta \rightarrow \frac{R}{\nu_v}, \qquad (95)
$$

$$
K_{sc} = \frac{2}{R^2} \frac{\lambda_c^3}{\lambda^3}; \qquad (96)
$$

$$
\lambda_c = \left(\frac{Q^2 r_0 g_0 R^2}{A\beta^2\gamma^5}\right)^{\frac{1}{3}}. \qquad (97)
$$

where λ_c is the critical lattice constant.
The string becomes unstable when the shift of the betatron tune is equal to half the distance from the original, dilute beam stopband, viz.:

$$
\delta\nu_v = \frac{1}{2}(\nu_v - \frac{m}{2}N_p) \qquad (98)
$$

$$
\Delta\nu_v < \delta\nu_v \text{stable motion} \qquad (99)
$$

$$
\Delta\nu_v = \delta\nu_v \text{limit of stability} \qquad (100)
$$

The string is stable in the vertical plane as long as the the ion separation λ is larger than the critical value, viz.:

$$
\lambda > \lambda_c\, c_v, \qquad (101)
$$

$$c_v = (\nu_v \delta_v)^{-\frac{1}{3}}. \tag{102}$$

Using the equations of motion in the horizontal plane:

$$\begin{aligned} x'' + (K_h - K_{sc}) &= h\,\delta, \\ \delta'' + 2K_{sc}\delta &= 2\gamma^2 K_{sc} h x. \end{aligned} \tag{103}$$

and pproximating the logitudinal tune by:

$$\nu_s = R\,\sqrt{2\,K_{sc}} \tag{104}$$

the limits of stability in the radial plane are derived in similar fashion.

References

1. H. Walther, Phase Transitions of Laser-Cooled Ions, in Advances in At. Mol. and Opt. Physics, **31**, p.137, (1993).
2. H. Walther, Atoms in Cavities and Traps, in Advances in At. Mol. and Opt. Physics, **32**,p. 379, (1994).
3. J.P. Schiffer and P. Kienle, Z. Phys. A - Átoms and Nuclei, **321**, 181 (1985).
4. J.P. Schiffer and A. Rahman, Feasibility of a Crystalline Condensed State in Cooled Ion Beams of a Storage Ring. PHY-5121-ME-88. Feb. 1988, Argonne National Laboratory.
5. Chiping Chen and Ronald c. Davidson, Phys. Rev. Lett. **72**, 2195 (1994), *and* Phys. Rev. E, **49**, 5679 (1994).
6. R.W. Hasse and J.P. Schiffer, Ann. of Phys. **203**, 419 (1990).
7. J.Wei, X.P.Li, A.M.Sessler, Crystaline Beam Ground State, BNL–52381, June 1993.
8. H. Thomas, G.E. Morfill, V. Demell, J Goree, B Feuerbacher, and D. Möhlmann, Phys. Rev. Lett. **73**, 652 (1994).
9. CRYSTAL. A Storage Ring for Crystalline Beams. Laboratori Nazionale di Legnaro (INFN), LNL-INFN(Rep) 80 (1994).
10. A.F. Haffmans, Confinement and Stability of Crystalline Beams in Storage Rings. Contribution to the Proceedings of the 6th Workshop on Advanced Accelerator Concepts, 1994.
11. A.F. Haffmans, D. Maletić and A.G. Ruggiero, Particle Motion in Crystalline Beams, BNL-60436 (1994).
12. A.F. Haffmans, D. Maletić and A.G. Ruggiero, Matrix Formulation of the Particle Motion in Crystalline Beams, Proceedings of the Fourth European Particle Accelerator Conference.

13. E.D. Courant and H.S. Snyder, Ann. of. Phys., **3**, 1-48, (1958).
14. John Guckenheimer and Philip Holmes, Nonlinear Oscillations, Dynamical Systems, and Bifurcations of Vector Fields, Springer-Verlag, New York, 1983.
15. J.A.G. Roberts and G.R.W. Quispel, Phys. Rep. 216, 63 (1992).
16. A.F. Haffmans, D. Maletić, and A.G. Ruggiero, Crystalline Beams: The Horizontal Zigzag, BNL- (1994).
17. A.G. Ruggiero, Demonstration of No Feasibility of a Crystalline Beam in a Betatron Magnet, BNL-49529 (1993).
18. A.F. Haffmans, D. Maletić, and A.G. Ruggiero, Crystalline Beams: The String, BNL-60619 (1994).
19. A.F. Haffmans, D. Maletić, and A.G. Ruggiero, Crystalline Beams: The Vertical Zigzag, BNL-60743 (1994).
20. A.F. Haffmans, D. Maletić, and A.G. Ruggiero, Study of the Effect of the Space Charge using Synch, BNL-60521 (1994).
21. A.F. Haffmans, D. Maletić and A.G. Ruggiero, Behaviour of Space-Charge Dominated Ion Beams in Storage Rings, Proceedings of the Fourth European Particle Accelerator Conference, (to be published).
22. A.G.Ruggiero, Nuclear Fusion of Protons with Boron, BNL-AD/AP-48, 1992. Also Proc. of 1993 PAC, Washington, vol.1 page 700. Also Il Nuovo Cimento, 106A, n.12, (1993), 1959.

Schottky Spectra and Crystalline Beams

D.V. Pestrikov
Budker Institute for Nuclear Physics,
630090 Novosibirsk, Russian Federation

Abstract

In this paper we review specific features of Schottky noise spectra of cold beam and Schottky noise measurements at NAP-M as a background for the crystalline beams physics.

1 Introduction

After upgrading in 1976 the cooling device at NAP-M, we have found that the efficiency of the beam cooling substantially exceeded that, which could be expected according to papers [1] and [2]. Apart from understanding of this fact [3], the enhancement of cooling demanded more careful measurements of the beam momentum spread (σ). Since typically experiments at NAP-M were carried out with a coasting proton beam, first we tried to use the method, based on the measurement of the decay rate ($1/\tau$) of the first harmonic of the revolution frequency (ω_0) in the beam current after the kicking out of the small fraction of the beam. The debunching rate was expected to be proportional to the revolution frequency spread of the beam ($1/\tau = \Delta\omega_0 = \omega_0\alpha_p\sigma/p$, where $p = Mv\gamma$ is the momentum of the particle and $\alpha_p = 1/\gamma^2 - 1/\gamma_{tr}^2$, $Mc^2\gamma_{tr}$ is the transition energy of the ring).

More close inspection of this process showed that in a low-intensity beam with strong cooling of particles ($\lambda \gg \Delta\omega_0$, λ is the momentum cooling decrement) the debunching occurs due to diffusion of particles along the beam. That results in the debunching rate (see in Ref. [4], or in Ref. [5]).

$$\frac{1}{\tau} \simeq \frac{(\Delta\omega_0)^2 + \Omega^2}{\lambda}, \quad \Omega \leq \lambda,$$

where Ω is the coherent frequency shift of the first harmonic. At high beam intensities the beam momentum spread and the debunching rate can increase due to contributions from the turbulent beam heating (see, for example, in Ref. [6], or in Ref.[7])

$$\left(\frac{\sigma}{p}\right)_f^2 \simeq \left(\frac{\sigma}{p}\right)_{in}^2 + C\frac{\theta_g}{2\pi}\frac{Ne^2\omega_0(Z/n)_{eff}}{2\pi pv\alpha_p}, \quad \frac{1}{\tau} \simeq \Delta\omega_0 > \lambda,$$

where θ_g is the azimuthal length of the gap in the beam, $(Z/n)_{eff}$ is an effective longitudinal coupling impedance, the subscripts f and in mark the values before and after the beam debunching, and the dimensionless factor C takes the value between 1 and 2. Such a possibility was also indicated experimentally by visible echoing of the signal from the pick-up electrode during debunching. Complicated interpretation of the results of measurements convinced us to look for an alternative method for reliable measurements of σ.

Such a method was found out in the measurements of the longitudinal Schottky noise of the coasting proton beam [8]. As a background that time we used the facts that in normal beams the width of the noise spectrum determines the momentum distribution function $[f(\Delta p)]$, while its power gives the number of particles of the beam ($P = N$). The first results of Schottky noise measurements at NAP-M showed [9, 10, 11] that main properties of the beam noise in a deeply cooled proton beam are very sensitive to collective interactions of particles. The interpretation of these measurements also occurs to be very nontrivial and finally resulted in the assumption of the formation in storage rings of the crystalline (or crystal-like) beams Refs [12, 6]. In this report we remind the reader main results of the Schottky noise measurements at NAP-M, discuss the properties of simplest crystalline geometries in a coasting beam and specific features of the Schottky noises in such crystalline beams.

2 Schottky Noise of Cooled Beams

First, we remind the reader basic properties of the longitudinal Schottky noise of the coasting beam, which can be expected within the framework of the perturbation theory (see, for example, in [5] for more detail). Due to thermal motion of particles a coasting beam produces noise signals, which can be detected by pick-up electrodes. These so-called Schottky noise signals are described by the correlation functions

$$K_n(\tau) = < \rho_n(\tau)\rho_n^*(0) >, \tag{1}$$

where

$$\rho_n(\tau) = \sum_{a=1}^{N} \exp[-in\theta_a(\tau)]$$

is a Fourier harmonic of the microscopic linear density of the beam, and the brackets $< \ldots >$ mean the statistical averaging over the beam fluctuations. The Fourier amplitude

$$K_n(\omega) = \int_{-\infty}^{\infty} d\tau K_n(\tau) e^{-i\omega\tau} \tag{2}$$

yields the Schottky noise spectrum, while the integral

$$P_n = \int_{-\infty}^{\infty} \frac{d\omega}{2\pi} K_n(\omega) \tag{3}$$

gives the Schottky noise power. For normal beam with low intensity direct calculations (see in Ref. [5]) yield

$$K_n(\omega) = K_n^{(0)}(\omega) = N \int_{-\infty}^{\infty} d\Delta p f(\Delta p)\delta(\omega - n\omega_0\alpha_p\Delta p/p), \quad P_n = N. \tag{4}$$

These simple relations gave the basis for non-destructive Schottky noise diagnostic of beams, enabling the measurements of both the momentum distribution function and the beam intensity [8].

Analytic formulae for Schottky spectral densities in the cooled beams generally depend on the collision integral, which is used to describe the beam cooling. In the simplest case, when both the longitudinal cooling rate λ and diffusion coefficient d (so that $\sigma^2 = d/2\lambda$) are constant, simple calculations (see in Ref.[5]) result in

$$K_n^{(0)}(\tau) = N\exp\left(-q\lambda|\tau| + q[1 - e^{-\lambda|\tau|}]\right), \tag{5}$$

and respectively

$$K_n^{(0)}(\omega) = \frac{2Ne^q}{\lambda}\sum_{l=0}^{\infty}\frac{(-q)^l}{l!}\frac{q+l}{(q+l)^2+z^2}, \quad z = \frac{\omega}{\lambda}, \quad q = \frac{n^2\Delta\omega^2}{\lambda^2}. \tag{6}$$

Although the power of this spectrum is the same as in Eq.(4) $P_n = K_n^{(0)}(\tau = 0) = N$, the shape of the spectrum may deviate very strongly from the beam distribution function. For example, if $q \ll 1$, which corresponds to the case of strong cooling, we can omit in the r.h.s. of Eq.(6) all higher powers of q and obtain

$$K_n^{(0)}(\omega) = 2N\frac{q\lambda}{\omega^2 + (q\lambda)^2}, \quad q \ll 1. \tag{7}$$

The width of this spectrum $\delta\omega = \pm n\Delta\omega(n\Delta\omega/\lambda)$ is substantially smaller than the width of the frequency distribution function ($n\Delta\omega$). Due to the beam cooling coherent fluctuations with low n are frozen in the beam. On the contrary, if $q \gg 1$, the main contribution to the integral over τ in Eq.(6) gives the region, where $\lambda\tau \leq 1/q$ and hence

$$-q\lambda\tau + q[1 - e^{-\lambda\tau}] \approx -q\frac{(\lambda\tau)^2}{2} + O([\lambda\tau]^3),$$

so that the integration over τ results in

$$K_n^{(0)}(\omega) = \frac{N}{\sqrt{2\pi}|n|\Delta\omega}\exp\left(-\frac{\omega^2}{2(n\Delta\omega)^2}\right). \tag{8}$$

The Schottky noise spectrum of an intense beam is affected by the collective interaction of particles. In typical conditions this interaction can be described in terms of the so-called dielectric constant of the beam ($\epsilon(n,\omega)$) in such a matter that the Fourier harmonics of $\rho_{n\omega}$ are modified according to the following equation

$$\rho_{n,\omega} = \frac{\rho_{n,\omega}^{(0)}}{\epsilon(n,\omega)}, \tag{9}$$

where $\rho_{n,\omega}^{(0)}$ describes coherent fluctuations neglecting collective interactions of particles. Analytic and others properties of the function $\epsilon(n,\omega)$, in general case, depend on the collision integral describing the beam cooling and heating (see, for instance, in Refs. [5] and [13]). For the case, when collisions can be described by the Fokker-Planck collision integral, the function $\epsilon(n,\omega)$ reads [4], [5]

$$\epsilon(n,\omega) = 1 + \zeta e^{q} \sum_{l=0}^{\infty} \frac{(-q)^l}{l!} \frac{q+l}{-iz+q+l}, \quad z = \omega/\lambda. \tag{10}$$

Here, $\zeta = \Omega_n^2/(n\Delta\omega)^2$,

$$\Omega_n^2 = n^2\omega_0^2 \frac{Ne^2\omega_0\alpha_p}{2\pi pv}\left(\frac{-iZ_n}{n}\right) \tag{11}$$

is the square of the coherent frequency shift, and Z_n is the so-called longitudinal coupling impedance. The eigenfrequencies of coherent oscillations of the beam can be found as the roots of the dispersion equation

$$\epsilon(n,\omega) = \epsilon'(n,\omega) + i\epsilon''(n,\omega) = 0. \tag{12}$$

Below transition energy of the ring and for non-relativistic ion beams the main contribution to the coupling impedance gives the space charge of the beam, when

$$\left(\frac{-iZ_n}{n}\right) = \frac{1}{v}[2\ln(b/a)+1], \quad |n| \ll R_0/b, \tag{13}$$

where b and a are the pipe and beam radii respectively. In this case, all the roots in Eq.(12) correspond to stable (and decaying) oscillations. However, due to energy loses in the beam surroundings the coupling impedance Z_n is a complex value ($Z_n = Z_n' + iZ_n''$). For that reason, the roots of Eq.(12) generally move in the upper half-plane of ω, when the beam current increases. For a given frequency spread the requirement $\mathrm{Im}(\omega) = 0$ and Eq.(12) determine the bean threshold current (N_{th}) and the coherent frequency shift on the threshold (ω_{th}). Near the threshold ($|N - N_{th}| \ll N_{th}$) the roots of the dispersion equation have the form

$$\begin{aligned} \omega - \omega_{th} &= \frac{N - N_{th}}{N_{th}} \frac{(\partial\epsilon'/\partial\omega)_{th}}{|(\partial\epsilon/\partial\omega)_{th}|^2}, \\ \delta = -\mathrm{Im}(\omega) &= -\frac{N - N_{th}}{N_{th}} \frac{(\partial\epsilon'/\partial\omega)_{th}}{|(\partial\epsilon/\partial\omega)_{th}|^2}. \end{aligned} \tag{14}$$

According to Eqs (1) and (9) the Schottky noise spectrum of intense beam is described by the following spectral density

$$K_n(\omega) = \frac{K_n^{(0)}(\omega)}{|\epsilon(n,\omega)|^2}, \tag{15}$$

where the function $K_n^{(0)}$ is determined in Eq.(6). Using Eq.(10) we can rewrite $K_n(\omega)$ using the only function $\epsilon(n,\omega)$

$$K_n(\omega) = \frac{2N}{\omega|\epsilon_n(\omega)|^2}\mathrm{Im}\left[\frac{\epsilon_n - 1}{\zeta}\right], \quad \zeta = \Omega_n^2/(n\Delta\omega)^2. \tag{16}$$

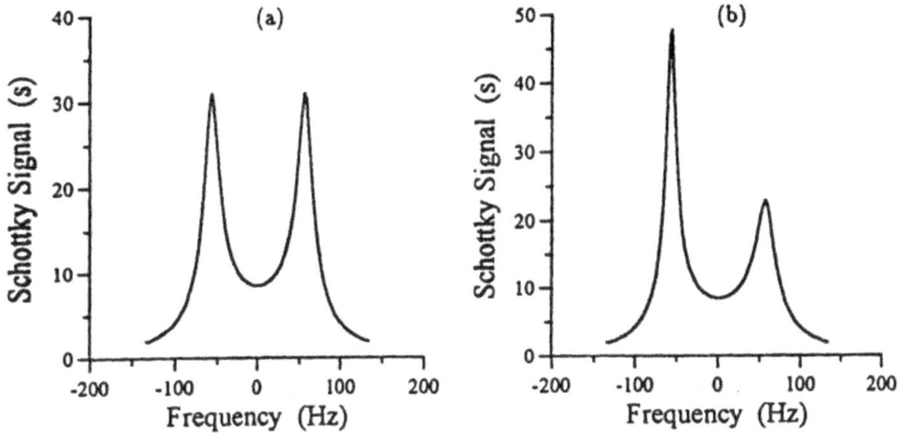

Figure 1: Schottky spectra $[\sqrt{K_n(\omega)}]$. Frequency is measured from the harmonic $n\omega_0/2\pi$, $N = 10^8$, $\lambda = 100$ 1/s, $\delta\omega = 1.$ 1/s, $\mathrm{Im}Z/n = 500$ Ohm, $n = 8$; a: $\mathrm{Re}(Z_n)/|n| = 0$ Ohm, b: $\mathrm{Re}(Z_n)/|n| = 50$ Ohm.

As is seen from Fig.1, collective interaction of particles can have strong effect on the shape of the Schottky spectra, if $|\Omega_n^2| \gg (n\Delta\omega)^2$, and $|\Omega_n^2| \gg \lambda^2$, when it results in essential suppression of the Schottky signal and in the splitting of the spectrum in two narrow peaks, shifted (up and down) from the harmonic $n\omega_0$ by the coherent frequency shift. These peaks describe the propagation along the beam of slow-decaying coherent modes. If energy loses in the beam surroundings are negligible, spectrum is symmetrical relative to $n\omega_0$ (Fig.1, (a)). However, if $Z_n' \neq 0$, the wave propagating up-stream the beam has lower decrement than the wave, propagating down-stream. In this case, the Schottky spectrum get an asymmetry as shown in Fig.1, (b).

Let us note, that the beam cooling strongly affects also the widths of the peaks in the Schottky spectra. If we take, for example, $\mathrm{Im}(\zeta) = 0$ and $\Omega_n > \lambda/2 \gg n\Delta\omega$, then initial collective modes of the cooled beam decay with decrements equal to $\lambda(1/2 + q)$, while for uncooled (but still very cold) beam these decrements are exponentially small

$$\delta_n = \sqrt{\pi/2}\frac{\Omega_n^4}{(|n|\Delta\omega)^3}\exp\left(-\frac{\Omega_n^2}{2n^2\Delta\omega^2}\right). \tag{17}$$

So that the peaks in the spectra of the strongly cooled beam are much wider than that for uncooled beam with the same longitudinal temperature.

The Schottky noise power can be easily found in the case, when $\mathrm{Im}\zeta = 0$ and when direct calculation results in (see, for example, in Ref. [5])

$$P_n = \frac{N}{1 + N/N_c}, \quad N_c = \left(\frac{\sigma}{p}\right)^2 \frac{2\pi p v \alpha_p}{e^2\omega_0(Z''/n)}. \tag{18}$$

This expression predicts strong suppression of the signal power for the strongly cooled, or for high intensity beams, when the noise power varies proportional to the longitudinal beam temperature

$$P_n \simeq N_c = \left(\frac{\sigma}{p}\right)^2 \frac{2\pi p v \alpha_p}{e^2\omega_0(Z''/n)}, \quad N \ll N_c \tag{19}$$

If the beam is potentially unstable ($Z'_n \neq 0$), simple analytic expressions for the function $P_n(N)$ can be obtained only near the thresholds of instability, where for the case of strongly cooled beam ($q \ll 1$) and $|\Omega_n| > |n|\Delta\omega$, we can write (for more detail see in Ref.[5], or in Appendix A)

$$P_n = \frac{N_c}{1 - N/N_{th}}, \quad N_{th} - N \ll N_{th}. \tag{20}$$

In the case of strong cooling ($\lambda \gg |n|\Delta\omega$), the threshold current (N_{th}) can essentially exceed N_c [5]

$$N_{th} \simeq \frac{2\pi p v}{e^2\omega_0\alpha_p}\left(\frac{\lambda}{\omega_0}\right)^2 \frac{Z''_n/n}{(Z'_n)^2} \gg N_c. \tag{21}$$

3 NAP-M

The proton synchrotron NAP-M (see in Fig.2) was initially designed as a prototype of the antiproton accumulator NAP and then was constructed especially to study the electron cooling. One of the important requirements to the lattice of the antiproton accumulator

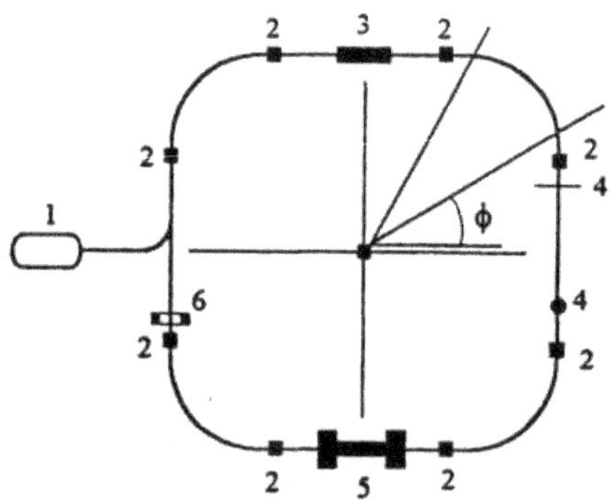

Figure 2: Schematic layout of the NAP-M. 1 – 1.5 MeV electrostatic accelerator; 2 – correcting magnet (provides dipole, quadrupole and sextupole fields); 3 – RF-station; 4 – the flying-wire beam size monitor; 5 – electron cooling device; 6 – Schottky pick-up.

NAP was the capability to accept antiproton beams with large momentum spreads ($\sigma/p \approx 5\%$). For that reason, in NAP-M we used the edge focusing with edges of dipoles directed in the center of the ring (for NAP the tangent of the edge angle was chosen to be $\tan\phi = 0.538$). However, during commissioning of the NAP-M it was found out that the ring operates better, if the working point corresponding to this edge angle is corrected. Such corrections could be done using 8 correcting magnets (items 2 in Fig.2) placed along the ring about 40 cm from each end of the dipoles. Each of these correctors could give vertical and horizontal dipole fields, quadrupole fields (to enable corrections within $\Delta\nu = \pm 0.1$) and sextupole fields.

After injection in NAP-M at 1.5 MeV, the proton beam prior to measurements was accelerated till the required energy (typically 65 MeV). Then, the RF-station was switched off, and after beam debunching the electron cooling device was switched on. Typical parameters of the NAP-M, obtained after commissioning, and some parameters of electron and proton beam are listed in the Tables 1, 2 and 3.

Table 1: Parameters of the NAP-M after commissioning.

Energy	1.5 - 150 MEV
Bending radius	300 cm
Length of the straight section	710 cm
Perimeter (Π)	4723 cm
Horizontal betatron tune (ν_x)	1.34
Vertical betatron tune (ν_z)	1.24
Horizontal β-function in the middle of the straight section (β_x)	319 cm
Vertical β-function in the middle of the straight section (β_z)	483 cm
Dispersion function in straight sections	536 cm
Momentum compaction factor (α)	0.8
Kinetic transition energy ($Mc^2[\gamma_{tr} - 1]$)	110 MeV
Aperture of the vacuum chamber in the dipoles ($h \times w$)	7 × 11 cm
Aperture of the vacuum chamber in the straight sections (ø)	7 cm
Average Vacuum	5×10^{-10} Torr

Table 2: Parameters of the electron cooling device.

Energy	35 KeV
Beam current	0.3 A
Beam radius	0.5 cm
Beam angular divergency	3×10^{-3}
Length of the cooling section	100 cm
Longitudinal magnetic field	0.1 T

Table 3: Parameters of the cooled proton beam.

Energy	65 MeV
Beam current	$\leq$ 50 mkA
Revolution frequency	2.248 MHz
Beam radius	0.005 cm
Beam angular divergency	1.25×10^{-5}
Beam momentum spread	1.53×10^{-6}

The transverse beam size was measured using the flying-wire monitor (item 4 in Fig.2) and measuring the spot size of hydrogen atom due to recombination of protons in the cooling region (see in Ref. [14] for more detail). The results of these measurements were used for the calculation of the transverse beam temperature. The longitudinal beam temperature was recalculated from the Schottky noise measurements as is described in the next section.

4 Schottky Noise Measurements

The beam noise spectra were measured at the 5-th and 8-th harmonics of the revolution frequency using the special integral pick-up electrode (item 6 in Fig.2). To enhance the sensitivity of measurements, the capacity of the pick-up C_d and the inductance L were chosen to form a resonance circuit. The Q-factor of this circuit was $Q = 500$. The shunt impedance of the circuit was 40 KOhm. The noises of amplifier were lower than 12 db. After the transformation into a digital code the beam signal, measured during 0.25 s, was transmitted to a computer and was processed using the FFT-code. The frequency band of the measured signal on the ADC input was equal to 1 kHz, the clock frequency of ADC was 2 kHz, the conversion error was 0.1%.

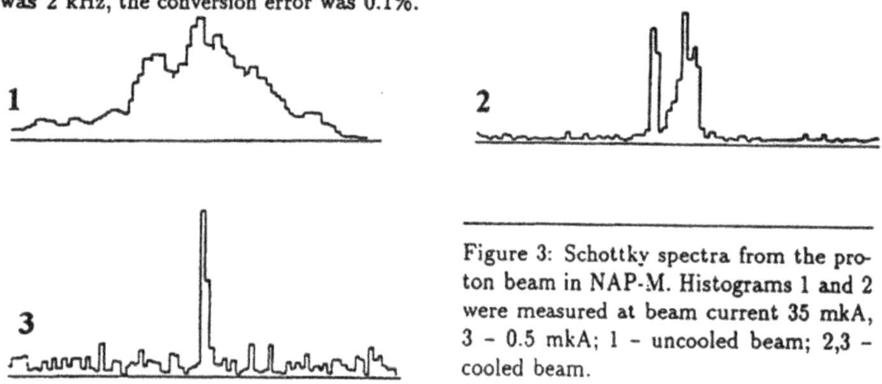

Figure 3: Schottky spectra from the proton beam in NAP-M. Histograms 1 and 2 were measured at beam current 35 mkA, 3 - 0.5 mkA; 1 - uncooled beam; 2,3 - cooled beam.

Typical spectra of the Schottky noise (see in Fig.3) clearly indicated strong effect of the beam cooling on the shape of these spectra. In particular, at large beam currents in the noise spectrum of the cooled beam appear two well pronounced peaks (see in Fig.3, (2)). The distance between these peaks was measured to evaluate the coherent frequency shift of the beam (for the 8th harmonics - full points in Fig.4). As expected, this distance decreases with a decrease in the beam current until becoming beyond the resolution of the monitor at small beam currents (Fig.3 (3), $I_p = Ne\omega_0/2\pi = 0.5$ mkA).

The linear fit of these data, which could be expected from Eq.(11), results in

$$\frac{\Omega_n^2}{2\pi n} \simeq 5.22 \times 10^{-7} N \quad [1/s^2]. \tag{22}$$

The slope in this dependence corresponds to the coupling impedance equal to $Z''/n \approx 503$ [Ohm], which is about 2.5 times lower than the value, which could be expected according to Eq.(13) ($Z/n \simeq 1200$ Ohm). It is difficult now to point out some particular element in NAP-M, which could be responsible for this effect. As is seen from Fig.4, the fitting curve is a little bit tilted relative data points. This probably indicates that the accuracy in the determination of the spectrum splitting in these measurements was not very good.

The Schottky noise power was calculated integrating the measured spectra over the frequency. It was found that the noise power of uncooled beam varies according to $P_n = N$.

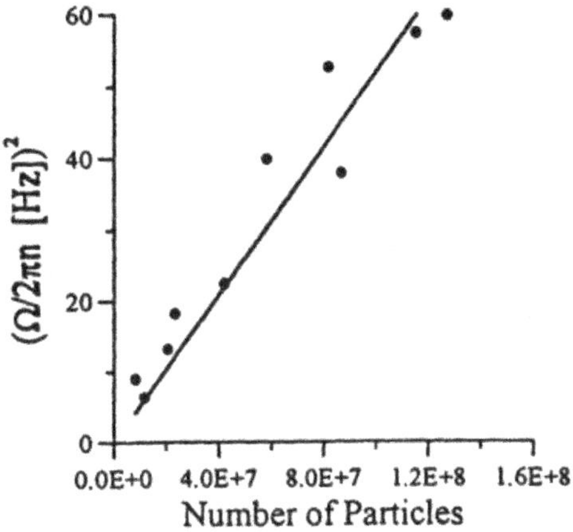

Figure 4: Dependence of the square of the spectrum splitting $[\Omega_n/(2\pi n)]$ on the number of particles in the beam. Full points present measurements [11], solid line - linear fit through origin.

The beam cooling resulted in strong suppression of the noise power. As is seen, however, from Fig.5 the noise power of the cooled beam is not a constant value. In paper [11], where we used Eq.(18) to treat these data, the visible increase in the powers of the spectra after $N \simeq 3 \times 10^7$ was interpreted as an indication of some threshold in the dependence of the equilibrium momentum spread ($P_n \propto \sigma^2$, $N \gg N_c$) on the beam current.

The observed blow-up of the noise power may also occur due to approaching to the threshold of the collective instability of the measured harmonic in the beam current (see in Eq.(20)). In order to inspect such a possibility we plotted the inverse noise power ($1/P_n$) vs. the beam intensity (the curve (b) in Fig.5). These data can be fitted well using a linear function

$$1/P_n = B - AN, \quad B \approx 9.23 \times 10^{-6}, \quad A \approx 8 \times 10^{-14}, \tag{23}$$

which corresponds to the following threshold number of particles

$$N_{th} = \frac{B}{A} \simeq 1.15 \times 10^8.$$

The parameters λ, Z' and the beam momentum spread (σ) can be calculated fitting $P(N)$ using Eq.(15). This yields

$$Z''/n \simeq 450\ \mathrm{Ohm}, \quad Z' \simeq 1500\ \mathrm{Ohm}, \quad \lambda \simeq 150\ 1/\mathrm{s}, \quad \frac{\sigma}{p} \simeq 1.2 \times 10^{-6}. \tag{24}$$

As is seen from Fig.6, the obtained value of Z' although being quite reasonable, contradicts to the observation in the Schottky spectra of two well-pronounced peaks. Since the spectrum splitting was seen in measurements starting at least from $N \simeq 10^7$ (see in Fig.4), we must reject the possibility to explain the observed increase in the noise power due to approaching to the threshold of instability of the 8th harmonics in the beam current. Adopting again that $P_n \propto \sigma^2$ we, however, obtain from Fig.5 and Eq.(23) that the variation of the beam momentum spread of the cooled beam in NAP-M with the beam intensity can be described using the following expression

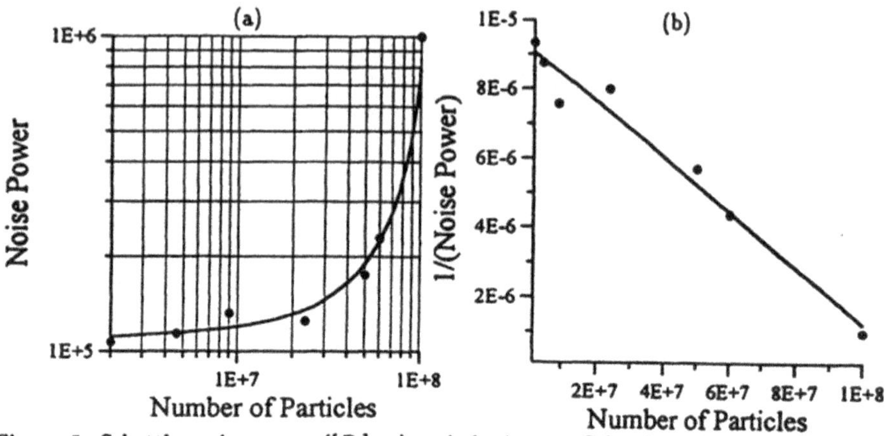

Figure 5: Schottky noise power ($[P_n]$: a) and the inverse Schottky noise power ($[1/P_n]$: b) vs. number of particles in the beam (full points - measurements [11]). Solid line in (b) presents the linear fit of the data using $1/P_n = B - AN$ with coefficients $B \simeq 9.23 \times 10^{-6}$ and $A = 7.97 \times 10^{-14}$, which results in $N_{th} \simeq 1.15 \times 10^{8}$.

$$\sigma^2 = \frac{\sigma_0^2}{1 - N/N_{th}}, \quad \frac{\sigma_0}{p} \simeq 1.4 \times 10^{-6}, \quad N_{th} \simeq 1.15 \times 10^{8}. \tag{25}$$

Note, that within the range of these measurements ($N \leq 10^8$) the fitting parameter N_{th} does not depend on the longitudinal beam temperature ($\propto \sigma^2$).

5 Beam ordering

The measured values of the transverse ($T_\perp$) and longitudinal (T) temperatures of the cooled proton beam in NAP-M differ by several orders of magnitude. In such conditions, for normal beams we could expect that the dependence of T on the beam current is determined by the intra-beam scattering (IBS). However, the dependence described in Eq.(25) strongly contradicts to predictions of the conventional IBS theory. In papers [12] and [6] it was assumed that the observed suppression of the IBS in very cold beams occurs due to a crystal-like ordering in longitudinal positions of protons, which may take place at very low longitudinal temperatures.

As was pointed out in Ref.[6], in very cold beams important modification of IBS comes from the selection of the collisions onto the fast and adiabatic, which is generic for the kinetics in external fields. If for example, τ_{col} is a collision time, colliding particles can change the energies of their betatron oscillations only in the case, when $\tau_{col}\omega_{x,z} \lesssim 1$. If $\tau_{col}\omega_{x,z} \gg 1$, collisions occur adiabatically slowly compared to betatron oscillations of particles and do not result in the energy exchange between of betatron oscillations of particles and their systematical longitudinal motion. The maximum impact parameter, corresponding to fast collisions

$$\rho_{max} = u/\omega_{x,z} = \alpha_p R_0(\sigma/p)/\nu_{x,z} \simeq 1.1 \times 10^{-4} \text{ cm}$$

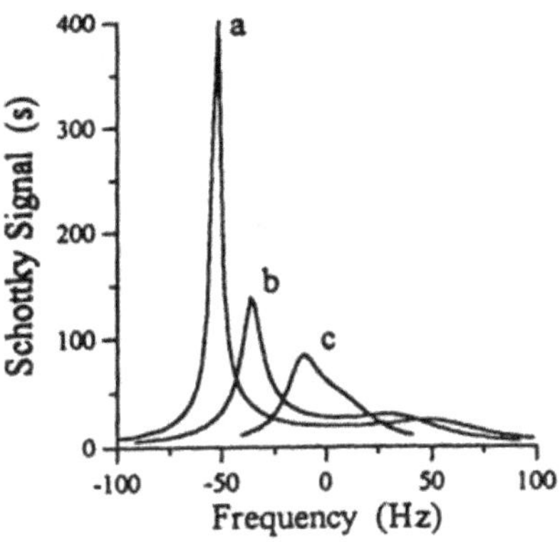

Figure 6: Schottky spectra near the 8th harmonic of the revolution frequency for the case, when $\sigma/p = 1.2\times10^{-6}$, $\lambda = 150$ 1/s and $Z' = 1500$ Ohm; a: $N = 10^8$, b: $N = 5\times10^7$, c: $N = 10^7$.

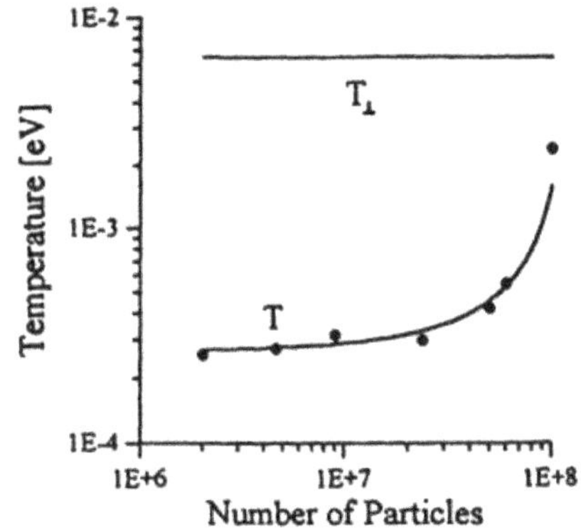

Figure 7: Dependence of longitudinal temperature ($T = \sigma^2/M$) on the number of particles in the beam. Recalculated from Fig.5. Horizontal line presents the transverse beam temperature recalculated from the measured beam radius [$T_\perp = pv\nu a^2/(2R_0\beta)$]

and $\rho_{min} \simeq e^2\beta^2/(Mv^2a^2) \simeq 1.5\times10^{-5}$ cm, yield too small value of the Coulomb logarithm for such collisions ($\ln[\rho_{max}/\rho_{min}] \simeq 2$) to make reliable results, obtained in the logarithmic approximation.

Adiabatic collisions occur as if the beam consists of macroparticles, spreaded over the region of betatron oscillations (transverse and longitudinal). In the coasting beam the mobility of these macroparticles is determined by the longitudinal temperature. The equilibrium in the beam of macroparticles is reached due to their collisions and corresponding energy exchange with coherent fluctuations in such a beam. If coherent oscillations of the beam are stable, only those of coherent modes, which have the damping times (τ_{coh}) comparable to the relaxation time of the beam (τ_r), can contribute significantly in the beam heating (cooling). Such a channel can dominate in the beam relaxation due to the Coulomb collisions, when the Debye length for longitudinal motion (r_D)

$$r_D = a\frac{\sigma}{p}\sqrt{\frac{\Pi pv\alpha_p}{Ne^2}}. \tag{26}$$

becomes comparable, or less than the average distance between particles in the beam of macroparticles $d = \Pi/N$. In this case, the correlations in the particle motions are extended till the interparticle distances, which results in their periodical spacing along the closed orbit with the period d. Although the calculation of the thermodynamic variables for such a beam with arbitrary temperature is very difficult, it seems that high-temperature and low-temperature limits fit each other well and agree with the following expression for the N-particle distribution function [6]:

$$f_N \propto \exp\left\{-\sum_{a=1}^{N}\left[\frac{\alpha_p \Delta p_a^2}{2MT} + \frac{n_D^2}{2}\left(\theta_a - \frac{2\pi a}{N}\right)^2\right]\right\}, \quad n_D = \frac{R_0}{r_D}, \tag{27}$$

which holds, if

$$\Gamma^{(1)} = \frac{2\pi n_D}{N} \gg 1. \tag{28}$$

The function f_N in Eq.(27) was calculated in the smoothed focusing approximation, when the variations of α_p along the closed orbit are ignored. The longitudinal coherent oscillations of a coasting beam in such a ring are stable for all temperatures provided that $\alpha_p > 0$. This means that forming of the chain with the decrease in the longitudinal temperature occurs gradually without sharp phase transitions. For parameters of NAP-M the condition (28) reads

$$\frac{\sigma}{p} \ll \frac{10^{-3}}{\sqrt{N}}.$$

Although it could hold at low beam currents (say, $N \leq 10^5$), but for most range of the beam intensities in the described Schottky noise measurements the parameter $\Gamma^{(1)}$ was small.

Subsequent study of the possibility of the forming in storage rings of the crystalline beams was mainly performed using the so-called molecular dynamics simulations (see, for example, in Refs [15] and [16]). It showed that in analogy with the case of the infinite and very cold one component plasmas, the phase transitions in focused beams start when its plasma coupling parameter $\Gamma \simeq (Qe)^2\varrho/T$ (here, ϱ is the density of the ion beam) exceeds the critical value $\Gamma_{th} \simeq 170$. Depending on its intensity the beam can form the string, zigzag and more complicated ordered structures around the closed orbit. However, it seems that the threshold for string ordering in these calculations agrees, in general, with the result of simplified calculation described above.

6 Stability of the Crystalline Strings

Like in normal beams, the intensity N of very cold crystalline beam ($\Gamma \gg 1$) can be limited by Coulomb repulsion of ions, which pushes them off the unperturbed closed orbit. As a result, after some threshold intensity [$N \geq N_{th}$] the ion oscillations become unstable. The effect of this instability on phase transitions of a beam moving in a straight channel with constant focusing and equal vertical and horizontal tunes was studied in Ref.[16]. The results of these studies showed that an increase in the beam intensity causes a sequence of the phase transitions from the string to a zigzag ordering, then from the zigzag to helix and to others more complicated structures.

However, these results cannot be applied directly to describe the beams in storage rings. The reason is that these phase transitions occur due to effect of the beam space charge on the symmetry of its equilibrium state. Roughly speaking, in a straight channel with constant focusing the first transition occurs when the beam space charge compensate the strength of the focussing. In the storage ring particles move along the closed orbits and for that reason, trajectories corresponding to the closed orbits must be periodic functions of s. This results in various space charge instabilities, which must be overcame to make phase transitions possible. For example, if the ring focusing is constant along the closed orbit, the beam space charge results in the so-called negative mass instability so that an equilibrium cannot be reached for a coasting beam (see, for instance, in Eq.(27), or in Refs [6] and [18]). On the contrary, in the ring with alternating gradient focusing the negative mass instability is avoided, if particle energies are below transition, but due to longitudinal modulation of betatron functions the beam space charge results in numerous instabilities of the betatron beam phase space [1]. Due to these instabilities the so-called Laslett tune shift (ε is the beam emittance)

$$\Delta\nu_L = \frac{Ne^2}{4\pi p v \gamma^2 \varepsilon}, \tag{29}$$

in real rings cannot significantly exceed some threshold value, which is about $\Delta\nu_L \simeq 0.1$.

Let us briefly discuss the effect of this space charge instability on the behaviour of crystalline beams in storage rings. The oscillations of ions are described by the following equations ($\gamma \simeq 1$):

$$\frac{d\mathbf{p}_m}{dt} = \frac{Qe}{c}\mathbf{v}_m \times \mathbf{H}(\mathbf{r}_m) + (Qe)^2 \sum_{m\neq n} \frac{\mathbf{r}_m - \mathbf{r}_n}{|\mathbf{r}_m - \mathbf{r}_n|^3}. \tag{30}$$

Here, $\mathbf{H}(\mathbf{r})$ is the magnetic field of the ring lattice,Qe is the charge of the ion. These equations describe both new closed orbits [2] and oscillations near these orbits. For the string geometry the particles are spaced along the closed orbit with the period $d = \Pi/N$. Then, if $x_m \ll d$, $z_m \ll d$ and $s_m \ll d$ are the offsets of the m-th ion, the linearization of Eqs(30) near the closed orbit yields:

$$z''_m + g_z(\theta)z_m = \frac{(Qe)^2}{M\omega_0^2 d^3}\sum_{m\neq n}\frac{z_m - z_n}{|m-n|^3}, \quad g_z = \frac{R_0^2}{R^2}n_B(\theta),$$

$$x''_m + g_x(\theta)x_m = \frac{R_0^2}{R}\frac{\Delta p_m}{p} + \frac{(Qe)^2}{M\omega_0^2 d^3}\sum_{m\neq n}\frac{x_m - x_n}{|m-n|^3}, \quad g_x = \frac{R_0^2}{R^2}[1 - n_B(\theta)], \tag{31}$$

$$\frac{\Delta p_m{}'}{p} = -\frac{2}{R_0}\frac{(Qe)^2}{M\omega_0^2 d^3}\sum_{m\neq n}\frac{s_m - s_n}{|m-n|^3}, \quad \frac{s'_m}{R_0} = -\frac{x_m}{R} + \frac{\Delta p_m}{p}.$$

Here, integers m and n mark the beam ions, $\theta = s/R_0$, $2\pi R_0$ is the closed orbit perimeter, n_B is the field index. Since for any particle in the beam $\mathbf{r}_m = \mathbf{r}_{m+N}$, we expect solutions

[1] First the interaction of the crystalline beam with the alternating gradient lattice and possible limitations were discussed in Ref.[17].

[2] In storage rings these closed orbits are periodical solutions of Eqs (30) with the lattice period. More generally, these trajectories are given by the fixed points of nonlinear resonances, excited by forces in r.h.s. of Eqs (30).

of Eqs(31) in the form

$$z_m = Z_l \exp(ik_l m - i\nu\theta), \quad x_m = X_l \exp(ik_l m - i\nu\theta),$$
$$s_m = S_l \exp(ik_l m - i\nu\theta), \quad (\Delta p/p)_m = \Delta_l \exp(ik_l m - i\nu\theta). \tag{32}$$

Here, $k_l = 2\pi l/N$, and $l = 1, 2, \ldots N-1$. Substituting these expressions in Eqs(31) we obtain the system of equations for amplitudes Z_l, X_l and S_l

$$Z_l'' + [g_z(\theta) - \Sigma_l]Z_l = 0, \quad \Sigma_l = \frac{2(Qe)^2}{M\omega_0^2 d^3}\sum_{n=1}^{\infty}\frac{1-\cos(k_l n)}{n^3},$$
$$X_l'' + [g_x(\theta) - \Sigma_l]X_l = \frac{R_0^2}{R}\Delta_l, \quad \Delta_l' = -2\Sigma_l\frac{S_l}{R_0}, \quad \frac{S_l'}{R_0} = -\frac{X_l}{R} + \Delta_l. \tag{33}$$

These ordinary linear equation with periodic coefficients can be solved using standard procedures. The only difference compared to the case of a normal beam is that for the string beam the rigidity coefficients in Eqs (33) depend on the mode-number (l) and since Eqs (33) are uncoupled in l, different numbers l specify different eigenmodes of the beam. Note also that according to Eqs(33) the mode with $l = 0$ is not affected by the Coulomb interaction. More generally, Eqs.(33) determines $3N/2$ β-functions, phase advances per the period of the lattice and tunes. These β-functions cover some band between the unperturbed $\beta(\theta)$ $[l = 0]$ and β-function of the so-called zigzag mode ($l = N/2$). The width of this band increases, when modes approach the border of the stability diagram. Since the highest Σ_l corresponds to the zigzag mode this mode becomes unstable earlier than others. As seen from Eqs.(33), an increase in the strength of the beam space charge (Σ_l) generally decreases rigidities of betatron and increases the rigidity of synchrotron oscillations in the string. For this reason, the tunes of vertical modes decrease with an increase in Σ_l and these oscillations lose stability when the tune reaches the nearest parametric resonance of betatron oscillations. The coupling of horizontal and longitudinal oscillations results in special dependences of tunes of these oscillations on the strength of the space charge. As is seen from Fig.8, the tune of the fast mode (horizontal) always increase, while the tune of the slow mode after passing the maximum tend to zero. Depending on the strength of the focusing, these modes can lose stability when either betatron oscillations reach the nearest betatron parametric resonance, or due to the instability of longitudinal oscillations. The last possibility is specific for the motion of particles along the closed orbit. In spite of high enough periodicity (the lattice of NAP-M consists of 4 periods) the maximum tune shifts and the threshold beam intensity are not high (according to data, shown in Fig.8, $\Delta\nu_{max} \simeq 0.2$ and $N_{th} \simeq 2.4 \times 10^6$). In any case, the phase transition string to zigzag in storage rings is accompanied by the envelope instability. Generally, the strengths of perturbations due to this instability can be decreased, if the variations of the lattice betatron functions are made as smooth as possible. The nonlinearity of the Coulomb force also may help the beam to survive during such a transition. Although the molecular dynamics simulations give very promising predictions (see, for example, in Ref.[19]), it seems that the understanding of the possibility of phase transitions in the lattices with alternating gradient focusing is still an open question.

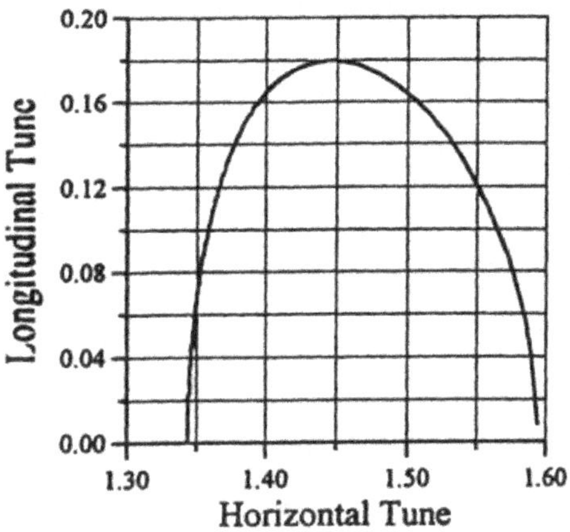

Figure 8: Working path $\nu_-(\nu_+)$ for the zigzag mode of small radial-longitudinal oscillations of the string crystalline configuration in NAP-M. Oscillations become unstable due to an instability of the slow zigzag mode (longitudinal), which occurs at $N_{th} \simeq 2.4 \times 10^6$.

7 Schottky Noise in the String

Solutions of Eqs (33) enable the calculation of various values, related to the Schottky noise of the string-beam. First, let us calculate $K_n(t) =< \rho_n(t)\rho_n^*(0) >$. For the sake of simplicity, we assume the smoothed focusing approximation ($R_0 \simeq R$, $\beta_x \simeq R/\nu_x$) and that harmonic number n is not very high. We write

$$\rho_n(t) = \sum_{a=1}^{N} e^{-in\theta_a(t)} \simeq -\frac{in}{R}\sum_{a=1}^{N} s_a(t)\exp\left(-\frac{2\pi ina}{N}\right) = \frac{-in}{R}\sqrt{N}S_n, \qquad (34)$$

where we used $\theta_a = (2\pi a/N) + (s_a/R)$, $|ns_a|/R \ll 1$ and the definition

$$S_n = \frac{1}{\sqrt{N}}\sum_{a=1}^{N} s_a(t)\exp\left(-\frac{2\pi ina}{N}\right). \qquad (35)$$

Now, taking as an independent variable $\theta = \omega_0 t$ and using Eq.(1), we obtain

$$K_n(\theta) = \frac{n^2}{R^2}N < S_n(\theta)S_n^*(0) > . \qquad (36)$$

In the smoothed focusing approximation, the value $S_n(\theta)$ can be calculated easily (see, for example, in Ref.[5], or in Appendix B)

$$S = -\frac{\nu_+}{Q_-}A_+\sin\psi_+ - \frac{\nu_-}{Q_+}A_-\sin\psi_-, \quad Q_\pm = \nu_x^2 - \Sigma - \nu_\pm^2. \qquad (37)$$

Here, subscripts + and − mark the fast and slow normal modes, $\psi'_\pm = \nu_\pm$, amplitudes $A_\pm$ are expressed in terms of the action variables of these modes

$$J_\pm = \nu_\pm\left(\frac{C}{\pm Q_\mp}\right)\frac{A_\pm^2}{2}, \quad C = \sqrt{(\nu_x^2 - 3\Sigma)^2 + 8\Sigma},$$

and subscript n is omitted for short. Substituting these expressions in Eq.(36) and using $(<\nu_\pm J_\pm>=T)$, we find

$$K_n(\theta)=\frac{NTn^2}{2pv\Sigma}\left[\frac{1+\chi}{2}\cos(\nu_-\theta)+\frac{1-\chi}{2}\cos(\nu_+\theta)\right],\quad \chi=\frac{\nu_x^2-3\Sigma}{\sqrt{(\nu_x^2-3\Sigma)^2+8\Sigma}}. \qquad (38)$$

Taking into account that for small harmonic numbers

$$\Sigma_n\simeq n^2\frac{Ne^2L}{\Pi pv},\quad L=\sum_{n=1}^{n_{max}}\frac{1}{n},\quad \pi n_{max}\ll N,$$

we rewrite the correlation function K_n like follows

$$K_n(\theta)=N_c\left[\frac{1+\chi}{2}\cos(\nu_-\theta)+\frac{1-\chi}{2}\cos(\nu_+\theta)\right],\quad N_c=\frac{NTn^2}{2pv\Sigma}\simeq\frac{\Pi T}{2e^2\Lambda}. \qquad (39)$$

Before discussion the spectral properties of the noise, corresponding to this equation, let us also calculate similar correlation function for the radial betatron oscillations $K_n^{(x)}(\theta)=<X_n(\theta)X_n^*(0)>$. Using Eqs(B.3), we obtain

$$K_n^{(x)}(\theta)=\frac{NT}{M\omega_0^2}\left[\frac{1+\chi}{\nu_+^2}\cos(\nu_+\theta)+\frac{1-\chi}{\nu_-^2}\cos(\nu_-\theta)\right]. \qquad (40)$$

The mixing of the radial and longitudinal spectra described by Eqs (39) and (40) is not a specific future of crystalline beams. Such a mixing will be seen in the spectra of both crystalline and normal beams and is caused by the motion of particles along the closed orbit, resulting in the radial-longitudinal coupling. These equations, however, show that for crystalline beam the approaching to the threshold of stability corresponding to a phase transition ($\nu_-\to 0$) is accompanied by the increase in betatron correlation function $[K_n^{(x)}(\theta)]$. Such a blow-up in betatron Schottky spectrum could be expected beforehand. The instability pushes particles off the old equilibrium orbit and changes the dimension of the equilibrium state of the beam (1d string →2d zigzag ordered beam).

8 Conclusion

The Schottky noise measurements at NAP-M clearly indicate the threshold [3] in the dependence of the beam noise power on its number of particles, which occurs about at $N=N_{th}\simeq 1.2\times 10^8$. This threshold does not correspond to an approaching the beam to the threshold of some coherent instability of the measured harmonics of the beam current and does not depend on the longitudinal beam temperature.

Very likely, this threshold indicates an effect of the particle correlations on the beam cooling. Although such an influence is specific for beams with very high phase space

[3] Fig.5 probably indicates one more threshold at $N_{th}\simeq 3\times 10^7$. For proton beam in NAP-M at this intensity the Laslett tune shift reaches $\Delta\nu_L\simeq 0.1$. Another possible change in conditions at this intensity was mentioned in [6]

density [4], these measurements do not give, however, clear indication of the transition of the beam in NAP-M in the crystalline state. This question still demands more experimental study, or reliable simulations.

The calculations of the stability of particle oscillations in the string geometry indicate the possibility of the formation of the string-ordered crystalline beam in NAP-M only at very low intensity $N \leq 2 \times 10^6$(see also in Refs [6] and [5]). For higher beam intensities more complicated crystalline structures must be expected [16].

The calculation of the Schottky spectra in the string-ordered beam predicts different dependences of the longitudinal and betatron correlation functions on the beam current near the threshold of stability of oscillations in the string (transition beam intensity). Simultaneous blow-up in the betatron spectrum and no blow-up of the corresponding lines in longitudinal spectrum is specific for the crystalline state of the beam and can be used as an indication of the string to zigzag transition.

N.S. Dikansky, V.V. Parkhomchuk and A.G. Ruggiero are kindly acknowledged for their stimulating discussions.

A Noise Power near Threshold of Instability

Let us calculate P_n near threshold of instability of coherent oscillations. In this region we can write

$$\epsilon(N,\omega) \simeq -\left(1-\frac{N}{N_{th}}\right)+\left(\frac{\partial\epsilon}{\partial\omega}\right)(\omega-\omega_{th}), \quad |N-N_{th}| \ll N,$$

and therefore

$$P_n = \int_{-\infty}^{\infty} \frac{d\omega}{2\pi}\frac{2N}{\omega|\epsilon(N,\omega)|^2}\mathrm{Im}\left\{\frac{\epsilon-1}{\zeta}\right\} \simeq \frac{N}{\omega_{th}|\partial\epsilon''/\partial\omega|_{th}(1-N/N_{th})}\frac{\zeta''}{|\zeta|^2}, \tag{A.1}$$

where we used Eqs(14). Now, assuming $q \ll 1$ we write

$$\epsilon = 1+q\zeta\left(\frac{i}{z}-\frac{1+iz}{1+z^2}\right), \quad z \gg \sqrt{q},$$

or

[4] For example, the cooling decrements (λ) of the laser-cooled atomic beam can be expressed in terms of the pair-correlation function $[K(r)]$ according to [20]

$$\lambda = \lambda^{(0)}\left[1+4\pi\varrho\int_0^{\infty} drrK(r)\right],$$

where $\lambda^{(0)}$ is the cooling decrements of a single atom. This expression shows that in crystalline beam, where $1+4\pi\varrho\int_0^{\infty} drrK(r) = 0$, holds $\lambda = 0$. Similar suppression the single particle electron cooling may take place in very cold and dense ion beams.

$$\epsilon'' = q\zeta' \left(\frac{1}{z} - \frac{z}{1+z^2}\right) - \frac{q\zeta''}{1+z^2}. \tag{A.2}$$

Then, the dispersion equation $\epsilon''(\omega) = 0$ results in $z_{th} = \zeta'/\zeta''$, while the calculation of $\partial\epsilon''/\partial\omega$ results in

$$\frac{\partial \epsilon''}{\partial z} = q\zeta' \left(-\frac{1}{z^2} - \frac{1}{1+z^2} + \frac{2z^2}{(1+z^2)^2}\right) + q\zeta'' \frac{2z}{(1+z^2)^2}.$$

Substituting in this equation z_{th}, we obtain

$$z_{th} \left(\frac{\partial \epsilon''}{\partial z}\right)_{th} = -\frac{q(\zeta'')^3}{|\zeta|^2},$$

which after substitution in Eq.(A.1) yields

$$P_n \simeq \frac{N}{q(\zeta'')^2(1 - N/N_{th})}.$$

Taking into account that at the threshold of instability $q(\zeta'')^2 \simeq \zeta'$ ($q \ll 1$), we obtain Eq.(20).

B Integrals of Motion in the String

For the sake of simplicity we use the smoothed focusing approximation. Then, equations from the 2nd and 3rd lines in Eqs(33) read

$$\begin{aligned} X_l'' + [\nu_x^2 - \Sigma_l]X_l &= I_l, \quad I_l = R\Delta_l, \\ I_l' = -2\Sigma_l S_l, &\quad S_l' = -X_l + I_l. \end{aligned} \tag{B.1}$$

These equations are made by the Hamiltonian (subscript l is omitted for short)

$$H = \frac{p^2 + (\nu_x^2 - \Sigma)X^2}{2} + \frac{I^2}{2} + \Sigma S^2 - XI, \quad p = X' \tag{B.2}$$

Direct solution of Eqs(B.1) results in the following expressions

$$\begin{aligned} X &= A_+ \cos\psi_+ + A_- \cos\psi_-, \quad p = -\nu_+ A_+ \sin\psi_+ - \nu_- A_- \sin\psi_-, \\ S &= -\frac{\nu_+}{Q_-} A_+ \sin\psi_+ - \frac{\nu_-}{Q_+} A_- \sin\psi_-, \quad Q_\pm = \nu_x^2 - \Sigma - \nu_\pm^2, \\ I &= Q_+ A_+ \cos\psi_+ + Q_- A_- \cos\psi_-, \quad \psi_\pm' = \nu_\pm, \\ w &= (w - 2\Sigma)(w + \Sigma + 1 - \nu_x^2), \quad w = \nu_\pm^2, \quad Q_+ Q_- = -2\Sigma, \end{aligned} \tag{B.3}$$

which generate canonical transformation from the variables (X, p, S, I) to the action-angle variables (J_+, ψ_+) and (J_-, ψ_-), so that

$$\begin{aligned} H &= \nu_+ J_+ + \nu_- J_-, \quad J_\pm = \nu_\pm \left(\frac{C}{\pm Q_\mp}\right)\frac{A_\pm^2}{2}, \\ A_\pm^2 &= \frac{(Q_\mp X - I)^2}{C^2} + \frac{(2\Sigma S + Q_\mp p)^2}{\nu_\pm^2 C^2}, \\ C &= \sqrt{(\nu_x^2 - 3\Sigma)^2 + 8\Sigma}. \end{aligned} \tag{B.4}$$

This canonical transformation enables numerous perturbation theory calculations for the string-ordered beams. If, for example, λ_x and λ_s are radial and longitudinal cooling decrements, simple calculations using

$$\frac{dJ_\pm}{dt} = -\lambda_x p \frac{\partial J_\pm}{\partial p} - \lambda_s I \frac{\partial J_\pm}{\partial I} = -\lambda_x p \frac{\partial X}{\partial \psi_\pm} - \lambda_s I \frac{\partial S}{\partial \psi_\pm}$$

result in

$$\frac{dJ_\pm}{dt} = -\Lambda_\pm J_\pm, \quad \Lambda_\pm = \Gamma \pm \frac{\delta}{2}\frac{\nu_x^2 - 3\Sigma}{\sqrt{(\nu_x^2 - 3\Sigma)^2 + 8\Sigma}}, \quad |\Lambda_\pm| \ll |\nu_\pm|,$$

where $\Gamma = (\lambda_x + \lambda_s)/2$ and $\delta = \lambda_x - \lambda_s$. Note, that the fast mode (J_+, ψ_+) has smaller cooling decrement $(\Lambda_+ \leq \Lambda_-)$, if $\lambda_s \geq \lambda_x$.

References

[1] G.I. Budker. Atomnaja Energia, **22**, p. 346, 1967.

[2] Ya.S. Derbenev, A.N. Skrinsky. Kinetics of Electron Cooling of Beams in Storage Rings of Heavy Particles. Preprint INP-225, Novosibirsk, 1968.

[3] Ya.S. Derbenev, A.N. Skrinsky. In Proc of the 10-th Internat. Conf. on High Energy Accel., v.1, p. 516, Serpukhov 1977; see also in Ya.S. Derbenev, A.N. Skrinsky. In Physics Revievs, ser. Soviet Physical Reviews, v. 3, p. 165, 1981.

[4] N.S. Dikansky, D.V. Pestrikov. In Proc. of the 6th All Union Particle Accelerator Conference, v. 1, p. 111, Dubna, 1979.

[5] N.S. Dikansky, D.V. Pestrikov. Physics of Intense Beams and Storage Rings. AIP PRESS, New York, 1994.

[6] N.S. Dikansky, D.V. Pestrikov. The Influence of the Ordering Effects on Relaxation of Coasting, Cold Beam in Storage Rings. Preprint INP 84-48, Novosibirsk, 1984; see also in Proc. of ECOOL 1984, p. 275, Kernforschungszentrum Karlsruhe, 1984; or in Sov. Journ. of Tech. Phys. **56**, N 2, .286, and N 3, p.505, 1986.

[7] V.V. Parkhomchuk. Experiments at NAP-M. These Proccedings.

[8] J. Borer, P. Bramham, et al. In Proc. of the 9-th Intern. Conf. on High Energy Accel., p. 53, Stanford, SLAC 1974.

[9] G.I. Budker, A.F. Bulushev, N.S. Dikansky, et al. In Proc. of the 10-th Internat. Conf. on High Energy Accel., v. 1, p. 498, Serpukhov 1977.

[10] V.V. Parkhomchuk, D.V.Pestrikov. Sov. Journ. of Tech. Phys., **50**, p. 1411, 1980.

[11] E.N. Dementiev, N.S. Dikansky, et al. Sov. Journ. of Tech. Phys., **50**, p. 1717, 1980.

[12] V.V. Parkhomchuk. In Proc. of ECOOL 1984, p. 71, Kernforschungszentrum Karlsruhe, 1984.

[13] S. Cocher and I. Hofmann. Part. Accel. **34**, 189, 1990.

[14] G.I. Budker et al. Particle Accelerators, **7**, p. 197, 1976.

[15] J.P. Schiffer. In Proc. of the Workshop on Crystalline Ion Beams, Wertheim, 1988, GSI-89-10, GSI, 1989, p2.

[16] R.W. Hasse, J.P. Schiffer. In Proc. of the Workshop on Crystalline Ion Beams, Wertheim, 1988, GSI-89-10, GSI, 1989, p33.

[17] I. Hofmann, J. Struckmeier. In Proc. of the Workshop on Crystalline Ion Beams, Wertheim, 1988, GSI-89-10, GSI, 1989, p140.

[18] A.G. Ruggiero. In Proc of the Workshop on Beam Cooling and Related Topics. CERN 94-03, p. 274, 1994.

[19] J. Wei, X. Li and A.M. Sessler. In Proc. of the Workshop on Beam Cooling and Related Topics. CERN 94-03, p. 366, 1994.

[20] Yu.L. Klimontovich. Statistical Physics. Nauka, 1982.

STUDIES OF CRYSTALLINE BEAMS AT LEGNARO

A.PISENT , A.BUROV#

INFN-Laboratori Nazionali di Legnaro, via Romea 4

I-35020 Legnaro (PD), Italy

ABSTRACT

Beam crystallization in a storage ring was studied theoretically (mainly in smooth approximation) and simulated in the lattice proposed for CSR. The effects of shear forces, envelope instability and periodicity breaking due to the presence of the electron cooling are discussed.

1. Introduction

At LNL (Laboratori Nazionali di Legnaro) the construction of an ion storage ring with the specific task of an experimental search for beam crystallization has recently been proposed . The project (CSR) is illustrated in ref .[1,2].

We shall here concentrate on some aspects of the beam physics studies done in these years at LNL. These studies have been oriented toward the determination of ring design criteria for a feasible crystallization experiment. Two main aspects have been considered and analyzed at different levels: the crystalline state structure and the evolution from ordinary to ordered beam.

The first aspect consists mainly in the determination of the geometry in which the cooled ions organize themselves (minimizing the potential energy), and the study of the coherent modes of these crystals. The comprehension of this part, achieved by analytic models and computer simulations is rather advanced.

The evolution of the beam during the cooling process is more complicated, involving different phenomena such as coherent instabilities and intrabeam scattering, and also the numerical approach is not easy. Our simulations show the relevance of the envelope instability and the importance of choosing a betatron phase advance per period lower than 90^0. Beam impedance evaluations for CSR ring have been initiated, and in [15] the two aspects of broad band impedance for non relativistic beams and electron

---

On leave from the Budker Institute of Nuclear Physics, 630090, Novosibirsk, Russia.

cooling impedance evaluation are considered. Finally some studies of the alignment and gradient error tolerances of our ring have been done.

2. Coulomb Crystal and Stability

In a Coulomb Crystal ions of the same charge are confined in a regular pattern[3,4,5]. The confinement in one direction (longitudinal, z) is guaranteed by the periodicity (or infinite extension) of the crystal, and in the other two directions (vertical y, horizontal x) by an external (linear) focusing.

In the rest frame of the crystal, that we suppose for the moment inertial, only electric fields are important both for focusing and particle-particle interaction, since relative velocity between ions is low.

In stationary condition, equilibrium requires that at each ion location

$$\vec{E} = \vec{0} \tag{1}$$

when adding the Coulomb and focusing fields. The stability of the equilibrium requires:

$$\frac{\partial E_x}{\partial x} > 0, \frac{\partial E_y}{\partial y} > 0, \frac{\partial E_z}{\partial z} > 0. \tag{2}$$

This last condition is clearly in contrast with

$$div\vec{E} = 0. \tag{3}$$

For this reason the equilibrium cannot be stationary and the external focusing must be time dependent (periodic). In a periodic focusing channel the equilibrium configuration is a periodic solution of the equations of motion, and the ions in the equilibrium positions outside the axis undergo a periodic displacement. In order to have a situation as close as possible to the time independent case, this displacement should be small. This is achieved when the period of the external focusing is short in comparison with the oscillation period of the crystal, and the external oscillating force can be replaced by its average.

Since the crystal frequencies are generally of the order of the zero space charge frequency or lower, there are the conditions for smooth approximation, i.e. phase advance per period much lower than π.

The same argument applies immediately to a beam traveling on a straight line (the transformation between inertial systems changes magnetic into electric fields) and alternating gradient quadrupoles are needed.

In a storage ring the rest frame of the reference particle is not inertial, and apparent forces seem to help, since in weak focusing conditions it is possible to have focusing in both vertical and horizontal directions. The Hamiltonian for a particle circulating in a horizontal ring is:

$$H(\theta) = \frac{1}{2}(p_x^2 + p_y^2 + p_z^2) - h\gamma x p_z + g_x \frac{x^2}{2} + g_y \frac{y^2}{2} + \xi V$$
$$V = \sum_i \frac{1}{r_i} \tag{4}$$

where $\theta = s/R$ is the independent parameter, $g_x = h^2 - R^2 K$ and $g_y = R^2 K$, $h = R/\rho(\theta)$, r_i is the distance to the i-th particle, R is the average ring radius, K the focusing strength . The space charge parameter is:

$$\xi = \frac{Z^2}{A} \frac{R^2 r_p}{d^3 \beta^2 \gamma^2} \tag{5}$$

with Z and A ionization and mass numbers, βc velocity of the reference particle, $r_p = 1.55\ 10^{-18}\ m$ classical proton radius, $d = 2\pi R\gamma/N$ average inter particle longitudinal distance, unit for all the lengths involved in the Hamiltonian; the momenta are measured in $p_0 d/R$ units .

If we consider K and ρ constant along the ring, the equations of motion become:

$$\begin{cases} x'' = \gamma p_z - Q_x^2 x - \xi \dfrac{\partial V}{\partial x} \\ y'' = -Q_y^2 y - \xi \dfrac{\partial V}{\partial y} \\ z' = p_z - \gamma x \\ p_z' = -\xi \dfrac{\partial V}{\partial z} \end{cases} \tag{6}$$

with:

$$Q_x^2 = 1 - R^2 K$$
$$Q_y^2 = R^2 K \tag{7}$$

One can have focusing in both transverse planes (weak focusing) if $0 < R^2 K < 1$. But longitudinally there are problems.

From the first equation, $x''=0$ for the beam baricenter and for an energy different from the reference if:

$$x = \gamma \frac{p_z}{Q_x^2} \tag{8}$$

and Eq. (6)a becomes:

$$z'' = \eta \gamma^2 \frac{\partial V}{\partial z}$$
$$\eta = \frac{1}{\gamma^2} - \frac{1}{Q_x^2} \tag{9}$$

Therefore it is possible to have (stationary) focusing and stability (at least for small ξ) in both the transverse planes , but the longitudinal stability requires $h<0$, i.e. $KR^2<1-\gamma^2<1$, inconsistent with the condition for the transverse stability. This condition was found by A.Sessler and J.Wei[4] and recognized as a negative mass instability by A.G.Ruggiero[6].

From another point of view we can say that also in this case the consequences of $div\bar{E} = 0$ cannot be circumvented*.

3. Ground States and coherent modes in a Ring

The Hamiltonian in smooth approximation is:

$$H = \frac{1}{2}(p_x^2 + p_y^2 + p_z^2) - \gamma x p_z + Q_x^2 \frac{x^2}{2} + Q_y^2 \frac{y^2}{2} + \xi V , \tag{10}$$

0where Q_x and Q_y are machine tunes. The equations of motion are the same as for weak focusing (Eq. (6)), but, since values are averaged, Q_x and Q_y are free.

To analyze the problem of ground state configurations, and coherent oscillations around those states, one has to write each coordinate as the sum of an equilibrium position plus a small deviation, namely:

$$\begin{cases} x_i = a_i + \delta x_i \\ y_i = b_i + \delta y_i \\ z_i = c_i + i + \delta z_i \end{cases} , \tag{11}$$

while the electric field can be linearized around the equilibrium position as:

$$\begin{aligned}
\frac{\partial V}{\partial x} &= -\sum_j \frac{a - a_j}{r_j^3} - \sum_j \left[\frac{1}{r_j^3} - 3\frac{(a - a_j)^2}{r_j^5} \right] [\delta x - \delta x_j] \\
\frac{\partial V}{\partial y} &= -\sum_j \frac{b - b_j}{r_j^3} - \sum_j \left[\frac{1}{r_j^3} - 3\frac{(b - b_j)^2}{r_j^5} \right] [\delta y - \delta y_j] \\
\frac{\partial V}{\partial z} &= -\sum_j \frac{c - c_j - j}{r_j^3} - \sum_j \left[\frac{1}{r_j^3} - 3\frac{(c - c_j - j)^2}{r_j^5} \right] [\delta z - \delta z_j]
\end{aligned} . \tag{12}$$

The equilibrium values (a_i, b_i, c_i) have to satisfy the equations of motion with l.h.s. equal to zero.

In particular eq. (6)c gives the condition

$$p_z = \gamma a , \tag{13}$$

---

*Note that these considerations are somewhat different from those about the longitudinal stability in the presence of an RF field. In this case infact $\frac{\partial^2 V}{\partial z^2}$ is always positive (repulsive Coulomb force) while for the RF field one can change the sign by choosing a different phase.

this means that at the equilibrium the longitudinal momentum must be proportional to the transverse displacement, that is nothing but the condition of equal revolution frequency for all the particles of the crystal.

Let's state this more clearly: the ring is not isochronous (the longitudinal stability of the crystal requires to operate below transition energy) but the particle distribution is isochronous. If we look at the p_z vs x plot after a successful simulation we can see a slope that is the signature of a crystalline ground state (see for example Fig. 5).

The other conditions for a_i, b_i, c_i read:

$$
\begin{aligned}
a &= \frac{\xi}{Q_x^2 - \gamma^2} \sum \frac{a - a_j}{r_j^3} \\
b &= \frac{\xi}{Q_y^2} \sum \frac{b - b_j}{r_j^3} \\
0 &= \sum \frac{c - c_j - j}{r_j^3}
\end{aligned} \quad . \tag{14}
$$

Due to the curvature the first two equations have different scaling parameters[7,8],

$$
\lambda_x = \left[\frac{3\xi}{2\left(Q_x^2 - \gamma^2\right)} \right]^{1/3}, \lambda_y = \left(\frac{3\xi}{2Q_y^2} \right)^{1/3} . \tag{15}
$$

Next step is the stability analysis. We consider the deviation from equilibrium configuration as a sum of coherent modes of oscillation; for each of them we can write:

$$
\begin{pmatrix} \delta x_j \\ \delta x_j \\ (\delta p_z)_j \end{pmatrix} = \begin{pmatrix} \delta x_q \\ \delta x_q \\ (\delta p_z)_q \end{pmatrix} e^{i(qj - \nu\theta)} , \tag{16}
$$

where $q = 2\pi k/N$ (k integer, $|k| < N/2$) is the mode angle, and $\delta p_z = p_z - \gamma a$ is the momentum deviation from equilibrium. Doing this substitution into the equations of motion one gets the following linear equations for the deviations from equilibrium:

$$
\begin{cases}
\delta x'' = \gamma \delta p_z - (Q_x^2 - \xi C_x)\delta x \\
\delta y'' = -(Q_y^2 - \xi C_y)\delta y \\
\delta p_z'' = -\xi C_y (\delta p_z - \gamma \delta x)
\end{cases} , \tag{17}
$$

where the subscript q is omitted and:

$$
\begin{aligned}
C_x &= 2\sum_{j\geq 1}\left(1-\cos(qj)\right)\left[\frac{1}{r_j^3}-\frac{12a^2\delta_j}{r_j^5}\right]\\
C_y &= 2\sum_{j\geq 1}\left(1-\cos(qj)\right)\left[\frac{1}{r_j^3}-\frac{12b^2\delta_j}{r_j^5}\right]\\
C_z &= 2\sum_{j\geq 1}\left(1-\cos(qj)\right)\left[\frac{3j^2}{r_j^5}-\frac{1}{r_j^3}\right]\\
\delta_j &= \begin{cases}1, j=odd\\ 0, j=even\end{cases}
\end{aligned}
\quad , \qquad (18)
$$

and the eigenfrequencies:

$$
\begin{aligned}
2\nu_\pm^2 &= Q_x^2 - \xi C_x + \xi C_z \pm \sqrt{(Q_x^2 - \xi C_x + \xi C_z)^2 - 4\xi C_z(Q_x^2 - \xi C_x - 1)}\\
\nu_y^2 &= Q_y^2 - \xi C_y
\end{aligned}
\quad . \qquad (19)
$$

The constants in the case of a string are for example: $C_z / 2 = C_x = C_y \equiv C_\perp$ with:

$$
C_\perp = 2\sum_{j\geq 1}\frac{1-\cos(qj)}{j^3}. \qquad (20)
$$

We have written a code that calculates the dispersion curves for the eigenmodes of the string and zig-zag. The information contained in the dispersion curves can be used for example to evaluate which modes can be excited by perturbations, such as lattice misalignments or periodic focusing strength variation.

4. Determination of thresholds

Another application of the dispersion relation is the calculation of the thresholds between one structure and the following one. Let's consider the transition from string to zig-zag.

The zig-zag will be formed in the transverse plane corresponding to a lower λ (let's suppose y). The zig-zag amplitude is determined by the Eq.(14)b:

$$
1=\frac{\xi}{Q_y^2}4\sum_{k=0}^{\infty}\frac{1}{[(2k+1)^2+4b^2]^{3/2}}, \qquad (21)
$$

and for $b=0$ the threshold is $\xi_1 = Q_y^2/4.2$, corresponding to λ=.709.

If we calculate the condition for the vertical π mode of the string to have 0 frequency one gets:

$$
0=Q_y^2-2\xi\sum_{k=1}^{\infty}\frac{1-\cos\pi k}{k^3}=Q_y^2-4\xi\sum_{k=1}^{\infty}\frac{1}{(1+2k)^3}, \qquad (22)
$$

and it is intuitive that this mode once unstable generates the zig zag.

The two methods give the same result, but in the first case we had to know the more complex structure, in the second case the simpler.

For the next step, the transition from zig-zag to a 3D structure, the advantage of the stability analysis is clear. The first approach requires the knowledge of the geometry of the 3D structure, and the comparison of the two potential energies (zig-zag and 3D structure) to determine the minimum. This has been done in [7,8] for rectilinear homogeneous and isotropic focusing, taking into account helixes and more complex structures; in the general case of a storage ring the problem is rather complicated even in smooth approximation.

Looking at the stability of the coherent modes of a zigzag it is much easier to find the transition ξ.

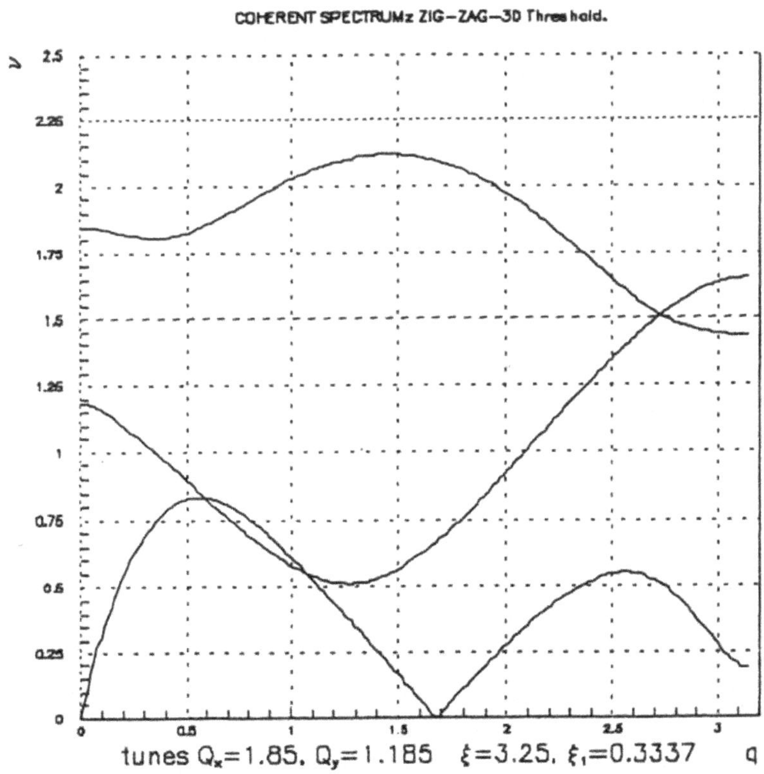

Fig 1 Coherent spectrum (ν_+,ν_- and ν_ν as a function of q, in rad) at zig-zag 3D structure transition. At zero mode all particles oscillate in phase, and the tunes are the same as for particles without Coulomb interaction. The mode q=*1.68* has *0* frequency and makes the structure unstable.

Indeed for CSR parameters (Q_x=*1.85*, Q_y=*1.185*) the ν_- mode q=*1.68* gets unstable when ξ=*3.25* (λ_x=*1.26*,λ_y=*1.51*) in excellent agreement with the simulations that we show later. In Fig. 1 the spectrum is shown, with the three branches determined by Eq. (19).

Finally it is interesting to observe that the single particle stability condition of Eq. (2) is a necessary but not sufficient condition for stability, since single particle coordinates are not normal coordinates of the system. Nevertheless the thresholds calculated in this way can be useful, since it can be proved that in this way transition ξ is always over estimated by less than a factor two.

For the string zig-zag threshold we have:

$$0 = Q_y^2 - 2\xi \sum_{k=1}^{\infty} \frac{1}{k^3}, \tag{23}$$

and $\xi_1 = Q_Y^2 / 2.4$, corresponding to λ=*0.855*. For the zig-zag 3D structure transition with CSR parameters the threshold is ξ=*5.7*, determined by the instability of the longitudinal oscillations.

5. Simulations

Multiparticle simulations have been performed with the code PARMT[9]. We use a subroutine where the space charge forces are calculated summing up the relative contributions from each particle in the cell and for identical distributions in the two adjacent cells. The cooling is simulated by a reduction of the divergence and momentum spread every turn.

The first application of those simulations is the determination of ground states. Cooling the beam adiabatically, starting from normal conditions (1πmmmrad emittance, Δp/p=1%) the particles organize themselves in the ground state. In doing this the cooling rate must be small enough to avoid the formation of metastable structures (local energy minima), but can still be considerably stronger than a realistic cooling.

We have first of all considered a rectilinear and smooth case (10^0 betatron phase advance for both planes) getting the same structures as in [7].

Then we have simulated CSR ring and classified the ground state structures λ=5 (Table 1). In Fig. 2 some examples are shown. All the 3D structures present the characteristic linear dependence of longitudinal velocity respect to transverse position as shown in Fig. 5.

Table 1 - Crystalline structures for the CSR working point[10].

	Horizontal	Vertical
Ring Length [m]	68.8	
Periodicity	8	
Tune	1.85	1.185
Beta Function [m]	6.7-10.4	5.1-8.4
Dispersion [m]	3.4-3.6	0
String	$0 < \lambda_x < 0.59$	$0 < \lambda_y < 0.71$
Zig - Zag	$0.64 < \lambda_x < 1.23$	$0.77 < \lambda_y < 1.48$
Rect. Tetra	$1.28 < \lambda_x < 1.37$	$1.54 < \lambda_y < 1.64$
Rhomb. Tetra	$1.61 < \lambda_x < 2.25$	$1.92 < \lambda_y < 2.7$
Twisted 4-Plet	$2.33 < \lambda_x < 2.41$	$2.79 < \lambda_y < 2.89$
Helix	$2.43 < \lambda_x < 3.64$	$2.92 < \lambda_y < 4.37$
Shell + String	$4.02 < \lambda_x < 4.82$	$4.82 < \lambda_y < 5.78$

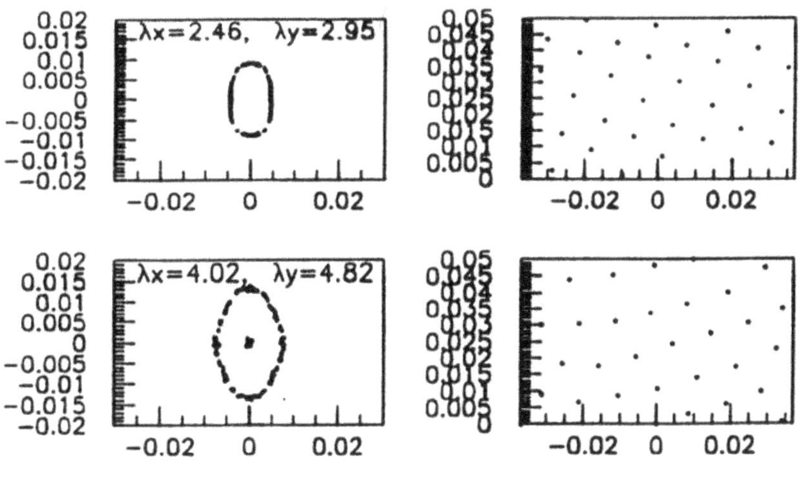

Fig.2 Crystalline configurations. On the left x-y plane projection, on the right particles on the outer shell plotted as if the mantle were the plane of the drawing.

6. Transverse and longitudinal breathing

Since the lattice is not perfectly smooth, the equilibrium ion positions transversally oscillate (or breath) passing from a focusing to a defocusing quadrupole. Moreover longitudinally the particles, that have all the same revolution frequency, oscillate respect to each other due to the presence of the drift spaces. In fact an external particle, with a higher linear velocity respect to a more internal one, will gain space in drift sections and lose in the bendings, so to have the same average angular velocity.

As already mentioned, the transverse breathing is lower for a smaller phase advance per period. The longitudinal breathing is lower for a smaller bending angle per period. To keep both those angles small is evidently costly, since it implies the construction of a ring with high periodicity. For this reason the lattice of CSR has been chosen so to be just "smooth enough".

In Fig. 3 the position of the particles on the mantle of a shell is plotted for four different locations in the lattice period. Both the longitudinal and the transverse oscillations can be seen.

These oscillations are appreciable but still considerably smaller than the crystal step size. Their period is the lattice period that is much shorter than the elastic period. For this reason they do not easily transfer energy to the crystal modes.

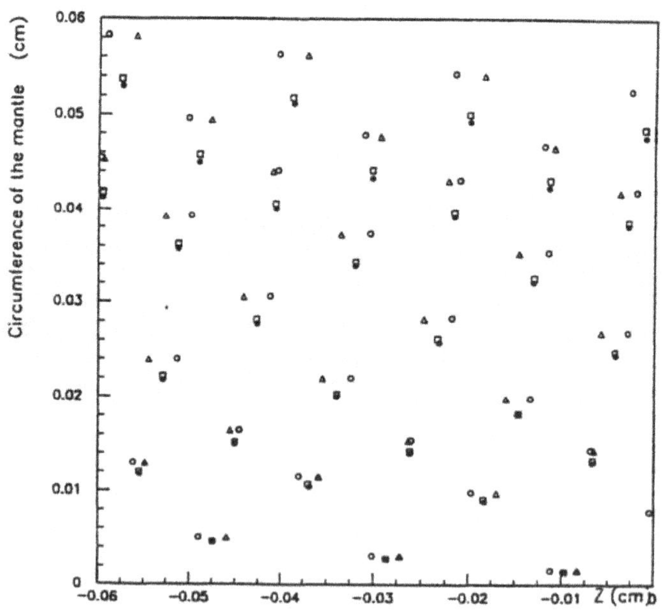

Fig 3. Relative positions of the particle of the outer shell for different locations in the focusing cell of CSR.

7. Shear forces and cooling

In the ground state all the particles have the same angular velocity, while a cooling device tries to give them the same linear velocity. This poses some limitations, and has suggested the use of special cooling devices to angular velocity equalization. Let's consider what happens with usual cooling.

In smooth approximation, and averaging the effect of the cooling, Eq. (6) gives for the equilibrium:

$$z' = p_z - \gamma x = 0$$
$$p_z' = -\xi \frac{\partial V}{\partial z} - \lambda_c p_z = 0 \tag{24}$$

where $\lambda_c>0$ is the cooling rate per radiant. The first equation states that at the equilibrium all particles have the same angular velocity, the second that the elastic shear force balances the cooling force. As a result the crystal is deformed. This effect for a horizontal zig-zag is sketched in Fig. 4.

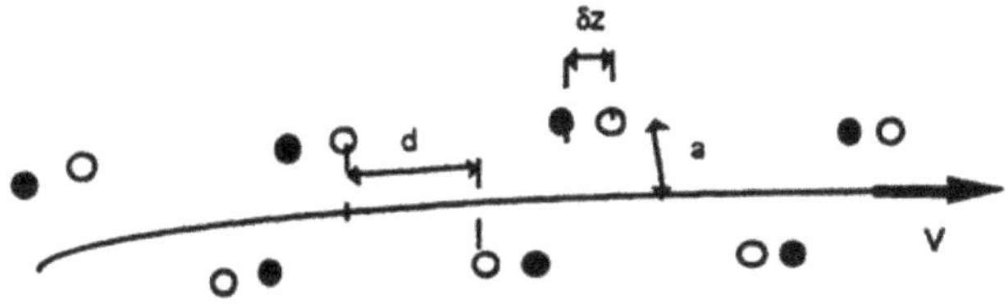

Fig. 4 Deformation of a horizontal zig-zag due to cooling (from white to black circles).

If the deformation δz is comparable with 1 the crystal is destroyed by the cooling. Linearizing the field around the equilibrium position this condition becomes:

$$1 >> |\delta z| = \frac{\lambda_c \gamma |a|}{\xi \frac{\partial^2 V}{\partial z^2}} \approx\propto \frac{\lambda_c \gamma}{\lambda^{3/2} Q^2} \tag{25}$$

where the last estimate is very rough, and considers a complex crystal as an homogeneous cylindrical charge distribution. We can see that for the same structure parameter a lattice with higher Q can better survive shear forces. The same conclusion was drawn by J. Shiffer[11]; nevertheless our argument is different, and in smooth approximation we do not have any temperature involved, while we can chose a lower cooling rate to preserve the crystal.

The limit is given by the necessity of having a cooling strong enough to overcome the intrinsic heating mechanisms of the crystal. In Fig 5 a two shells structure obtained in CSR with linear velocity cooling ($\lambda_x=12.5$, $\lambda_y=15.$, $2\pi\lambda_c=.09$) is shown. May be that for bigger structures angular velocity cooling or higher Q will be needed.

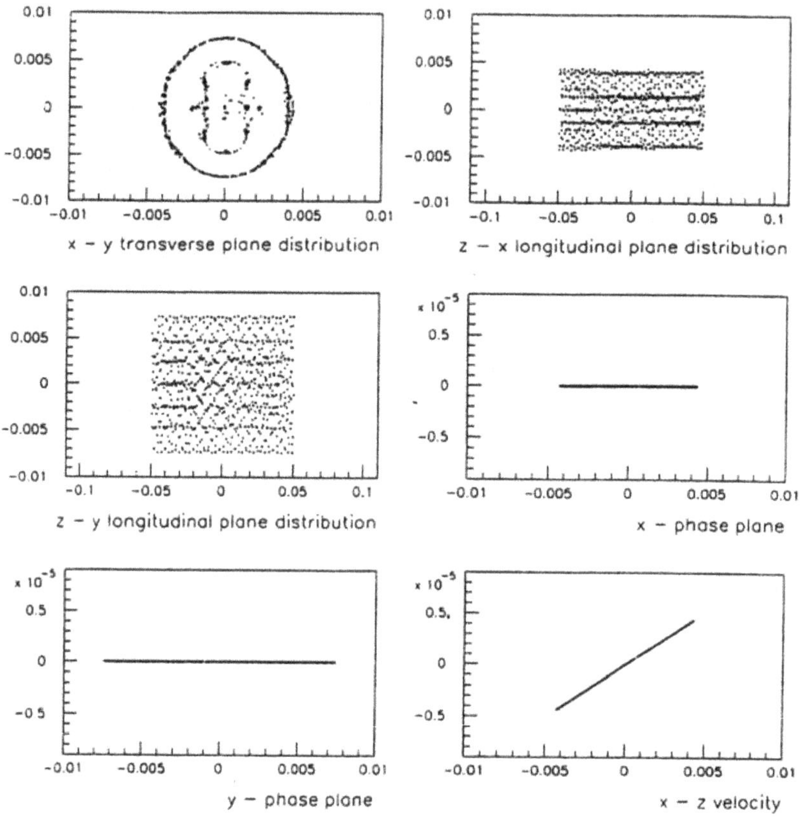

Fig. 5 Two cells structure(λ_x=12.5, λ_y=15.,$2\pi\lambda_c$=.09).; all lengths are in cm, all angles in rad. The bottom right figure shows the velocity gradient as a function of the transverse coordinate, signature of any crystal.

8. Envelope instability.

The current circulating in the ring for the various structures shown is relatively low. For example for Li_7^+ (that can be laser cooled) 10^8 particles accumulated at β=0.07 correspond to λ_x=21 and λ_y=26 (complicated multishell structures) . At injection (1π *mm mrad* normalized emittance) this corresponds to a Laslett tune shift of just 10^{-3}.

During the cooling process from this situation to a crystalline beam the tune has to go to zero and various parameter regions have to be crossed, where space charge and intra beam scattering (IBS) become important. Beams with significant space charge (relative tune shift of some fraction) can suffer coherent instabilities, such as the envelope instability: the envelope oscillation tune per period (depressed by space charge defocusing) can fall near a parametric (half integer) resonance and the envelope oscillation can become unstable with a growth time shorter than for IBS. It has been analytically proved (for restrictive hypothesis) and numerically shown that the instability is avoided if the zero current betatron phase advance per period σ_0 is smaller than 90^0 [12,13].

In the case of CSR there is a cooling that can compete with the instability growth rate. In Fig. 6 is shown the result of the simulations for *Li* (parameters mentioned before); the minimum cooling rate necessary to reach[14]:

$$\Gamma = \frac{(Ze)^2}{akT} = 1, \tag{26}$$

has been plotted as a function of σ_0; a is Wigner Size radius ($\approx\lambda$d), Γ=1 is considered a typical reference value on the way towards ordering. Simulations are done with PARMT using the Poisson solver routine.

As foreseen by the theory, there is a steep jump around the 90^0; for this reason, and keeping in mind the limitations of present cooling systems, we assumed as a design criteria for CSR lattice $\sigma_0<90^0$. This condition is not contradictory with the requirement of smoothness $\sigma_0<<\pi$, but adds a threshold value.

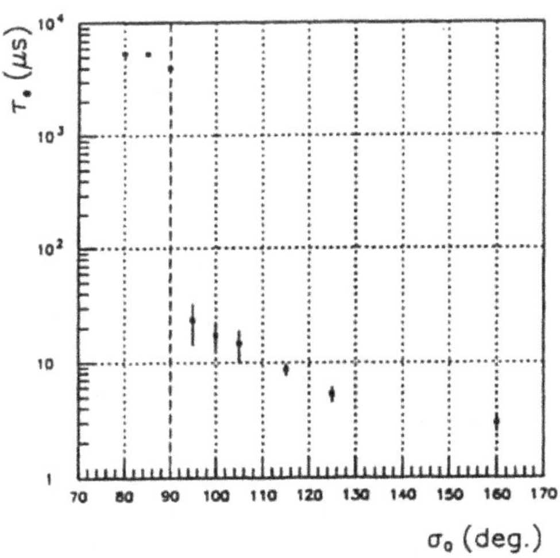

Fig. 6 Maximum cooling time necessary to reach the condition Γ=*1* as a function of the zero current phase advance per cell. The steep jump at 90^0 is due to envelope instability.

9. Periodicity Breaking due to the electron cooling

Error studies for CSR have been initiated, with the simulation of quadrupoles and dipoles misalignments of *0.05 mm*, rotations of *0.2 mrad* and field errors of 10^{-4}.

In parallel we begin to consider the effect of a simple thin lens, representing the electron cooler. The electrons indeed have a focusing effect that can be considered as a thin lens of focal length:

$$\frac{1}{f}=\frac{2\pi Znr_p l_c}{A\beta^2}, \tag{27}$$

where $l_c=1m$ is the length of the cooler, $n=10^8\ cm^{-3}$ is the electron density. For our *Li* beam $f=36\ m$. The focusing strength of the cooler is almost doubled by the focusing of the solenoid.

The periodicity braking due to this effect is overwhelming respect to gradient errors in the quadrupoles. The effective periodicity of the lattice is lowered to one, and instabilities related to parametric resonances can occur .

In Fig. 7 are shown the cooling curves ($1/\Gamma$ vs number of turns) for different f, for a vertical zig-zag in CSR ring. The beam is again *Li*, $\lambda_y=0.86$ $2\pi\lambda_c=.05$ ($\tau=65\mu s$) . The parameter f increases regularly by $10\ m$ from one curve to the next one, while the Γ saturation zones are denser around three values, roughly 1, 100 and 2000. This effect is relevant since the crystallization is prevented by our perturbation even with a very optimistic cooling. The plateaux in the curves correspond to the condition:

$$\lambda_c = 2\Lambda(\Gamma), \tag{28}$$

where the second term represents the time constant of the heating mechanism, function of the order parameter Γ.

In general a Mathieu equation will be driven:

$$Y'' + \left(\nu^2 + 2G\cos m\vartheta\right)Y = 0 \tag{29}$$

with

$$\nu = \frac{m}{2} + \varepsilon \qquad , \qquad G = R / 2\pi f$$

with m integer, ε distance from the closest half integer resonance. The growth rate at resonance is $\Lambda_{inst} = G/m$. At different beam temperature Y can have different meanings:

- for Y single particle position ν is the (space charge depressed) tune,
- for Y envelope dimension ν is the envelope tune,
- for Y crystal coherent mode amplitude ν is the tune.

By equating this growth rate with the cooling rate we get the threshold in terms of f:

$$f = \frac{1}{m}\frac{R}{\pi\lambda_c}. \tag{30}$$

In the case of Fig. 7 the fundamental mode is at $f=400m$; the agreement is rather poor since we could interpret the data with a fundamental $f=70m$ and a second resonance $f=35m$. With the simple argument here presented we overestimate the growth rate by about a factor 6. In other words the heating rate is mitigated respect to the instability (linear) growth rate by some mechanism, that can be different for space charge dominated, IBS dominated and crystallized beams. Investigations on this point are in progress.

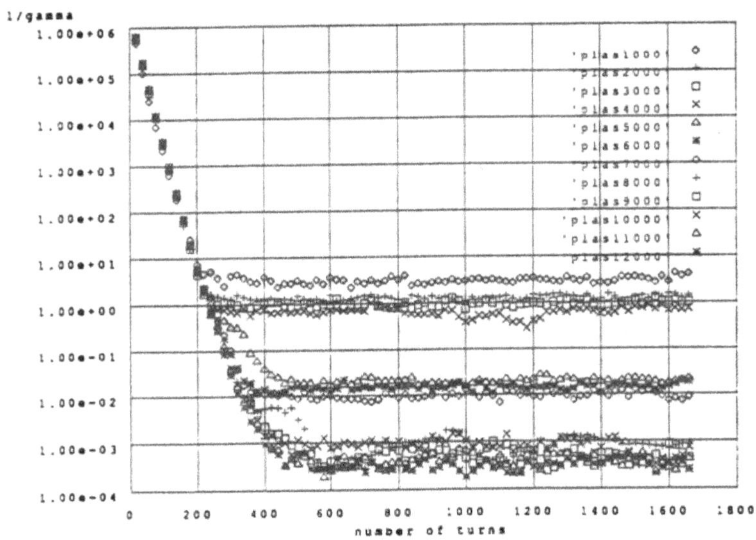

Fig.7 Cooling curves in CSR perturbed by a thin lens of different focal lengths: from 10 to 120 m. 10 m step.

10. Conclusions

Ground states have been investigated both analytically (for string and zig zag) and with multi-particle simulations. In 3D structures all ions move with the same revolution frequency (intuitive idea of a circulating crystal).

The development of ordered structures requires smooth focusing (small phase advance per period). The overcoming of envelope instability sets the phase advance to less than 90^0. A high horizontal tune is necessary to limit the crystal deformation due to the equilibrium between cooling and shear elastic forces. Therefore a perfect machine for crystallization has high periodicity and low phase advance per period. Drift spaces and costs requirements push in the opposite direction. A ring with periodicity 8 and phase advances 83.3^0 and 53.3^0 has been selected for CSR. In this lattice structures up to two shells have been simulated.

Errors and the electron cooling focusing break the periodicity and can cause problems that are under investigation.

11. Acknowledgements.

This paper summarizes the work done within CSR group with the help of many people that we all thank: B.Yang, S.Gustafsson (Computer simulations), I.Hofmann, L.Tecchio, G.Bisoffi, G.Lamanna, P.Lenisa, V.Variale (Envelope instability studies), T.Clauser, M.Rutigliano, V.Stagno (Error tolerances studies). A.G.Ruggiero initiated ground states configurations studies at LNL and has supported us with very effective discussions.

12. References

1. L.Tecchio et al. this workshop.
2. G.Bisoffi et al. "*Crystal Storage Ring: conceptual design*",LNL-INFN(REP) 97/95.
3. Rahman and J.P. Schiffer; Phys. Rev. Lett. **57** (1986) 1133.
4. J.Wei, X.P. Li and A.M. Sessler; Crystalline Beam Ground State; BNL-52381, June 1993.
5. J.Wei, Xiao-Ping Li and A. Sessler; Phys. Rev. Lett. **73** (1994) 3089.
6. A.G.Ruggiero; *Demonstration of No Feasibility of a Crystalline Beam in a Betatron Magnet*; BNL-49529, September 1993.
7. Hasse and J.P. Schiffer; Ann. of Phys. **203** (1990) 419.
8. Hasse; Phys. Rev. Lett. **67** (1991) 600.
9. A.Burov, A. Pisent, L. Tecchio and B. Yang, "*Crystallized Ion Beams in a Storage Ring*" (Accepted by Nuovo Cimento D)
10. J. Struckmeier and I Hoffman, GSI-ESR-87-03, Darmstadt 1987.
11. J.Shiffer this workshop.
12. I.Hofmann, L.J. Laslett, L. Smith and I. Haber; Particle Acc. **13** (1983) 145.
13. I.Hofmann and J. Struckmeier; Proc. of the Workshop on Crystalline Ion Beams, Wertheim, Germany, October 1988, Report GSI 89-10 (1989), 140.
14. B.Yang, G. Lamanna, G. Bisoffi, P. Lenisa, L. Tecchio and V. Variale "*The Blow-up Rate of the Envelope Instability of Cold Ion Beams in a Storage Ring*" (Accepted by Physics of Plasmas)
15. V.G.Vaccaro and A.Burov this workshop

Intrabeam Scattering in Dense and Cold Beams*

M. SEURER, Q. SPREITER AND C. TOEPPFER
Institut für Theoretische Physik II, Universität Erlangen
D-91058 Erlangen, Germany

ABSTRACT

We investigate the dynamics of intense, dense heavy ion beams. Our main interest lies in the various processes causing emittance changes. We extend the well known envelope equations for periodic quadrupole channels by introducing a dynamics for the emittances. The additional equations account for intrabeam scattering. We further derive a generalization of the envelope equations to the case of storage rings, where bending magnets and thus dispersion are present. The model is supplemented by means of molecular dynamics (MD) computer simulations. The main result of these simulations are: the extraction of relaxation rates in a regime, where common theories cannot be applied and the theoretical verification of indirect beam cooling.

1 Introduction

Since the first speculations about crystalline beams in storage rings by Schiffer and Kienle ten years ago[1] much theoretical work was done on the problem of stable transport of such beams in storage rings. Up to now such systems of so-called Coulomb crystals were only observed in ion traps[2]. In storage rings, crystallization could not yet be observed except the case of a simple onedimensional linear chain in the NAP-M in Novosibirsk[3] and in the ESR in Darmstadt[4]. A major problem on the way towards crystalline beams are the large intrabeam scattering rates in the gas-liquid regime preceding the transition to a Coulomb crystal. Intrabeam scattering causes emittance growth which inhibits further cooling.

In the case of periodic quadrupole channels intrabeam scattering in conjunction with envelope oscillations lead to emittance growth. This can easily be deduced by a simple model which we outline shortly in Section 2. The model yields an extension of the well known envelope equations of Lapostolle and Sacherer by introducing a

*supported by the Bundesministerium für Bildung und Forschung (BMBF), Gesellschaft für Schwerionenforschung, Darmstadt, the European Community and the Studienstiftung des deutschen Volkes

dynamics for the emittances. In this model an additional parameter, a relaxation rate, has to be introduced. The concept of envelope equations can also be applied to storage rings with bending magnets. This we will show in Section 3. In the parameter region of the gas-liquid transition, where scattering processes become very important, molecular dynamics (MD) computer simulations are an appropriate technique to study the dynamics of the beam. We outline shortly our simulation methods in Section 4. From these simulations we have extracted relaxation rates which we will discuss in Section 5. After that we present in Section 6 our simulation results concerning indirect transverse cooling of an ion beam in the TSR at the MPI Heidelberg. Finally, the dependence of the emittance growth rates on different values of phase advances per unit cell is discussed in Section 7.

2 Envelope equations in periodic quadrupole channels

High brilliances and high intensities require special techniques to understand the beam dynamics. The simple single particle dynamics is insufficient because of the space charge. The Coulomb repulsion between the particles contributes an additional force which must be taken into account. If the density is homogeneous and the beam cross section is an ellipse with half axes in the horizontal x- and vertical y-direction, then the space charge force acting on a particle is a linear function of the distances x and y of a particle from the center of the beam. But the proportionality factor of this force is a function of the density which in turn is a function of the beam dimensions. It is therefore insufficient to investigate the single particle dynamics since the single particle dynamics depends on the collective dynamics. This problem was solved by Lapostolle and Sacherer in 1971[5, 6] who developed the envelope equations. They considered the rms-dimensions of the beam and found differential equations for the beam envelopes including the effect of the space charge. In the rest frame of the beam and for a coasting beam these envelope equations take the form:

$$\ddot{\overline{x}} = a_{\mathrm{ext},x}\overline{x} + \frac{n_z Z^2 e^2}{4\pi\varepsilon_0 m(\overline{x}+\overline{y})} + \frac{\varepsilon_x^2}{\overline{x}^3}. \tag{1}$$

Here, n_z is the particle line density, Z and m are the ion charge and mass, respectively, and $a_{\mathrm{ext},x}$ is a time dependent external force due to the focusing structure. $\overline{x} = \sqrt{\langle xx \rangle}$ is the envelope in radial direction where the brackets denote averaging over the whole phase-space:

$$\langle xx \rangle = \int \mathrm{d}x\mathrm{d}y\mathrm{d}v_x\mathrm{d}v_y\mathrm{d}v_z\, xx f(x,y,v_x,v_y,v_z),$$

where f is the normalized particle distribution function in phase-space. The symbol

$$\varepsilon_x^2 = \det\begin{pmatrix} \langle xx \rangle & \langle xv_x \rangle \\ \langle xv_x \rangle & \langle v_x v_x \rangle \end{pmatrix} = \langle xx \rangle\langle v_x v_x \rangle - \langle xv_x \rangle^2 \tag{2}$$

is the rms-emittance of the beam. The first term in Eq. 1 is the external force, the second is the space charge contribution and the third is a pressure term due to the temperature motion of the ions. The equation for the vertical y-direction is obtained by interchanging x and y in Eq. 1. In the case of bunched beams the structure of the envelope equations is the same as in the case of coasting beams, only the space charge term is more complicated:

$$\ddot{\overline{x}} = a_{\mathrm{ext},x}\overline{x} + \frac{NZ^2e^2}{20\sqrt{5}\pi\varepsilon_0 m\overline{x}^2} g\left(\frac{\overline{y}}{\overline{x}}, \frac{\overline{z}}{\overline{x}}\right) + \frac{\varepsilon_x^2}{\overline{x}^3}, \tag{3}$$

where $g(\alpha, \beta)$ is the elliptic integral of the second kind:

$$g(\alpha, \beta) = \frac{3}{2}\int_0^\infty \mathrm{d}s(1+s)^{-3/2}(\alpha^2+s)^{-1/2}(\beta^2+s)^{-1/2}.$$

Here, N is the number of particles in the bunch. Again, one obtains the equations for the vertical y- and the longitudinal z-direction by interchanging x, y and z. There are two main drawbacks of this model. The first one is that it does not include dispersion effects which are present in storage rings. An extension of the envelope equations for a coasting beam to structures including bending magnets is presented in the following section. The second drawback is that the emittances in this equation have to be provided as an external parameter. Normally they are treated as constant. But in high intensity beams there can be considerable emittance growth due to space charge homogenization, instabilities and intrabeam scattering. The problems of space charge homogenization and instabilities are treated elsewhere[7, 8, 9, 10], here we will focus on the effects of intrabeam scattering. The mechanism causing emittance growth in periodic quadrupole channels is the relaxation of a temperature anisotropy. To show this, we establish a relation between the beam temperature in x-direction T_x, the emittance ε_x and the envelope $\overline{x}$:

$$T_x = \frac{m\varepsilon_x^2}{k_B\overline{x}^2}. \tag{4}$$

Since the emittance is approximately constant in a periodic quadrupole channel, it is evident that variations in envelopes cause variations in temperature. Typically, these variations of the envelopes are with opposite phase in x- and y-direction, thus causing a large temperature anisotropy. We make a simple temperature relaxation ansatz, substitute the temperatures by the emittances with the aid of Eq. 4 and gain additional equations governing the time evolution of the emittances[11, 12]:

$$\frac{\mathrm{d}}{\mathrm{d}t}\varepsilon_x^2 = \tau_c^{-1}\left(2\frac{\varepsilon_x^2}{\overline{x}^2} - \frac{\varepsilon_y^2}{\overline{y}^2} - \langle v_z v_z\rangle\right)\overline{x}^2. \tag{5}$$

The results for ε_y^2 and $\langle v_z v_z\rangle$ are derived analogously. These are the results for coasting beams, the results for bunched beams are derived if $\langle v_z v_z\rangle$ is replaced by $\varepsilon_z^2/\overline{z}^2$. Here,

τ_c^{-1} is a relaxation rate which is itself a function of the beam temperature and the particle density. The relaxation rate τ_c^{-1} is the only parameter in the model. Before discussing this parameter in Sec. 5 we show how envelope equations can be established for beams in storage rings where bending magnets are present.

3 Envelope equations in storage rings

To develop envelope equations we first establish the single particle equations of motion in a bending magnet in the rest frame of the beam. In order to get these equations of motion we performed the following steps[13].

1. Propagation of the particles in the actual system of reference by the time step dt.
2. Boost of all particle coordinates from this system of reference to the laboratory system.
3. Rotation of the coordinate system by the angle $\frac{\gamma\beta c}{R}dt$. Here γ and β are the usual relativistic factors, c is the speed of light and R is the bending radius.
4. Boost of all particles to the co-propagating system of reference.
5. Synchronization of particle clocks

In each of these steps we only kept terms up to linear order in dt. This procedure leads to the equations of motion in a pure bending magnet. The full equations of motion arising from this scheme are exact. For the following considerations we keep only the terms up to linear order in x, y, z, v_x, v_y and v_z, with which we denote the particle positions and velocities in the rest frame of the beam. This approximation is valid since the nonlinear terms are small as they contain additional factors x/R or $v/\gamma\beta c$. The coordinate system is oriented such that the unit vector in x-direction $\mathbf{e}_x$ points to the center of the bending circle, $\mathbf{e}_z$ points in the direction of beam propagation and $\mathbf{e}_y$ is perpendicular on $\mathbf{e}_x$ and $\mathbf{e}_z$ in such a way that $\mathbf{e}_x$, $\mathbf{e}_y$ and $\mathbf{e}_z$ form a right-handed coordinate system.

$$\frac{\mathrm{d}}{\mathrm{d}t}x = v_x \tag{6}$$

$$\frac{\mathrm{d}}{\mathrm{d}t}y = v_y \tag{7}$$

$$\frac{\mathrm{d}}{\mathrm{d}t}z = v_z + \frac{\gamma^2\beta c}{R}x \tag{8}$$

$$\frac{\mathrm{d}}{\mathrm{d}t}v_x = a_{sc,x} - \frac{\gamma^2\beta c}{R}v_z - \frac{\gamma^2\beta^2c^2}{R^2}x \tag{9}$$

$$\frac{\mathrm{d}}{\mathrm{d}t} v_y = a_{\mathrm{sc},y} \tag{10}$$

$$\frac{\mathrm{d}}{\mathrm{d}t} v_z = a_{\mathrm{sc},z} \tag{11}$$

$a_{\mathrm{sc},x}$, $a_{\mathrm{sc},y}$ and $a_{\mathrm{sc},z}$ denote the acceleration due to the interparticle forces. After establishing the single particle equations of motion we will focus now on the derivation of envelope equations. It is evident that nothing changes in the vertical direction: the dynamics of the y-direction is decoupled from the dynamics of all other directions. In the accelerator plane the dynamics is fully coupled. But since we apply periodic boundary conditions in z-direction of beam propagation, the distribution function is a function of x, y, v_x, v_y and v_z only and we can omit the dynamics of the z-coordinate. Thus there remain only the x-, v_x- and v_z-coordinate as candidates for the envelope equations in the accelerator plane. The set of all second moments that can be constructed from these coordinates is $\langle xx \rangle$, $\langle xv_x \rangle$, $\langle xv_z \rangle$, $\langle v_x v_x \rangle$, $\langle v_x v_z \rangle$ and $\langle v_z v_z \rangle$. The velocity in z-direction, i.e. the particle longitudinal momentum is unchanged during the passage through a bending magnet. Therefore the rms velocity spread $\langle v_z v_z \rangle$ does not change, which means that $\langle v_z v_z \rangle$ is a constant of motion. The second constant of motion we found is the generalized emittance in x-v_x-v_z-space:

$$\varepsilon_{\mathrm{rms}}^3 = \det \begin{pmatrix} \langle xx \rangle & \langle xv_x \rangle & \langle xv_z \rangle \\ \langle xv_x \rangle & \langle v_x v_x \rangle & \langle v_x v_z \rangle \\ \langle xv_z \rangle & \langle v_x v_z \rangle & \langle v_z v_z \rangle \end{pmatrix}. \tag{12}$$

The conservation of $\varepsilon_{\mathrm{rms}}^3$ is established by straightforward calculations making use of Eqs. 6 - 11 under the assumptions that $a_{\mathrm{sc},x}$ is a linear function of x and that a_z vanishes. Despite these reduction of variables by exploiting the constants of motion we end up with four second moments which have a time dependence. The dynamics is summarized in the following generalized envelope equations:

$$\frac{\mathrm{d}}{\mathrm{d}t} \overline{x} = \xi \overline{x} \tag{13}$$

$$\frac{\mathrm{d}}{\mathrm{d}t} \xi = \frac{\varepsilon_{\mathrm{rms}}^3 + \left(\langle x v_z \rangle \xi - \langle v_x v_z \rangle \right)^2 \overline{x}^2}{\overline{x}^4 \langle v_z v_z \rangle - \overline{x}^2 \langle x v_z \rangle^2} + a_{\mathrm{ext},x} + \frac{n_z Z^2 e^2}{4\pi\varepsilon_0 m (\overline{x} + \overline{y}) \overline{x}} - \frac{\gamma^2 \beta^2 c^2}{R^2} - \frac{\gamma^2 \beta c}{R} \frac{\langle x v_z \rangle}{\overline{x}^2} - \xi^2 \tag{14}$$

$$\frac{\mathrm{d}}{\mathrm{d}t} \langle x v_z \rangle = \langle v_x v_z \rangle \tag{15}$$

$$\frac{\mathrm{d}}{\mathrm{d}t} \langle v_x v_z \rangle = a_{\mathrm{ext},x} \langle x v_z \rangle + \frac{n_z Z^2 e^2}{4\pi\varepsilon_0 m (\overline{x} + \overline{y}) \overline{x}} \langle x v_z \rangle - \frac{\gamma^2 \beta^2 c^2}{R^2} \langle x v_z \rangle - \frac{\gamma^2 \beta c}{R} \langle v_z v_z \rangle \tag{16}$$

$$\frac{\mathrm{d}}{\mathrm{d}t} \overline{y} = \eta \overline{y} \tag{17}$$

$$\frac{\mathrm{d}}{\mathrm{d}t} \eta = a_{\mathrm{ext},y} + \frac{\varepsilon_y^2}{\overline{y}^4} + \frac{n_z Z^2 e^2}{4\pi\varepsilon_0 m (\overline{x} + \overline{y}) \overline{y}} - \eta^2. \tag{18}$$

We note that Eqs. 13 and 14 cannot be summarized in form of a second order differential equation, since the dynamics of the intermediate variable ξ, which is related to the logarithmic time derivative of the envelope, depends on ξ itself. The first term in Eq. 14 accounts for the pressure and corresponds to a term $\varepsilon_x^2/\overline{x}^4$ in the case of a periodic quadrupole channel. We note that the rms-emittance ε_x defined in Sec. 2 is no longer a constant of motion in a bending magnet, but in a matched beam it becomes a periodic function of the lattice. The second term in Eq. 14 is again the external focusing due to the quadrupoles, the third term is due to the space charge. The fourth term reflects the focusing properties of a pure bending magnet, the fifth term is due to the velocity spread and is typically defocusing. These two terms vanish in the case of a straight section. The last term is of merely algebraic origin. These equations form a set of six coupled differential equations which have to be solved simultaneously. We are especially looking for solutions which are matched, i.e. which are periodic with the lattice period in all variables. The search is done by a simple iteration process: We solve numerically the differential equations 13 - 18 over one accelerator period. As initial values of a new run we take a weighted arithmetic mean of the initial and the end value of the last run. The starting point for each run is chosen as a symmetry point of the lattice. Thus we have $\xi = 0$, since ξ depends on the α-function[14] and the α-function vanishes at a symmetry point. As starting values for the iteration we take the envelope $\overline{x}$ resulting from the single particle beam dynamics and we take $\langle x v_z \rangle = \langle v_x v_z \rangle = 0$. The method converges rapidly after 50 – 100 iterations. The matching procedure enables us to provide our MD computer programs with reasonable starting data.

4 The molecular dynamics simulation technique

In a regime were collisions between beam particles become frequent and correlations between the beam particles get larger it is convenient to apply the molecular dynamics (MD) simulation method. This method is based on the numerical solution of the equations of motion. Typically, two types of forces are acting on the beam particles: the external force exerted by the focusing structure and the internal force due to the Coulomb interaction of the beam particles. In molecular dynamics simulations the internal force is calculated by summing up the exact Coulomb interaction forces of all pairs of particles in the beam. Thus all correlation effects are taken into account correctly. This is in contrast to other simulation techniques like particle-in-cell (PIC) simulations[15]. Here, only the mean field contributions are taken into account. The differences between the molecular dynamics method and two types of particle-in-cell methods are summarized in Tab. 1. As mentioned above, the internal force is calculated by summing up the forces between all possible pairs of beam particles. Thus, the numerical effort scales as $\sim N^2$ where N is the number of particles in the simulation box. This scaling law sets an upper bound to the size of the physical system,

Table 1: The table shows the application range of different simulation methods. d is the typical wavelength of the phenomenon which has to be investigated. $r_{WS} = \sqrt[3]{4\pi n/3}$ is the Wigner Seitz radius and $\lambda_D = \sqrt{\varepsilon_0 k_B T / n_0 Z^2 e^2}$ is the Debye length. A '+'-sign indicates that the simulation technique is a good tool to investigate the phenomenon, a '-'-sign denotes that the simulation technique is not appropriate to investigate the phenomenon.

	$d < r_{WS}$ hard collisions	$r_{WS} < d < \lambda_D$ soft collisions, short wavelength collective modes	$d > \lambda_D$ collisionless, long wavelength collective modes
Molecular Dynamics	+	+	up to some λ_D (limited by box size)
PIC with microparticles (Vlasov-Poisson)	-	+	up to some λ_D (limited by box size)
PIC with macroparticles (Vlasov-Poisson)	-	0	+

which can be simulated. So, the molecular dynamics method is not appropriate to study long wavelength phenomena, which is its main drawback. To simulate coasting beams we apply periodic boundary conditions in z-direction of beam propagation. The Coulomb forces are calculated with a modified Ewald method[16]. In the case of a bunched beam we use a hierarchical tree code[17] that reduces the numerical effort to nearly $\sim N \log N$. Therefore it is possible to simulate 2500 beam particles within reasonable time. For the numerical integration of the equations of motion we use a 4th order predictor-corrector method or a 5th order Runge-Kutta method with adaptable stepsize[18, 19]. The particle positions and velocities are initialized such that their rms-values equal the values obtained by the matching procedure discussed above.

5 Relaxation rates in the gas-liquid transition regime

One might expect that the gas-liquid transition manifests itself in the transport coefficients. We measure temperature relaxation rates in our MD simulations. To extract the relaxation rates we use two different methods. The first method is to simulate the beam, measure the emittance growth rates and fit the relaxation rate in the phenomenological model discussed above such that the emittance growth rates in the MD-simulation and in the phenomenological model are equal. The second method is to cool the beam longitudinally until a dynamical equilibrium between external cooling and internal heating is reached. Then one can calculate the relaxation rate and the plasma parameter from the cooling rate, the emittances and the envelopes. Fig. 1 shows the dependence of the relaxation rate τ_c^{-1} on the plasma parameter

$$\Gamma = \frac{Z^2 e^2}{4\pi\varepsilon_0 k_B T} \sqrt[3]{\frac{3}{4\pi n}} \tag{19}$$

for bunched beams and coasting beams with different numbers of particles per unit length. The plasma parameter is a measure of the coupling strength of a plasma. Low plasma parameters correspond to weak coupling, high plasma parameters to strong coupling. The time scale is measured with respect to the typical time scale of the plasma, the plasma frequency $\omega_p = \sqrt{\frac{Z^2 e^2 n}{\varepsilon_0 m}}$. The behaviour is similar for all cases: for small Γ the relaxation rate increases as predicted by the kinetic theory[20]: $\tau_c^{-1} \propto \omega_p \Gamma^{3/2} \ln\Lambda$. Here, $\ln\Lambda$ is the Coulomb logarithm which is due to the introduction of cut-offs at small and large scattering angles. At $\Gamma \approx 0.3$, deviations from the kinetic value appear. The relaxation rate measured in the MD-simulations is lower than the kinetic value. At $\Gamma \approx 0.7$ a flat maximum is observed. For Γ-values larger than 1.0 the relaxation rate starts to drop. The decrease can be fitted by $\tau_c^{-1} \propto \omega_p \Gamma^{-3/2}$. The simulations confirm that in this regime the relaxation rate is independent of the density of the beam particles. It only depends on the temperature of the beam. The decrease in the relaxation rate is the precursor of the transition of the plasma from gaseous through a liquid to a crystalline state. This was also observed in simulations by Wei et al., who mainly investigated beam crystallization[21] and in laser cooling experiments in ion traps[22]. The different Γ-dependence of the relaxation rate in the kinetic and in the liquid regime is also clearly visible in the time development of the emittances of beams which are not cooled. Then, the emittance is expected to grow according to the differential equation

$$\frac{\mathrm{d}}{\mathrm{d}t}\varepsilon = \frac{1}{\tau_c}\alpha_s \varepsilon$$

where α_s is a constant which only depends on the focusing structure. The growth is not exponential as one might suppose, since the relaxation rate τ_c^{-1} depends itself on the emittance. In the regime, where the kinetic theory is still valid, the emittance is expected to grow as $\varepsilon \propto t^\beta$, where $\beta = 1/3$ in the case of a bunched beam and $1/3 \leq \beta \leq 1/2$ in the case of a coasting beam[12]. Fig. 2 shows the time development of the emittance of a beam where the initial value of the plasma parameter was $\Gamma = 0.18$. The dashed line represents the expected behaviour according to the kinetic theory which is valid as the plasma parameter is sufficiently small. As the beam gets hotter the relaxation rate gets smaller and thus also the growth rate. As shown in Fig. 3 the time development of the emittance changes when one starts the simulation at $\Gamma = 1.8$ which is beyond the maximum of the relaxation rate at $\Gamma = 0.7$. Now, the growth rate increases as the beam gets hotter. This curve cannot be fitted by the kinetic model. The curvature of the curve is positive. We denote this behaviour as anti-kinetic.

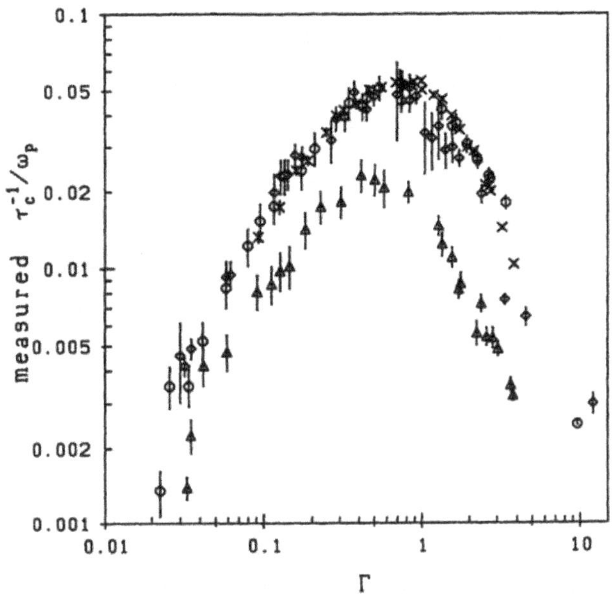

Figure 1: The figure shows the dependence of the collision rate τ_c^{-1} scaled by the plasma frequency on the plasma parameter Γ in a bunched beam (diamonds) and in a coasting beam. For the coasting beam the numbers of particles per mean particle distance were $n_z = 60$ (circles) and $n_z = 13$ (crosses) and $n_z = 2.5$ (triangles). The data for bunched beams are measured by the first, the data for coasting beams are measured by the second method described in Sec. 5. The error bars are due to the statistical fluctuations of runs with different initial particle configurations.

6 Indirect Cooling

The most promising cooling method to overcome the maximum of the heating rate in the gas-liquid regime seems to be laser cooling. By means of this cooling method very high cooling rates are obtainable. The drawback of laser cooling is that it is only applicable to the longitudinal beam direction, the transverse degrees of freedom remain uncooled. A possible scenario to cool also the transverse directions is indirect cooling. Due to the temperature exchange through intrabeam scattering between the hot transverse directions and the cold longitudinal direction also the transverse directions can be cooled. We verified this by the simulation of a $^{202}Hg^+$ coasting beam of an energy of 0.81 MeV/u in the lattice of the Heidelberg test storage ring (TSR). (It should be noted, however, that such a beam would require unrealistic high magnetic fields of $B = 13$T in the bending magnets.) The particle line density was $n_z = 1 \cdot 10^6 \mathrm{m}^{-1}$. As initial values for the temperatures we chose $T = 500$K in all

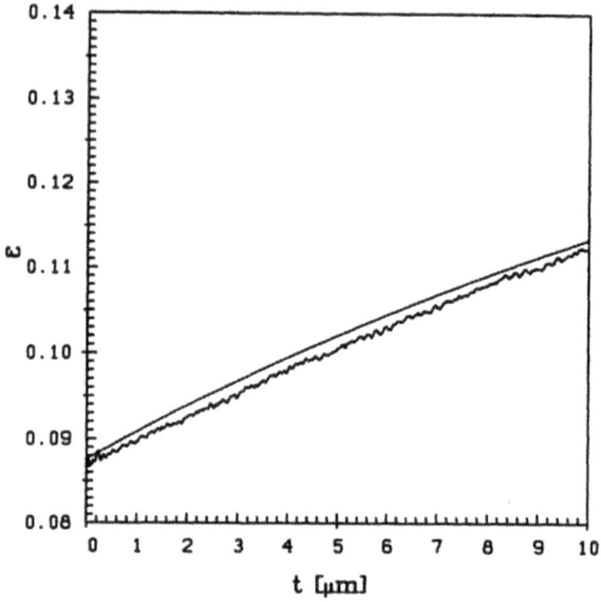

Figure 2: The time development of the emittance of the beam where the initial value of the plasma parameter was $\Gamma = 0.18$. The full line are the MD data, the dashed line the data expected from the kinetic theory.

directions. The cooling rate was $\Lambda_c = 10^6 \mathrm{s}^{-1}$. We achieved the cooling by rescaling the particles' longitudinal velocities, measured in the rest frame of the beam, by a constant factor such that $\mathrm{d}T_z/\mathrm{d}t = -\Lambda_c T_z$. Fig. 4 shows the time evolution of the longitudinal temperature T_z in this simulation. The scale of the ordinate is logarithmic. As a consequence of the very high cooling rate of $\Lambda_c = 10^6 \mathrm{s}^{-1}$ the longitudinal temperature initially drops rapidly from $T_z = 500\mathrm{K}$ to a value of $T_z = 0.2\mathrm{K}$. At this temperature a dynamical equilibrium between the external cooling and the heating due to the temperature exchange with the hot transverse directions is established. In Fig. 5 the time development of the transversal emittances in this simulation is shown. There is no steep decrease in the emittances as in the case of the longitudinal temperatures. Instead of that, the emittances decrease slowly as a consequence of the slow heat exchange with the cool longitudinal direction. As the beam gets colder and denser the intrabeam scattering rates increase. These enhanced intrabeam scattering rates shift the dynamical equilibrium between the external cooling and the internal heating of the longitudinal directions; the longitudinal temperature increases again to a value of $T_z = 2\mathrm{K}$ as can be seen again in Fig. 4. The overall dynamical equilibrium is reached after about 10ms. After this time neither the longitudinal temperature nor the transversal

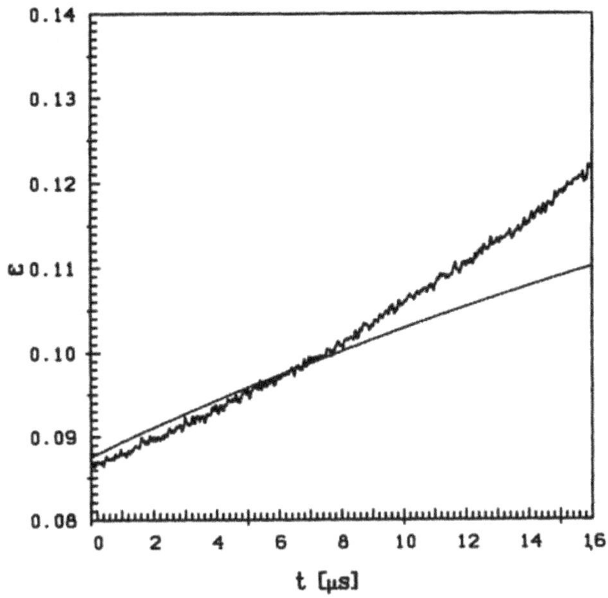

Figure 3: The time development of the emittance of the beam where the initial value of the plasma parameter was $\Gamma = 1.8$. The full line are the MD data, the dashed line the data expected from the kinetic theory with an adjusted initial relaxation rate.

emittances change anymore. The final state, however, is not the state of a Coulomb crystal. The beam remains in the low-Γ regime. Our simulation results confirm measurements of indirect transverse cooling at the TSR in Heidelberg[23]. It should be noted that there are proposals to accelerate the temperature exchange between the longitudinal and the transverse direction. The first proposal is the application of a coupling cavity to speed up the heat exchange[24]. The second proposal is to take advantage from a collective anisotropy instability by which the beam can be cooled if it is sufficiently space charge dominated in the longitudinal direction[25]. It is not certain, however, that this method will work in the regime where intrabeam scattering becomes severe: in this regime collective phenomena are strongly damped and thus indirect cooling may be more efficient to cool the beam transversally. In our simulations, this collective anisotropy instability was not observed. This may be due to our rather short simulation boxes which set an upper bound to the wavelength of the phenomena that can be studied.

From these simulation results we conclude further that the TSR lattice is not appropriate to achieve crystalline beams. We have therefore also simulated a simple FODO lattice with a phase advance of $\sigma_0 = 60°$ (without space charge). The starting

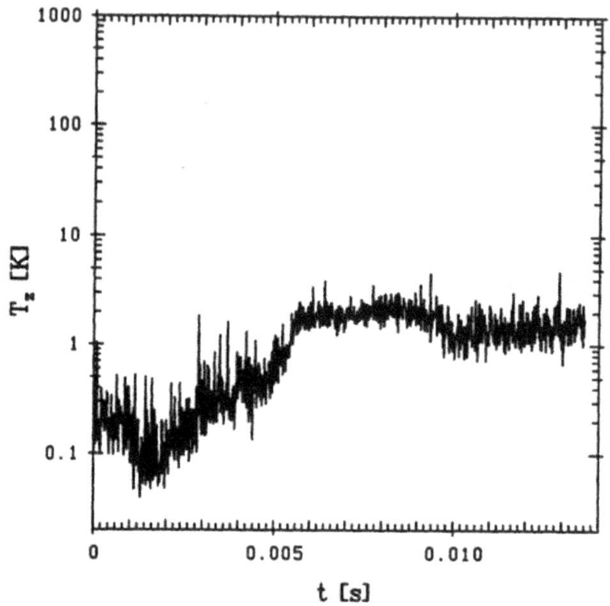

Figure 4: Time development of the longitudinal temperature in the simulation of a $^{202}Hg^+$-beam in the TSR. The scale of the ordinate is logarithmic.

temperature was again $T = 500$K in all directions, the ion charge was $m = 40$u. The particle line density was $n_z = 1.1 \cdot 10^6 \mathrm{m}^{-1}$, the applied cooling rate was $\Lambda_c = 6 \cdot 10^5 \mathrm{s}^{-1}$. These parameters corresponded to an initial plasma parameter of $\Gamma = 0.2$ which is below the plasma parameter at which the heating rate has its maximum. Fig. 6 shows the time development of the longitudinal temperature T_z during this simulation. At the initial stage during the first 10μs the temperature drops rapidly from $T_z = 500$K to $T_z = 100$K. Then, the decrease slows down as the beam is in the regime of maximum heating rates. After about 70μs the decrease becomes steeper again. This is due to the decrease in the heating rate after the maximum is overcome. The simulation was interrupted after about 200μs. At this point the longitudinal temperature had dropped to a value of around $T_z = 10$mK, corresponding to a longitudinal plasma parameter of $\Gamma = 5000$. It is clearly visible from Fig. 7 that the cooling also worked in the transversal direction. In this figure, the time development of the emittance is shown. The emittance drops from its initial value to a value which is about 1/20 of the initial value. At this level the decrease of the emittance stops. It should be noted that at such low emittances, where the plasma parameter is very high, the rms-emittance is no longer a correct measure of the beam temperature. The temperature definition given in Sec. 2 is no longer valid, since it was established under the premise of a

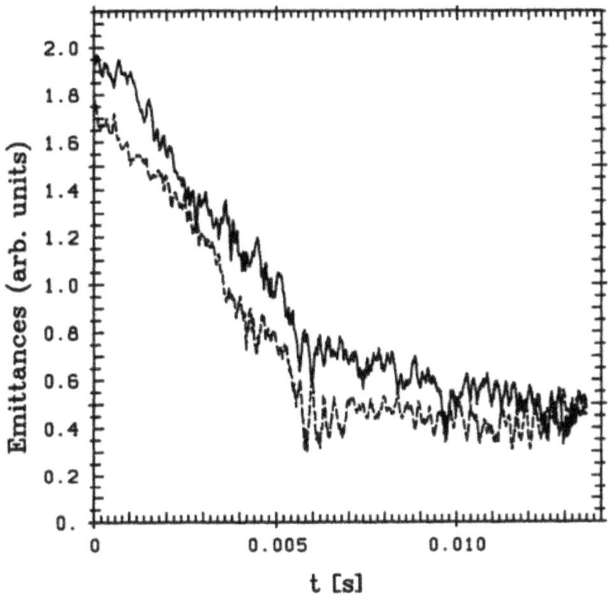

Figure 5: The time evolution of the horizontal emittance (full line) and the vertical emittance (dashed line) in the simulation of a $^{202}Hg^{+}$-beam in the TSR

gaseous or a liquid state without long-range correlations. The phase-space projection on the x-y plane revealed that the final state of this beam was a Coulomb crystal with four shells. Thus in this simulations a cooling rate of $\Lambda_c = 6 \cdot 10^5$ was sufficient to overcome the maximum in the heating rate and reach a crystalline beam. The cooling rate of $\Lambda_c = 6 \cdot 10^5 s^{-1}$ seems to be high, but one should keep in mind that our simulation was done with a beam of ion charge $Z = 10$. In a beam with ion charge $Z = 1$ much lower cooling rates will be sufficient.

7 Dependence of emittance growth rates on the phase advance per cell

The feasibility of indirect cooling is not the only problem faced in the search for crystalline beams. Another problem is the internal heating which is due to the focusing structure. Two aspects have to be taken into account:

1. In a crystalline beam the interparticle forces have to be stronger than the shearing forces in the bending magnets in order to withstand these forces. Strong interparticle forces are accomplished by low ion distances in the Coulomb crystal, i.e. high particle densities. Low shearing forces can be achieved by large

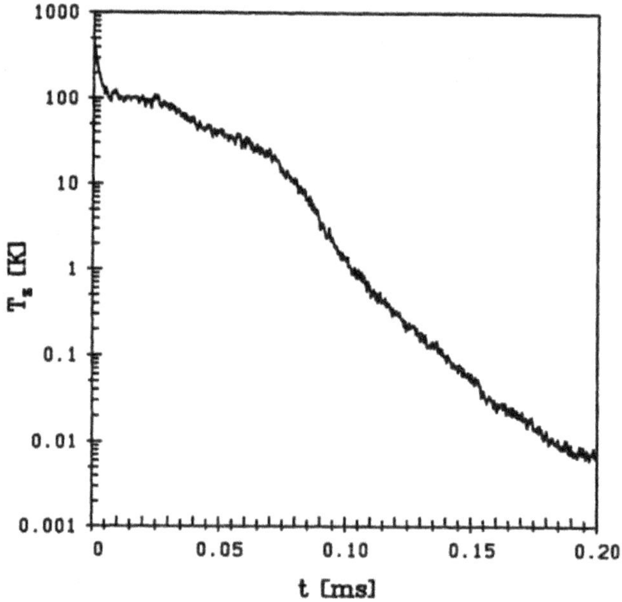

Figure 6: The time evolution of the longitudinal temperature in the simulation of a ion beam with ion charge $Z = 10$ and ion mass $m = 40$u. The focusing structure was a simple FODO-lattice with a phase advance without space charge of $\sigma_0 = 60^o$.

bending radii, which increases the costs of the accelerator facility, and small beam diameters, since the shearing force is proportional to the displacement x from the design orbit as can be seen from Eq. 8. Thus, both strong interparticle forces and small shearing forces can be obtained by high beam densities, which implies strong focusing and thus a high phase advance per cell.

2. The maximum applicable cooling rate of each specific ion is limited. As was pointed out in Sec. 2 the internal heating rate increases with Γ until a maximum at $\Gamma \approx 0.7$. Γ istself increases as the temperature decreases and the density increases. Thus, if indirect cooling of the transverse degrees of freedom proceeds, also the internal heating rate becomes larger. If the heating rate reaches the external cooling rate before its maximum value at $\Gamma \approx 0.7$ is overcome, there is a dynamical equilibrium between external cooling and internal heating. In this case beam crystallization cannot be reached.

Since the plasma frequency ω_p and the plasma parameter Γ both increase with the density, a high beam density also implies a high collision rate. Thus, from an intra-beam scattering point of view beams with rather low phase advances per unit cell

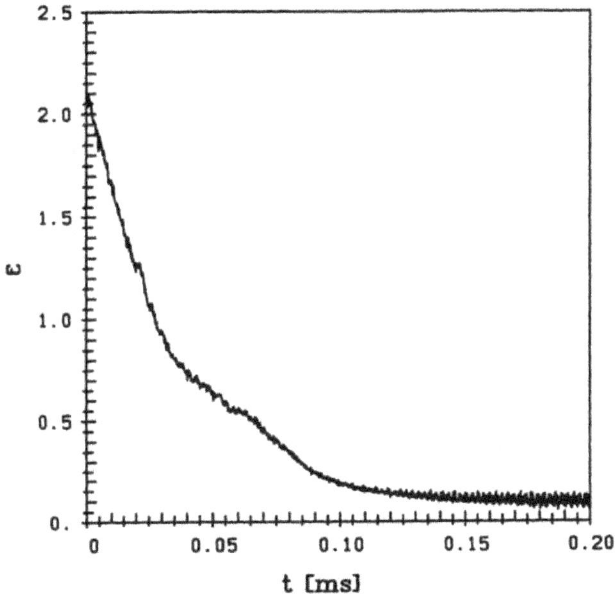

Figure 7: The time evolution of the emittance. The parameters were the same as in Fig. 6

are preferable. Since for a stable beam transport high phase advances are required, one has to design the focusing structure very carefully such that intrabeam scattering rates are not too large so that stable Coulomb crystals can be achieved. The increase of scattering rates with the focusing strength is also visible in our MD simulation results. We have simulated a beam in a simple FODO lattice with 5° bending magnets inserted between the quadrupoles. In each of the simulations the geometry of the lattice was the same but the phase advances and thus the focusing strengths were different. We made simulations for phase advances per unit cell of $\sigma_0 = 60^o, 90^o, 120^o$ and 150°. For each run we measured the growth rates. Fig. 8 shows the results for the emittance growth rates in the MD simulations (circles). The emittance growth rates increase with increasing phase advance per cell. The simulations were made in a parameter range, where the kinetic theory still delivers correct results, i.e. the region of low Γ-values. We have also calculated the kinetic value of the relaxation rates from the beam dimensions. These values are also plotted in Fig. 8 (triangles). It is clearly visible that the emittance growth rates increase even at a steeper pace than the relaxation rates. This can be understood as follows: the stronger focusing requires a larger ratio between the maximal and the minimal value of the beam envelopes. Therefore, the temperature anisotropy is also enhanced according to Eq. 4. A larger temperature anisotropy also increases the entropy gain due to the relaxation. Thus,

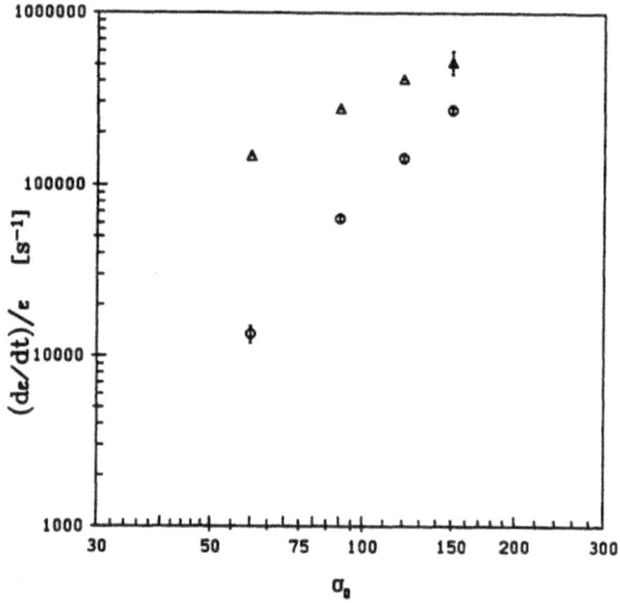

Figure 8: The figure shows the normalized emittance growth rates (circles) and the kinetic value of the relaxation rate (triangles) in dependency of the phase advance σ_0 of the lattice.

the average emittance increase is also enlarged. The conclusion is that a higher phase advance not only yields a higher emittance growth rate due to the enhancement of relaxation rates, but that lattices with higher phase advance show in general higher emittance growth rates beyond an increase of relaxation rates.

8 Conclusion

In high-intensity beams it is mandatory to take into account both the effects of space charge and intrabeam scattering. We have shown that the envelope equations in a periodic quadrupole channel can be extended to account for the effects of intrabeam scattering. The parameter of the model, we developed, a relaxation rate, shows a maximum at $\Gamma \approx 0.7$. The envelope equations can also be extended to include the effects of bending magnets. Then, the rms-emittance is no longer a constant of motion, but a generalized emittance can be shown to be a constant of motion. The problem of matching a space charge dominated beam to a specific accelerator structure can be solved rapidly. These equations have still to be extended to account also for the effects of intrabeam scattering. First results of molecular dynamics simulations of the

TSR-lattice at the MPI in Heidelberg confirm experimental results from this facility that indirect cooling of the transverse directions of a heavy ion beam is feasible. The studies on the dependency of intrabeam scattering effects on the focusing strength imply that a dedicated storage ring for beam crystallization has to be developed very carefully to guarantee both low intrabeam scattering rates and the transport of stable Coulomb crystals.

9 Acknowledgement

We acknowledge valuable discussions with P.-G. Reinhard and R. Grimm, H.-J. Miesner and D. Habs from the MPI Heidelberg. We thank the HLRZ Jülich and the RRZE Erlangen for providing computation time.

10 References

1. J.P. Schiffer and P. Kienle, *Z. Phys.*, **A321** (1985) 181.

2. F. Diedrich, E. Peik, J.M. Chen, W. Quint and H. Walther, *Phys. Rev. Lett.*, **59** (1987) 2935.

3. V.V. Parkomchuk, *ECOOL 1984*, Kernforschungszentrum Karlsruhe (1984) 71-83.

4. M. Steck, to be published in the *Proceedings of the First European Conference on Atomic Physics with stored Highly Charged Ions*, Heidelberg, 1995.

5. P.M. Lapostolle, *IEEE Trans. Nucl. Sci.* NS-18 (1971) 1101-1104.

6. F.J. Sacherer, *IEEE Trans. Nucl. Sci.* **NS-18** (1971) 1105-1107.

7. J. Struckmeier, J. Klabunde and M. Reiser, *Part. Acc.* **15**, (1984) 47-65.

8. T.P. Wangler, K.R. Crandall, R.S. Mills and M. Reiser, *IEEE Trans. Nucl. Sci.* **NS-32** (1985) 2196-2200.

9. O.A. Anderson, *Part. Acc.* **21** (1987) 197-226.

10. I. Hofmann, L.J. Laslett, L. Smith and J. Haber, *Part. Acc.* **13** (1983) 145-178.

11. M. Seurer, P.-G. Reinhard and C. Toepffer, *Nucl. Instr. and Meth.* **A 333** (1994) 286-294.

12. Q. Spreiter, M. Seurer and C. Toepffer, *Nucl. Instr. and Meth.* **A 364** (1995) 239-242.

13. C. Møller, *The Theory of Relativity* (Clarendon Press, Oxford, 1952).

14. E.D. Courant and H.S. Snyder, *Ann. Phys. (N.Y.)* **3** (1958) 1-48.

15. M.P. Allen, D.J. Tildesley, *Computer simulation of liquids*, (Clarendon Press, Oxford, 1989), ch. 3.

16. P.P. Ewald, *Ann. Physik* **64** (1921) 253.

17. J.E. Barnes and P. Hut, *Nature,* **324** (1986) 446.

18. W.H. Press, S.A. Teukolsky, W.T. Vetterling and P.P. Flannery, *Numerical Recipes in C* (Cambridge University Press, Cambridgh, 1992).

19. E. Hairer, S.P. Nørsett and G. Wanner, *Solving Ordinary Differential Equations* (Springer Verlag, Berlin, 1987).

20. S. Ichimaru, *Statistical Plasma Physics, Vol. 1*, (Addison-Wesley, New York, 1992), ch. 8.

21. J. Wei, X.-P. Li and A.M. Sessler, *BNL Report, Brookhaven National Laboratory,* **BNL-52439** (October 1994).

22. R. Blümel, C. Kappler, W. Quint and H. Walther, *Phys. Rev.* **A 40** (1989) 808-823.

23. H.-J. Miesner, M. Grieser, R. Grimm, D. Habs, B. Hochadel, C. Horvath, M. Madert, D. Schwalm, B. Wanner and A. Wolf, in: *Proceedings of the 14th Int. Conf. on Atomic Physics*, Boulder, 1994.

24. H. Okamoto, A.M. Sessler and D. Möhl, *Phys. Rev. Lett.*, **72** (1994) 3977.

25. I. Hofmann and R.W. Hasse, this conference proceedings.

Crystalline Beams in the Crystal Storage Ring at Legnaro [a]

A.F.Haffmans, D.Maletić, A.G.Ruggiero
Brookhaven National Laboratory, Upton, NY 11973, USA

Abstract

A computer program was developed in Mathematica[7] software package to examine if a particular accelerator lattice can support creation and maintain existence of the several ground states of a crystalline beam. The program is based on methods described earlier[1-6]. Results of this program, applied to the proposed CRYSTAL Storage Ring in Legnaro, Italy[8], are shown. This ring supports all of the three examined ground states, that is string, zigzag and helix.

1 Introduction

A program has been developed to examine if a given accelerator lattice can support creation and maintain existance of a several ground states of the crystalline beam, particularly, string, zigzag and helix states. The program is based on both analytical and matrix methods, described earlier[1-6] Comparison of the results of these two methods are shown here. Purpose of this exercise is to test the quality of our analytical model in different tune ranges. This can help us to use simple analytical equations, when appropriate, saving long time needed for the matrix manipulations in the matrix method.

Crystalline beam structures for the Legnaro ring are examined in the detail at the working-point tunes[8]. Also, these tunes were changed to some other values, by varying of the quadrupole gradient strenghts. Changes in the appearance of the crystalline beam structures, due to the tune variation, were noticed.

Limit	Matrix	Analytical
String Stability	$\lambda = 76.00$	$\lambda = 76.29$
String Bifurcation Direction	Vertical	Vertical
Zigzag confinement	$\lambda = 183.96$	
Zigzag Stability	$\lambda = 67.48$	$\lambda = 66.08$
	$\alpha = 1.55$	$\alpha = 1.60$
String-to-Zigzag Transformation	$\lambda_s = 82$	
	$\lambda_{zz} = 164$	
Helix Confinement	$\lambda = 151.39$	
Helix Stability (Analytical)	-	$\lambda = 102.78$
	-	$\alpha_h = 0.85$
	-	$\alpha_v = 1.54$
Zigzag-to-Helix Transformation	$\lambda_{zz} = 70$	
	$\lambda_h = 140$	
$\Delta\lambda = \pm 0.5 \quad \Delta\alpha = \pm 0.025$		

Table 1: Results for Legnaro ring. $\lambda_c = 63.4$. λ in $[\mu m]$ units.

The Program is written using the Mathematica[7] program package. We wish this program to be a valuable designing tool for new accelerator lattices supposed to facilitate crystalline beam structures.

2 Analytical Method

Conditions for the confinement and stability of the string, zigzag and helix states of a crystalline beam were developed earlier[1-5]. It was proven that constant gradient accelerators can not support crystalline beam, so such machines were not considered here, but only ones with alternating gradient.

2.1 The String

The only confinement condition for the string state is that the centrifugal force and the space–charge repulsion are balanced by the Lorentz force of the magnetic field. This condition must also be satisfied by zigzag and helix structures.

When the longitudinal ion spacing λ is reduced, the space–charge forces increase and the betatron tunes decrease. The string will become unstable when the edge of a stopband induced by the space–charge is encountered by vertical, ν_v, horizontal, ν_h, or longitudinal, ν_s, tune. Also, planar instability can appear due to the coupling: $\nu_h \pm \nu_s = m_\pm N_p$. In terms of the particle spacing, instability limit is given by

$$\lambda > \lambda_c c_i, \quad (i = v, h, s, c) \tag{1}$$

where

$$c_v = (\nu_v \delta \nu_v)^{-\frac{1}{3}} \tag{2}$$

$$c_h = (\nu_h \delta \nu_h)^{-\frac{1}{3}} \tag{3}$$

$$c_s = (\frac{4}{N_p})^{\frac{2}{3}} \tag{4}$$

$$c_c = (\nu_h q)^{-\frac{2}{3}} \tag{5}$$

and λ_c is the critical lattice constant

$$\lambda_c = (\frac{Q^2 r_0 g_0 R^2}{A\beta^2\gamma^5}) \tag{6}$$

N_p is the periodicity of the ring and q is the smallest positive value of

$$q = 1 \pm \sqrt{1 + \frac{\nu_h - m_\pm N_p}{\nu_h}} \tag{7}$$

POINT	G_f	G_d	ν_x	ν_y
0	4.41	-5.5195	1.907	1.085
1	2.898	-4.496	1.80	1.20
2	2.432	-4.461	1.6625	1.3875
3	1.117	3.912	1.525	1.525
4	0.986	-4.299	1.3875	1.6625
5	6.996	-7.266	2.1	0.9

Table 2: Moving the working point in Legnaro ring by changing quadrupole gradients

2.2 The Zigzag

Approaching the resonance destroyes the string, while is not enough to trigger the formation of the zigzag. For the creation of the zigazag, the periodicity of variation of the zigzag width a or height b need to be a harmonic of the storage ring period, yielding upper limit for λ

$$\lambda < 2.41\lambda_c\bar{c}_i, \quad (i = h, v) \tag{8}$$

where $\bar{c}_h$ differs from the c_h shown for the string because the beam width a and the momentum spread δ must be proportional to each other for the horizontal zigzag to exist, causing kinematic correction to appear

$$\bar{c}_h = [\nu_h(\delta\nu_h - \frac{\gamma^2}{2\nu_h})]^{-\frac{1}{3}} \tag{9}$$

If zigzag appears, it will develop with its aspect ratio $\alpha = a/\gamma\ \lambda$ and ion spacing related by the

$$\frac{\lambda^3}{f_3(\alpha)} = \frac{2}{g_0}\lambda_c^3\bar{c}_i^3, \quad (i = h, v) \tag{10}$$

where

$$f_n(\alpha, \tau) = \sum_i [\alpha^2 + (i - \tau)^2]^{-\frac{n}{2}} \tag{11}$$

Stability condition for the zigzag appears due to the same reason as one for the string. It can be shown[3-4] that it is

$$\frac{\eta_i}{f_3(\alpha)} < \frac{2}{g_0}(\frac{c_j}{c_i})^3, \quad (i = v, h, s, c) \tag{12}$$

where j denotes v or hz depending on the vertical or horizontal zigzag in question. The form factors η are defined as follows

$$\eta_{h,v} = 1 + \frac{f_3(\alpha)}{g_0} \tag{13}$$

$$\eta_e = \eta_s = \eta_c = 1 + \frac{f_3(\alpha) - 1.5\alpha^2 f_5(\alpha)}{g_0} \tag{14}$$

2.3 The Helix

Again, the strongest limit will superseed all others as the stability limit. If at the same time, the longitudinal force is equal to zero and the transverse confinement condition, equivalent to the one for the zigzag, is satisfied, zigzag, which have reached stability limit, will evolve into the helix structure. The upper limit for ion spacing in the helix is[5]

$$\lambda_u = (\psi_{h,v}^{(i)})^{\frac{1}{3}} \lambda_c \bar{c}_{h,v} \tag{15}$$

where $\psi_{h,v}^{(i)}$ are the initial values of the envelope form factors

$$\psi_h = \frac{2}{g_o}[f_3(\alpha_h, 0, \tau_{12}) + f_3(\alpha_h, \alpha_v, \tau_{13})] \tag{16}$$

$$\psi_v = \frac{2}{g_o}[f_3(\alpha_h, \alpha_v, \tau_{13}) + f_3(0, \alpha_v, \tau_{14})] \tag{17}$$

Initial aspect ratio for helix is determined by the horizontal and vertical tunes of the particular lattice through relation

$$2(\frac{\bar{c}_h}{\bar{c}_v})^3 = 1 + \frac{f_3(0, 1/2)}{f_3(\alpha^{(i)}), 1/4)} \tag{18}$$

and further development is governed by the relation between α_h, α_v and τ, for example when the helix originates from a horizontal zigzag

$$\frac{f_3(\alpha_h, \alpha_v, \tau + \frac{1}{2}) + f_3(0, \alpha_v, \frac{1}{2})}{f_3(\alpha_h, 0, \tau) + f_3(\alpha_h, \alpha_v, \tau + \frac{1}{2})} = (\frac{\overline{c_h}}{\overline{c_v}})^3 \tag{19}$$

Stability condition can be found with the same procedure as for the other structures, and is written as

$$\frac{\eta_i(\alpha, \tau)}{\psi_j(\alpha, \tau)} < \frac{2}{g_o}(\frac{c_j}{c_i})^3 \tag{20}$$

where

$$\alpha^2 = \alpha_h^2 + \alpha_v^2 \tag{21}$$

and $j = h, v$ while $i = h, v, s, c$. For definition of form factors η, please see our earlier paper[5]. There are eight separate stability equations described by the relation above. The one which needs the largest coefficient α to remain correct creates the ultimate limit of stability for the helix structure. Note that relation between α and τ figuring in the stability condition depends on the origin of the helix structure. However, the stability condition itself does not depend on the origin of the helix.

3 Matrix Method

Each magnet component of the accelerator lattice can be subdivided in a large number of sections where the parameters could be approximated as constants. For each of these sections we can associate 6×6 transfer matrix describing the propagation of the vector $(a, a', b, b', \sigma, \delta)$ through it. A storage ring is defined by the sequence of magnet elements. The total transfer matrix $\mathbf{M}$ per revolution is obtained by multiplication of all the matrices in given

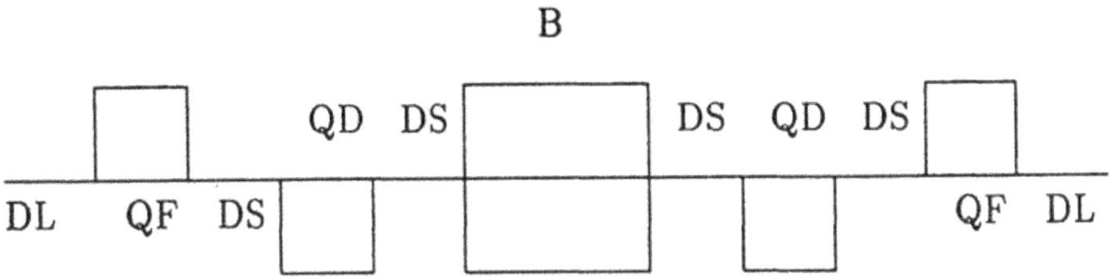

Figure 1: One period of the CRYSTAL ring at the Legnaro

order. Examination of the eigenvalues of the total transfer matrix allows us to determine the stability of the motion.

To find a closed periodic solution, we start at a waist. After guessing the initial phase space coordinates, we propagate that vector by sequential matrix multiplications corresponding to the storage ring we examine. We change the initial guess until $det(\mathbf{M}-\mathbf{I})=0$.

Space charge effects are introduced by the method developed in our earlier papers[1-6]. It is also treated as a sequence of kicks along the trajectory, at evenly distributed fixed points s_k. In the proximity of the kick the equation of motion reduces to

$$y'' = \frac{K_0}{\beta_v}\delta(s - s_k) \tag{22}$$

$$K_0 = K \times L_k \tag{23}$$

where L_k is the interaction length replaced by the kick.

Pt.	Mth.	λ_s^s	Dir.	λ_{zz}^s	α_{zz}^s	λ_h^c	α_h^s	α_v^s	λ_h^s
0	A	75.69	Ver.	65.52	1.60	151.39	0.85	1.58	102.78
0	M	75.69	Ver.	65.52	1.60	-	-	-	-
1	A	70.78	Ver.	61.28	1.60	166.10	1.10	1.53	94.41
1	M	70.78	Ver.	61.28	1.60	-	-	-	-
2	A	64.25	Ver.	78.16	1.00	189.71	1.502	1.401	87.04
2	M	67.25	Rad.	56.21	1.65	-	-	-	-
3	A	60.34	Ver.	65.91	1.50	162.50	1.53	1.00	97.99
3	M	75.34	Rad.	61.87	1.65	-	-	-	-
4	A	64.26	Rad.	71.28	1.60	No Helix			
4	M	84.26	Rad.	71.28	1.60	No Helix			
5	A	85.74	Ver.	74.23	1.60	No Helix			
5	M	85.74	Ver.	72.72	1.65	No Helix			

Table 3: Results for Crystal Beam structures in the Legnaro ring, when working-point is moved from its original value

4 Proposed Crystal Storage Ring at Legnaro

Legnaro CRYSTAL ring[8] has periodicity 8. Each period has a mirror symmetry and is shown in Fig. 1. The following values are assigned to the ring parameters:

$$B\rho = 4.2Tm \tag{24}$$

$$G_f = 4.41T/m \tag{25}$$

$$G_d = -5.5195T/m \tag{26}$$

$$L_{DL} = 2.0m \tag{27}$$

$$L_{DS} = 0.4m \tag{28}$$

$$L_Q = 0.125m \tag{29}$$

$$L_B = 1.0014m \tag{30}$$

These parameters produce tunes:

$$\nu_H = 1.907m \tag{31}$$

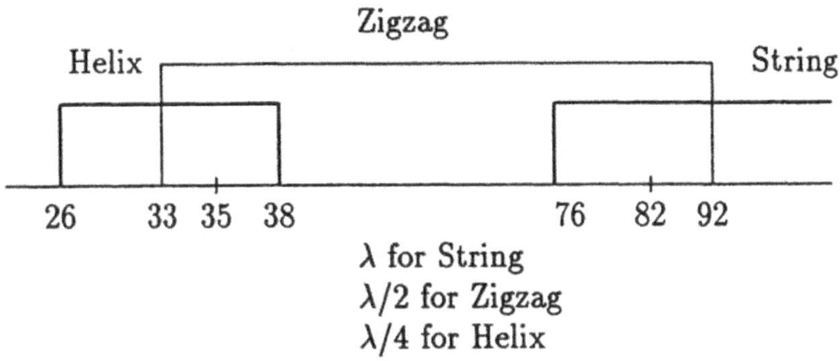

Figure 2: Areas of stability of crystalline beam structures in the CRYSTAL ring. Transition points are also marked. For the convinience axis was graded to show ion spacing of the string.

$$\nu_V = 1.085m \tag{32}$$

We assume that stored ions are gold, with charge state $Q = 51$ and $A = 197$.

5 Results

Tables 1 – 3 describe results of the Mathematica[7] program applied to the lattice of the proposed CRYSTAL storage ring in Legnaro. Ion spacing λ is shown in units of μm. Aspect ratio α is dimensionless.

Table 1 gives the areas of stability for each of the three examined crystalline beam structures. These results are shown at Fig. 2. graphicly. We assumed that the beam started as a string with a large ion-spacing. Increase in the ion density (ex. by adding the ions to the ring) would cause decrease of the ion spacing and eventually transformation of the crystalline beam from the one ground state to the other. Note that the ion spacing is doubled at the each bifurcation (ex. string with the $\lambda = 50$ becomes zigzag with

$\lambda = 100$). String and the zigzag areas of stability in the CRYSTAL ring overlap each other, as well as the zigzag and helix ones. This represents the opportunity for the "lower" structure to "evolve" into the "higher" one, before it is destroyed by the instability. Transition points shown in the Table I (String – to – zigzag and Zigzag – to – Helix) correspond to the points at which potentials of the two structures become equal. We expect that the bifurcation should occure at those points, with the appearance of the more energy favorable structure.

Developement of the string structure is the simple one. Its ion spacing simply decreases as ion concentration increases. Developement of the zigzag structure is a little more complicated. Expected change in the aspect ratio as the function of the ion spacing for the zigzag in Legnaro ring is shown in the Fig. 3. Similarly, Fig. 4 and 5 show the same type of the development for the horizontal and the vertical aspect ratio of the helix in the Legnaro ring. Both zigzag and helix developement are shown within the areas of stability, together with the boundaries of existence (between transition points) for the each structure.

We also examined how the crystalline beam structures will be affected in the CRYSTAL ring if we shift its working point by changing the quadrupole gradients. We have used SYNCH[9] program to find 5 other working points for the Legnaro ring. Appropriate quadrupole strenghts and tune values are shown in the Table 2. In the Table 3. we have shown how this working point change affects stability areas of the crystalline beam structures. Important changes were noticed, in particular disapearance of the helix structure under some conditions (see Table 3.).

Table 3 also serves as a tool of comparison between the Analytical and the Matrix Method. Matrix method, beeing accurate without any approximations, is the standard to which analytical one is beeing compared. One can notice that analytical model behaves well for very different horizontal and vertical tune values.

However, it is not good for the equal tunes.

6 Conclusion

We expect the results shown here to serve as guidelines for the design of a storage ring. Our results show that, if there is an effective way to cool the ion beam in he CRYSTAL ring, one can expect indeed it to harbor at least the three mentioned ground states of crystalline beam. In other words, the designers of the CRYSTAL ring have one more confirmation of the usefullness of their machine for the exploration of crystalline beams.

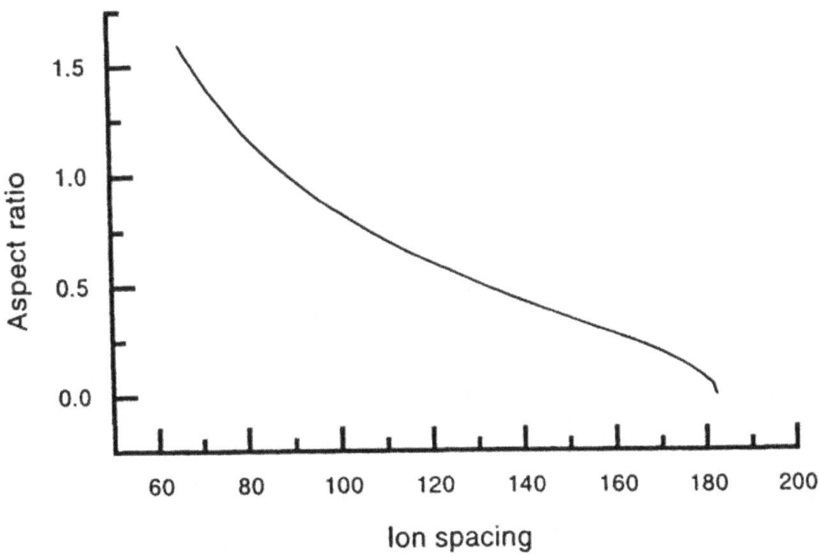

Figure 3: Aspect ratio α asunction of the ion spacing λ for the zigzag structure in the CRYSTAL ring

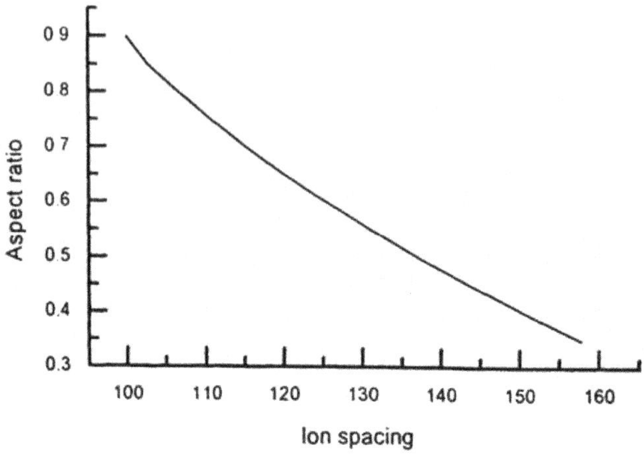

Figure 4: Aspect ratio α_H as function of the ion spacing λ for the helix structure in the CRYSTAL ring

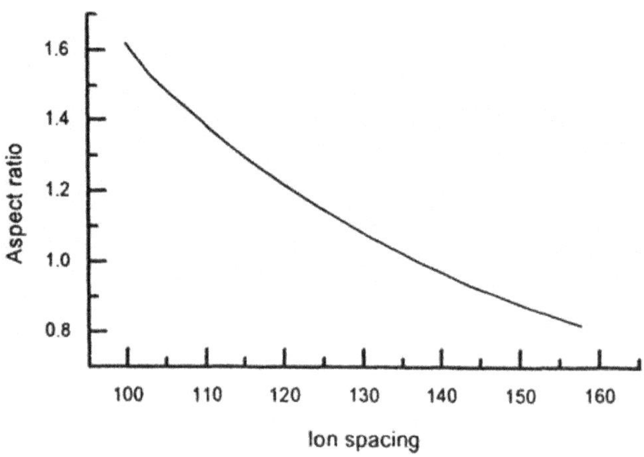

Figure 5: Aspect ratio α_V as function of the ion spacing λ for the helix structure in the CRYSTAL ring

Reference

1. A.F. Haffmans, D. Maletić and A.G. Ruggiero, *Particle Motion in Crystalline Beams, Informal Report*, **BNL-60436** , April 1994.
2. A.F.Haffmans, D.Maletić and A.G.Ruggiero, *Crystalline Beams: The String, Informal Report* , **BNL-60619** , April 1994.
3. A.F.Haffmans, D.Maletić, and A.G.Ruggiero, *Crystaline Beams: The Vertical Zigzag, Informal report* , **BNL-60743** , May 1994.
4. A.F.Haffmans, D.Maletić and A.G.Ruggiero, *Crystaline Beams: The Horizontal Zigzag, Informal report* , **BNL-60876** , May 1994.
5. A.F.Haffmans, D.Maletić A.G.Ruggiero, *Crystalline Beams: The Helix,* BNL Informal report, in preparation, December 1995.
6. A.F. Haffmans, D. Maletić and A.G. Ruggiero, *Study of the Effect of the Space Charge using SYNCH, Informal Report* , **BNL-60521** , April 1994.
7. S. Wolfram, *Mathematica* , Addison-Wesley, New York, 1991.
8. L.Tekkio, *Erice Workshop Proccedings* , 1996.
9. E.D.Courant, A.A.Garen and A.S.Kenney, *A Users Guide to SYNCH* , **BNL** , Upton, 1985.

Space Charge Instability in Storage Rings With Circular Closed Orbits

D.V. Pestrikov
Budker Institute of Nuclear Physics,
630090 Novosibirsk, Russian Federation

Abstract

In this report we show that in storage ring with circular closed orbits the space charge force does not result in the envelope instability of linear betatron oscillations. The reason is that without longitudinal modulation of the lattice focusing the space charge does not result itself in spontaneous modulations of periodical β-functions, which therefore are constant along the orbit. In these conditions, the only space charge instability of a coasting beam in such a ring can be the negative mass instability.

1. The Coulomb repulsion of particles is known as limiting the beam intensity in low-energy rings. This is a manifold phenomenon, which results in the blow-up of the beam sizes due to increase in the beam envelopes and due to excitation of nonlinear resonances of particle oscillations. The goal of this report is to remind the reader that for this instability the crucial point is the modulation of betatron functions along the ring. Since the instability occurs due to forces depending on relative distance between the particles in the beam, without such a modulation the harmonics of the interaction energy of particles in oscillation phases remain constant along the orbit. It can be shown (see, for instance, in Ref.[1]) that strengths of the space charge resonances are proportional to azimuthal harmonics of β-functions of the ring. It means that in a storage ring with circular closed orbits, where the focusing is constant, all these strengths will vanish. In this case, the beam sizes still blow-up due to decrease in the strength of the ring focusing, but it is not accompanied by an instability of betatron oscillations. This fact can be important when choosing a model for analytic, or computer simulation study of the space charge dominated beams.

Another important (but maybe more technical) point is a difference in the motion of particles in a storage ring and in unclosed channel. In the storage ring the solutions of the envelope equations (see, for instance, in Refs [2] and [3]) must be periodic functions with the lattice period, or with the period 2π. This requirement strongly restricts the class of solutions of envelope equations, which may describe β-functions of the ring. In computer simulations such periodic solutions must be found by direct tracking of particle motion over a lattice period and by finding initial conditions, which ensure the periodicity.

2. The perturbations of linear betatron oscillations due to space charge forces of the beam are described by the following equations

$$z'' + [g_z + \Delta g_z]z = 0, \quad x'' + [g_x + \Delta g_x]x = 0. \tag{1}$$

Here,

$$g_z = \frac{R_0^2}{R^2(\theta)} n(\theta), \quad g_x = \frac{R_0^2}{R^2(\theta)}[1 - n(\theta)],$$

$n(\theta)$ is the field index, $\Pi = 2\pi R_0$ is the perimeter of the orbit, $\theta = s/R_0$, and s is the path along the orbit. For a coasting beam with a Gaussian distribution in transverse coordinates

$$\varrho(x, z) = \frac{N}{2\pi\sigma_x\sigma_z\Pi} \exp\left(-\frac{x^2}{2\sigma_x^2} - \frac{z^2}{2\sigma_z^2}\right) \tag{2}$$

the functions Δg_z and Δg_x read

$$\Delta g_z = -\frac{Ne^2R_0}{\pi\gamma^2 pv\sigma_z(\sigma_z + \sigma_x)}, \quad \Delta g_x = -\frac{Ne^2R_0}{\pi\gamma^2 pv\sigma_x(\sigma_z + \sigma_x)}. \tag{3}$$

The r.m.s. beam sizes σ_x and σ_z are expressed in terms of the beam emittances $\epsilon_{x,z}$ and beam momentum spread $(\Delta p/p)$ using

$$\sigma_z = \sqrt{\beta_z\epsilon_z}, \quad \sigma_x = \sqrt{\beta_x\epsilon_x + \eta^2(\Delta p/p)^2}.$$

Here, $\beta_{x,z}$ are the β-functions of the horizontal and vertical betatron oscillations, $\eta(\theta)$ is the dispersion function of the ring. For the sake of simplicity, below we assume $\beta_x\epsilon_x \gg \eta^2(\Delta p/p)^2$, what may occur, for instance, in a strongly cooled beam. Then, $\sigma_x \approx \sqrt{\beta_x\epsilon_x}$.

Defining the so-called Floquet functions $f_{x,z} = \sqrt{\beta_{x,z}/R_0}$, we obtain from Eqs(1) usual envelope equations (see, for instance, in Refs [2] and [3])

$$f_x'' + g_x f_x = \frac{1}{f_x^3} + \frac{2\Delta Q_x}{f_z r_z + f_x r_x}, \quad \Delta Q_{x,z} = \frac{Ne^2}{2\pi\gamma^2 pv\sqrt{\epsilon_{xz}}(\sqrt{\epsilon_z} + \sqrt{\epsilon_x})}$$

$$f_z'' + g_z f_z = \frac{1}{f_z^3} + \frac{2\Delta Q_z}{f_z r_z + f_x r_x}, \quad r_{x,z} = \frac{\sqrt{\epsilon_{x,z}}}{\sqrt{\epsilon_x} + \sqrt{\epsilon_z}}. \tag{4}$$

3. Very complicated nonlinear equations (4) usually must be solved numerically. In this report we consider a special case, when closed orbits are circular ($R(\theta) = R_0$) and therefore,

$$g_x = 1 - n = \nu_x^2, \quad g_z = n = \nu_z^2.$$

Then, Eqs(4) can be rewritten in the form

$$f_x'' + \nu_x^2 f_x = \frac{1}{f_x^3} + \frac{2\Delta Q_x}{f_z r_z + f_x r_x}, \quad f_z'' + \nu_z^2 f_z = \frac{1}{f_z^3} + \frac{2\Delta Q_z}{f_z r_z + f_x r_x}. \tag{5}$$

In contrast to the case of transport channel, in storage rings solutions of these equations must be periodic functions of θ:

$$f_x(\theta + 2\pi) = f_x(\theta), \quad f_z(\theta + 2\pi) = f_z(\theta).$$

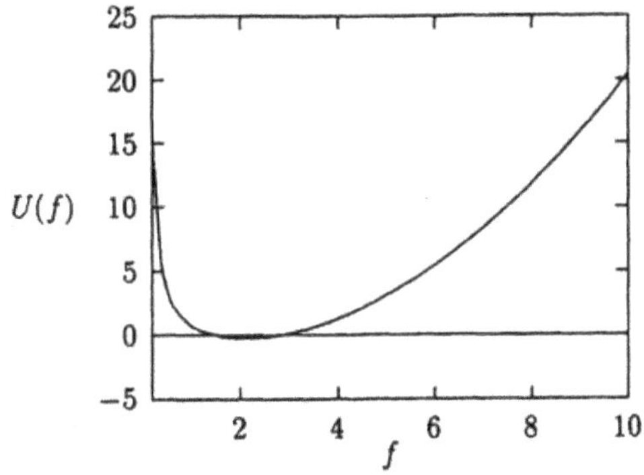

Figure 1: An example of the effective potential well for envelope equation (6) vs f; $\nu = 1/\sqrt{2}$, $\Delta\nu_L = 1$.

This requirement strongly restricts the number of solutions, which can describe β-functions. In order to simplify calculations, below we consider a special case, when $\epsilon_x = \epsilon_z = \epsilon$, $\nu_x = \nu_z = \nu$ and therefore, $f_x = f_z = f$, while ΔQ becomes equal to the Laslett tune shift

$$\Delta Q = \Delta\nu_L = \frac{Ne^2}{4\pi\gamma^2 p v \epsilon}.$$

In our case, the function f obeys the following one-dimensional equation

$$f'' + \nu^2 f = \frac{1}{f^3} + \frac{2\Delta\nu_L}{f}, \tag{6}$$

which can be obtained using the Hamiltonian:

$$h = \frac{p^2}{2} + U(f), \quad U(f) = \frac{\nu^2 f^2}{2} + \frac{1}{2f^2} - 2\Delta\nu_L \ln f, \quad p = f'. \tag{7}$$

As is seen from Fig.1, the effective potential well $U(f)$ has a minimum, which describes oscillations of $f(\theta)$. The period of these oscillations reads

$$\theta_T = \sqrt{2}\int_{f_1}^{f_2} \frac{df}{\sqrt{h - U(f)}}, \quad U(f_{1,2}) = h. \tag{8}$$

Solutions of Eq.(6) will be unique and periodic with the ring period, if for some h and $\Delta\nu_L$ the period θ_T reaches the values $2\pi m$ ($m = 1, 2, \ldots$). An example in Fig.2 shows that it never occurs except the case, when $h = U_{min}$ and hence, $f = const$. The value of this constant β-function can be found from the equilibrium equation:

$$\nu^2 f = \frac{1}{f^3} + \frac{2\Delta\nu_L}{f}.$$

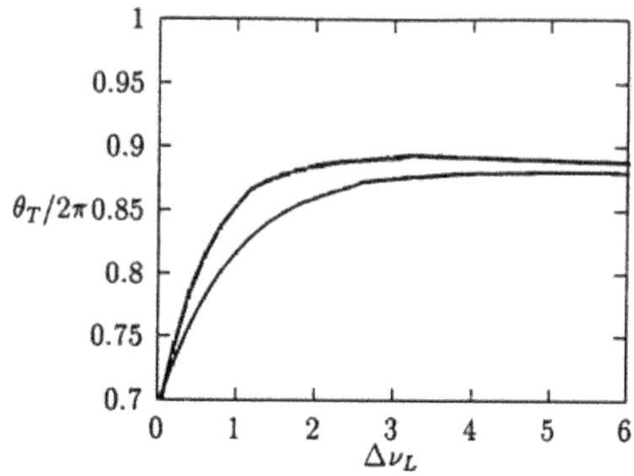

Figure 2: Dependence of the period of beatings of β-function (θ_T) on the Laslett tune shift. Upper curve: $h = 1.5$, lover curve: $h = 3$; $\nu = 1/\sqrt{2}$.

Note, that this equation has roots for any $\Delta\nu_L$, which means that for circular closed orbits the Coulomb repulsion never results in instability of linear betatron oscillations. The value of the β-function can be found using:

$$\beta = \beta_0 \left[\sqrt{1 + \left(\frac{\Delta\nu_L}{\nu}\right)^2} + \frac{\Delta\nu_L}{\nu} \right], \quad \beta_0 = R/\nu. \tag{9}$$

According to this equation, even if beam emittances remain constant, the increase in $\beta(\Delta\nu_L)$ will result in the blow-up of transverse beam sizes with an increase in the beam current. A growth of the beam emittances with the beam current, generally, will give faster than linear increase in the squares of the transverse beam sizes.

References

[1] N.S. Dikansky, D.V. Pestrikov. Physics of Intense Beams and Storage Rings. AIP PRESS, New York, 1994.

[2] F. Sacherer. IEEE Trans. on Nucl. Science, NS-18, p. 1105, 1971.

[3] S. Machida Ph.D. Thesis. University of Tokyo, 1990.

Particle Trapping with an RF Cavity

ALEX CHAO
Stanford Linear Accelerator Center, Stanford, CA 94309, USA

It is in principle possible to trap charged particles in an rf cavity. For simplicity, the cavity is considered to be a closed rectangular box with three dimensions ℓ_x, ℓ_y and ℓ_z, and the cavity mode is supposed to provide

$$\begin{aligned}
E_x &= E_{x0} \sin\frac{\pi x}{\ell_x} \cos\frac{\pi y}{\ell_y} \cos\frac{\pi z}{\ell_z} e^{-i\omega t} \\
E_y &= E_{y0} \cos\frac{\pi x}{\ell_x} \sin\frac{\pi y}{\ell_y} \cos\frac{\pi z}{\ell_z} e^{-i\omega t} \\
E_z &= E_{z0} \cos\frac{\pi x}{\ell_x} \cos\frac{\pi y}{\ell_y} \sin\frac{\pi z}{\ell_z} e^{-i\omega t} \\
B_x &= \frac{ic\pi}{\omega}\left(\frac{E_{z0}}{\ell_y} - \frac{E_{y0}}{\ell_z}\right) \cos\frac{\pi x}{\ell_x} \sin\frac{\pi y}{\ell_y} \sin\frac{\pi z}{\ell_z} e^{-i\omega t} \\
B_y &= \frac{ic\pi}{\omega}\left(\frac{E_{x0}}{\ell_z} - \frac{E_{z0}}{\ell_x}\right) \sin\frac{\pi x}{\ell_x} \cos\frac{\pi y}{\ell_y} \sin\frac{\pi z}{\ell_z} e^{-i\omega t} \\
B_z &= \frac{ic\pi}{\omega}\left(\frac{E_{y0}}{\ell_x} - \frac{E_{x0}}{\ell_y}\right) \sin\frac{\pi x}{\ell_x} \sin\frac{\pi y}{\ell_y} \cos\frac{\pi z}{\ell_z} e^{-i\omega t},
\end{aligned} \tag{1}$$

where ω is the mode frequency given by

$$\omega = \pi c \sqrt{\frac{1}{\ell_x^2} + \frac{1}{\ell_y^2} + \frac{1}{\ell_z^2}} \tag{2}$$

In this note we calculate the trapping mechanism of the field pattern (1). The trapping occurs for either sign of the charge of the particles.

Maxwell equations also requires for this rf mode to obey

$$\frac{E_{x0}}{\ell_x} + \frac{E_{y0}}{\ell_y} + \frac{E_{z0}}{\ell_z} = 0 \tag{3}$$

The mode pattern (1) traps particles around the node point $(0,0,0)$. Such a trapping mechanism is of course considered favorable for a particle trap. On the other hand, it may also happen that undesirable stray particles are trapped at nodes of an rf device, which then perturbs the normal operation of an accelerator. Such trappings are then undesirable and are to be avoided.

Consider the motion of a cold particle near the center of the cavity $x \ll \ell_x, y \ll \ell_y, z \ll \ell_z$. The field seen by this particle is approximately

$$\begin{aligned}
\vec{E} &\approx \pi e^{-i\omega t}\left[\frac{E_{x0}}{\ell_x} x\hat{x} + \frac{E_{y0}}{\ell_y} y\hat{y} + \frac{E_{z0}}{\ell_z} z\hat{z}\right] \\
\vec{B} &\approx 0
\end{aligned} \tag{4}$$

The magnetic field is negligible not only because it is second order in x, y, z, but also because the magnetic force is weak for nonrelativistic particles. For what follows, we replace $e^{-i\omega t}$ by $\cos \omega t$. In this work, we do not consider how the particles are injected into the cavity, or how the particles can be cooled by some means.

The equation of motion of a particle with rest mass M and charge Q is

$$M\ddot{x} = QE_x \approx Q\frac{E_{x0}}{\ell_x} x \cos \omega t \tag{5}$$

and similarly for the y- and z-motions. The three dimensions are decoupled, and by a change of variable from t to $v = \frac{1}{2}\omega t$, they read

$$\frac{d^2 u}{dv^2} + (\beta \cos 2v)u = 0 \tag{6}$$

For the x-motion, for example, we have

$$\beta = -\frac{4QE_{x0}}{M\ell_x \omega^2} \tag{7}$$

Particle trapping is then based on the alternating focusing nature of Eq.(6). Indeed, the alternating sign.of $\cos 2v$ provides an opportunity for the u-motion to be focussed regardless of the sign of β. Although the field can not focus all three dimensions at a given instant of time,The values of β for the three dimensions can not all have the same sign, as shown by an inspection of Eq.(3). the alternating rf field may still provide net focusings to all three dimensions, thus trap the particles.

Equation (6) is a special case of the Mathieu's equation. In order for the u-motion to be stable, β and ω must satisfy some condition. The stability boundary is determined by the existence of periodic solutions of $u(v)$ of the types

$$u(v) = \sum_{m=0}^{\infty} A_m \left\{ \begin{matrix} \sin \\ \cos \end{matrix} \right\} (2m+1)v \tag{8}$$

Following the standard Mathieu's equation analysis,[1] it can be shown that for the periodic solutions (8) to exist, the value of β must satisfy

$$2 \pm \beta = \cfrac{\beta^2}{2 \cdot 3^2 - \cfrac{\beta^2}{2 \cdot 5^2 - \cfrac{\beta^2}{2 \cdot 7^2 - \cdots}}} \tag{9}$$

where + sign is for the existence of the sin solution, while the − sign is for the cos solution. A numerical calculation gives the results $\beta = \mp 1.816$. We thus find that the stability condition is

$$|\beta| < 1.816 \tag{10}$$

---

a*

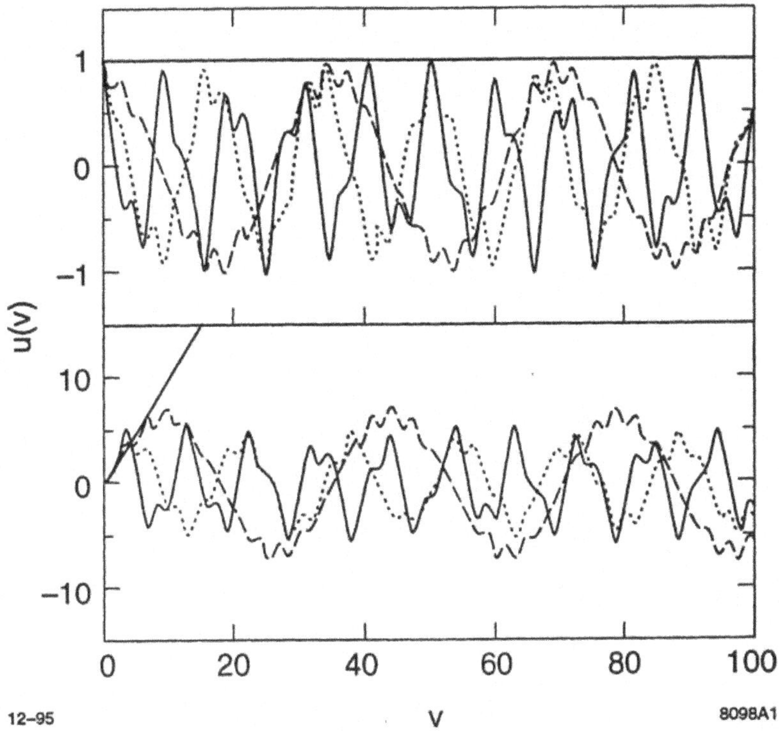

Fig. 1. Particle motion in an rf trap.

Condition (10) can be written as

$$QE_0\ell \leq 4.3Mc^2 \tag{11}$$

where E_0 is the cavity field strength, ℓ is the size of the cavity. This condition is equivalent to requiring the energy gained from the cavity electric field over the distance of the cavity dimension must not render the particle relativistic. Such a condition is easily satisfied in practice.

We can track the motion of a particle using Eq.(6). Figure 1(a) shows the results for tracking a particle with initial conditions $u(0) = 1$ and $\frac{du}{dv}(0) = 0$ for 4 values of $\beta = 0$ (solid curve), 0.5 (dashed curve), 1 (dotted curve), and 1.5 (solid curve), for the range of v from 0 to 100. The motion is stable for these values of β. Figure 1(b) is the same with the initial conditions $u(0) = 0$ and $\frac{du}{dv}(0) = 1$. A strong focusing occurs when $\beta \sim 1$, away from 0 or the threshold value of 1.816.

Figures 2(a) and 2(b) show the particle motion near the threshold $\beta = 1.816$. The two values of β are 1.814 (solid curve) and 1.818 (dotted curve). The transition from

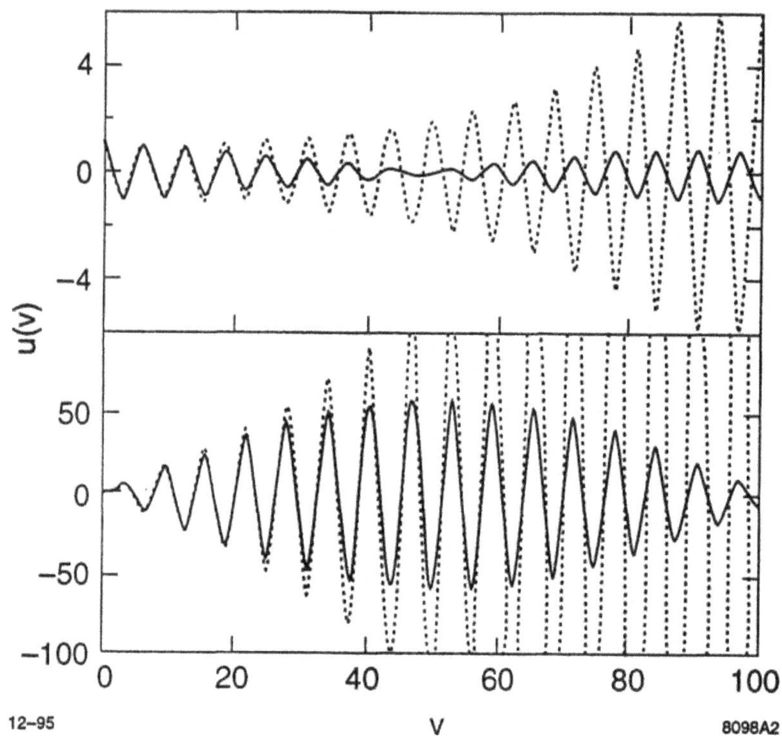

Fig. 2. Particle motion in an rf trap near the instability threshold.

stable to unstable motions is rather sudden.

It may be instructive to examine the qualitative behavior of the particle motion using a simplified model instead of the Mathieu's equaton analysis. To do so, consider the replacement

$$\beta \cos 2v \to \begin{cases} |\beta| & \text{when } \beta \cos 2v > 0 \\ -|\beta| & \text{when } \beta \cos 2v < 0 \end{cases} \tag{12}$$

In other words, replace the $\beta \cos 2v$ focusing by a piecewise constant alternating focusing. We expect to obtain a stability condition similar to Eq.(10) but with somewhat different numerical coefficients.

With the replacement (12), the particle motion over one period of time $v_{\text{final}} - v_{\text{initial}} = \pi$ is described by a matrix map

$$\begin{bmatrix} u \\ \frac{1}{\sqrt{|\beta|}} \frac{du}{dv} \end{bmatrix}_{\text{final}} = T \begin{bmatrix} u \\ \frac{1}{\sqrt{|\beta|}} \frac{du}{dv} \end{bmatrix}_{\text{initial}} \tag{13}$$

where

$$T = \begin{bmatrix} \cos(\frac{\pi}{2}\sqrt{|\beta|}) & \sin(\frac{\pi}{2}\sqrt{|\beta|}) \\ -\sin(\frac{\pi}{2}\sqrt{|\beta|}) & \cos(\frac{\pi}{2}\sqrt{|\beta|}) \end{bmatrix} \begin{bmatrix} \cosh(\frac{\pi}{2}\sqrt{|\beta|}) & \sinh(\frac{\pi}{2}\sqrt{|\beta|}) \\ \sinh(\frac{\pi}{2}\sqrt{|\beta|}) & \cosh(\frac{\pi}{2}\sqrt{|\beta|}) \end{bmatrix} \tag{14}$$

Stability of motion is determined by the condition $|\text{trace}(T)| \leq 2$, or

$$|\cos(\frac{\pi}{2}\sqrt{|\beta|})\cosh(\frac{\pi}{2}\sqrt{|\beta|})| \leq 1 \tag{15}$$

Like the Mathieu's equation, there are infinite number of stable bands of values of β. The stable band near $\beta = 0$ has the largest width, and it spans the region

$$|\beta| < 1.425 \tag{16}$$

Equation (16) is to be compared with the exact result, Eq.(10).

The model (12) can also be used to estimate the oscillation frequencies of the particle motion in the rf cavity trap. In the stable region, the oscillation frequency is determined by

$$f = \frac{\omega}{2\pi}\frac{1}{2\pi}\cos^{-1}[\cos(\frac{\pi}{2}\sqrt{|\beta|})\cosh(\frac{\pi}{2}\sqrt{|\beta|})] \tag{17}$$

If $|\beta| \leq 1$, we have

$$f \approx \frac{\omega|\beta|}{16\sqrt{3}} \tag{18}$$

The three oscillation frequencies are therefore

$$f_x \approx \frac{QE_{x0}}{4\sqrt{3}M\ell_x\omega}, \qquad \text{similarly for } f_y, f_z \tag{19}$$

The period V of oscillation can be read off from Figs. 1 and 2. This gives $V \approx 35$, 17, 10, 104 for $\beta = 0.5$, 1, 1.5 and 1.814, respectively. These periods can be compared with the approximate expression (19), which gives $V \approx \frac{8\sqrt{3}}{\beta}$. For $\beta = 0.5$, 1, 1.5 and 1.814, this gives $V = 28$, 14, 9, and 7.6. The result for $\beta = 1.814$ is way off, but the other values are reasonable.

Take $\ell_x = \ell_y = \ell_z = 30$ cm, Eq.(2) gives $\omega = 0.54 \times 10^{10}\ \text{s}^{-1}$ (0.86 GHz).[2] Consider antiprotons, we have $Q = e$, $Mc^2 = 938$ MeV. Let the cavity field be such that $E_{x0} = E_{y0} = 100$ KV/cm and, from Eq.(3), $E_{z0} = -200$ KV/cm. We have $|\beta_x| = |\beta_y| = \frac{1}{2}|\beta_z| = 0.28 \times 10^{-4}$, easily satisfying the stability condition (10). Using Eq.(20), we can also estimate the oscillation frequencies to be $f_x = f_y = \frac{1}{2}f_z = 85$ KHz.

The rf trap mentioned in the above example can in principle be used to trap cold electrons or positrons. We then have $|\beta_x| = |\beta_y| = \frac{1}{2}|\beta_z| = 0.051$, and the particles are trapped with $f_x = f_y = \frac{1}{2}f_z = 160$ MHz.

The map (14) has the Courant-Snyder invariant

$$J = (1 - \cos^2 \cosh^2)u^2 + [u \sin \sinh + \frac{du}{dv}(\cos \sinh + \sin \cosh)]^2 \tag{20}$$

where the arguments of sin, cos, sinh, and cosh functions are $\frac{\pi}{2}\sqrt{|\beta|}$. When $|\beta| \ll 1$, we have the invariant[3]

$$\frac{J}{\pi^2|\beta|} \approx (\frac{du}{dv})^2 + \frac{\pi}{2}\sqrt{|\beta|}u\frac{du}{dv} + \frac{\pi^2}{12}|\beta|u^2 + \mathcal{O}(|\beta|^2) \tag{21}$$

The invariant (21) allows an estimate of the particle oscillation amplitude given the initial conditions $u(0)$ and $\frac{du}{dv}(0)$. For example, when $u(0) = 0$, Eq.(21) gives a maximum oscillation amplitude

$$u_{\max} \approx \frac{4}{\pi}\sqrt{\frac{3}{|\beta|}}\frac{du}{dv}(0) \tag{22}$$

Equation (22) can be compared with the tracking results shown in Fig.1(b).

If a cloud of particles are trapped, and if the rms kinetic energy is K, the rms size of the particle cloud is

$$\sigma_x = \frac{4}{\pi}\sqrt{\frac{K\ell_x}{QE_{0x}}}, \qquad \text{and similarly for } \sigma_y, \sigma_z \tag{23}$$

Note the result (23) does not depend explicitly on the mass of the particle. Take $K = 1$ KeV, then the previous numerical examples (for both antiprotons and positrons) give $\sigma_x = \sigma_y = \sqrt{2}\sigma_z = 7$ mm. Space charge effects however have been ignored in Eq.(23).

I would like to thank Sandro Ruggiero and Paul Channel for their helpful comments.

1. References

1. Morse and Feshbach, *Methods of Theoretical Physics*, McGraw-Hill, 1978.

2. One may want to avoid this highly degenerate mode by having $\ell_{x,y,z}$ slightly different from one another, but here we just want a crude numerical example.

3. Carried out to higher orders, we have $\mathcal{O}(|\beta|^2) = -\frac{\pi^4}{240}|\beta|^2(\frac{du}{dv})^2 - \frac{\pi^5}{720}|\beta|^{5/2}u\frac{du}{dv} - \frac{\pi^6}{5040}|\beta|^3u^2 + \mathcal{O}(|\beta|^{7/2})$.

IV. EXPERIMENTS

WORKING GROUP ON EXPERIMENTS AND SPECIAL SESSION ON ELECTRON COOLING

D. Reistad

The Svedberg Laboratory, Box 533, S-751 21 Uppsala, Sweden

ABSTRACT

The Working Group on Experiments took place on Sunday, 19 November. There were discussions on the NAP-M and ESR experimental results on possible one-dimensional beams, on intra-beam scattering and its interplay with three-dimensional laser cooling, on experiments with space-charge dominated beams performed at the Indiana Cooler, on experimental results on local density enhancement around ions during electron cooling performed at the TSR, tune shift measurements, and operation with isochronous condition. A new proposal for arranging coupling between the transverse and the longitudinal degrees of freedom was put forward. During a Special Session on Electron Cooling on Thursday, 16 November, there was a discussion on what the minimum electron beam temperature is, and if crystallization can be achieved with electron cooling only, also on neutralization of the electron beam, on expanded electron beams, and on "electron heating."

1. How do we interpret the NAP-M and ESR results on possible one-dimensional beams?

Observations that may be explained by assuming ordering in the beam have been made at the NAP-M ring of Novosibirsk and at the ESR ring of Darmstadt. These are the only two particle beam cooler rings, where such observations have been made. The results from Novosibirsk and Darmstadt were reported by Parkhomchuk and Steck respectively.

At the NAP-M, a "magnesium-jet beam profile monitor" could be used both to measure the vertical profile and the intensity of the beam. It could also be used to measure the transverse cooling time[1-2]. A flying-wire monitor was used to measure the transverse profile of cooled beams, when the resolution of the magnesium-jet beam profile monitor was not sufficient. There was also a monitor to observe the power of the Schottky noise from the coasting beam.

Measurements were performed with a beam of 2×10^7 65 MeV protons, which was cooled with electrons so that the transverse beam radius was about 0.1 mm and the momentum spread was 10^{-6}.

Several years after that the measurements were made, it was realized that the observed cooling time and equilibrium $\Delta p/p$ contradicted standard intra-beam scattering theory by two orders of magnitude. This can be explained by assuming ordering in the beam (reducing the coupling between the longitudinal and transverse motion).

At the ESR[3-4], many different heavy ion beams (from C^{6+} to U^{92+}) have been injected and stacked, and cooled with electrons. The beam size during cooling has been measured with a moveable scraper with a resolution of 0.1 mm. The width and the power of a Schottky band was measured as the number of stored ions decreased. It was found that for heavy ions $(Z > 20)$ the power of the Schottky band directly reflects the number of particles and that the width of the Schottky band reflects the momentum spread through $\Delta p/p = \eta^{-1}\Delta f/f$. No collective beam behaviour was observed. This is explained by the strong heating by intra-beam scattering, which is present for heavy ions.

The absence of Schottky noise suppression was checked by injecting a beam current which was sufficient to be measured with the DC current transformer (about 20 μA). The beam current was subsequently measured with this transformer until the signal disappeared. Thus the beam lifetime became known, and the intensity was known independently of the Schottky noise. The Schottky power always agreed with the value determined from the lifetime. From time to time the electron current was switched off or changed. The noise power without cooling agreed with the noise power of the cooled beam. In case of Schottky noise suppression there should have been a considerable difference between the signals from the cooled and uncooled beams.

The measurements were made with a wide range of the number of ions; from a single ion for the heaviest ions and $10^2 - 10^3$ ions for the light ones up to $10^7 - 10^9$. For all ions from Ar^{18+} and up, a sudden jump in the width of the Schottky spectrum was found to occur when the number of ions reached a value of a few thousand. It is conjectured that this may be evidence for ordering in the beams.

Parkhomchuk compared the conditions at NAP-M and at the ESR. If one calculates the coupling parameter Γ as the ratio between the potential energy associated with the *transverse* beam radius $a_\perp$ and the effective *longitudinal* temperature $E_k(\Delta p/p)^2\eta$, where E_k is the (total) kinetic energy of the ions[5], one finds values of the coupling parameter close to unity in both cases. It is not obvious, however, that it is appropriate for ordering in the beam to calculate Γ in this way.

The inter-particle spacing at NAP-M and at the ESR were very different. 2×10^7 protons in the 47 m NAP-M corresponds to an average longitudinal spacing of only 2.4 μm, thus two orders of magnitude *less* than the observed transverse beam size. On the other hand, the 2×10^3 ions in the 108 m ESR are 5 cm apart, two orders of magnitude *more* than the transverse beam size.

Steck pointed out, that the relative longitudinal motion between the ions is very slow compared to the betatron oscillations. Thus, the ions are effectively behaving as 'disks,' blocking the way for each other.

Sessler and Wei have attempted molecular dynamics (MD) simulations of the NAP-M experiment. They found that below a threshold density, N_c, the beam does behave like a 1-D chain of 'disks', with one proton in each. The 'disks' pass each other only occasionally. Above N_c, the longitudinal oscillation (coupled to transverse motion) destroys the disk structure. They found from their simulations however, that $N_c < 1.5\times10^5$, thus, they could not confirm the NAP-M experience, in which ordering was observed to set in with as many as 2×10^7 protons circulating in the ring.

A problem, which was pointed out by Sessler, is that the long-range phenomena, that are responsible for the Schottky noise, can not be predicted by MD simulations. New techniques are required and have to be developed, in order to predict the characteristics of the Schottky noise, which will be generated by ordered and near-ordered ion beams.

In conclusion, one has to say that the NAP-M observations seem indirect[6], and that it is therefore very valuable that ordering phenomena have now been observed more convincingly at another ring. It should now be an important task for the theorists to do molecular dynamics simulations and other theoretical work to completely understand the ESR beams, and it should also be a challenge to some of the other rings to carefully repeat and improve the NAP-M and ESR experiments.

2. Do we completely understand intra-beam scattering in a gaseous beam?

Miesner and Grimm discussed the relation between intra-beam scattering and three-dimensional laser cooling:

The pronounced two-component nature of the longitudinal velocity distribution has very important and interesting consequences for the three-dimensional laser cooling process. As found at the TSR, a main part of the laser-cooled beam can be much colder than one would naïvely expect from the simplified intrabeam scattering theory that assumes Gaussian distributions.

We know that the key to understand this is to distinguish between two different types of collisions in the beam, frequent, soft collisions leading to a diffusive heating and rare, hard collisions leading to large sudden velocity changes.

We certainly do understand intrabeam scattering in the sense that we can write down Piwinski's general formulas and integrals (they are of course OK and they somehow contain the complete physics). We would like to obtain an intuitive model for the cooling physics that takes these two subdistributions into account.

In this sense, we are trying two find answers to two main questions:

(i) In how far can we still extract useful information out of the simplified theory using Gaussian distributions ? Our results indicate that, if the right amount of thermal energy is dissipated by the laser, the shape of the longitudinal distribution plays no rôle for the indirect transverse cooling effect.

(ii) Is it possible to modify the standard theory somehow on the basis of a simple two-component model?

3. Experiments with space-charge dominated beam

Lee reported on experiments on space-charge dominated beams performed at the IUCF Cooler Ring, where the bunch length is roughly proportional to the third power of the beam current.

He did however remind us, that the relative importance of the space charge force to the emittance deserves careful studies.

Information on the relative importance of space charge and $\varepsilon_{zz'}^2/z_m^3$ (where $\varepsilon_{zz'}$ is the longitudinal emittance and $2z_m$ is the bunch length) can be obtained by fitting measured bunch length data to the solution of the longitudinal envelope equation,

$$k_{z0}^2 z_m - \frac{K_L}{z_m^2} - \frac{\varepsilon_{zz'}^2}{z_m^3} = 0$$

where k_{z0} is the longitudinal focusing force and $K_L = \frac{3}{2} g \frac{N r_0}{\beta^2 \gamma^3}$ where N is the number of particles in the bunch, g is a geometry factor, and r_0 is the classical electron radius.

An approximate solution, exact when $K_L = 0$ or when $\varepsilon_{zz'} = 0$ and with maximum error of 3 % when both terms are of the same magnitude is[7]

$$z_m = \left(\frac{K_L}{k_{z0}^2} + \frac{\varepsilon_{zz'}^{3/2}}{k_{z0}^{3/2}} \right)^{1/3}.$$

The data obtained at IUCF show that the ratio between the space charge and the emittance terms is about 30-40 %.

Lee also discussed some experimental results with a double rf. system[8-10], used at the IUCF Cooler to alleviate the space charge problem by reducing the peak current.

In one of their experiments, the beam intensity was increased a factor four with $V = V_1 \sin(h\omega_0 t) + V_2 \sin(2h\omega_0 t)$, $V_2 = -V_1/2$.

4. Experimental results on local electron density enhancement around ions during electron cooling

Wolf discussed experimental results from the TSR on local electron density enhancement around ions during electron cooling. The local density can be estimated from the cooling decrement $\alpha = -F/Mv$ and also from the rate of electron-ion recombination during electron cooling. These methods show good agreement over the ion charge range investigated, from $Z = 1$ to $Z = 20$. Local enhancements up to 70 % have been observed at $E_{\text{c.m.}} < 10^{-3}$ eV, insensitive to $T_{e\perp}$.

The TSR results seem to be confirmed by MD simulations performed by Zwicknagel and reported by Toepffer. Zwicknagel found in his MD simulations electron densities near ions, which for large coupling and low velocity are much larger than those of the linearized Debye- Hückel theory. There are two new features:

(i) The polarization is so strong, that the nonlinear shielding length becomes larger than the Debye length.

(ii) Due to to multiparticle collisions many electrons become trapped into high-lying Rydberg states. This enhances the radiative recombination rate.

An important consequence for electron coooling is that for low ion velocity, $v_i < v_{e\parallel}$, and large coupling, $Z\Gamma^{\frac{3}{2}} \geq 1$, nonlinear shielding leads to a different scaling of the drag coefficient,

$$\frac{F_{\text{cool}}}{v_i} \propto Z^2 \underset{\substack{\text{non-linear}\\\text{shielding}}}{\Rightarrow} Z^{1.5}.$$

Toepffer also showed first results by Spreiter for a magnetized, anisotropic plasma. These need, however, further improvement of the statistics. In dependence on the magnetic field there apparently are two regimes, depending on the ratio of the Larmor radius to the inter-electron distance. They strongly urge measurements exploring the dependence of observables on the magnetic field.

5. Tune shift measurements

Chanel reported on measurements made with a quadrupolar pickup on LEAR[11].

A pickup can be made sensitive to oscillations in the shape of the circulating beam. Thus, it can be used to study quadrupolar instabilities (such as those created by the presence of ions in an antiproton beam for example). By exciting the beam with a quadrupolar kicker and measuring the response with a quadrupolar pickup, i.e. by performing a quadrupolar transfer function measurement, it is possible to determine the degree of stability of the beam. In this case, the transfer function provides information on the working point of the particles as a function of their betatronic amplitude, thus the incoherent detuning due to the space charge.

The (horizontal and vertical) quadrupolar tunes $\nu_{2,x}$ and $\nu_{2,z}$ obtained in presence of space charge have been derived by Hardt for the case of elliptical beams[12].

$$\nu_{2,x} = 2\left(\nu_{0,x} - \Delta\nu_{i,x}\left(\frac{3}{4} - \frac{1}{4}\frac{a_x/a_z}{\left(1 + a_x/a_z\right)}\right)\right),$$

$$\nu_{2,z} = 2\left(\nu_{0,z} - \Delta\nu_{i,z}\left(\frac{3}{4} - \frac{1}{4}\frac{a_z/a_x}{\left(1 + a_z/a_x\right)}\right)\right),$$

where a_x/a_z is the ratio between the horizontal and vertical beam sizes; for the evaluation of the measurements Chanel assumed that

$$a_x/a_z = \sqrt{\nu_{0,z}/\nu_{0,x}}\,.$$

$\nu_{0,x}$ and $\nu_{0,z}$ are the single-particle (dipolar, i.e. "normal") tunes, and $\Delta\nu_{i,x}$ and $\Delta\nu_{i,z}$ are the incoherent tune shifts, to be determined by the measurement.

Measurements have been done on electron cooled beams while varying the number of particles N, the electron current I_e and the tune to test effects of resonances.

During the measurements f_{rev}, $\nu_{0,x}$, $\nu_{0,z}$, $\nu_{2,x}$, $\nu_{2,z}$, and the rms. beam size (measured with the beam ionization profile monitor) were recorded.

The measurements have allowed an evaluation of the behavior of the beam under the effects of strong space-charge and intra-beam scattering. Provided that the beams are well aligned in the electron cooler, the equilibrium between cooling and intra-beam scattering seems to be as predicted by the Möhl-Giannini intra-beam scattering codes[13].

In the vicinity of a resonance the beam distribution becomes uniform. This is not only seen with the beam ionization profile monitors, but also from the width of the quadrupolar beam transfer function signal.

6. Operating with isochronous condition

Lee remarked that the crystallization presented by all calculations is equivalent to the attainment of the isochronous condition (transition energy). All particles march like "soldiers in an army" in the storage ring. Each particle has its own momentum-dependent closed orbit and performs small betatron oscillations around it. Therefore, particle stability at transition energy is an important topic for the "crystal-beam community." Questions such as what is the effect of impedance, noise, rf bucket, nonlinear momentum compaction factor, etc require study.

Steck pointed out that the ESR has been operated with a coasting beam right on the transition energy without problems.

7. Proposal for a static mixing device

Schiffer presented an idea by Caspers for using a Wien filter to obtain mixing of the transverse and the longitudinal phase planes for laser cooling. The idea is to use an electric and a magnetic field orthogonal to each other and to the beam, so that the electric and the magnetic forces cancel for a given $\beta = v/c$.

Any Δp would lead to a deflection, but also any transverse momentum (not position since the fields should be homogeneous) orthogonal to the magnetic field would be converted into $\Delta p/p$.

One would need two such devices, one with horizontal magnetic field and one with vertical magnetic field. These would form a passive structure with homogeneous fields, it would therefore not give rise to filamentation or heating.

8. How to analyze high-order envelope instabilities?

Lee pointed out that higher order envelope resonances can be important. The envelope tune of a cylindrically symmetric beam is bounded by $\sqrt{2}\nu_{\text{particle}} \leq \nu_{\text{envelope}} \leq 2\nu_{\text{particle}}$, where ν_{particle} is the particle betatron tune[14-15]. The Mathieu instability corresponds to $2\nu_{\text{envelope}}$ = integer. Thus perturbations arising from external sources or from intrinsic envelope mismatch can provide envelope and particle instabilities. These cause halo formation in intense charge beam transport systems.

Besides the linear instability, $3\nu_{\text{envelope}}$ can also cause problem to the crystallization of very cold beams.

Lee urged careful studies on the higher order envelope instabilities using molecular dynamics simulations.

Characteristic features of parametric resonances can be found in ref. 16.

9. Special session on electron cooling

There was a special session on electron cooling on Thursday, 16 November. Some of the points that were discussed are briefly summarized below.

9.1 What is the minimum electron beam temperature, and could crystallization be achieved with electron cooling only?

The minimum possible electron beam temperature is determined by the intra-beam scattering among the electrons to be of the order of $\frac{e^2}{4\pi\varepsilon_0} n_e^{\frac{1}{3}}$. The required temperature of the ions to reach crystallization ($\Gamma \approx 200$) is $\sim \dfrac{1}{200}\dfrac{Z^2 e^2}{4\pi\varepsilon_0} n_i^{\frac{1}{3}}$. In a typical situation for crystallization, n_e and n_i are of the same order. Achieving crystallization of a proton beam by electron cooling is thus not possible. On the other hand, the factor Z^2 would seem to help for heavy ions, and achieving crystallisation through electron cooling might seem possible.

Lebedev made however a comparison between electron cooling and intra-beam scattering, and concluded that since intrabeam scattering grows as Z^4 while the friction force only grows as Z^2 it is more difficult to cool a beam of highly charged ions through the maximum of intrabeam scattering.

Thus, one would have to conclude that crystallization will not be achieved with electron cooling only.

9.2 Neutralization of the electron beam

Neutralization of the electron beam seems useful, since it gives a better match between the electron and ion velocity distributions. Quite good neutralization has been achieved at LEAR[17]. For some reason however, neutralization has not improved the electron cooling times. Another un-resolved question is the mechanism of escape of the ions across the potential barrier.

9.3 Expanded electron beams

Measurements of drag force with expanded electron beams have been performed at CRYRING[18-19] and at the TSR[20-21]. Although the electron beam according to models ought to be magnetized, and the transverse electron temperature therefore should not be important for the drag force, these two rings do show the best normalized drag forces.

For obtaining crystallized beams, it may be important to have modest magnetic fields in the electron coolers in order not to perturb the crystals too much. The electrons may therefore not be totally magnetized, and electron temperature may indeed be an important issue. Combining a GaAs photocathode[22] and expansion of the electron beam would seem to be the way to produce the lowest possible temperature of the electron beam. Adiabatic acceleration[23] was also discussed, it does not however seem to be obvious that it can be used in practice to obtain cold electron beams for cooling.

9.4 What is electron heating?

Merging a stored proton or ion beam with electrons does not only have a beneficial influence on the ion beam. One usually unwanted effect of the electrons is recombination. The electron beam is also producing an additional impedance to the stored beam, thus driving instabilities in intense beams[24-25]. In certain cases, particularly if the stored beam has a larger size than the electron beam, the non-linear electrical field from the electrons may drive resonances in the beam[26-27].

10. Conclusion

Fascinating linear ordering phenomena have been observed at the NAP-M and even more convincingly at the ESR. It is important that the other cooler rings also look for ordering, and that a theoretical effort is undertaken to understand the long-range plasma effects directly responsible for producing the Schottky noise effects, which have been seen.

Many difficult problems remain to be solved before 3-D crystallization in storage rings will be achieved; such crystallized beams, like the Eloisatron, represent utopia.

11. Acknowledgments

Acknowledgments are due to everybody who took part in the working group and in the special session. Special thanks for help with this manuscript to Michel Chanel, Håkan Danared, Rudi Grimm, Valery Lebedev, S.Y Lee, Hans-Joachim Miesner, Vasily Parkhomchuk, Andy Sessler, Markus Steck, and Christian Toepffer.

References

1. V.V. Parkhomchuk, this workshop.
2. G.I. Budker et al., *Particle Accelerators* **7** (1976) 197.
3. M. Steck, this workshop.
4. M. Steck et al., *1st Euroconference on Atomic Physics with Stored Highly Charged Ions*, Heidelberg, March 1995, proceedings to be published in *Hyperfine Interactions*.
5. S. Chattopadhyay, *CERN* 84-11.
6. D. Habs, *MPI* H – 1987 – V 10, Max-Planck-Institut für Kernphysik, Heidelberg.
7. M. Reiser, *Theory and Design of Charged Particle Beams*, Wiley, 1994, page 425-426.
8. S.Y. Lee et al., *Phys. Rev.* **E 49** (1994) 5717.
9. J.Y. Liu et al., *Phys. Rev.* **E 50** (1994) R3349.
10. J.Y. Liu et al., *Particle Accelerators* **49** (1995) 221.
11. M. Chanel, *Mesures avec pick-up quadripolaire à LEAR*, unpublished, 1995.
12. W. Hardt, *On the incoherent space-charge limit for elliptic beams*, unpublished, 1966.

13. R. Giannini, D. Möhl, *CERN PS/AR Note* 92-22.
14. S.Y. Lee, A. Riabkov, *Phys. Rev.* **E 51** (1995) 1609.
15. A. Riabko et al., *Phys. Rev.* **E 51** (1994) 3529.
16. S.Y. Lee, *AIP Proceedings* **326** (1995) 13.
17. D. Möhl, this workshop.
18. H. Danared et al., *Phys. Rev. Letters* **72** (1994) 3775.
19. H. Danared, this workshop.
20. M. Grieser et al., *Proc. Fourth European Particle Accelerator Conference*, London, U.K., 27 June-1 July 1994, 518.
21. D. Habs, this workshop.
22. R. Calabrese, L. Tecchio, *Il Nuovo Cimento* **104A** (1991) 1127.
23. A.V. Aleksandrov et al., *Phys. Rev.* **A46** (1992) 6628.
24. A.V. Burov, *Electron Cooler Impedances*, unpublished, 1995.
25. V. Parkhomchuk, D. Pestrikov, *CERN* 94-03, 327.
26. D. Reistad et al., *CERN* 94-03, 183.
27. T. Tanabe et al., *CERN* 94-03, 312.

THE LEGNARO CRYSTAL PROPOSAL

L. TECCHIO
Department of Experimental Physics University of Torino and INFN, Torino, Italy.

G. BISOFFI, G. CIULLO, A. DAINELLI, S. GUSTAFSSON
M.F. MOISIO, A. PISENT, M. POGGI, B. YANG*
Laboratori Nazionali di Legnaro, Legnaro, Italy
**LABEN spa, Vimodrone (Mi), Italy*

A. BUROV, N.S. DIKANSKY, B. GRISHANOV, V.V. PARKHOMCHUK
Budker Institute of Nuclear Physics, Novosibirsk, Russia

R. CALABRESE, V. GUIDI, P. LENISA, M. PILATI
Department of Physics University of Ferrara and INFN, Ferrara, Italy

T. CLAUSER, G. LAMANNA, M. RUTIGLIANO, V. STAGNO, V. VARIALE
Department of Physics University of Bari and INFN, Bari, Italy

V.G. VACCARO
Department of Physics University of Napoli and INFN, Napoli, Italy

E. MARIOTTI, L. MOI
Department of Physics University of Siena and INFN Ferrara, Italy

ABSTRACT

The Conceptual Design of a low-energy heavy-ion storage ring, called CRYSTAL, proposed for the experimental demonstration of Crystalline Beams is presented.

The design of the project has been optimized for the study of crystalline beams; but provisions have also been made for carrying out experiments of different nature in nuclear, atomic and molecular physics. The Tandem-ALPI heavy-ion facilities of the Laboratori Nazionali di Legnaro is assumed as the injector for the storage ring. For the Crystalline mode the possibility to inject particles from an indipendent low energy source is also considered.

The storage ring has a circumference of *70 meters*, large enough to allow a maximum magnetic rigidity of *3.8 Tm*. It has a periodicity of eight, made of eight dipole magnets surrounded by doublet quadrupoles, sextupoles and a drift space free of magnets *3.4 meters* long.

The demonstration of Crystalline Beams is feasible with very effective electron and laser cooling and requires dedicated diagnostics.

The existence of stable ground states for Crystalline Beams has been demonstrated by computer simulations. The basic structures are the helices and twisted multiplets. The ground state configurations depend on the focusing parameters of the storage ring. The transition thresholds from one ground state to another depend on the focusing parameters and particles density.

1. Introduction

After the successful exploitation of electron cooling in several heavy ion storage rings the possibility of generating Crystalline Ion Beams, i.e. beams with particles located at fixed positions relative to one another, has gained the interest of particle accelerator physicists[1].

New cooling methods like laser cooling[2], give further opportunity to reach ultra cold system of particles necessary for the state transition to the Crystalline configuration.

Crystalline Beams will give knowledge on a completely new research field, reaching new heights in precision measurements and techniques. The applicative potentialities of Crystalline Beams justifies a careful investigation on this subject.

A part from the pioneering work[3] performed by E.N. Dementiev *et al.* in 1980, only recently[4] some indications of longitudinal order in a storage ring has been detected, for very low intensity ion beams.

During the last decade computer simulations[5,6,7] have been carried out, their results showed various kinds of Crystalline structures in an equally focused straight channel. The properties of Crystalline ion beams were studied both analytically and numerically in a model where the beam was considered at rest in a pipe with longitudinally homogeneous and transversely isotropic external focusing.

The predicted Crystalline structures were experimentally achieved in a small ring-like ion trap by employing laser cooling[8,9].

Only recently the feasibility of Crystalline beams in realistic storage rings were demonstrated[10,11]. The particle trajectory curvature and the inhomogeneity of AG focusing were taken into account in deriving the Hamiltonian[12,13]. This Hamiltonian was used to study the existence and properties of Crystalline Beams. Periodic ground states of a Crystalline Beam were achieved and are very similar to the ones experimentally observed in the ion traps. Also other authors demonstrate[13,14], in an independent way that only in an alternating gradient storage ring the Crystalline Beams may exist. In was further found that Crystalline structures can be created at any density if the beam energy is lower than the transition energy of the storage ring. Above transition energy the negative mass instability prevent the formation of Crystalline Beams[13,14]. Furthermore, the total energy of the system is not a constant of the motion and heat will transfer into the system. If the cooling is not efficient enough the increase of temperature will influence the particle correlation and the Crystalline Beam will reach its melting point.

Numerical calculations was performed by simulating a realistic storage ring[15] and the ground state of Crystalline Beams were achieved. Although applied to a specific case, simulations have a general validity and can be useful for any storage ring.

Electron and laser cooling techniques are applied to cool down the ion beams to reach Crystalline states. In spite of the viscous heating by the shear induced in the bending fields, simulations indicates that even high density Crystalline Beams do exist in a storage ring. The particles undergo oscillations due to the lattice function variations (breathing), but their amplitudes are small for a smooth machine[10].

A serious limitation of transverse beam temperature, which prevent the formation of Crystalline structures, is the main result of two phenomena appearing in the transport of very intense beam of particles: the envelope instability and the intrabeam scattering. Both phenomena have been numerically investigated. The simulations show that the heating of the beam due to envelope instability[10] is even stronger than the heating due to the intrabeam scattering. By an independent analysis of both phenomena the required conditions to avoid envelope instability have been found. The effect of intrabeam scattering cannot be avoided, but only compensated by the cooling and reduced by the smoothness of the lattice.

2. Envelope Instability of a Space Charge Dominated Beam

The transport of intense and cold ion beams through a storage ring has some common features with beams in heavy ion inertial fusion, for which several critical aspects have been revealed since the 1950s. Originally envelope equations are derived by Kapchinskij and Vladimirskij based on the Vlasov equation[16].

The envelope equations are self-consistent since the space charge distribution is exactly of Kapchinskij-Vladimirskij type. Analytical and numerical studies[17] are made by I. Hofmann *et. al.* demonstrating that the K-V distribution in a periodic focusing channel is subject to various instabilities. The main conclusion of their work is that the phase advance of a single particle, σ_0, in the cell should not exceed 90^0 to avoid the second-order instability. Struckmeier and Reiser[18] studied the behavior of mismatched intense charged-particle beams in periodic transport channels and found

found two mismatch oscillations, one in which the amplitude of oscillations in the two transverse directions are "180^0 out of phase" and one in which they are "in phase". Still they confirm that $\sigma_0 < 90^0$ is a safe condition with regard to envelope perturbations for mismatched beams. For the instability of cold and space charge dominated ion beams in a storage ring, Hofmann and Struckmeier[19] demonstrated the existence of such an effect by computer simulation.

Based on electron and laser cooling, longitudinal beam temperature in the *mK* range has been achieved experimentally in a storage ring[20], but still the transverse temperatures are about several thousand Kelvin[21]. Besides experimental efforts to develop fast transverse cooling techniques, the need for an advance in understanding the transport of space charge dominated beams in a storage ring has been intensified as well. The envelope instability can be considered as an effect that prevent to achieve low transverse temperatures.

Being the beams, in our case, always under frictional cooling forces, a quantitative knowledge of the instability growing rates is of particular interest.

To study the envelope instability, the beam transport has been simulated by using the PARMT code[22,23]. Simulations were performed for a general storage ring and the results are of general usefulness and scalable to any storage ring. In order to study the instability a semi-Gaussian particle distribution is considered. For all simulations the beam consists of 10^8 ions of $^9Be^+$ with $\beta=0.07$. The initial emittances are chosen with a space charge phase advance of $\sigma=32^0$, which should cause envelope instability. To study separately only the envelope instability, excluding the contribution of the intrabeam scattering on the emittance growth, the simulations have been made with a self-consistent calculation of space charge forces by means of a Poisson solver.

According to *Ref.* 17, multipolarity of the unstable oscillations depends on the single phase advance per cell σ_0. Simulations were done for three different values of the phase advance σ_0: (i) $\sigma_0 = 60^0$, (ii) $\sigma_0 = 90^0$, (iii) $\sigma_0 = 125^0$. It is confirmed that $\sigma_0 < 90^0$ is a safe condition for the envelope mode[10,24].

A quantitative estimation of the blow-up rate of the envelope instability is of great interest for general understanding. The onset condition of envelope instability depends on the single particle phase advance of a storage ring σ_0 and on the space charge dominated phase advance σ, which is reduced because of the cooling. To simulate quantitatively the grow time of the envelope instability the beam is chosen with an initial emittance of $10^{-6}\pi$ *mrad* and is cooled down to a plasma parameter $\Gamma=1$. The cooling process is simulated by reducing the emittances of a fixed quantity, once every revolution. The blow up rate of the envelope instability is deduced from the required cooling forces to obtain the desired plasma parameter. The results of simulations are shown in *Fig. 1*, for different σ_0. The drastic change of the curve behavior for $\sigma_0 > 90^0$ clearly indicates transverse heating from the envelope modes. For a phase advance $\sigma_0 < 90^0$ the beam can be cooled to reach $\Gamma=1$ with a minimum e-folding cooling time of *5.8 ms*. This cooling time represent a lower limit, in fact longer cooling time were not explored because they requires very long computation time. In addition, the use of longer cooling time can also produce not correct results because of numerical errors, which are affecting the calculations involving small numbers. For $\sigma_0 > 90^0$ the envelope instability produces three order of magnitude faster blow up rate, which is in the *ms* range. The simulations shows that as a coherent effect, the heating due to the envelope instability is even stronger than the heating due to the intrabeam scattering.

Compared with the transverse cooling time of the best experimental results, the blow up time of the envelope instability is very fast in the present storage ring and the transverse beam temperature can be hardly decreased below a certain value. Therefore, the results of simulations suggest that a storage ring aimed to crystallize ion beams has to fulfill the condition that $\sigma_0 < 90^0$.

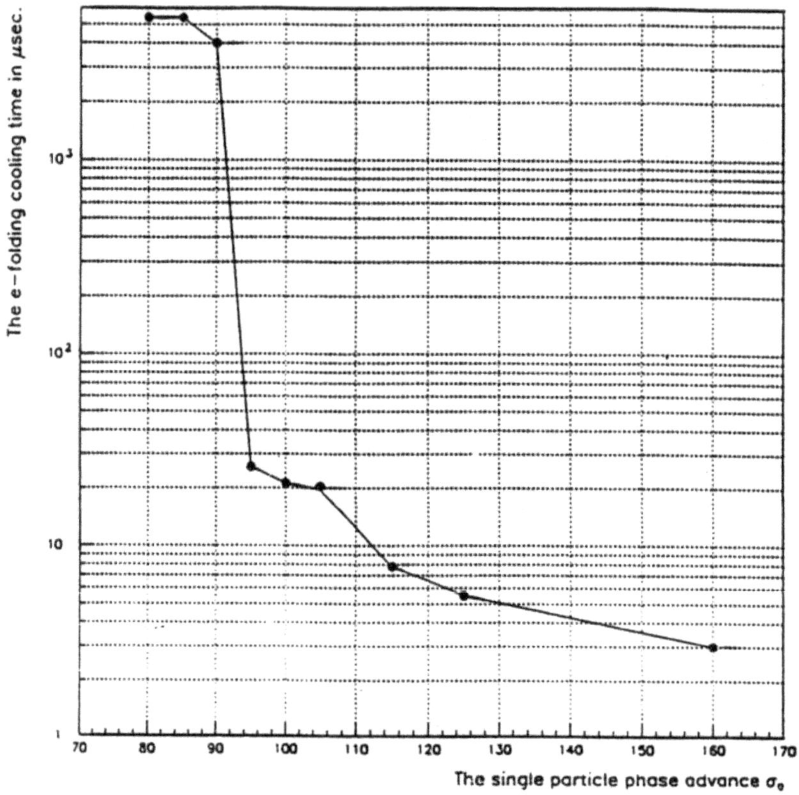

Fig. 1 - The e-folding cooling time required to compensate the blow up rate of the envelope instability, in order to obtain a beam with plasma parameter $\Gamma=1$, in function of the single particle phase advance σ_0.

3. The Storage Ring: Lattice and General Layout

The overall size of the storage ring has been determined by some general considerations such as the largest specific kinetic energy of the ions which have to be stored, the matching with the heavy ion injector and, most important, the requirements for Crystalline Beams. The maximum specific energy has been fixed at *50 MeV/u* for *Gold* ions *(51+)*, corresponding to a magnetic rigidity of *3.8 Tm*. The storage ring develop a total length of *70 m*. The maximum specific energy has been chosen in connection with the use of the storage ring for other purposes, namely nuclear, atomic and molecular physics. The magnetic rigidity of *3.8 Tm* places the CSR Storage Ring in a complementary position with respect to the existing similar facilities.

The shape and symmetry of the ring has been determined by a compromise between an ideal machine for crystallization purposes, where the beam is preserved from external perturbations, and the need of several straight sections where to accomodate injection, extraction, acceleration, cooling and diagnostic devices.

It has been proven[25] that the minimization of the intrabeam scattering blow up is related to the minimization of the first derivatives of the horizontal and vertical β–functions and of the dispersion function η' along the machine. The simulations performed with the computer code MAD[26] show that high periodicity of the lattice and short drift sections are actually preferable. As a matter of fact, the electron cooler device needs a rather long drift space and therefore set up the length of all the drift sections.

In addition, the demand of single particle phase advance $\sigma_0 < 90^0$, in order to avoid envelope instability, imposes already a minimum periodicity to the ring; the eightfold symmetry is the minimum geometry that satisfy this requirement.

The needs for many straight sections, combined with the high periodicity required to obtain Crystalline Beams and to avoid crossing systematic betatron resonances, suggests the choice of an eightfold symmetry (see *Fig. 2*). For the given periodicity of the lattice, the choice of the betatron tunes has been done to stay far from the following low-order systematic resonances:

$$Q_x = i, \qquad Q_y = j$$
$$2Q_x = k, \qquad 2Q_y = l$$
$$Q_x + Q_y = m, \qquad Q_x - Q_y = n$$

with *i, j, k, l, m* and *n* natural integers.
Several sets of values can satisfy these requirements and one can change between different working points and use the most suitable for the actual experiment. Although the storage ring can be used for different applications, here the working point directly connected to the formation of Crystalline Beam structures only, will be discussed.

The magnetic lattice consists of eight 45° bending magnets, *32* quadrupoles and *32* sextupoles. The dipoles are sector magnets with zero entrance and exit angles; their length along the reference orbit is *2 m*. Only one type of quadrupole is used; their length is *0.25 m* and are assembled in four families for the machine operating in the normal mode. To operate the machine for Crystalline Beams only two families of quadrupole are required. The bending magnets have a vertical aperture of *80 mm*, corresponding to a vertical acceptance of *50 π mm mrad*; while the quadrupoles have a bore radius of *65 mm*. Two families of sextupoles have been considered to correct the horizontal and vertical chromaticities. The sextupoles have the same bore radius of quadrupoles and their length is *0.20 m*.

The drift space between the quadrupole-sextupole and quadrupoles is *0.1 m* and *0.4 m*, respectively; while the length of the straight sections is *3.4 m*. These latter will house the injection and extraction systems, electron and laser cooling devices, RF cavity, diagnostic devices and correcting magnets.

The transition energy $\gamma_{tr} = 1.76$ is a large value which always keeps the machine operating well below it.

A summary of the lattice and cell parameters is given in *Tables I* and *II*.

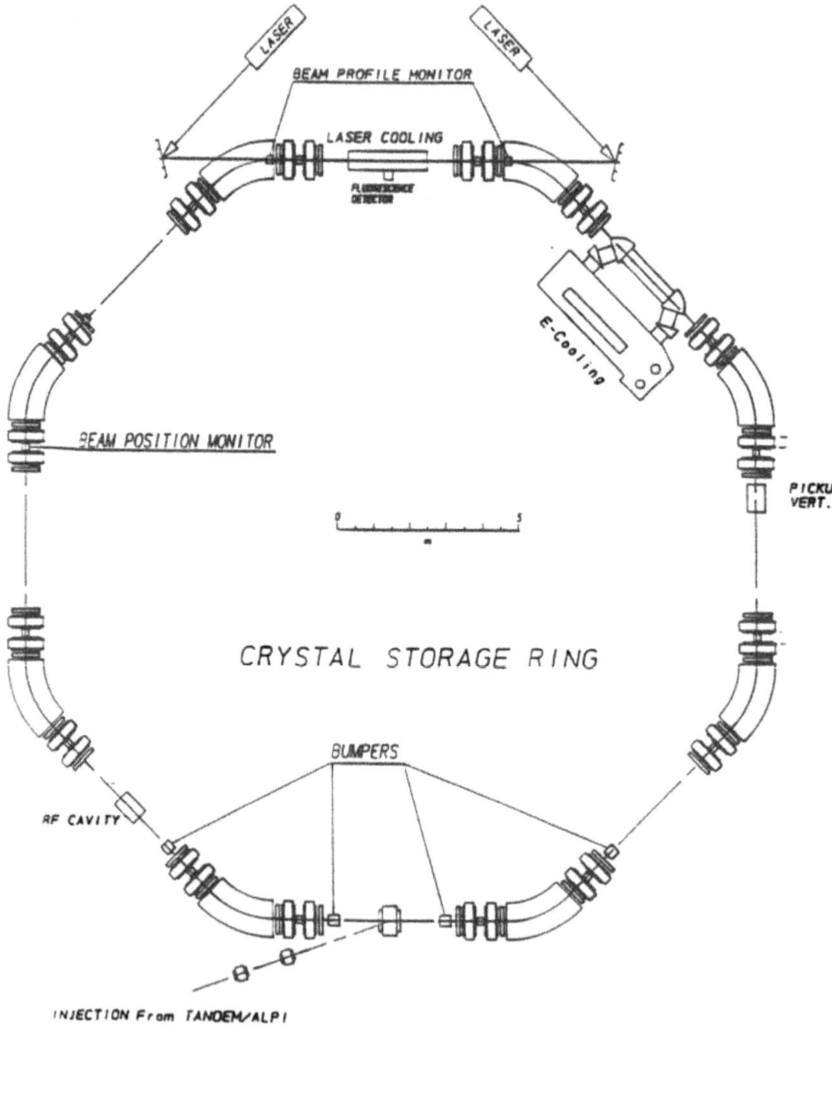

Fig. 2 - Layout of the CRYSTAL Storage Ring.

Table I - Lattice Parameters of the CRYSTAL Storage Ring.

Description	Crystalline Mode
Ring Length [m]	68.8
Lattice Periodicity	8
Magnetic Rigidity [Tm]	3.8
Bending Radius [m]	2.55
Phase Advance per Period:	
Horizontal	83^0
Vertical	53^0
Transition γ	1.76
Betatron Tunes:	
Q_x	1.85
Q_y	1.185
Natural Chromaticity:	
ξ_x	-0.72
ξ_y	0.21
β_x, max [m]	6.74
β_y, max [m]	10.4
Dispersion max [m]	3.6
Momentum Compaction	0.32

The positioning of the magnets requires special care. The following alignment *rms* errors are considered to be within the present state of the art: *0.1 mm* for both dipole and quadrupole lateral displacement and *0.2 mrad* for axial rotation. In addition, an integrated dipole field error of the order of 10^{-4} *rms* is required. This cause an uncorrected orbit distorsion with an expected value less than *0.8 mm*.

To keep the maximum acceptance of the ring, the closed orbit distorsion has to be minimized. The compensation can be done by correcting dipoles properly located. In addition to the correcting dipoles, extra horizontal compensation can be obtained from backleg windings on the bending magnets.

3.1 The Working Point for Crystalline Beams

The accumulation of an ordered structure is the result of an equilibrium between the interparticle Coulomb force and the external focusing system. The effect of different lattices on the rate of intrabeam scattering (IBS) of a previously cooled ion beam was investigated with the computer code INTRABS[27]. In comparison, the features of the beam itself (type of ion, its charge state, velocity, initial values of ε_x, ε_y and $\Delta p/p$ - typical of an already cooled beam) were left unchanged, and only the lattice was varied.

Since, by definition, at equilibrium the compression of the six-dimensional phase space provided by the cooler and averaged on one turn cancels the phase space growth due to intrabeam scattering, it results that a lower IBS growth rate (expressed in our case as the increase rate of the transverse emittance and the longitudinal momentum width) poses less stringent requirements on the

Table II - Cell Parameters of the CRYSTAL Storage Ring.

Description	Crystalline Mode
PERIOD Length [m]	8.6
Dipoles:	
Dipole Length [m]	2.0
Bending Angle	45^0
Bending Radius [m]	2.55
Vertical Gap [cm]	8
Drifts:	
Drift Length [m]	3.4
Quadrupoles:	
QF Length [m]	0.25
Strength [m^{-2}]	0.4
Gradient [T/m]	0.48
QD Length [m]	0.25
Strength [m^{-2}]	-0.6
Gradient [T/m]	-0.72
Bore Radius [cm]	6.5

cooling system performance and is therefore favourable.

The working point for Crystalline Beams is characterized by smooth β–functions in order to decrease as much as possible the intrabeam scattering. Furthermore, crystallization requires low phase advance per super-period[17] and hence low tunes.

When the beam crystallize the tune shrinks towards zero (not necessarily for a string configuration) due to very strong space charge defocusing. It is favourable to be below the strong integer resonances $Q_x = Q_y = 2$ since these resonances are difficult to pass.

The working point for Crystalline Beams is determined optimizing the storage ring from the point of view of intrabeam scattering, as discussed above. Only such a working point allows to decrease the beam temperature sufficiently to obtain Crystalline structures.

The lattice functions $\beta_x(s)$, $\beta_y(s)$ and $D_x(s)$ for such a configuration are plotted in *Fig. 3*. The tune for the Crystalline working point is $Q_x = 1.85$, $Q_y = 1.185$. There are more resonances in the vicinity of this working point. However, almost all systematic resonances are driven by skew fields (n_y is odd) and thus less dangerous.

3.2 The Injection and Extraction Systems

Heavy ions will be injected into the CRYSTAL Storage Ring using the existing facility at LNL: the XTU Tandem with ALPI superconducting post-accelerator. The most efficient way of injecting into a storage ring is the pulsed mode, where the XTU Tandem-ALPI provides short beam pulses of heavy ions with peak current, giving good quality beams with small transverse emittance and momentum spread. With this scheme, it is possible to obtain beams with a wide range of mass numbers and sufficiently large charge states produced by stripping targets at proper location[28].

If required, it will also be possible to inject into the CSR directly from the Tandem, bypassing ALPI.

The ions usefull for laser cooling are $^7Li^+$, $^9Be^+$ and $^{24}Mg^+$, for which the Tandem has a very poor production efficiency. To produce such kinds of ions, an indipendent low energy source, followed by a small electrostatic accelerator, can easily be installed closed to the injection, in the CSR area.

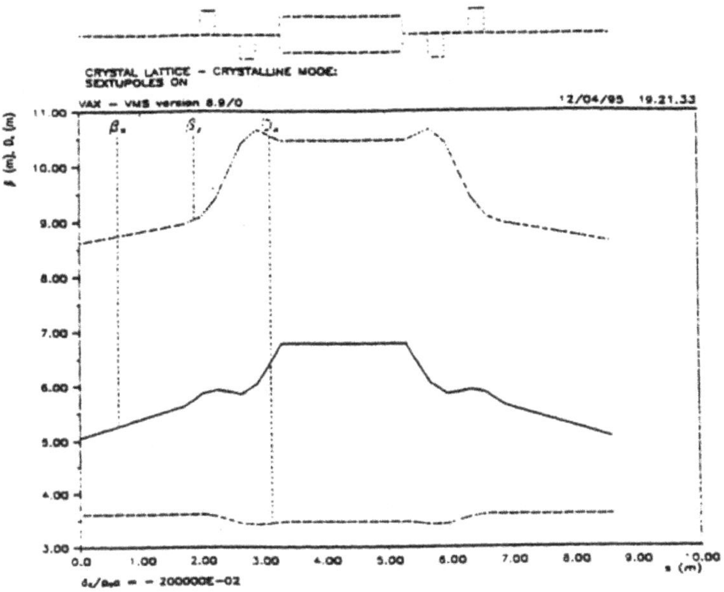

Fig. 3- The Lattice Functions for the Crystalline Mode.

The most suitable source is a CHORDIS type and is chosen to provide a low emittance beam ($< 1\pi$ *mm mrad*) and small initial energy spread ($< 10^{-4}$), with typical current intensity of a few tens of *mA*. The beam is accelerated electrostatically by a precision high voltage accelerator. The accelerator can operate at energies from *50* to *400 KeV*. The ions are mass separated in a *45* degree analyzing magnet and then pass through the injection beam line.

An estimation of the injection parameters for different ion species are summarized in *Tab. III*.

Beam injection of heavy ions take place in one of the straight sections as shown in *Fig. 2*.

Table III - Parameters from Injection.

Ion Species	Li	Be	Mg	S	I	Au
Q	1	1	1	16	40	51
A	7	9	24	32	127	197
β=v/c	0.0117	0.0091	0.0034	0.185	0.11	0.099
Bρ [Tm]	0.25	0.25	0.25	1.17	1.135	1.196
Kin. Energy [MeV/u]	0.4	0.35	0.13	16.4	6.11	4.59
Number of Bunches	1	1	1	6	9	11
N_{inj} 10^7/bunch	200	200	200	2.1	0.5	0.375

For the demonstration of Crystalline Beams, the CHORDIS source can produce intense enough beams of $^7Li^+$, $^9Be^+$ and $^{24}Mg^+$ that a single turn injection probably will be sufficient for the required intensities. The intensities of ion beams attainable out of a Tandem-ALPI complex might be not enough and it is necessary to adopt a multiturn injection scheme.

The same kicker magnet, which deflect the beam in at the injection, can also deflect the beam out at the moment of extraction.

3.3 The Magnet System

The Storage Ring lattice consists of *8* periods with *16* horizontally focusing quadrupoles (*QF*) and *16* horizontally defocusing quadrupoles (*QD*) and *8* bending dipoles. In addition, *32* sextupoles are also inserted to correct the chromaticity.

The Dipoles are *C*-type to get a vacuum chamber with offtakes for pumps in addition to the extentions for alignement.

Given the nominal field of *1.5 T* and a gap height of *80 mm*, special attention will be paid to the dimensioning of the magnet circuit in order to avoid iron saturation that could cause some field distortion. The iron has to be characterized by good saturation properties. To obtain a good uniformity ($\Delta B/B_0 < 10^{-4}$) at different field values ($B_{min}= 0.1\ T$, $B_{max}= 1.5\ T$) in the vacuum pipe region it has been necessary to increase the pole width.

The coils will be made up with *10.5 x 10.5* mm^2 water cooled hollow conductors (central cooling hole *7 mm* diameter). Every coil is arranged in six pancakes, with *24 turns* and four layers each. All the pancakes are electrically in serie and the cooling circuits are parallel-connected; the selected maximum current density is *4.86* A/mm^2. The total dissipated power in the dipole system is *410.4 kW*. The main parameters of the dipoles are listed in *Table IV*.

The dipoles have been designed for DC operation mode. If a AC operation mode is required they must be modified by using laminated core (lamination thickness: *0.35 mm*) to minimize iron current losses. In order to minimize the total stored energy and therefore the peak power it could be necessary also to reduce the pole width to the detriment of the homogeneous field area width.

Table IV - Basic Dipole Parameters.

Parameter	Symbol	Unit	Values
Type			Curved C
Number of Magnets			8
Max. nominal Field	B_{max}	T	1.5
Bending Angle	q	Degrees	45
Bending Radius	r	m	2.55
Geometrical Length	L_{dip}	m	2.0
Gap Height	g	mm	80
Current	I	A	340
Voltage	V	V	151
Power	P	kW	51.3
Total Weight		ton	21

The quadrupoles in the Storage Ring have a length *0.25 m* and a maximum design gradient of *12.3 T/m*, consistent with the magnetic rigidity.

The maximum field gradient of *12.3 T/m* gives a maximum pole tip field of *0.8 T*. Because of the high field it is necessary to avoid iron saturation by shaping properly the return joke. The quadrupoles , being four-fold symmetric, will be cut in four identical pieces.

The coils will be made up with *9.5 x 9.5* mm^2 water cooled hollow conductors (central cooling hole *4.5 mm* diameter). Every coil is arranged in *60* turns and twelve layers. All the coils are electrically in series and the cooling circuits are parallel-connected; the selected maximum current density is *4.72* A/mm^2. The total dissipated power in the quadrupole system is *270 kW*. *Table V* lists the fundamental parameters.

The quadrupoles have been designed for DC operation mode. If a AC operation mode is required they must be modified by using laminated core (lamination thickness: *0.35 mm*) to minimize iron current losses and slightly changing the yoke shape to minimize the total stored energy.

The proposed solution for the Q-pole power supplies has to provide two power supply busses, for each one of the QF and QD quadrupoles (two families). The quadrupoles of each family are powered in series. All the dipoles are independently powered.

The DC power supplies are conventional and their main parameters are listed in *Table VI*; a digital programmed mode of operation is necessary for a possible slowly pulsed operation mode, in this case suitable power supplies must be provided.

In order to mantain a stable machine operation, mainly at the Crystalline mode, the ripple of the power supplies has to be better than 10^{-5}.

Table V - Basic Quadrupole Parameters.

Parameter	Symbol	Unit	Values
Number of Magnets			32
Max. Field Gradient	G_{max}	T/m	12.3
Max Pole Tip Field	G_p	T	0.8
Geometrical Length	L_{quad}	mm	250
Aperture Diameter	g	mm	130
Number of Turns/pole	Np		60
Current	I	A	350
Voltage	V	V	24
Power	P	kW	8.4
Total Weight		kg	750

Table VI - Total Power Supplies Requirements

Parameters	Unit	Values
N. of Dipole Power Supplies		8
Total Output Power	kW	411
Total Input Power	kVA	514
N. of Quads Power Supplies		2
Total Output Power	kW	270
Total Input Power	kVA	338

3.4 Fringe Fields, Misalignments and Field Errors

The design of a realistic storage ring demands an accurate investigation about the effects produced by fringe fields, misalignments and field errors in bending magnets and quadrupoles, that act as perturbations on the beam. These perturbations lead to a distorsion of the closed orbit and produce resonance lines, which should be crossed by the beam under cooling process.

The formation of Crystalline structures in ion beams under perturbated conditions have been studied by using the computer code PARMT. In the simulations a real cooling factor (at least for the string case) has been used and any kind of misalignments, field errors, linear fringe fields and periodicity perturbation have been taken into account.

For simulations a beam of $^9Be^+$ with $\beta = 0.07$ and initial emittance of $1\ \pi\ mm\ mrad$ was considered. These simulations have a general validity, in fact similar results are obtained by using different kind of ions. The cooling effect was taken into account by reducing the beam divergences by a given factor, after each revolution. A reduction factor of *1.003*, corresponding to an effective cooling time of *1 ms*, is used for simple Crystalline structures (string and zig-zag). Since the computation time increases quadratically with the number of ions, simulations of *3D* structures demand huge computing time (or unphysical cooling factor) and are here not considered. The results obtained after 10^4 revolutions are shown.

The computer code PARMT take into account fringe fields of both quadrupoles and dipoles. The field gradient $g(s)$ on the axis of a quadrupole field is calculated with the following ansatz:

$$g(s) = g_0 sin^2\left(\frac{1}{2}\pi s / L_R\right); 0 \le s \le L_R$$

where L_R is the length of the fringe fields. The magnetic field can be written as a power series containing only derivatives of the function $g(s)$ as coefficients[29]. The calculation of the particle transformations through the fringe fields is carried out by numerical integration of a system of differential equations. In the simulations $L_R = 5\ cm$ has been used, which is realistic representation of well-designed quadrupoles.

The fringe field of dipoles is calculated more simply through a thin matrix formalism that takes into account the edge effects in bending magnets by inserting the fringe field term y:

$$\psi = K_1\left(\frac{g_a}{\rho_0}\right)\left(\frac{1+\sin^2\alpha}{\cos\alpha}\right)\left[1 - K_1K_2\left(\frac{g_a}{\rho_0}\right)\tan\alpha\right]$$

where a is the angle of rotation of pole face, g_a is the gap of magnet, r_0 the bending radius of central trajectory, K_1 and K_2 as usual are 0.35 and 4.0 respectively.

The fringe fields in the quadrupoles produce a small orbit displacement; while the fringe fields in the dipoles contribute very slightly.

Misalignment

In general, misalignment effects can be taken into account by applying an angular deflection on the beam particles, that is a change in particle divergence. The particle divergence changes, due to any type of shifts and angular rotations of the optical elements, can be estimated separately and then added. So in calculations, after each bending magnet and quadrupole, changes of particle divergences have been introduced randomly in both horizontal and vertical planes along the ring circumference. As maximum shifts are chosen values of *0.1 mm* and as maximum rotations *0.2 mrad*. These kinds of misalignments produce two different maximum divergence changes for the two family of quadrupoles used in CSR. In fact a maximum divergence change of *0.045 mrad* in both transverse planes has been evaluated for the focusing quadrupoles and *0.073 mrad* for the defocusing one. The maximum divergence changes added after the bending magnet due to misalignments have been evaluated as *0.08 mrad* in the horizontal plane and as *0.25 mrad* in vertical plane. The divergence values are chosen randomly between the given maximum values and then added along the storage ring.

The misalignments produce a closed orbit distortion that does not prevent the formation of Crystalline Beams.

To determine the influence on the closed orbit distorsion due to quadrupole and dipole misalignments, thier contributions have been separately studied. The misalignments of the dipoles are more effective than those on quadrupoles; the maximum allowable tolerances for the formation of Crystalline Beams have been found to be of *2 mm* shifts and *4 mrad* rotations for quadrupoles and of *1.5 mm* shifts and *2.5 mrad* rotations for dipoles.

Field Errors

The effects of field errors on the Crystalline Beams have also been included in the code. In all the quadrupoles and dipoles the field errors have been taken into account by adding a relative error to the values of both quadrupole strengths and curvature radii. In the simulations a set of field errors, corresponding to the *0.01 %* of the designed values, are randomly chosen. The resulting closed orbit distorsion is negligible in comparison with the distorsions induced by misalignments. By increasing the relative field errors up to *0.2 %* on both dipole and quadrupole elements still Crystalline Beams are obtained, but with an increases of the closed orbit distorsion.

Periodicity Perturbations

To have the possibility to correct the chromaticity a series of sextupoles are placed along the ring. Furthermore, a solenoid is inserted in a drift section to apply the electron cooling. These elements break the symmetry in the ring and then are considered as periodicity perturbations. In this case sextupoles are placed symmetrically along the ring, then, it is only the solenoid used for the electron cooling that breaks the symmetry. The sextupoles and the solenoid have to be taken into account in the simulations because they can become a source of heating for beam particles. The simulation results have shown that Crystalline Beams are still obtained but with an unlike closed orbit distortion and this is mainly due to the sextupoles effect. The solenoid insertion produces almost no effect.

In summary, all the computer simulations have shown that in a realistic lattice, which include linear fringe fields, misalignments, field errors and periodicity perturbations, Crystalline structures are still obtained. In all computer simulations, for the realistic lattice the following tolerances are assumed: misalignments *0.1 mm* shifts and *0.2 mrad* rotations, field errors *0.01 %*.

3.5 Beam Instabilities

A variety of instabilities (transverse and longitudinal) can occur in a storage ring. The existing literature[30] discuss quite extensively all of them.

The most important instabilities to be considered are the transverse ones, since they lead to immediate particle losses. When the beam emittance is reduced, the space-charge shift the working point to approach a resonance, which leads to a fast transverse beam growth and loss.

Coulomb scattering between particles in the beam can couple one phase plane to another and transform the common longitudinal motion of the particles into transverse energy spread.

The Landau damping of the beam is suppressed by the cooling and it cannot always compensate the blow-up of the ion beam.

The cooling of the momentum spread to very low values can also produce longitudinal instabilities. They are mainly related to the resistivity of the walls surrounding the beam.

Reactions between the ion beam, the residual gas and the electron beam also cause important beam losses in a storage ring.

Non-linear, single-particle phenomena and collective effects have been studied in connection with their implications on the design of the storage ring lattice.

All these phenomena were quite extensively studied in a previous report[28]. Here, instability phenomena connected to the ring impedance, that more strongly influence the formation of Crystalline Beams, are discussed.

The performances of a storage ring, especially in the case of very cold and intense beams, is limited by coherent instabilities. Ions generate an electromagnetic field that acts back on the beam after interaction with its surroundings. If a perturbation exists, that could be enhanced by such a field causing an unstable process.

The parameter that takes into account beam-surrounding interaction is the coupling impedance. It is therefore very important to estimate the transverse and longitudinal coupling impedances in order to evaluate the stability limits.

Measurements carried out show that impedances can be modeled as follows[31]:

A) - the resistive wall component;

B) - the parasitic low $Q \approx 1$ resonators with broad band component such as: 1) bellows, 2) vacuum pump ports, 3) kicker and septum, 4) vacuum chamber cross-section variation, 5) diagnostic devices like beam profile monitors and, 6) electron beam impedance (electron cooler);

C) - the parasitic high $Q >> 1$ resonators with narrow band component (higher longitudinal mode in cavities);

D) - the space charge impedance.

In addition, the impedance contribution of bellows and strip line beam position monitors have to be taken into account. Electron beam impedance also has to be taken into account.

The estimations show that longitudinal and transverse instabilities can be damped by the cooling process. In fact, the calculated growth time of instabilities are never less than the cooling time available from electron or laser cooling.

For the transverse case, the coherent motion for lower mode number can be damped by a feed-back system[31].

On the contrary, the growth times of instabilities caused by electron beam impedance are several orders of magnitude greater than growth time due to the resistive wall or space charge impedance. Especially for the longitudinal instability, very short cooling times are required for high charge state ions.

3.6 The Cooling System

Very efficient phase-space cooling is required to attain crystalline ion beam i.e. beams with very small temperature (in the rest frame system), compared with the average Coulomb interaction energy. Two techniques of cooling are applied to the CRYSTAL Storage Ring: electron cooling and laser cooling (see *Fig. 2).*

Electron and/or laser cooling are needed to attain Crystalline Beams since they have features which complement each other.

Electron cooling provides a phase-space contraction for an ion beam, through the overlapping of an ultracold electron beam, continuously renewed. Coulombian scattering forces between ions and electrons equalize the temperatures of the beams and allow to attain a relatively low momentum spread $\Delta p_p/p = 10^{-5} \div 10^{-6}$. Notice that assuming an electron beam longitudinal temperature of $e^2 n_e^{1/3}$. the ion plasma parameter after cooling will be $\Gamma_l = Q^2 (n_i/n_e)^{1/3}$. If the electron and ion densities are of the same order of magnitude, then the crytical value for Cristalline Beams of $\Gamma_l = 170$ is reached when $Q^2 > 13$.

Laser cooling acts by virtue of momentum exchange between an ion beam and laser light, whose frequency is chosen in correspondence with high values of the electric dipole for the stored ions. Photon excitation and subsequent re-emission causes, in principle, longitudinal momentum $\Delta p_p/p$ even three orders of magnitudine lower than in the electron cooling case. This latter method has a working regime that recommends a precooled ion beam; the former can therefore be considered as complementary to the laser cooling. Furthermore, in the present storage rings, laser cooling acts only along the longitudinal axis, so that transverse cooling is bound up to electron cooling efficiency. To increase the cooling efficiency in the transverse planes a new method for transverse laser cooling is under developement. In addition, laser cooling of bunched beams is a new and very promising techniques for longitudinal phase space compression. With this method also the indirect cooling of transverse dimensions can effectively be increased. It is also clear that the laser cooling makes a decisive selection on the kind of ion to be employed. For these ion beams, electron cooling reduces emittance and momentum spread without decreasing the ion current, allowing the application of laser cooling on a dense ion beam with still enhanced efficiency.

Electron Cooling

The electron cooling is a well known and tested technique for cooling ions in storage rings[32]. The efficiency of the method is strictly related to the electron beam parameters. A typical electron cooling device suitable for CSR is discussed and the design parameters are reported in *Table VII.*

In general, a cooling device may be divided in three sections:

1) electron gun, where the electrons are produced and accelerated;
2) drift section, where the electron beam are merged with ion beams at the same relative velocity, after being bent by a toroid;
3) collector, which has to recoils electrons without disturbing the electron and ion optics.

Table VII - Electron Cooler Design Parameters.

Parameters	Symbol	Value
Electron Energy Range [keV]	E_e	1-20
Electron Current [A]	I_e	0.1 -1
Cathode diameter [mm]	f_c	40
Gun Perveance [mP]	P_g	2
Transverse Electron Temperature [eV]	T_e	0.1
Length of Cooling Section [m]	l	1
Magnetic Field [T]	B_0	0.2
Collector Perveance [mP]	P_C	5
Relative Current Loss	I_{loss}/I	$<10^{-5}$
Dissipated Collector Power [W]	W_C	100
Average Vacuum Pressure [Torr]	P_{gas}	10^{-10}

Electrons are emitted by resistively heating a flat disk thermionic cathode. The cathode is surrounded by a Pierce electrode[33] and followed by *7* ring shaped electrodes. The immersed-flow configuration (magnetized electron) compensates the space-charge blow-up and transverse energy heating due to optic errors. The gun has a perveance of *2.3* μP. The extraction by Pierce optics compensates at the first stage the space-charge effect, the first electrode provides a fixed current extraction. The second electrode supplies an electric field which compensates the defocusing effects of the anode hole. After the extraction the electron may experience the acceleration fields of the following electrodes in the Pierce optics, adiabatic acceleration[34] or resonant acceleration[35].

The different accelerating methods improve the parameters related to the beam temperature, therefore they influence the cooling time of the ion beams.

After extraction and acceleration the electron beam experiences the bending magnetic field provided by a toroids and dipoles. At the junction between solenoid and toroid the magnetic field is reduced and needs correction in order to avoid transverse heating in the boundary of the beam. Then the electrons are merged with the ion beam, which has a lower initial diameter; big care needs the alignment between the beams. The cooling time for different kind of ions, at the injected energies, calculated by using the standard formula[36] are shown in *Table VIII.*

Table VIII - Cooling Time for Different Ions.

	Li	Be	Mg	I	Au
Z	1	1	1	40	51
A	7	9	24	127	297
β	0.07	0.07	0.07	0.11	0.1
γ	1.0025	1.0025	1.0025	1.006	1.005
Energy [keV]	1.28	1.28	1.28	3.1	2.6
τ [ms]	29	42	112	2	1.2

In the low temperature range ($T_i < T_e$), the contribution to the cooling time is mostly due to the electron temperature. Then the accelerating methods may improve the feature of electron cooling. By adiabatic acceleration[34] it is possible to reduce the longitudinal temperature and suppress the transverse-longitudinal relaxation by a strong magnetic field. The adiabatic expansion provides a reduction of the transverse beam temperature of the electron beam[37]. Both methods contributes to enhance the cooler performances.

The cooling section affects the ion beam optics. The solenoidal confining magnetic field induces a rotation of the beam around its axis. This effect provides a coupling between the two transverse plane, as result there is lowering of the betatron oscillation limit and improving the cooling force for small beam emittance. Instead, toroids give a component of magnetic field perpendicular to the ion beam, requiring correction dipoles to compensate the diverging effects.

Furthermore, the electron beams acts on the ion beam like an electrostatic lens, i.e. β-function variation in the cooling section produces a tune shift. For a constant perveance gun the tune is independent from the electron energy.

After the interaction the electrons pass through a solenoid, and are decelerated and captured by a collector. The collector potential is close to the cathode one. The electrons are decelerated in order to dissipate lower power and to prevent the secondary electron emission. But lower the collector potential higher the probability to have the reflection backwards of primary or secondary electrons. Primary and secondary electrons, accelerated back, excite the transverse energy of the beam in their travel forwards and backwards in the cooling section.

The parameters required for the collector are: perveance of $5\ \mu P$, current loss lower than 10^{-5} and the power dissipated $100\ W$. These features are accomplished by applying a magnetic mirror that spread on the collector surface the electron beam and reflect the electrons which may travels in the opposite direction[38].

Laser Cooling

Standard laser cooling of ion beams is a very powerful method to reach very low longitudinal temperatures[39]. Its basic principles and techniques are well known. It is the result of velocity-selective transfer of photon momentum from a laser beam to a moving ion. The laser cooling technique is effective if the selected ion has a strong dipole transition from a stable or metastable state to an excited one. Obviously the required wavelength should be accessible to the laser technique. The ion species untill now used in storage rings are: $^{7}Li^{+}$, $^{9}Be^{+}$ and $^{24}Mg^{+}$.

One possible technique for longitudinal laser cooling consists in superimposing two counterpropagating laser beams on the ion beam in one of the straight sections of the ring. The frequencies of the two lasers are set, after the different Doppler shifts have been taken into account, to the opposite sides of the Doppler linewidth of the ion. The two laser frequencies are then swept across the Doppler width. In this way the velocity distribution of the ions is compressed to the centre of the Doppler line. The suitable set-up concerning the longitudinal laser cooling for CSR has been extensively discussed in *Ref.* 28. In the following, more attention is dedicated to a new proposal for the transverse laser cooling.

Transverse Cooling

In order to overcome the problem of the short interaction time between ions and light when the beams are mutually orthogonal, is proposed a scheme where almost all the ions stay on resonance with the laser during the full time they cross the laser beam, i.e. they are on resonance regardless of their velocities. This can be achieved considering a suitable broad-band laser whose spectrum is so large as to cover about one half of the Doppler profile of the ions. In this way no laser tuning is needed and the ions experience the radiation pressure each time they cross the laser beam, independently of the perturbations they have undergone in the storage ring. A second important point is that the optical pumping problems due to the ground state level structure are avoided when using a broad-band laser.

Let us assume that the cooling laser is directed along the transverse x-direction; all the ions whose velocity component v_x is negative are on resonance with the red detuned laser and, therefore, are under cooling. If the laser beam is reflected backward through a mirror, then also the ions with positive velocity component will be on resonance. The same argument holds true for cooling of the other transverse degree of freedom provided that another laser beam is used.

As the interaction time is very short (i.e. of the order of few ns) the ions undergo a pulsed excitation which impose the calculation of the probability p of excitation to the upper level at each turn of the ion about the storage ring. In a first approximation, p is a function of the interaction time $t_{int}=L/v_L$ (where L is the length of the ion beam-laser overlapping region and v_L the ion longitudinal velocity) and of the Rabi frequency Ω_R:

$$p = \sin^2\left(\frac{\Omega_R t_{int}}{2}\right)$$

By introducing the excited level lifetime τ and the saturation parameter $S=I/I_S$, where I is the laser intensity and $I_S = \pi hc/\lambda^3\tau$ is the saturation intensity (λ being the laser wavelength, c the speed of light and h the Planck constant), the Rabi frequency becomes

$$\Omega_R = \frac{1}{\tau}\sqrt{\frac{S}{2}}$$

Let v_p denote the initial transverse velocity of ions. With the above assumptions the number of turns, which are necessary to set an ion with zero transverse velocity, is given by the number of photons the ions has to absorb divided by the number of photons absorbed at each crossing with the laser beam. This holds:

$$n = \frac{M v_p \lambda}{hp}$$

where M is the ion mass. The cooling time T_{cool} is n times the revolution period, $T = L_S/v_L$ (L_S being the ring length). T_{cool} depends essentially on two correlated parameters, the interaction time t_{int} and the saturation parameter S. Then T_{cool} holds:

$$T_{cool} = \frac{v_p}{v_L}\frac{M L_S \lambda}{h}\left\{\frac{1}{\sin\sqrt{\dfrac{L W_0 \lambda^3}{8\pi q D h c \tau v_L^2}}}\right\}^2$$

where D is the transverse size of the overlapping region, W_0 the laser power and q the number of longitudinal modes of the laser.

In order to demonstrate the feasibility of the proposed method we discuss the case of $^7Li^+$ and $^{24}Mg^+$ ions confined in the ASTRID Storage Ring at Aarhus[40], the main parameters of which are reported in *Table IX*. The chosen values of β will be obtainable also at CSR. In *Table IX* the cooling times for a fixed values of W_0 are given, assuming $L=5$ *cm* and $D=1$ *mm*. The laser powers are *1.5 W* at *548 nm* for cooling of $^7Li^+$ and *100 mW* at *280 nm* for $^{24}Mg^+$. The first one is obtainable from commercially available dye-lasers. A comment is necessary as concerns the UV laser. A power of about *100 mW* at *280 nm* in continuous operation has already been obtained by frequency doubling a dye-laser in an external cavity[41]. No reference instead exists in literature about the possibility of doubling a multi-mode laser as the one we suggest to use. A careful theoretical and experimental investigation is therefore requested about this point.

The proposed technique can be considered as complementary to the indirect cooling, to be used both together with indirect cooling itself for high density beams and alone for low density beams. As concerns the first case, the technique can be employed to complete the cooling process once indirect cooling has brought the ion beam to such low temperatures where intrabeam scattering is more effective. On the other hand, however, the high densities necessary for indirect cooling process are no more requested for this direct cooling. The diffusion process due to intrabeam scattering, which is the main source of heating for high density beams, can be overcome by the use of very low density beams. Working with a mumber of particles of the order of 10^5 - 10^6 the heating rate due to intrabeam scattering is negligible and the main sources of heating are the collisions with the residual gas and the statistical nature of the spontaneous force. From *Table IX* it emerges that the technique allows to subtract from the system about *200 K* in one second: this value is sensitively larger than the expected heating rates due to the two processes mentioned above.

The minimum achievable temperature is bound to the frequency detuning between resonance frequency (center of Doppler profile) and that of the sharp edge $\Delta\nu_0$:

$$T_{\min} = \frac{1}{k_B}\frac{mc^2}{8\ln 2}\left(\frac{\Delta\nu_0}{\nu_0}\right)^2$$

The ultimate transverse temperature is also affected by the precision of the *90°* angle between laser and ion beams. If ϕ is the (small) deviation of this angle from *90°*, the *1st* order Doppler shift is:

$$\nu_l' = \nu_l\,(1+\beta\phi)$$

Table IX - Main Parameters for Transverse Ion Cooling.

Parameter	$^7Li^+$	$^{24}Mg^+$
lower state	2s $^3S_1$ (F=5/2)	3s $^2S_{1/2}$
upper state	2p $^3P_2$ (F=7/2)	3p $^2P_{1/2}$
lifetime of lower state (s)	50	stable
lifetime of upper state τ (ns)	43	3.5
transition wavelength λ (nm)	548	280
natural linewidth $\Delta\nu_H$ (MHz)	3.7	45
total laser linewidth $\Delta\nu_L$ (GHz)	1.9	2.1
laser cavity length L_C (m)	6.8**	3.7
number of longitudinal modes q	270**	47
laser intensity per mode* I (mW/cm^2)	11.1	4.2
total laser power W_0 (W)	1.5	0.1
saturation intensity I_S (mW/cm^2)	8.8	810
saturation parameter $S=I/I_S$	1.26	5.2×10^{-3}
interaction time t_{int} (ns)	30	56
excitation probability p	0.075	0.16
transverse velocity v_P (m/s)	1.1×10^3	5.7×10^2
Doppler cooling limit (mK)	0.089	1.08

* The laser intensity is calculated by assuming that the laser beam is focused to a rectangular spot, with L=5 cm and D=1 mm.

** The number of longitudinal modes and the laser cavity length is calculated by considering also the power broadening and the short interaction time.

Taking this effect into account, the detuning from resonance of the sharp edge side of the laser profile becomes:

$$\Delta\nu = \Delta\nu_O + \nu_l \beta \phi$$

This means that, due to this small deviation, the frequency of the laser can be moved either towards the blue region, giving origin to heating process, or more towards red, corresponding to a smaller possible reduction of the Doppler profile of the ion beam.

For the practical situation of ASTRID, the angle ϕ must be smaller than *0.36 mrad* in order to have $\Delta\nu - \Delta\nu_O < 1\ GHz$. A realistic value for alignment precision is *1 mrad*. The effect can be compensated in experiments by properly varying the laser frequency in order to drive back the displacement of the laser frequency to its reference value $\Delta\nu_O$.

Another source of error is the unavoidable divergence of a gaussian laser beam $\theta = \frac{\lambda}{\pi\omega_L}$. This effect has the following consequences in the cylindrical focalization system (*5 cm x 2 mm*). In x - y plane, where the waist is smaller the divergence is about *1 mrad*, taking into account the initial transverse velocity of the ions, a starting Doppler shift of less than *5 MHz* has to be expected, this value is going to decrease as the cooling proceeds. In x - z plane the divergence of the beam is much less (< *0.02 mrad*), due to its larger waist, accounting for the longitudinal velocity of the ions the expected Doppler shift is less than *100 MHz*. This can cause a small heating in the longitudinal direction, where in any case is counteracted by the strong laser cooling force.

Laser Characteristics

We shall now discuss how it is possible to produce a laser whose frequency spectrum suites the requirements for transverse laser cooling as indicated above. The "white-light" atom cooling has been proposed by Moi[42] and successfully achieved by many groups[43,44,45,46], which adopted different solutions for the practical realization of the broad-band laser.

Here is proposed to realize the "white" cooling by means of "long-cavity" multi-mode dye laser with a total linewidth $\Delta\nu_\Lambda$ and a frequency separation between the longitudinal modes $\Delta\nu_M = c/2L_c$, L_c being the laser cavity length. If the laser is resonant with a two level atom or ion, whose homogeneous absorption linewidth is $\Delta\nu_H$, then the "white-light" condition is reached when $\Delta\nu_M < \Delta\nu_H$ and $\Delta\nu_L > \Delta\nu_\Delta$, with $\Delta\nu_D$ = Doppler linewidth. All the ions are instantaneously on resonance, regardless of their speed, and all the ions are submitted to the radiation pressure during all the time they cross the laser beam. Standard dye lasers have a Dn_M of the order of *300-400 MHz*, i.e. too large for most cases (for example $\Delta\nu_H$ for sodium is *10 MHz*). Therefore the length of the laser cavity L_c has to be prolonged in order to afford every longitudinal mode to be sufficiently close to its neighbors, according to the equation:

$$L_C = c/2\Delta\nu_H = \pi\, c\, \tau$$

Such kind of laser has been realized[47] by Liang et al.and by Moi and collaborators[48] and laser cooling of a *Na* beam has been experimentally observed[43] In this case the laser cavity length was $L_c = 15\ m$.

In *Table IX* the length of the laser cavities for transverse laser cooling of the two selected ions are reported. The laser cavity for $^7Li^+$ should be $L_c = 40\ m$ but, when the power broadening of the absorption line and the line broadening due to transient interaction time are taken into consideration, L_c is reduced to less than *7 m*. In order to cool an ion beam with a transverse temperature of about *500 K*, the bandwidth of the laser should be of the order of *2 GHz* (see *Table IX*).

Sharp edge Problem

One of the problems of white-lasers, whose bandwidth shows, in general, a Gaussian profile, is the lack of a natural sharp edge in the frequency domain, which limits the achievable lowest temperature. Many solutions can be adopted as already checked through experimental investigation, e.g. see 43,49. Here a possible approach is given as an example. If the laser cavity length is reduced, the frequency mode separation becomes larger than the homogeneous linewidth of the ions and the cooling process stops because the automatic shift from one node to the next one does not work any more. But, if the laser power density W_L is high enough to broaden the atomic line, the condition of white light can be again established. W_L can be chosen in such a way that only the central modes satisfy the white light condition, while all others do not. The ions are decelerated, switching from one mode to the next, until the mode intensity is large enough, then suddenly the cooling stops.

3.7 The Vacuum System

In many cases the lifetime of a beam is limited by reactions with residual gas (see chapter *6* in *Ref.* 28). The most important effects are the loss, and the capture, of electrons by scattering on residual gas. Single and multiple Coulomb scattering are also limiting the ion beam lifetime.

In order to avoid considerable beam losses, the storage ring is designed for operation in the 10^{-12} *Torr* region. This vacuum level can be achieved and mantained in a *2.5 mm* thick stainless steel vacuum chamber. The bending magnets aperture give a good conductance over all their length. Dipole magnets are connected by a cylindrical pipe of *13 cm* radius which extends through the quadrupoles. The vacuum chamber in the straight sections is also cylindrical with the same radius of *13 cm*. To reach such a vacuum level all equipment inside the vacuum has to be designed and treated very carefully. Only stainless steel and alluminum oxide has to be used as well as copper seals in all flanges. In order to remove mainly H_2 hidden inside the materials, the whole equipment has to be heated to $900\,^0C$ before the installation. In addition after the installation the system should be baked to $300\,^0C$ with simultaneous pumping.

The pumping system makes use of conventional pumps for ultrahigh vacuum, namely cryogenic and ion pumps with the addition of non-evaporable getters. Vacuum ports are located in each of the bending magnets, with an aperture of about *10 cm* to which ion pumps are attached. Four cryogenic pumps would be sufficient for all the system.

Vacuum pressure gauges can be located in the wall facing the vacuum port, in principle at every location, though probably one reading per arc and one per straight section will be sufficient. One gas analizer is necessary to determine the chemical composition of the residual gas.

Vacuum valves have to be located with the goal to provide easy access and operation and to leave sections of the beam pipe under vacuum when specific locations are open at atmospheric pressure for maintenance or repair. Considering the size of the ring and its geometry, it is sufficient to provide high quality remotely controlled vacuum valves in each straight section as well as at each end of the electron cooling device.

4. Diagnostic of Ordered Beams

Besides several diagnostic methods to detect order in a beam of particles[50], a new fluorescence-based method is proposed to detect ordering in a one-dimensional ion beam.

A pulsed laser, resonant with the traveling ions, is split in two parts, which simultaneously cross the ion beam at right angle at two nearby positions along the storage ring. This laser-to-ion crossing area is followed by four photomultipliers which detect the photons emitted by the ions that have previously been excited by the laser beams. The signals recorded by the photomultipliers are analyzed when one laser beam is moved with respect to the other one. In the absence of ordering, no correlation in fluorescence signals should be recorded while changing the relative distance of the two laser beams. On the contrary, if a string were obtained as a result of cooling, a strong correlation between the signals should be observed. Suppose that one of the four photomultipliers detects the fluorescence of an ion excited by one of the laser beams. If one of the other three photomultipliers detects a simultaneous fluorescence signal, it means that the other laser has interacted with another ion in the string, in turn indicating that the distance between the two ion-to-laser crossing points is an integer multiple of the string interparticle spacing. Then, by slightly moving the second laser beam, the correlation signal should vanish. A sort of periodical dependence on the distance between the laser beams should appear, whose contrast will mainly depend on the detection efficiency.

The diagnostics is conceived with the aim of detecting ordering within a string of ions in a storage ring. Typical values of the interparticle spacing for the string configuration lie between $s=10$-$100\ \mu m$; for the following we shall assume $s=50\ \mu m$. At non-zero temperature, ions are expected to oscillate both in the transverse and longitudinal directions (incoherent motion). Moreover, for a string longitudinal oscillations of equilibrium positions can also occur (coherent motion) through long-range waves. For this diagnostic system the distance between the two laser beams can be chosen to match interparticle spacing of the string. In this case, the effect of coherent motion is ineffective as the nearest-neighbor spacing is relatively uniform. The effect of long wavelength on interparticle oscillations is negligible for a relative distance of a few lattice steps. An analytical evaluation of this effect can be found in Ref. 51. Short wavelengths mostly affect the fluctuations in interparticle spacing, δs. Based on *Ref.* 51, a rough estimate holds: $<\delta s^2>/s^2 r 1/\Gamma$, where Γ is the plasma parameter for the ion beam. When the ion beam is being cooled δs becomes even lower than s; as an example, at $\Gamma=100$, $\delta s=5\ \mu m$. If the distance between the laser beams were larger than one interparticle spacing, the correlation signal should become progressively weaker.

The configuration of the diagnostic device can also compensate for a possible influence of transverse oscillations. Since the laser beams cross the string perpendicularly and their spots are small enough that they do not overlap, the ions suffering transverse oscillations in the direction of the laser can always be resonant, irrespective of their coordinates along that axis. Ion oscillations in direction orthogonal to both the laser and the string could in principle move the target ion outside the laser beam spot. To avoid this effect, and the consequent less efficiency, the laser beam can be focused by a cylindrical lens. This optical element can be arranged to produce a focal segment orthogonal to the directions of both the laser and the string. In this way the locations where the laser beams impinge on the string are two thin regions; these can be as wide as several hundreds of microns in one dimension without overlap between them. Transverse oscillations of a very cold ion beam are expected to be lower than this value.

A single ion of the beam interacts with one of the laser beams for very short time, t_{int}. The probability of excitation to the upper level $P(t_{int})$ is:

$$P(t_{int}) = \sin^2(\omega_R t_{int}/2)$$

where ω_R is Rabi frequency for the transition under consideration, which is proportional to the square root of the laser intensity. Thus, a reasonable interaction probability requests a sufficiently high laser intensity to make up for the short interaction time. However, this is not a problem since pulsed laser are several times as intense as the saturation intensity and a good excitation probability $(\sim 1/2)$ can be achieved with existing lasers. Moreover the focusing lens increases the intensity of the laser beam at the interaction point.

5. The String and Zig-zag Cases

In a previous report[52] the existence of ground states in the CRYSTAL Storage Ring has been demonstrated. The ground states have been obtained by cooling a hot beam until reaching Crystalline structures. To avoid the huge computing time demanded to obtain such a complex structures an unrealistic cooling factor was used. To demonstrate the possibility to reach Crystalline Beams using a realistic cooling factor the simple case of string and zig-zag configurations were investigated.

Numerical simulations were done by using the computer code PARMT. A typical laser cooled $^9Be^+$ ion beam is considered; the initial condition at the injection are the following: $\varepsilon = 1\pi\ mm\ mrad$, $\Delta p/p = 10^{-4}$, $\beta = 0.07$, $N = 2.7x10^5$ - $3x10^6$. The cooling factor used in the simulation is $K = 1.0004$ for the string and $K = 1.003$ for the zig-zag, corresponding to *11 ms* and *1 ms*, respectively. Compared with the experimental values for the longitudinal laser cooling, the cooling factors are quite close to the realistic ones. The difference is that here the same cooling rate is applied in the longitudinal direction as well as in the transverse one. The simulations are done for the realistic storage ring and take also into account misalignments, linear fringe fields, orbit perturbations, field errors, etc.

Fig. 4 shows both the results relative to a string and a zig-zag configuration, while *Table X* summarizes the main parameters of the simulation.

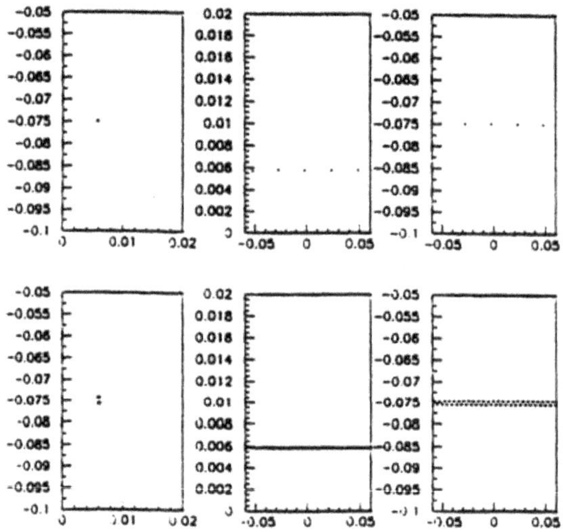

Fig. 4 - Simulations of string and zig-zag configurations, for the realistic storage ring. The ions are $^9Be^+$ with $\beta = 0.07$. The dimensions are in centimeters.

For the string a plasma parameter $\Gamma = 11$ has been obtained, which corresponds to a beam temperature *of 1.8 mK.*

The interparticle distance results to be *260 μm*.. The critical ion spacing for a string is $L_0 = 75$ *μm*, corresponding to about 10^6 circulating ions. At the critical ion spacing the string configuration becomes unstable and a new Crystalline structure (zig-zag) appear.

Being the focalization predominant in the horizontal plane the resulting structure is a vertical zig-zag. The simulation for the zig-zag gives a plasma parameter $\Gamma = 46$, which corresponds to a beam temperature of *7.6 mK*. The interparticle distance results of *30 μm* longitudinally and of *8 μm* transversally.

Notice that the zig-zag configuration, as well as the string one, is not affected by the curvature when traveling across the dipole magnets.

Similar results can be obtained simulating different ion species. The $^9Be^+$, as well as $^7Li^+$ and $^{24}Mg^+$, are candidates for Crystalline Beams. Laser cooling associated to other cooling techniques (indirect cooling, RF cooling, etc.) offers a good possibility to attain Crystalline Beams.

Table X - Main Parameters from the Simulation of String and Zig-zag Configurations.

Parameter	String	Zig-zag
Number of Ions	2.7×10^5	3×10^6
Initial Emittan. [π mm mrad]	1	1
Reduction Factor (cooling)	1.0004	1.003
Long. Interpart. Dist. [µm]	260	30
Trans. Interpart. Dist. [µm]	-	8
Plasma Parameter Γ	11	46
Beam Temperature [mK]	1.8	7.6

6. Shear Forces and Breathing

The attainment of Crystalline states for a cooled beam in a storage ring can be prevented by the viscous heating due to the shear induced in the dipoles. Shear will transfer some of the energy of coherent motion to the thermal energy and may prevent reaching the low temperatures required for formation of Crystalline states. The solution at the equilibrium for the third equation of motion[52] impose the condition of equal revolution frequency for all the particles of the crystal, which is the condition for a Crystalline Beam to survive into a general and realistic lattice. In fact, in the numerical simulations no continous shear slippage was observed, particles only underwent longitudinal and transverse oscillations with the ring lattice periodicity. While the mentioned solution is a direct consequence of smooth approximation (the equation of motion are time-independent), in the simulations the general time-depending equations are used. Also in this case a stable and periodical solution can be found. This solution has a periodicity corresponding to the lattice cell of the magnetic structure. An example of such behavior is shown in *Fig. 5*, where a Crystallized shell is shown at different positions within a cell of the CRYSTAL Storage Ring. The Crystalline Beam can travel along the storage ring without continous slippage if it has a proper value of the momentum gradient $\partial p_z / \partial x$. In fact, all the structures achieved in the simulations have the gradient shown in *Fig. 6*. The same results were found[53] also by *A. Sessler and al.*; the authors explain the results as that the crystal "takes up the difference" between constant angular velocity and constant linear velocity by adjusting the spacing between particles, i.e. by converting potential energy into kinetic energy.

Only on average, the Coulomb repulsion of the space charge dominated beam is balanced by the external confining forces. An alternation of the focusing gradient and the bending field along the circumference imposes a variation of the beam sizes with the lattice periodicity. This correspond to a "breathing" of the beam and can be analytically study. A comparison between the analytical results and the PARMT simulations for a crystallized shell is shown in *Fig. 7*. The breathing is rather small because of the smoothness of the CRYSTAL Storage Ring: $\Delta a_x / a_x = 11\%, \Delta a_y / a_y = 9.5\%$.

7. Conclusions

Phase transitions in a confined ionic system have been found by computer simulations[5,6] and later experimentally confirmed in ionic traps[8,9]. In the case of the ion traps, the simulations of ground states have been done considering a longitudinal homogeneous and transversally isotropic straight linear focusing channel. For a anisotropic focusing trap, where the transverse focusing is the same in both directions and the longitudinal focusing is different from the transverse ones, phase transitions have been studied as a function of the anisotropy[54].

Particle motion in a storage ring is intrinsically more complicated because of the particle angular velocity and the AG focusing.

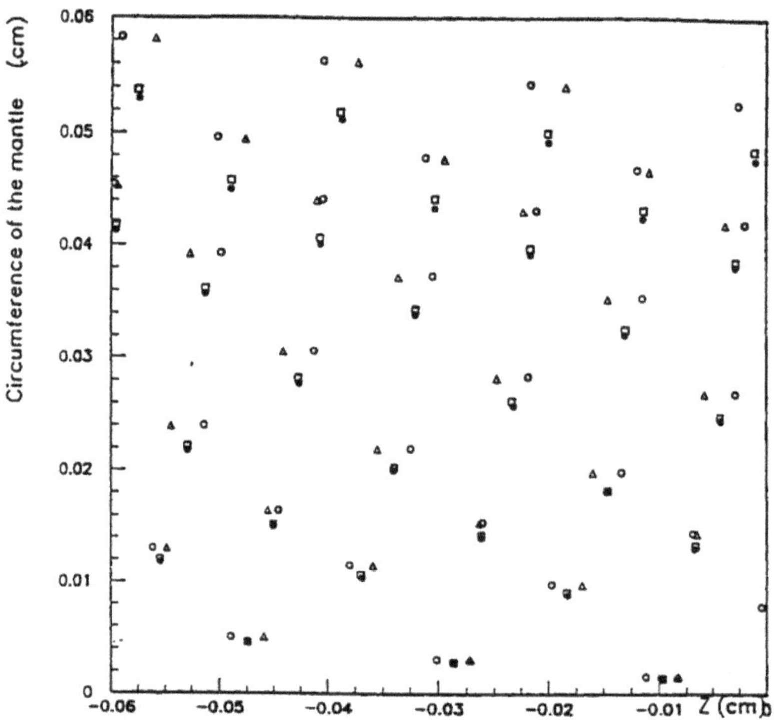

Fig. 5 - Shell configuration in a periodic cell of the CRYSTAL Storage Ring. The particles are projected onto the mantle at different positions in the cell: ● - middle of the drift space, ▵ - entrance of the dipole, ○ - exit of the dipole, ❑ - middle of the next drift section.

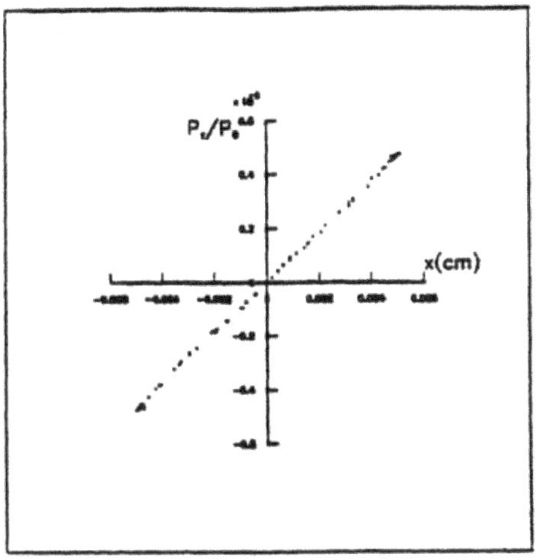

Fig. 6 - Longitudinal momenta p_z in the reference frame versus the horizontal coordinate of particles in a steady Crystalline state of a cooled beam. The momenta are given in units of the reference particle momentum p_0.

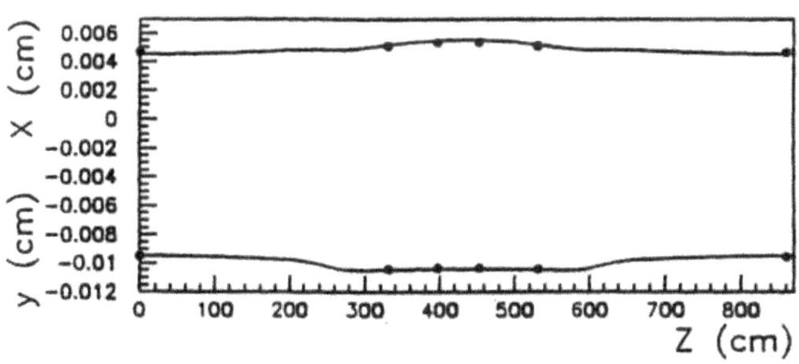

FIG. 7 - Breathing: comparison between the analytical solutions (solid curve) and the PARMT results (dots).

On the basis of this knowledge the phase transitions of cold ion beams in a storage ring are numerically simulated. The ground states are determined in function of the focusing parameters.

In analogy with the simulations in ion traps, the ground states of Crystalline Beams in a storage ring are determined by solving the equations of motion derived from the general Hamiltonian[12,53].

Respect to the case studied by Hasse and Schiffer[6] we can derive analogous structure parameters λ_x, λ_y but different in the two planes. In the case of a zero curvature and an isotropic focusing these parameters turn to the parameters introduced in *Ref.* 6.

Both the structure parameters are proportional to the linear density, with a constant that depends on focusing, main energy and ion kind. Every ground state configuration is defined in a range of λ_x, λ_y and therefore in a range of density. The extreme values (critical values) of such range correspond to the stability limits. When the limit is reached the ground state in examination became unstable and a new structure appears. At very low densities the particles are aligned along the direction of motion and at higher densities more complicated structures appear.

The ground states have been investigated by means of the computer code PARMT. As an example, a beam of $^9Be^+$ with $\beta = 0.07$ and initial emittance $\varepsilon = 1\pi\ mm\ mrad$ was studied. The results are very general, and can be extended to every ion species of different charge state and energy. For completeness the lattice perturbations, i.e., misalignments, linear fringe fields, etc., are also included in the simulation.

Ground states can be described in function of the linear density and structure parameters λ_x, λ_y so that the phase transitions for different Crystalline structures are defined only by the machine tunes and the particle linear density. The basic structures are the helix and twisted multiplets; some of them are displayed in *Fig. 8 and 9*, for different linear densities. *Fig. 9* shows the side view of a typical structure where the particles from the outer shell are plotted as if the mantle were the plane of the drawing. This picture clearly shows the hexagon patterns that are typical of Crystalline structures.

The Crystalline structures and the transition thresholds achieved in the traps well reappear in the storage ring when $\lambda_x = \lambda_y$. Notice that the condition for isotropic focusing ($\lambda_x = \lambda_y$) in a storage ring is $Q_x^2 - \gamma^2 = Q_y^2$, instead of $Q_x^2 = Q_y^2$ as in a trap.

In all cases considered, at very low linear density the particles are always strung out along the z-axis as a string, then increasing the density a zig-zag appears and so on, up to obtain *3D*-structures.

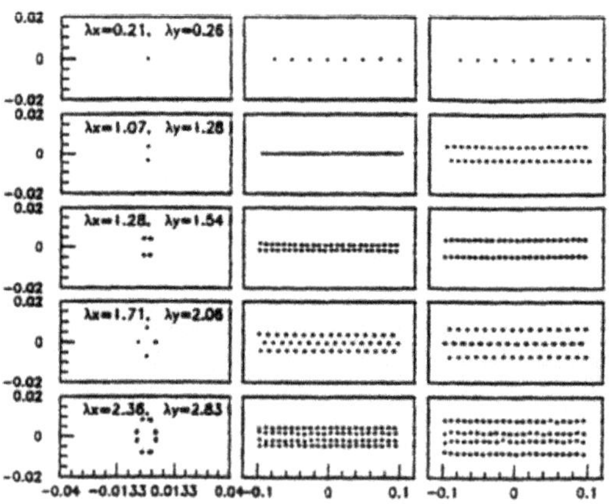

Fig. 8 - Crystalline configurations for different values of the structure parameters λ_x, λ_y. From left to right, projection of particles onto the x - y, z - x and z - y planes are shown for five different Crystalline structures. All dimensions are in centimeters.

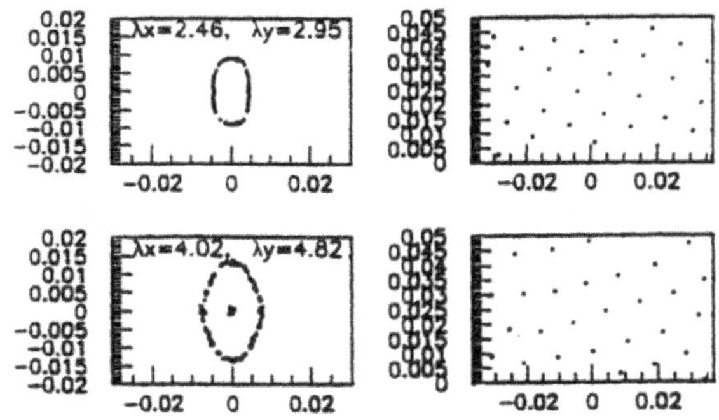

Fig. 9 - Crystalline configurations for different values of the structure parameters λ_x, λ_y. On the left is an x - y plane projection and on the right the particles from the outer shell are plotted as if the mantle were the plane of the drawing. All dimensions are in centimeters.

References

1. J.P. Schiffer and P. Kienle, Zeitschr. f. Phys. A321.
2. T. Haensch and H. Schawlow, Opt. Com. 13 (1975) 68.
3. E.N. Dementiev, N.S. Dikansky, A.S. Medvedko, V.V. Parkhomchuk, D.V.Pestrikov; Sov. Phys. Tech. Phys. 25 (1980) 1001.
4. M. Steck et al.; Suppression of Intrabeam Scattering in Cooled Heavy Ion Beams; this Workshop.
5. A. Rahman and J.P. Schiffer; Phys. Rev. Lett. 57 (1986) 1133.
6 R.W. Hasse and J.P. Schiffer; Ann. of Phys. 203 (1990) 419.
7. R.W. Hasse; Phys. Rev. Lett. 67 (1991) 600.
8. G. Birkl, S. Kassner and H. Walther; Nature 357 (1992) 310.
9. I. Waki, S. Kassner, G. Birkl and H. Walther; Phys. Rev. Lett. 68 (1992) 2007.
10. G. Bisoffi et al.; The Crystal Storage Ring Project at Legnaro; Presented at the European Particle Accelerator Conference EPAC 94, London June 27-July 1, 1994.
11. X.P. Li, A.M. Sessler and J. Wei; Crystalline Beam in a Storage Ring: How Long Can It Last?; Presented at the European Particle Accelerator Conference EPAC 94, London June 27-July 1, 1994.
12. A.V. Burov et al.; Crystallized Ion Beams in a Storage Ring; Accepted for Publication into Nuovo Cimento.
13. J. Wei, X.P. Li and A.M. Sessler; Crystalline Beam Ground State; BNL-52381, June 1993.
14. A.G. Ruggiero; Demonstration of No Feasibility of a Crystalline Beam in a Betatron Magnet; BNL-49529, September 1993.

15. L. Tecchio et al. (CSR Collaboration); CRYSTAL: A Storage Ring for Crystalline Beams: LNL-INFN (REP) 80/94.
16. I.M. Kapchinskij and V.V. Vladimirskij; Proceedings of International Conference on High Energy Accelerators, CERN, Geneva (1959), 27.
17. I. Hofmann, L.J. Laslett, L. Smith and I. Haber; Particle Acc. 13 (1983) 145.
18. J. Struckmeier and M. Reiser; Part. Acc. 14 (1983) 227.
19. I. Hofmann and J. Struckmeier; Proc. of the Workshop on Crystalline Ion Beams, Wertheim, Germany, October 1988, Report GSI 89-10 (1989), 140.
20. J.S. Hangst et al.; Phys. Rev. Lett. 67 (1991) 1238.
21. B. Hochadel et al.; Proceedings of the Workshop on Beam Cooling and Related Topics, Montreux, 4-8 October 1993, CERN 94-03, 198.
22. G. Boicount and J. Merson, Los Alamos National Laboratory, Report LA-UR-90-127, January 1990.
23. J. Struckmeier and I Hoffman, GSI-ESR-87-03, Darmstadt 1987.
24. B.Yang et al.; The Blow-up Rate of the Envelope Instability of a Cold Beam in a Storage Ring, Accepted for publication in Physics of Plasma.
25. J.D. Bjorken and S.K. Mtingwa; Part. Acc. 13 (1983) 115.
26. H. Grote; The MAD Program CERN/SL/90-13 (AP).
27. M. Conte and M. Martini; Part. Acc. 17 (1985) 1.
R. Giannini; INTRABS Computer Code, CERN PS 1993.
28. L. Tecchio et al. (CSR Collaboration); CRYSTAL: A Storage Ring for Crystalline Beams; LNL-INFN (REP) 80/94.
29. K. Steffen; High Energy Beam Optics, Wiley, New York.
30. J.L. Laclare; Proc. CERN Accelerator School, CERN 85-19.
31. J.L. Laclare; Proc. of Fifth General Accelerator Physics Course, Jyvaskyla (1992), pg. 349 and pg. 385.
32. H.Poth; Phys.Rep. 196 (1990) 135.
33. J.R. Pierce; Theory and Design of Electron Beams, (ed. D. Van Nostrand Company, Princeton).
34. A.V. Aleksandrov et al.; Phys. Rev. A 46 (1992) 6628.
35. V.I Kudelainen et al.; Sov. Phys. Techn. Phys 16 (1972) 1821.
36. N.S.Dikansky et al.; Report INP 88-61.
37. H. Danared et al.; Phys. Rev. Lett. 72 (1994) 3775.
38. A.N. Sharapa and A.V. Shemyakin; Nucl. Instrum. Meth. A 351 (1994) 295.
39. T. Hänsch and A. Schawlow, Opt. Com. 13 (1975) 68.
40. J.S. Nielsen et al.; Proc. of the Workshop on Beam Cooling and Related Topics, Yell. Rep.CERN 94-03 p. 339 and references therein.
41. J.S. Nielsen to be published.
42. L. Moi, Opt. Comm. 50 (1984) 349.
43. S. Gozzini et al.; Appl. Phys. B 54 (1992) 428.
44. P. Strohmeier et al.; J. Opt. Comm. 73 (1989) 451.
45. M. Zhu et al.; Phys. Rev. Lett. 67 (1991) 46.
46. I.C.M. Littler et al.; Zeit. Phys. D 18 (1991) 307.
47. J. Liang, L. Moi, C. Fabre; Optics Comm. 52 (1984) 131.
48. L. Moi, M. Allegrini, P. Bicchi, S. Gozzini, J.H. Xu, SPIE 701 (1987) 176.
49. J. Hoffnagle; Opt. Lett. 13 (1988) 102.
50. L. Tecchio et al. (CSR Collaboration); CRYSTAL Storage Ring: Conceptual Design Report, LNL-INFN (REP) 97/95.
51. L.D. Landau and E.M. Lifshitz, Statistical Physics (Pergamon Press, London, 1958), p. 482.
52. A. Pisent and A. Burov; Studies of Crystalline Beams at Legnaro, this Workshop.
53. J. Wei, Xiao-Ping Li and A. Sessler; Phys. Rev. Lett. 73 (1994) 3089.
54. J.P. Schiffer; Phys. Rev. Lett. 70 (1993) 818.

A Thermal Wave Model for Electromagnetic Traps

Renato Fedele
Universita' di Napoli and Istituto Nazionale di Fisica Nucleare, Napoli, Italy

Rosa Poggiani
Universita' di Pisa and Istituto Nazionale di Fisica Nucleare, Pisa, Italy

Gabriele Torelli
Universita' di Pisa and Istituto Nazionale di Fisica Nucleare, Pisa, Italy

ABSTRACT

The Thermal Wave Model can be used to study the equilibrium condition in electromagnetic traps; its reliability has been proved in a simple case. The final aim is to study the equilibrium condition of a $\bar{p}e^{+}$ plasma suitable for Antihydrogen formation.

1 Introduction

In the last years some interest has been devoted to the formation of Antihydrogen atoms. The theoretical reasons for such an interest are largely discussed e.g. by Charlton et al., who also looks at the technical possibility to obtain a useful formation rate. Very recently Antihydrogen formation in flight has been announced by a LEAR collaboration [2]. This positive result demonstrates that Antihydrogen exists, but does not give us a method to produce Antihydrogen at a reasonable rate and in conditions suitable for further experimentation, since the Antihydrogen atoms are produced at an energy of ~1.7 GeV.

The measurements to be performed on Antihydrogen atoms, i.e. spectroscopy to check CPT invariance and gravitational measurements to check gravity on antimatter [1], require stable or metastable atoms at thermal or epithermal energy and therefore large efforts have been engaged to develop the techniques to produce low energy Antihydrogen atoms at high rate.

2 Antihydrogen Formation

The possible reactions to produce Antihydrogen have been discussed in the recent years [3] and here we will just summarize the advantages and difficulties of each one of them.

The fundamental process is the radiative two bodies recombination:

$$\bar{p} + e^{+} \longrightarrow \bar{H} + h\nu \tag{1}$$

where the Antihydrogen is prevalently formed in low n quantum states, but the rate per antiproton:

$$\Gamma_1 = 6 \times 10^{-11} T^{-\frac{1}{2}} n_e \quad s^{-1} \tag{2}$$

is extremely low for the positron densities $n_e(cm^{-3})$ and temperatures $T(K)$ that can be achieved. The laser stimulation can induce the formation of Antihydrogen atoms in a specific quantum state and can increase the total rate by two orders of magnitude.

The competing process is the three bodies recombination when a second positron in the interaction volume provides for the energy–momentum balance:

$$\bar{p} + e^{+} + e^{+} \longrightarrow \bar{H} + e^{+} \tag{3}$$

This rate is quite high and dramatically dependent on the temperature:

$$\Gamma_2 = 3.8 \times 10^{-9} T^{-\frac{9}{2}} n_e^2 \quad s^{-1} \tag{4}$$

but unfortunately the Antihydrogen atoms are formed in very high Rydberg states. To be available for further experimentation the atoms must decay to a quantum state with a binding energy substantially higher than the thermal energy of the positron plasma. This probability has not yet been evaluated.

Some attempt has been done to check both recombination processes using protons and electrons in nested [4] and in combined [5] traps. Actually the three bodies process is dominating in the conditions of the tests, but Hydrogen formation has never been observed. To justify the negative results a bad overlap in space or time of the electrons and protons clouds has been claimed, but a very low probability to decay to a detectable metastable state can also be a reasonable explanation.

The third and promising method to form Antihydrogen is based on the use of positronium [6]:

$$\bar{p} + e^+e^- \longrightarrow \bar{H} + e^- \qquad (5)$$

A keV energy antiproton beam traverses a positronium cloud evaporating from the surface of a target hit on the other side by a positron beam. The rate is high and is growing with the fourth power of the positronium quantum number. A laser excitation of the positronium cloud can allow to get a very high formation rate and to partially select the Antihydrogen final quantum number [7].

Anyway this third process is based on a different technology and hence is out of the aims of this note that is intended to discuss the dynamics of the plasma in a trap to obtain some more insight about the Antihydrogen formation in a mixed plasma of positrons and antiprotons.

3 Electromagnetic Traps

A trap is a cylindrical symmetric device formed by three electrodes shaped as conjugate rotational hyperboloids (one central ring electrode and two side cap electrodes) to form an hyperbolic trap. A plane of symmetry cuts the axis in the central point of the trap. DC and RF potentials applied between the central and the side electrodes provide for the confinement of particles of both signs (Paul trap). A strong uniform magnetic field parallel to the symmetry axis increases the trapping stability (combined trap). The device using DC potential only is called Penning trap. When the magnetic field is present the electrodes can be cylinder segments, the central one usually longer than the side ones, forming a cylindrical segmented trap. In this geometry DC and RF potentials provide for the longitudinal confinement, while the radial one is performed by the magnetic field [8]. The particle dynamics in these devices is very well known in the classical [9] or quantum [10] context as far as the space charge effects can be neglected. This condition is no more satisfied when the Debye length:

$$\lambda_D = \left(\frac{T}{4\pi q^2 n}\right)^{\frac{1}{2}} \qquad (6)$$

is shorter than the bunch dimension; here and in the following the temperature T will be measured in energy unit, q will be the elementary charge and n the particle density.

When the space charge cannot be neglected, the classical approach is able to describe the equilibrium conditions for any plasma composed by several different species of particles with charges of the same sign [11,12]. In a plasma composed by particles of opposite charges, like the antiproton–positron one for the Antihydrogen formation, the repulsive force due to the space–charge is cancelled and the canonical angular momentum conservation does not guarantee any more the long term radial confinement of the plasma [13]. We need a more sophisticated approach to know if there is a stable equilibrium condition and what will be the values of the particle density and temperature for a cold trapped neutral plasma.

4 Thermal–Wave Model

The *Thermal-Wave Model* (TWM) is a new approach to the beam dynamics proposed in 1991 by R.Fedele and G.Miele [14]. It is based on the formal analogy between the electromagnetic beam optics, the non–relativistic quantum mechanics and the relativistic–charged–particles beam optics in paraxial approximation. The validity of the model has been proved by several successful applications to linear and non–linear problems in transverse and longitudinal relativistic beam dynamics [15]. The TWM describes the particle beam by a transverse density distribution function $\rho(r,z,\phi)$ proportional to $|\Psi|^2$. The wavefunction Ψ obeys to a two–dimensional Schroedinger–like equation where:

- the beam emittance ϵ plays the role of the quantum constant $\hbar$;
- the potential $u(r,z,\phi)$ is the potential seen by the beam, i.e. the one generated by the machine structure, by the beam itself and by its interaction with the chamber walls etc...;
- the coordinates are referred to the reference trajectory in paraxial approximation;
- the longitudinal coordinate z plays the role of the time: $z = ct$.

The Schroedinger–like equation in a dimensionless form can be written as:

$$i\epsilon\frac{\partial}{\partial z}\Psi = -\frac{\epsilon^2}{2}\nabla_t^2\Psi + U(r,z,\phi)\Psi \tag{7}$$

where $U(r,z,\phi) = \frac{u(r,z,\phi)}{m_0\gamma\beta^2c^2}$ is the dimensionless potential, m_0 the rest mass.

If the wave function Ψ is properly normalized ($||\Psi||^2 = 1$) and N is the total number of beam particles, the transverse particle number density, i.e. the transverse beam profile, is given by:

$$\rho(r,z,\phi) = N|\Psi(r,z,\phi)|^2 \tag{8}$$

Usually the beam equilibrium condition corresponds to the lowest possible averaged energy of the system and therefore to the solution of the Schroedinger–like equation with the lowest quantum number.

This model allows to transfer the whole of our knowledge about the Schroedinger equation to the charged–particle beam optics.

5 The TWM for Electromagnetic Traps

Inside an electromagnetic trap non–relativistic particles move in a three–dimensional space. Moreover the reference trajectory is reduced to the central point of the trap, this means that the coordinates used in the paraxial approximation and the laboratory coordinates coincide.

Non–relativistic particles in a three–dimensional space in a coordinate system at rest still satisfy the basic hypothesis of the TWM. The last condition to use the paraxial approximation is that the beam divergence around the reference trajectory must be small and it is not obvious what this condition means in the trap where the reference trajectory is a point.

Looking at the use of the paraxial approximation in the transverse beam dynamics it can be seen that *small divergence* or *small angle* really means that the transverse potential energy of the particles and therefore their transverse kinetic energy must be small compared to the particle energy, i.e. that the transverse motion in the moving reference frame must be non–relativistic. This condition is fully satisfied in a trap where the particle kinetic energy is lower than a few eV against a minimal rest energy of 0.5 MeV for the electron.

The possibility to use the TWM for electromagnetic traps needs to be verified and a simple way is to use the TWM to calculate a trial solution in a trap and to compare the results with the ones of classical analysis.

5.1 Classical solution

A very accurate and complete calculation of the equilibrium conditions for a cylindrical plasma column confined by a strong magnetic field parallel to the cylinder axis is reported in Ref. 16. It is based on statistical mechanics and includes the description of the transport of the column to the thermal equilibrium. To reproduce the full calculation using the TWM is out of the aims of this paper; as a preliminary test on the TWM validity the particle–number distribution calculated with both methods will be compared.

Following the classical solution the particle–number density can be written as:

$$n(r) = n_0 \exp(\frac{q\phi(r)}{T} - \alpha r^2) \tag{9}$$

where:

- n_0 is the central density;
- $\alpha = \frac{m_0}{2T}(\Omega\omega - \omega^2)$;
- Ω is the cyclotron angular speed $\frac{qB}{m_0}$;
- B is the magnetic field;
- ω is the coherent rotation angular speed;
- $\phi(r)$ depends on the electric potential distribution.

If the potential is generated only by the plasma itself $\phi(r)$ can be obtained by the numerical solution of the Poisson equation and the result is that $\phi(r)$ cancels the quadratic dependence due to the second term up to some value of the radius and then goes to zero in a few Debye lengths.

5.2 TWM calculation

We will use some approximations when applying the TWM to the above situation:

- we will not directly deal with the magnetic field and the related cyclotron motion, but we will just retain its major mechanical effect, i.e. we will reduce the effect of the magnetic field to a confining harmonic force $f = k'r$, where $k' = qB\omega$;
- we will assume that the particle number density is constant up to a radius r^*, i.e. the particle self-field completely cancels the confining potential.

With these approximations the relevant variable is the radius and the dimensionless Schroedinger-like equation can be written as:

$$i\epsilon\frac{\partial}{\partial s}\Psi = -\frac{\epsilon^2}{2}\nabla_t^2\Psi + U\Psi \tag{10}$$

where:

- $s = ct$ plays the role of time;
- $U = constant = U_0$ for $0 \leq r \leq r^*$,

- $U = \frac{1}{2}kr^2$ for $r \geq r^*$;
- $k = \frac{k'}{m_0c^2}$ is the dimensionless confining strength;
- $\epsilon = \frac{\epsilon'}{m_0c}$ is the dimensionless plasma radial emittance;
- $\epsilon' = \frac{r^*}{\sqrt{m_0T}}$ is the plasma radial emittance i.e. the product of the plasma radial dimension r^* times the thermal momentum.

The equation is easily solved in both intervals; by imposing the continuity of the potential U and of the wave function Ψ (but not of their derivatives) for $r = r^*$ the wave function can be written as:

$$\Psi(r,t) = A\exp(-\frac{iU_0s}{\epsilon})\begin{cases} \sqrt{\frac{\delta}{3\pi}} & 0 \leq r \leq r^* \\ \sqrt{\frac{\delta}{3\pi}}\exp(1 - \frac{\delta r^2}{2}) & r \geq r^* \end{cases} \tag{11}$$

where $\delta = \frac{\sqrt{k}}{\epsilon} = \frac{m_0c^2}{2}\frac{k}{T}$, $(r^*)^2 = \frac{2}{\delta}$ and Ψ has been normalized to 1 on the whole plane:

$$||\Psi||^2 = 2\pi[\int_0^{r^*} rdr|\Psi|^2 + \int_{r^*}^{\infty} rdr|\Psi|^2] = 2\pi A^2(1+\frac{1}{2})\exp(-2) = 1 \tag{12}$$

Therefore $A = \frac{\epsilon}{\sqrt{3\pi}}$.

The particle–number density is given by the product of $|\Psi|^2$ times the total number of particles N. Using the relative contributions of the central core ($\frac{2}{3}$) and of the tail ($\frac{1}{3}$) to the normalization the particle number density in the central core can be calculated:

$$n_0 = \frac{2}{3}\frac{N}{\pi(r^*)^2} = \frac{N\delta}{3\pi} \tag{13}$$

In our approximation the particle-number density can be written as:

$$n(r) = n_0\begin{cases} 1 & 0 \leq r \leq r^* \\ \exp(2 - \delta r^2) & r \geq r^* \end{cases} \tag{14}$$

and depends on three parameters: the plasma central density n_0, the temperature T and the confining potential strength k through the ratio $\delta = \frac{m_0c^2}{2}\frac{k}{T}$.

5.3 Analysis of the results

The results of the calculations of the particle density are quite similar; moreover the δ coefficient can be written as:

$$\delta = \frac{m_0c^2}{2}\frac{k}{T} = \frac{m_0c^2}{2T}\frac{qB\omega}{m_0c^2} = \frac{m_0}{2T}\frac{qB}{Tm_0}\omega = \frac{m_0}{2T}\Omega\omega \tag{15}$$

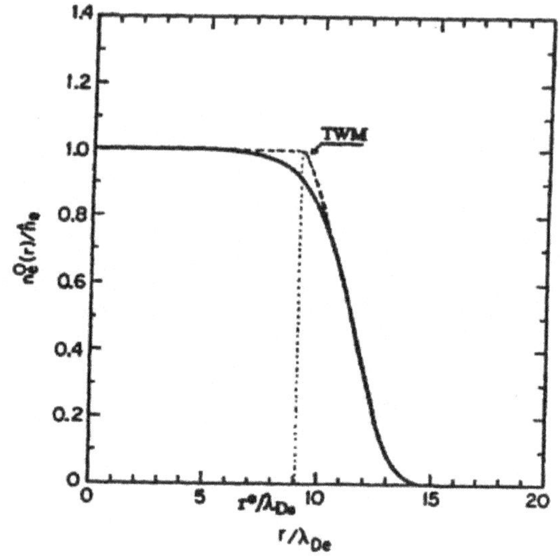

Figure 1: Comparison of the classical and of the TWM solutions

quite analogous to the term α of the classical calculation; the centrifugal part ω^2 is obviously not included in our approximations. The two results are displayed in Fig. 1, where the classical results has been taken from Ref. 11.

6 Conclusions

The *Thermal Wave Model* has been used to calculate the particle–number density in a plasma column, with very crude approximations. The results agree quite well, taking into account the used approximations, with the results of the classical analysis.

The use of the TWM for a detailed description of the distribution functions of plasmas in traps seems to be promising.

7 References

1. M. Charlton et al., *Phys. Reports* **241** (1994) 65
2. G. Baur et al., Phys. Lett. **B368** (1996) 251
3. see *Proceedings of the Antihydrogen Workshop*, Munich, 30–31 July 1992, Hyperf. Inter. **76** (1993)
4. W. Quint et al., Hyperf. Inter. **76** (1993) 181
5. T. W. Haensch et al., to be published
6. B. I. Deutsch. Hyperf. Inter. **44** (1988) 271
7. B. I. Deutsch, Hyperf. Inter. **76** (1993) 153
8. H. G. Dehmelt et al., Adv. At. Mol. Phys. **3** (1967) 53; D. J. Wineland et al., Adv. At. Mol. Phys. **19** (1983) 135; W. Paul, Rev. Mod. Phys. **62** (1990) 531
9. G. Z. Li and G. Werth, Phys. Scripta **46** (1992) 587
10. G. Z. Li, Z. Phys. **D10** (1988) 451
11. R. C. Davidson, *An Introduction to the Physics of Nonneutral Plasma*, Addison–Wesley (1990)
12. T. M. O'Neill, Phys. Fluids **24** (1981) 447
13. D. J. Wineland, *Proc. of the Cooling, Condensation and Storage of Hydrogen Cluster Ions Workshop*, SRI, Menlo Park, CA, USA, January 1987, pg. 181
14. R. Fedele and G. Miele, Il Nuovo Cimento **13D** (1991) 527
15. R. Fedele et al., Phys. Lett. **A179** (1993) 407; R. Fedele et al., Phys. Lett. **A185** (1994) 93; R. Fedele et al., Phys. Rev. **A45** (1992) 4045; R. Fedele et al., Phys. Rev. **A46** (1992) 6634
16. T. M. O'Neill and C. F. Driscoll, Phys. Fluids **22** (1979) 266; S. A. Prasad and T. M. O'Neill, Phys. Fluids **22** (1979) 278

THE STATUS OF LASER COOLING EXPERIMENTS IN ASTRID

J.S. HANGST, M. DREWSEN, A. LABRADOR, N. MADSEN, AND J.S. NIELSEN
Institute of Physics and Astronomy, Aarhus University, DK-8000 Aarhus, Denmark

P. SHI
Microelectronics Center, Danish Technical University, DK-2800 Lyngby, Denmark

O. POULSEN
Danish Research Ministry, DK-1260 Copenhagen, Denmark

J.P. SCHIFFER
Argonne National Laboratory, Argonne, IL, 60439 USA

V. LEBEDEV
CEBAF, Newport News, VA, 23606 USA

ABSTRACT

We briefly describe the current status of laser cooling experiments in the ASTRID storage ring. Open questions and future plans for addressing them are also discussed.

1. Review of Previous Results

The theory of laser cooling and the detailed results of our experiments have been treated elsewhere (see the references at the end of this article) and will not be repeated here. In the following we give a brief summary of experience at ASTRID.

Laser cooling of low energy (100 keV) beams has been underway at ASTRID since 1990. The goal of these experiments is to produce the coldest, densest possible stored beams and to study their behavior. The first work[1] was with $^7Li^+$ ions, chosen because the cooling transition is accessible with commercially available dye lasers. The disadvantages of using this two-component beam, in which only a small percentage of ions are in the coolable, metastable state, have been well documented[2]. Nonetheless, much was learned about laser cooling and fluorescence diagnostics in storage rings using the Li ion, both at Aarhus and at Heidelberg. Since 1993 we have exclusively used $^{24}Mg^+$ ions and Aarhus-developed laser frequency-doubling cavities[3] to produce the required ultraviolet light at about 280 nm. The Mg ion cooling transition, based in the ground state and having a lifetime of 3.5 ns and a Doppler temperature of about 1 mK, is well suited to storage ring cooling experiments.

Our current experimental setup is shown in Figure 1. Two different methods of longitudinal laser cooling have been employed. The first makes use of two lasers, one counterpropagating and one copropagating with the ion beam. The two lasers are frequency tunable, and are scanned in frequency to collect ions from a "hot" distribution ($\delta p/p \approx 10^{-3}$) into a "cold" one ($\delta p/p \approx 10^{-5}$). This technique is used on coasting (i.e., unbunched or DC) beams. It is with this technique that the interesting coherent behavior of

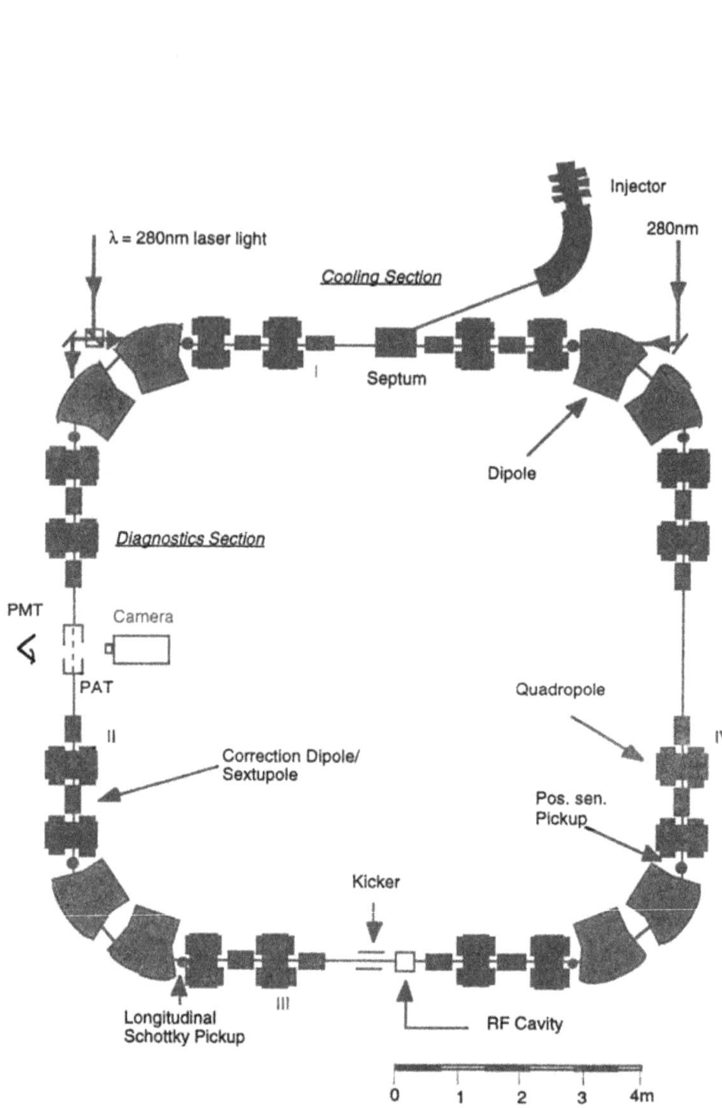

Figure 1. The ASTRID storage ring, illustrating laser and diagnostic systems.

the laser cooled beam[4] has been observed and studied with a longitudinal Schottky pickup. The unusual velocity distributions created by the short-ranged laser force lead to complex longitudinal plasma wave phenomena in the beam. The latest results and interpretation of these phenomena appear in these proceedings[5].

The second technique utilizes a standard radiofrequency cavity to bunch the beam, and a single, tunable laser to damp the synchrotron oscillations[6]. This technique has led to the observation of spacecharge dominated bunches in which the incoherent synchrotron oscillations appear to be suppressed by the interparticle forces. Such cold and dense bunches may offer another system in which crystallization might be possible. The stability of these bunches needs to be studied in detail. This form of laser cooling has the additional advantage of creating very well-defined particle bunches for experiments, and it has already been employed in an experiment to test special relativity in Heidelberg[7]. It has also been suggested that this technique might be useful for cooling intense beams of heavy, singly-charged ions for use in heavy ion inertial confinement fusion[8].

Of paramount importance in all of our experiments is the ability to directly measure the beam's velocity distribution by monitoring the laser fluorescence signal. This gives an unambiguous characterization of the longitudinal phase space in the coasting beam case, and, together with the bunch length measurements, completely characterizes the longitudinal phase space in the bunched beam case. However, we have not until now had a system which can accurately characterize the transverse phase space, i.e., a beam profile monitor or similar diagnostic device. Lack of such a system has led to unanswered questions about the behavior of the beam in both the bunched beam case and the coasting beam experiments on plasma waves. Since the Coulomb force couples all degrees of freedom in a spacecharge dominated beam, one cannot really disentangle longitudinal and transverse motions - all three phase space planes must be characterized. In Heidelberg, a profile monitor based upon ionization of the residual gas in the vacuum chamber has been used with great success, including the exciting results on sympathetic cooling presented at this workshop. The lower energy ions used in ASTRID make such a device impractical, due to the large electric fields used to collect the electrons onto the channelplate detector.

2. Transverse Diagnostics by Direct Imaging

In Aarhus we have chosen to develop a profile monitor based on imaging of the fluorescence light emitted by the ions. Such techniques have been used with great success in ion traps, as is illustrated by Professor Walther's contribution to these proceedings. Individual laser cooled ions, nearly at rest, are easily imaged in traps. A schematic diagram of the ASTRID system is shown in Figure 2. This system has been constructed, and the first tests with beam will begin at the end of 1995.

The heart of the system is an ultrasensitive, UV-enhanced, liquid nitrogen cooled CCD camera. Light from the stored ions is imaged onto the CCD chip by a system of lenses. For ease of handling, the lens system is external to the vacuum chamber. Both the lenses and the CCD camera are moveable, to allow for varying the optical magnification. A UV transmitting filter blocks out the visible light. The system can be oriented to measure either the horizontal or the vertical beam size. Not shown in Figure 2 are moveable apertures upstream and downstream of the camera section. These are used to stop the stray laser photons which scatter from the entrance and exit windows, and to aid in the overlap of the ion and laser beam. Also not shown is a moveable target which can be inserted into the vacuum chamber in order to image the laser beam profile at the proper longitudinal position. The laser beam passes through a beam expander before entering the vacuum chamber, allowing for adjustment of the laser beam diameter at the camera section.

As with any such optical system, the trick is to maximize the number of photons collected and minimize the noise. High laser power, beam current, and solid angle collected are good for the signal, as is high quantum efficiency of the detector (about 30% for our chip). Noise comes from stray UV photons, dark current in the CCD chip, and the noise inherent in reading out the chip. As is the case with the velocity diagnostics employing laser fluorescence, the light from the beam must be accumulated over many injection cycles in order to obtain good statistics. The triggered mechanical shutter allows one to study a particular time window in the cycle while accumulating over many cycles. The cryogenic cooling reduces the dark current to completely negligible levels, allowing for long accumulation times.

The CCD chip employed is a 2.54 cm square and has 24 micron square pixels. Since only one transverse beam dimension is being characterized, the pixels can be binned in the longitudinal direction. The ultimate resolution of the system is determined by the

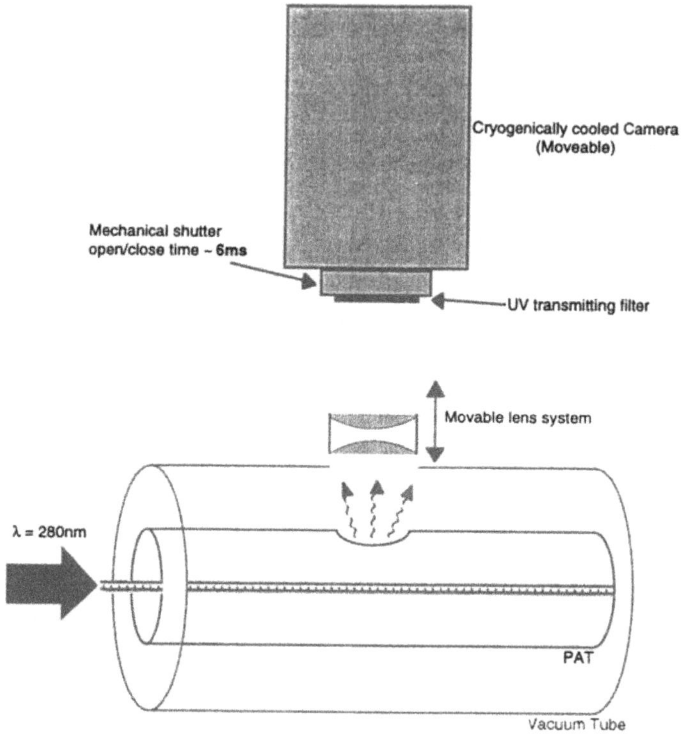

Figure 2. Schematic diagram of the imaging system for beam profile measurement. The post-acceleration tube (PAT) enables one to address a chosen velocity class by putting a voltage on the tube.

magnification (adjustable between 1 and 3 for the current system) and by the aberrations in the lens. There is a tradeoff between using large aperture lenses to maximize the light collected and using small apertures to minimize aberrations. Typical interparticle or inter-shell spacing for crystalline beams in ASTRID is about 50 microns, so, although we can't resolve individual high velocity ions, such a system can, in principle, resolve shell structure in a crystalline ion beam.

It is hoped that this system will have the necessary resolution and sensitivity to very accurately characterize the transverse phase space of the laser cooled beam in ASTRID. The first priority will be to try to duplicate, and hopefully learn to optimize, the sympathetic transverse cooling observed in the TSR - both with bunched and coasting beam. It is clear that the future of crystalline beams rests heavily on the ability to obtain efficient transverse cooling as well as the existing longitudinal cooling. If we can one day produce a crystalline beam, diagnosing its existence may be as easy as taking a photograph!

3. Acknowledgements

This work was supported by the Danish National Research Foundation through the Aarhus Center for Advanced Physics, and by the U.S. Department of Energy, Nuclear Physics Division, under Contract W-31-109-Eng-38.

4. References

1. J.S. Hangst *et al.*, *Phys. Rev. Lett.* **67**, 1238 (1991).
2. J.S. Hangst, Ph.D. Thesis, The University of Chicago, (1992), published as *Argonne Report* ANL/PHY-93/1.
3. J.S. Nielsen, *Optics Letters* **20**, 840 (1995).
4. J.S. Hangst *et al.*, Phys. Rev. Lett. **74**, 86 (1995).
5. V. Lebedev *et al.*, this volume.
6. J.S. Hangst *et al.*, *Phys. Rev. Lett.* **74**, 4432 (1995).
7. R. Grimm, private communication.
8. D. Habs, private communication.

Review of electron cooling investigation at the Novosibirsk INP

V. Parkhomchuk
Budker Institute of Nuclear Physics, Novosibirsk, 630090, Russia

Abstract

The history of discovery and the first experiments with electron cooling in Novosibirsk are presented. The question not finished on NAP-M and not clear up to now are presented for discussion at workshop.

1 Introduction

The electron cooling method was suggested by G. Budker in the middle sixties. The original idea of electron cooling published in 1966 [1]. The experiments were carried out on the special installation called NAP-M fig.1. The design activities at project NAP-M was started at November 1971 and first operation with proton beam was in September 1973. The first experiment with both beams electron and proton was started in May 1974. At this experiment was archived good result [2] very closed to theoretical prediction usual two component plasma heat exchange. But the main news about electron cooling would be find after few years experimental and theoretical investigation. In the course of the experiments carried out at NAP-M facility it was discovered that the time required for the cooling, expected to be several seconds, in fact turned out to be 0.1 sec. Such an abrupt increase in cooling efficiency was the result of a combined effect of two factors:first of all, the presence of the longitudinal magnetic field in cooling section and, secondly the electrostatic acceleration of the electron beam, which is accompanied by a considerable decrease in the longitudinal velocity spread of electrons. The magnetic field 'magnetizes' the transverse electron motion, and as result the cooling particles interact with a cool Larmor circle, but not with a hot free electron [3]. The effective temperature a Larmor circle only 1° K but a free electrons have temperature more them 2000° K! A temperature for the particles' longitudinal motion about 1^oK was obtained for proton beam with an energy of 65 MeV. Such a low temperature coursed some ordering in the particle position

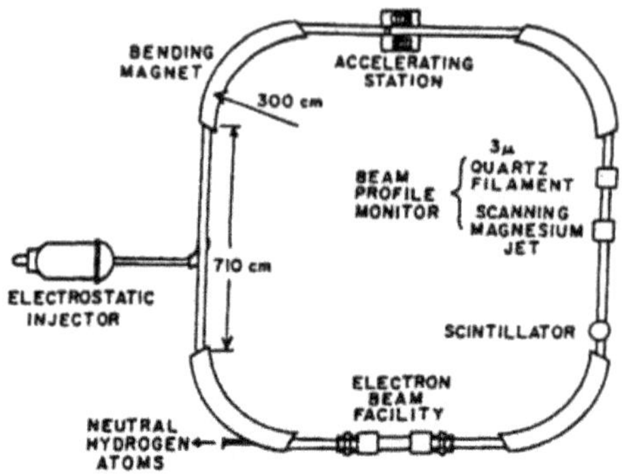

Fig. 1 - The NAP-M storage ring layout.

Figure 1: NAP-M

in the storage ring. The class of phenomena discovered aroused so much interest that the authors specifically called the process 'fast electron cooling'.

For a detailed study of the cooling under the condition of strong magnetization,an installation with a field of 4 kG and very good field homogeneity, was built. The experiments performed at this facility in the regime of single transmission of an ion through an electron beam were able to show an essential difference in the friction force for positively and negatively charge particles. I try to report more information about this investigation mainly concentrated on not clear up to now questions.

2 The fast electron cooling

After first successful experiments on NAP-M cooler was made modernization of magnetic field homogeneity and the electron gun designs. The main problem at initial version of electron cooler 'EPOHA' was a built-in ion pumps. The elements of this ions pumps was situated very close to electron beam and the fast neutral atoms from pumps can pass electron beam. The energy of this atoms was few keV and after ionizations by electron beam ions came to the walls of vacuum chamber. But the ionizations electron storage inside electron beam and the space charge electron beam increase too high. At time the first experiments with both beams very small electron beam immediately killed proton beam. After few days not successful experiments was founded that switch off internally ions pumps help to save proton beam.

For the study of influence the deviation of electron velocity on output the electron gun was installed the two plates which can made transverse electric field. This field exited the traverse Larmor rotation of electrons came out the electron gun. The result of measuring of the cooling time are shown in fig 2a.

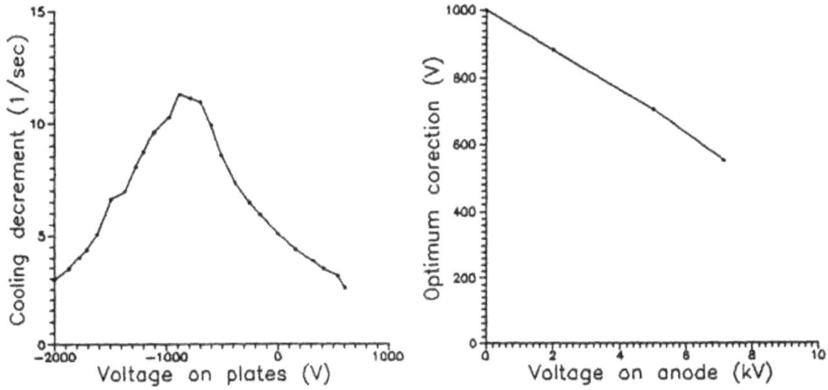

Figure 2: a-The inverses cooling time vs voltage on the plates for exiting Larmor rotation of electron beam,energy 65 MeV,electron current 0.3A. b-The optimum Larmor rotation correction vs voltage on the electron gun anode

It was very surprise for us that best cooling was obtain not for zero voltage on this plates. It means we have same misalignments at the electron gun and action this plates compensated this motion of electron beam. For testing this hypothesis we measured optimum for cooling voltage versus voltage on the gun anode fig.2b. Increasing the anode voltage decrease different potential cathode-anode and decrease amplitude of transverse motion as shown in fig.2b.

The measurement of cooling time vs amplitude of betatron motion are shown in fig.3. The maximum transverses decrement was very sensitive to the electron gun voltage ripples. For having the stable result we improved the high voltage power supply. The voltage ripple the cathode was not more them 10^{-5}. After improved all power supply we had maximum decrement more them 20 1/sec or cooling time less 0.05sec.

3 The equilibrium transverse and longitudinal beam momentum spread

For understanding of beam the equilibrium transverse and longitudinal momentum spread several methods were used. The most often we used the magnesium jet monitor. The thin jet crossed the proton beam and secondary ionization electron are collected by an electrostatic field on luminophor screen. The signal from photomultiplier which had optical contact with this screen was proportional density of proton current at position of the jet. The signal scanning the magnesium jet across the vertical beam direction shown on fig.3a.

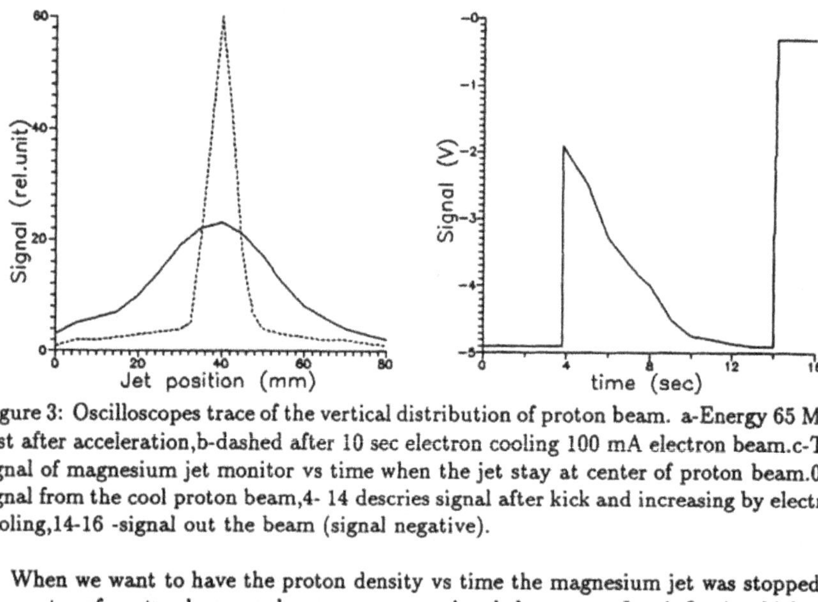

Figure 3: Oscilloscopes trace of the vertical distribution of proton beam. a-Energy 65 MeV just after acceleration,b-dashed after 10 sec electron cooling 100 mA electron beam.c-The signal of magnesium jet monitor vs time when the jet stay at center of proton beam.0-4-signal from the cool proton beam,4- 14 descries signal after kick and increasing by electron cooling,14-16 -signal out the beam (signal negative).

When we want to have the proton density vs time the magnesium jet was stopped in the center of proton beam and we measure as signal decreases after inflection kick and again return to old level at time of electron cooling. In the end of cooling we moved the magnesium jet out of proton beam for measure zero level of signal fig 4b. This result was obtain in 1995 and cooling time was near 3 sec. After cooling the resolution of this method was not enough and we used the flitting quartz wire [4] when thin 3mkm quartz wire pass the proton beam with velocities near 3m/sec and signal of electron emission from wire was proportional densities the proton beam. The result of measuring the beam diameter for fast cooling regime are shown on fig.4. The beam size extremely small and turn shift for this beam can be up to 0.05 as shown in fig 4b. The transverse momentum spread

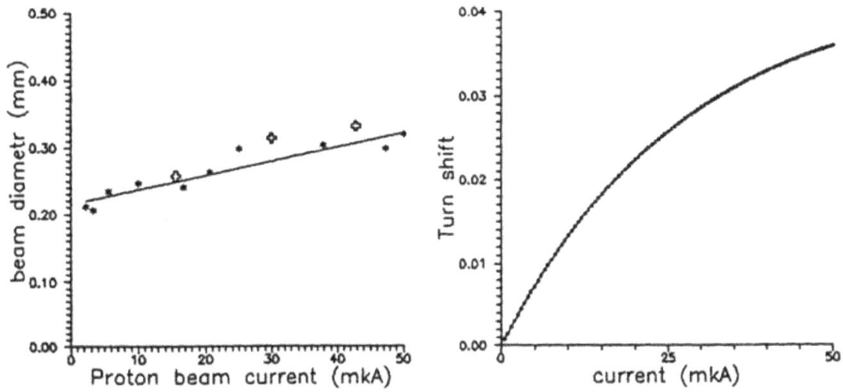

Figure 4: a-The beam diameter vs proton current. Energy 65 MeV,Electron beam current 0.3A. *measure by quartz wire (3mkm), +measure by hydrogen neutral atoms.b-The turn shift vs the beam current

for the small current was $\Delta p_{\perp}/p = a/\beta_{\perp} = 1.4 \cdot 10^{-5}$ and the transverse temperature 0.01eV or 100° K. The measuring of the longitudinal momentum spread of cooled proton beam was very hard problems. In first experiment we try used the information about the debunching time of proton beam. The debunching time of proton beam with a gap was measured versus the beam intensity using the method of knocking out a fraction of the beam (near 0.1* circumference) fig 5a.

At our report 1978 was used the relation between debunching rate τ^{-1} and the beam momentum spread:

$$\tau^{-1} = \frac{(\omega_0 \frac{\Delta p}{p}(\frac{1}{\gamma^2} - \frac{1}{\gamma_t^2}))^2}{\lambda_s}, \tag{1}$$

where λ_s-the longiutudinal cooling rate. In this model the momentum spread should be proportional of beam intensity at the power 1/3. But late analyses to show that in our condition the contribution of momentum spread at debanching time less them the contribution of the coherent space charge effects. On fig.5a. line with points show the

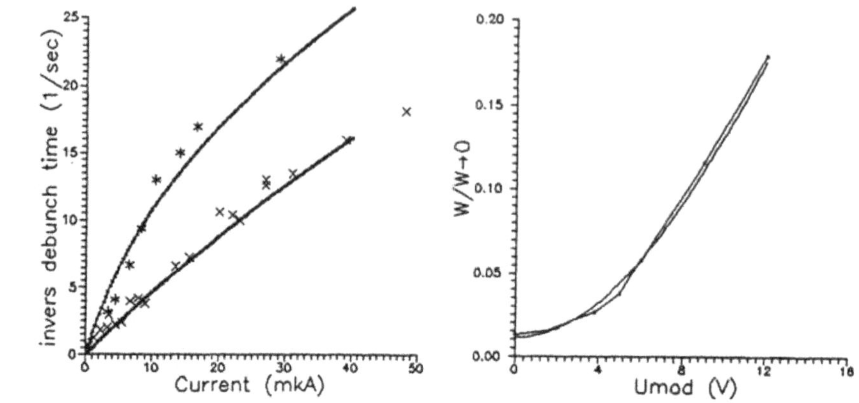

Figure 5: a-The inverses debanching time of proton beam versus the proton beam current. Upper date for the electron beam current 100 mA, lower one 300mA. (Energy 65MeV). The pointed lines is fit to expression (4)and ($\lambda_s = 20 and 70 1/s$). b-The ratio Schottky noise the cooled proton beam to signal not cooled beam versus amplitude of modulation. ($I_e = 300mA$)

calculating the debunching time at the wave space charge along beam:

$$\tau^{-1} = \frac{0.55\Omega_1^2}{\sqrt{\Omega_1^2 + \lambda_s^2}} \tag{2}$$

where Ω_1-coherent shift:

$$\Omega_n^2 = \frac{Ne^2n^2\omega_0 d\omega_0/dpZ}{2\pi R} \tag{3}$$

N-number of particles,e-electron charge,$\omega_0 = V_0/R$,Z-impedance of vacuum chamber. As you can easy see description of the debanching is not sensitive to the momentum spread. This fit is made to zero momentum spread!

For more direct measurement of the momentum spread we start measure the Schottky noise the proton beam. At first experiments we detected this signal for uncooled the proton beam. But after cooling this signal disappear and apparatus noise of the spectrum analyzers was at hundreds times more the beam signal. In this moment we cannot understood the reason of this effect. The Shottky noise is result of the voltage induced in pick-up electrodes motion of each proton. The proton circulating at the storage ring should induce in pick-up electrodes the periodical short pulses. If the motion of each protons is not correlated the spectral power from each proton is added to total Schottky signal. This experiment stimulated development the theory of the Schottky noise cool beam [5] and development new apparatus for measure Schottky noise [6] (1979). We understood that the signal to decrease so stronger by correlation at motion of protons. The energy of space charge fluctuation was too high for so small the beam temperature. As result on large distance scale the beam becomes more homogenous them it is possible from statistical limit $\pm\sqrt{N}$. The experiment shown that for small current the momentum spread was $\Delta p/p \approx 1.1 \cdot 10^{-6}$ or longitudinal temperature 1° K. The questions of Schottky noise result will be discusses at report Pestrikov D.V.. I want to show one interesting result from my point of view. In fig 5b are shown ratio of the Schottky power of cooling beam to the Schottky power uncooled beam (at both cases the proton current was 40 mkA). The fit was made at assumptions that the temperature of proton beam is equal the longitudinal temperature of electron beam:

$$T_p = T_e = T_0 + \frac{(eU_m * k)^2}{4E_k} \tag{4}$$

where T_0-the longitudinal temperature of electrons, E_k -kinetic energy of electrons, U_m-amplitude of modulation,k- factor of fitting. The pointed line in fig 5b is plotted for electron beam temperature 2.5° K and k=2.1. Why this result is wonderful for me. The intrabeam scattering for proton beam are equal to:

$$\frac{(dp_s/p)^2}{dt} = \frac{4\pi e^4 n_p L n_c}{v_\perp p^2}, \tag{5}$$

$n_p = 3.5 \cdot 10^7 1/cm^3$-the proton density,$Ln_c \approx 5$ -Colomb logarithm,$v_\perp = 1.410^5 cm/s$-the transverse velocity of the protons,p=Mv-the longitudinal momentum. For this parameters the longitudinal heating should be $\frac{(dp/p)^2}{dt} = 1.2 \cdot 10^{-6}$ 1/sec. It means for the cooling decrements 70 1/sec the equilibrium momentum spread should be $1.3 \cdot 10^{-4}$ but we had $\Delta p_s/p = 210^{-6}$ (2.5° K). How to suppress at foure order magnitude the power intrabeam heating? Our hypothesis was that the correlation at position the proton that suppressed the Schottky noise at the large distance (low frequency) are continued to smallest the distance between individual protons. At this case the longitudinal ordering have place and the intrabeam scattering should be suppressed [7].

4 The neutralization space charge electron beam

The fast electron cooling very sensitive to homogenous motion electron on the cooling section. Our measuring show that it is possible to have high enough cooling rate only

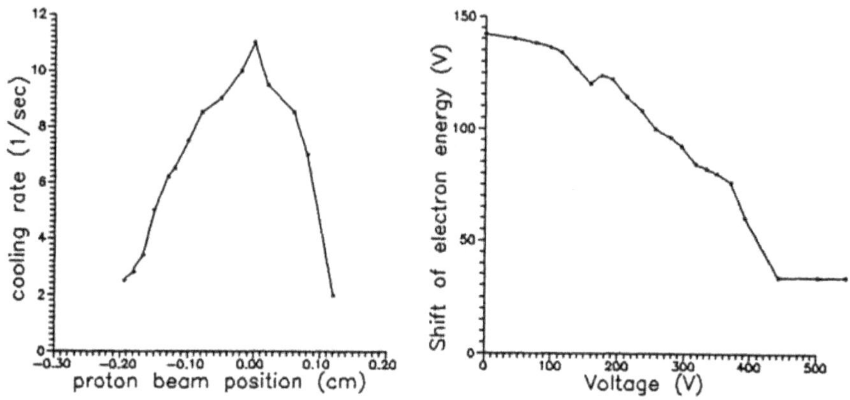

Figure 6: a-The cooling rate versus the proton beam position. b-The change optimum voltage on the electron gun versus the voltage feed to the ions traps electrodes ($I_e = 300mA$)

at very narrow region at electron beam. In fig 6a are shown the cooling rate versus the horizontal displacement of proton beam from optimum position.

The displacement less them 1mm coused decreasing decrement. For testing the influence the space charge field we try to neutralized the electron beam charge. For this on the special ion trap electrodes feeds the voltage for prevent the ions go out the cooling section. The ions storage on cooling section and neutralized the space charge electron beam. It effects on the values of the cathode potential for which the proton beam position was same. The changing of the cathode voltage for keeping the proton position are shown on fig 6b. As easy to see the neutralization looks perfectly. But cooling this neutralized electron beam was not too much improved. The first off all the coherent instability by interaction proton beams was happened with the proton current more them 5 mkA. But we not funded too large improvement at cooling and for small proton current.

The maximum currents for the electron beam with large length could be limited by the emergence of beam-drift instability in the electron-ion plasma. In an ultra-high vacuum ($10^{-8} - 10^{-10}$Torr) conditions it is possible to have perfect space charge neutralization but it seems same sort of art (not clear science)[8].

5 Single pass cooling experiment

For testing the ideas of fast cooling was constructed the single pass equipment[9]. This equipment had the electrostatic accelerator (energy 830keV) H^- ions,the solenoid with very homogenous magnetic field ($\Delta B/B = 10^{-5}$) from 1 to 4 kG, the electrostatic ions energy spectrometer. The change of ion charge sign is achieved with a special magnesium target at the solenoid entrance where the double ionization of the hydrogen negative ions occurs. The electron beam is formed in the electron gun placed in the magnetic field of the solenoid and it is transported along the solenoid magnetic field to an electron collector. The interaction of ions with the electron beam results in a variation of their energy and transverse velocities and this changing detected by electrostatic spectrometer. As example of the changing of ions energy in fig 6a shown the datc measuring for H^- and H^+ (electron current 3 mA,magnetic field 5 kG). The friction force for H^+ at few times more. With an increasing in electron current the friction force grows up and attains its maximum value at a current 8 mA (for B=4kG). This limitation are connected with grows up the longitudinal electron beam temperature.

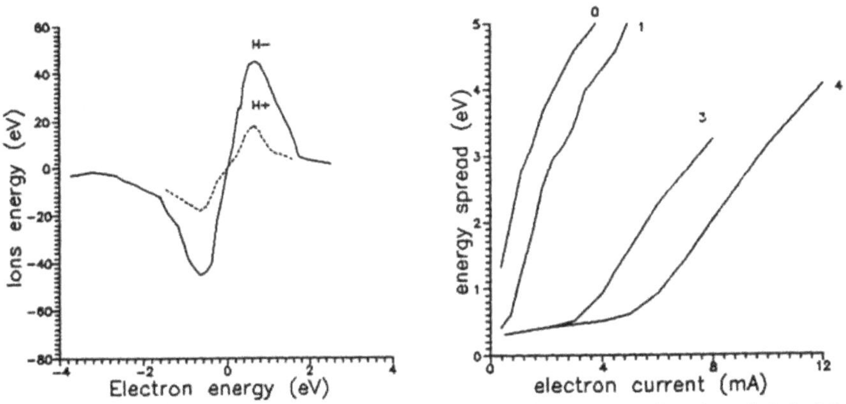

Figure 7: a-The friction force versus of electron energy (losses at cooling length 2.4m) for H^+ and H^- (B=4kG,I_e=3mA
. b-the energy spread of electron beam after passing of cooling section versus the electron current.

In fig6b shown the energy spread electron beam versus the electron current for different the magnetic field 0,1,3,4 kG. For B=0 the date are culculeted by expresion:

$$\frac{dT_s}{ds} = \frac{\pi e^3 j L_c}{W}\sqrt{\frac{m}{T_\perp}} \tag{6}$$

where s-longitudinal coordinates,L_c-Coulomb logarithm, the electron energy spread is $\Delta E = \sqrt{2WT_s}$. The suppressions of intrabeam scattering by the magnetic field can be describe the empirical formula:

$$\frac{dT_s}{ds} = \frac{\pi e^3 j L_c}{W}\sqrt{\frac{m}{T_\perp}} exp(-\frac{2.8e^2}{\rho_\perp(e^2 n_e^{1/3} + T_s)}) \tag{7}$$

where $\rho_\perp = \sqrt{2T_\perp mc^2}/(eB)$-the Larmor radius of electrons,T_s initial is $T_s = T_c^2/(2W) + 2e^2 n^{1/3}$. The maximum the friction force achieved for ion velocity value equal to: $v_m = \sqrt{2T_s/m}$.

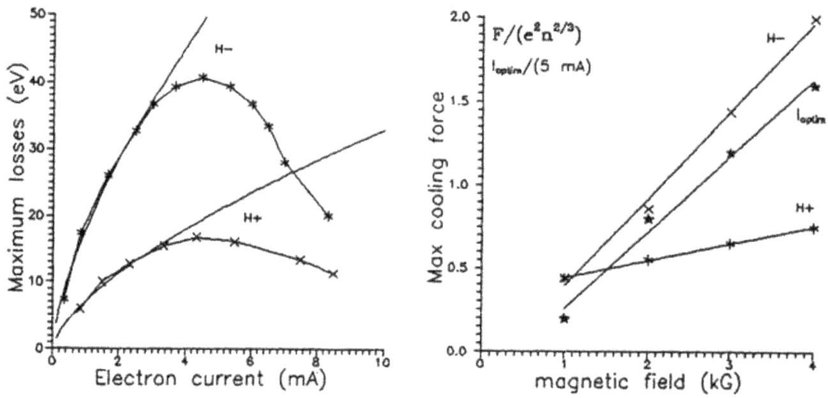

Figure 8: a-The maximum friction force losses versus of electron current (losses at cooling length 2.4m) for H^+ and H^- (B=3kG
.Solid line are obtained with expression $F_{max} = ke^2 n_e^{2/3}$, k=1.82 for H^- and k=0.72 for H^+.b-the maximum force and the optimum electron beam currents versus the solenoid magnetic field.

The maximum friction force measurements shown in fig8a. It is clear shown the optimum the electron beam current. In fig8b shown the maximum force (at units $e^2 \cdot n^{1/3}$ and optimum electron beam current versus the magnetic field.

6 Conclusion

Using the experimental information or empirical formulas it is possible to find the maximum cooling rate for the storage ring for study the crystallizing beams. It will be very interesting to try cool the negative ions or antiprotons. The minimum temperature for negative charge ions at few times less. The fast electron cooling have huge potential for this applications. The new ideas the produces the extremely cool electron beams by using lasser gun will open a lot of effects. The systems with the adeabatic cooling the transverse temperature (by magnetic field expantion) have problems with heating longitudional motion electron and seems not too perspective.

References

[1] Budker G.I. 1967. Efficient method for damping of particle oscillation in proton and antiproton storage ring Atomnaya Ehergiya 22, 346-8.

[2] Budker G.I.,Derbenev Ya.S.,Dikansky N.S.,Kudelainen V.I.,Meshkov I.N., Parkhomchuk V.V.,Pestrikov D.V.,Sukhina B.N.,Skrinsky A.N. 1975. Experiment on electron cooling,IEEE Trans. Nucl. Sci. NS-22,2093-7.

[3] Dikansky N.S,Kononov V.I,Kudelainen V.I,Meshkov I.N,Parkhomchuk V.V, Pestrikov D.V,Skrinsky A.N,Sukhina B.N, 1978. The study of fast electron cooling. Poc. 6 ALL-Union Meeting on Acceleration of Charged Particles (Dubna,1978)(Moscow, Nauka).

[4] Budker G.I,Dikansky N.S, Kudelainen V.I,Meshkov I.N,Parkhomchuk V.V, Pestrikov D.V,Skrinsky A.N,Sukhina B.N, 1978. Experimental study of electron cooling,1976,Particle Accelerarion 7 197-211.

[5] Parkhomchuk V.V.,Pestrikov D.V.Thermal noise of intensive beam at the storage ring. JTP 50 1411-18.

[6] Dementiev E.I.,Dikansky N.S,Medvedko A.S,Parkhomchuk V.V,Pestrikov D.V. 1980. Measurement of thermal noise of a proton beam at the NAP-M storage ring. JTP 50 1717-29.

[7] Parkhomchuk V.V. 1984. Physics of fast electron cooling. Proc.Workshop on Electron cooling and Related Applications,1984) ed. H.Poth (Karlsruhe:KfK).

[8] Burov A.V, Kudelainen V.I,Lebedev V.A,Parkhomchuk V.V,Sery A.A,Shiltsev V.D, Experimental Investigation of an electron beam in compensated state,Preprint 1993,CERN/PS 93-03(AR).

[9] Dikansky N.S., Kudelainen V.I,Lebedev V.A,Meshkov I.N, Parkhomchuk V.V, Sery A.A,Skrinsky A.N.,Sukhina B.N. Ultimate possibilities of electron cooling.Preprint INP 88-61 1988.

Measurements of the nonlinear electron cooling force at the IUCF Cooler Ring

S.Y. Lee, M. Bai, M. Ball, B. Brabson, J. Budnick, A. Chao[1], C.M. Chu, G. East, B. Hamilton, X. Kang, J.Y. Liu, K.Y. Ng[2], A. Pei, A. Riabko, T. Sloan, L. Wang, Y. Yan[1], X. Zhao

Indiana University Cyclotron facility, Bloomington IN 47405, USA

[1] *MS26, SLAC, Box 4349, Stanford, CA 94309*

[2] *Fermilab, P.O. Box 500, Batavia, Illinois 60510*

When the relative velocity between the proton beam and the cooling electrons is larger than the velocity spread of cooling electrons, protons are found to damp to limit cycle attractor arising from the nonlinear electron cooling force. This method is applied to measure the nonlinear cooling force at the relative velocity between the longitudinal and the transverse velocity spreads of cooling electrons. We found that the cooling force exhibits double peaks confirming the two component theory. Detailed analysis can be used to determine the electron cooling force more precisely.

1 Introduction

Since its invention by G.I. Budkar in 1966, *electron cooling*[1] has been widely applied to many storage rings for atomic, nuclear, particle[2], and accelerator physics research[3,4,5,6,7,8]. In particular, the electron cooling is seriously considered as the pre-cooler for crystal beams. Thus a better understanding of electron cooling would be beneficial to the research of crystal beams.

Recently, we have reported an experimental observation of the Hopf bifurcation in the synchrotron phase space when the relative velocity between the proton beam and the cooling electrons is greater than a threshold value[4,5,6]. We found that the threshold of bifurcation was related to the "temperature" or equivalently the longitudinal rms velocity spread of cooling electrons seen by protons. In these experiments, we found that the bifurcation could be asymmetric with respect to the relative velocity between the two beams, and the possible existence of the double peak in the cooling force. The asymmetry can arise from the combination of nonzero dispersion functions and the space charge depression of cooling electrons, the effect can be investigated by proper electron beam alignment. Thus the threshold of Hopf bifurcation will become symmetric for a properly aligned cooling system. This will also affect the amplitude of limit cycle attractor. Further experimental studies on Hopf bifurcation by varying the electron beam direction relative to the proton beam showed that the effective temperature was lowered by a factor of 3 and the resulting drag force exhibited characteristics of magnetized cooling[6]. However, the range of relative velocity between the proton beam and cooling electrons was limited due to the limitation of dynamical aperature for the off-momentum beam. This difficulty can be resolved by changing the energy of cooling electrons. This paper reports results of our new measurements where the energy of cooling electrons was varied.

The IUCF Cooler Ring [2] is a hexagonal shaped storage ring with a circumference of 86.8 m. The experiment was done with a 45 MeV proton beam injected and then stored with a 10 s cycle time. After 5 s from the start of the cycle, the 6 dimensional phase space coordinates were digitized at 10-revolution intervals for 16384 points[7,8]. The nominal rf cavity frequency was 1.03168 MHz with the harmonic number $h = 1$. The beam was typically a single bunch of about 4×10^9 protons with a typical length of about 100 ns FWHM with the rf peak voltage being about 41 V. For the experimental results reported in this paper, the rf voltage was varied between 40 V and 128 V. Since measurements of longitudinal motion were being made, the phase lock feedback loop for the rf, which is normally on, was switched off.

2 Synchrotron motion with damping

Let (ϕ, δ) be conjugate synchrotron phase space coordinates, where ϕ is the phase of a particle relative to the rf cavity voltage, and $\delta = \frac{\Delta p}{p}$ is the fractional momentum deviation of the particle from that of the synchronous particle. The difference equations describing the synchrotron motion are

$$\delta_{n+1} = \delta_n - \frac{2\pi\nu_s^2}{h\eta}(\sin\phi_n - \sin\phi_s) - f(\delta_n) \qquad (1)$$

$$\phi_{n+1} = \phi_n + 2\pi h\eta\delta_{n+1}, \qquad (2)$$

where $\eta = -0.86$ is the phase slip factor for the IUCF Cooler Ring at 45 MeV proton kinetic energy, ϕ_s is the phase of the synchronous particle, which for a stored beam is $0°$, h is the harmonic number, $f(\delta)$ is the damping force resulting from electron cooling, and the subscripts n refer to the revolution number. The small amplitude synchrotron tune ν_s at the zero synchronous phase is related to the rf cavity voltage V_{rf} by $\nu_s = \sqrt{h|\eta|eV_{rf}/2\pi E\beta^2}$, where $e, \beta c$, and E are respectively the charge, the speed, and the total energy of the proton. The angular synchrotron frequency is given by $\omega_s = \omega_0\nu_s$, where ω_0 is the angular revolution frequency.

The damping force $f(\delta)$ produced by the electron cooling is the result of a statistical exchange of energy in Coulombic collisions between the protons and relatively cold electrons as they travel together in the accelerator. In practice, electron cooling in synchrotrons is normally done in relatively short straight sections due to cost and space limitations. At IUCF the electron beam is mixed with the proton beam for distance of only 2.2 m or about 2.5% of the accelerator circumference. The electron beam radius is about 1.27 cm and the cathode temperature is about 1300°K or $kT_{\rm cath} = 0.11$ eV, where k is the Boltzmann constant. The maximum electron beam current is 4 A. The electron beam current used in this experiment was varied up to 1.5 A.

Assuming an electron beam with an isotropic phase space Maxwellian velocity distribution, the damping force in the non-magnetized binary collision theory is given by

$$f(\delta) = \frac{4\pi\alpha\Delta_{e\|}}{\omega_0} g(\zeta), \tag{3}$$

with a kinematic factor given by $g(\zeta) = g_0(\zeta)$ with

$$g_0(\zeta) = \frac{3\sqrt{\pi}}{4\zeta^2}\left[\mathrm{erf}(\zeta) - \frac{2\zeta}{\sqrt{\pi}}e^{-\zeta^2}\right], \tag{4}$$

where α is the $1/e$ damping rate for small relative velocities, $\Delta_{e\|} = \sigma_e/\beta c$ with σ_e as the rms velocity of cooling electrons, and $\zeta = (\delta - \delta_e)/\Delta_{e\|}$ with δ_e as the fractional momentum deviation of the proton traveling at the velocity of cooling electrons with respect to the synchronous particle of the rf cavity. From our previous measurements[4,5], we have $\alpha \approx 40$ s^{-1} and $\Delta_{e\|} \approx 3 \times 10^{-4}$ for the IUCF electron cooling system for a proton kinetic energy of 45 MeV. In our later experiments with optimal electron beam alignment, the $\Delta_{e\|}$ was found to be reduced by a factor of 0.6. This led to a reduction of the effective temperature by a factor of 3.

The effective temperature of the electron beam is related to the rms electron velocity spread by

$$kT_{\|\rm eff} = \frac{1}{2}m_e\sigma_e^2 = \frac{1}{2}m_e\beta^2c^2\Delta_{e\|}^2. \tag{5}$$

Because of the adiabatic acceleration, the longitudinal effective electron temperature is much smaller than that of the cathode temperature. Since there is no adiabatic damping in the transverse phase space, the effective transverse temperature remains 0.11 eV, which gives an equivalent momentum deviation of $\Delta_{e\perp} = 2.2 \times 10^{-3}$. For comparison, the bucket height of the synchrotron phase space at a rf voltage of 128 V is about 1.0×10^{-3}.

Since the cooling drag force of Eq. (4) is maximum at $|\delta - \delta_e| \approx \Delta_{e\|}$, it is evident that the center of the rf phase space will become unstable when $|\delta_e| > \Delta_{e\|}$[4,5]. The stability is ensured only if the damping effect of synchrotron motion with $\delta > 0$ has been exactly canceled by the excitation of synchrotron motion in the region $\delta < 0$, or vice versa. The balance between the damping and the excitation leads to the limit cycle attractor.

Since the threshold of Hopf bifurcation is related to the rms velocity spread of cooling electrons along the proton trajectory, the asymmetric Hopf bifurcation can arise from different velocity spreads seen by protons having a different momentum closed orbit. The increase in the rms velocity spread can result from misaligned proton beam in combination with the space charge depression of the electron beam given by

$$E(r) = E_{ax} + n_e \pi r^2 r_e m_e c^2, \quad E_{ax} = eU_{cath} - n_e \pi r_0^2 r_e m_e c^2 [1 + 2\ln \frac{r_{tube}}{r_0}], \quad (6)$$

where n_e is the electron beam density, r_0 is the electron beam radius, r_{tube} is the radius of the vacuum chamber in the acceleration column (see Eq. (98) of Ref. [1]). Thus the mis-alignment of the cooling electron beam can contribute to the asymmetry in the Hopf bifurcation, which was commonly observed in our earlier experiments. For a properly aligned electron beam, space charge depression will shift the relative velocity without affecting the velocity spread.

Since electrons having identical speed as those of protons have a much smaller momentum rigidity, the electron beam alignment can easily be achieved by superimposing a horizontal or a vertical dipole field to the solenoidal field. A method to obtain an "optimal" alignment is described as follows: When the rf frequency is shifted, the beam, which is originally at the center of the rf bucket (i.e., $\delta = 0$ and $\phi = 0$), will be dragged away from the origin and undergo synchrotron oscillations. If the damping force were linear over the entire range of v_{rel}, the proton beam would damp to a new fixed point attractor (FPA) ϕ_{FP}, where

$$\phi_{FP} \approx \frac{2\alpha}{\omega_0 \nu_s} \frac{h\eta\delta_e}{\nu_s}. \quad (7)$$

The resulting ϕ_{FP} is the synchronous phase angle ϕ_s of the beam.

3 Linear cooling force

In order to permit us to precisely measure the synchronous phase angle ϕ_{FP} shown in Eq. (7), we choose a small rf cavity voltage with $V_{rf} = 10$ V, where the corresponding synchrotron tune is 1.25×10^{-4} at $h = 1$. The rf synchronous phase angle is measured by stepping away from the reference frequency by 200 Hz [9]. Figure 1 shows the drag force, which is $eV_{rf} \sin \phi_{FP}$, vs the horizontal and vertical dipole currents (in Digital Analog Converter or DAC units), where the lower plot gives the drag force as a function of the vertical dipole strength V_{sol} with $H_{sol} = 160$ DAC. The maximum drag force is found to be about 1.8

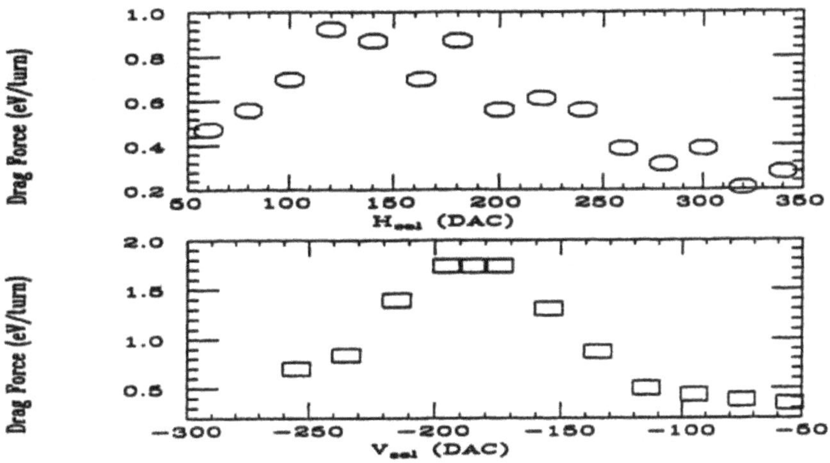

Figure 1: The measured damping drag force is plotted vs the horizontal and the vertical electron beam alignments in the upper and lower plots respectively. The lower plot shows the drag force vs V_{sol} with an optimal $H_{sol} = 160$ DAC value obtained from the upper plot. The DAC (Digital Analog Converter) unit in the coordinate is proportional to the current applied to the dipole coil winding.

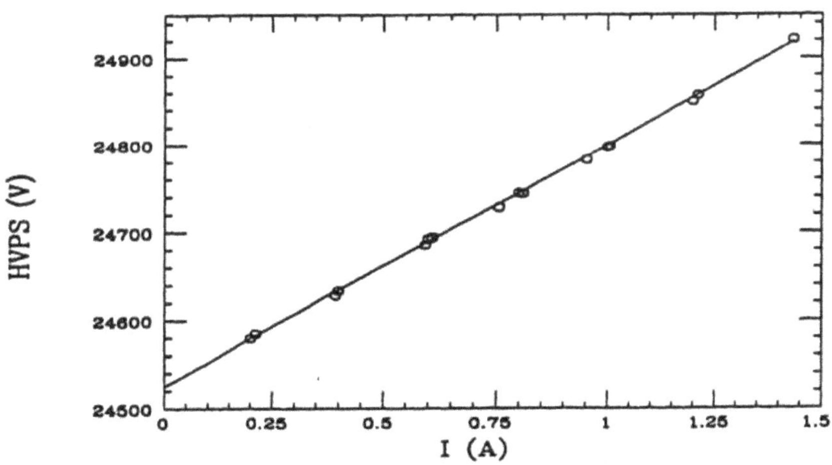

Figure 2: The measured electron beam energy on axis as a function of the current of cooling electrons is compared with Eq. (6).

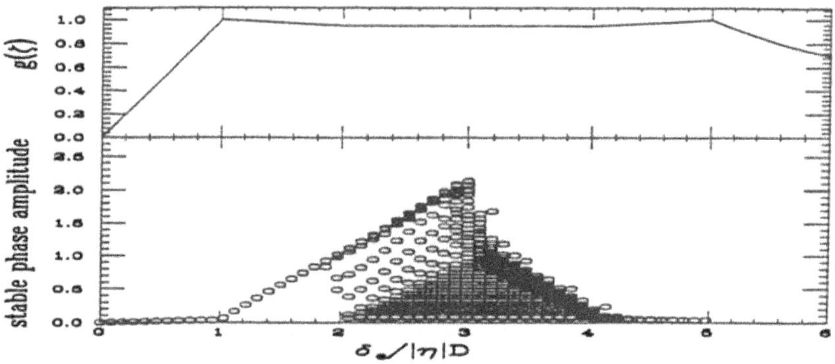

Figure 3: The lower plot shows the attractor solution and the stable synchrotron amplitude obtained from numerical simulations vs $\delta_e/|\eta|D$. The upper plot shows the corresponding kinematic function of Eq.(8) with parameters $\zeta_1 = 1, S_1 = -.05, \zeta_2 = 2, S_2 = 0, \zeta_3 = 4, S_3 = 0.05$, and $\zeta_4 = 5$ vs $\zeta = (\delta - \delta_e)/D$.

eV/turn. From Eq. (7), we obtain $\alpha = 45\ s^{-1}$, which agrees well with the value of 40 s^{-1} obtained from an earlier measurement by using the harmonic modulation to the HVPS (see Sec. III A of Ref. [5]).

Once the cooling rate is optimized, the energy of cooling electrons can be identified as E_{ax}. Figure 2 shows the measured E_{ax} as a function of the current of cooling electrons. The straight line is the theoretical curve with a geometric factor 2.74, which corresponds to $r_{tube} = 2.4r_0$.

4 Nonlinear cooling force

In the presence of magnetized cooling, the cooling drag force can be divided into two components, where the adiabatic component is maximum at the rms longitudinal velocity spread of cooling electrons and the non-magnetized component is peaked at the transverse rms velocity spread of cooling electrons. Thus we simulate the magnetized cooling with the following simplified drag force model represented by straight lines:

$$g(\zeta) = \begin{cases} \zeta & \text{if } 0 \leq \zeta < \zeta_1 \\ \zeta_1 + S_1(\zeta - \zeta_1) & \text{if } \zeta_1 \leq \zeta < \zeta_2 \\ \zeta_1 + S_1(\zeta_2 - \zeta_1) + S_2(\zeta - \zeta_2) & \text{if } \zeta_2 \leq \zeta < \zeta_3 \\ \zeta_1 + S_1(\zeta_2 - \zeta_1) + S_2(\zeta_3 - \zeta_2) + S_3(\zeta - \zeta_3) & \text{if } \zeta_3 \leq \zeta < \zeta_4 \\ [\zeta_1 + S_1(\zeta_2 - \zeta_1) + S_2(\zeta_3 - \zeta_2) + S_3(\zeta_4 - \zeta_3)]\zeta_4^2/\zeta^2 & \text{if } \zeta_4 \leq \zeta \end{cases} \tag{8}$$

with $g(-\zeta) = -g(\zeta)$, $\zeta = (\delta - \delta_e)/D$, and $D = 3 \times 10^{-4}$. Figure 3 shows the final equilibrium phase amplitudes of 40 particles, which are initially uniformly distributed along the $\delta \geq 0$ axis by solving Eq. (1) and (2) with $\nu_s = 0.0003, \eta = -.86$. The corresponding drag force model of Eq. (8) with parameters $\zeta_1 = 1, S_1 = -0.05, \zeta_2 = 2, S_2 = 0, \zeta_3 = 4, S_3 = 0.05$, and $\zeta_4 = 5$ is shown in the upper plot. The following observations can be drawn:

1. $S_1 \leq 0$ in the region $\zeta \in [1, 2)$ produces a limit cycle attractor (LCA) for $\delta_e/|\eta|D \geq 1$.
2. $S_2 = 0$ in the region $\zeta \in [2, 4]$ produces a center *damping-free* core for $\delta_e/|\eta|D \in [2, 4]$, where the center core and the LCA coexist. The limit cycle attractor becomes discontinuous at $\delta_e/|\eta|D \approx 3$.
3. The positive drag force slope of $S_3 = 0.05$ in the region [4,5] revitalizes the fixed point attractor (FPA) for $\delta_e/|\eta|D \in [4, 5]$;
4. For $\delta_e/|\eta|D > 5$, the synchrotron motion becomes unstable, which has resulted from a higher excitation rate of synchrotron oscillations at $\delta \leq 0$ than the damping of synchrotron motion at $\delta > 0$.

Thus an essential signature for a double peak in the electron cooling force is the re-appearance of FPA inside the LCA.

4.1 Possible hint of FPA inside the LCA

One method to investigate the Hopf bifurcation due to the nonlinear damping force is to displace the proton velocity from the electron velocity by changing the rf cavity frequency. Let f_0 be the rf cavity frequency when the synchronous proton travels at the same velocity as that of cooling electrons. At a new rf frequency f, the fractional momentum deviation, δ_e, of the electron beam from the proton beam becomes

$$\delta_e = \frac{(f - f_0)}{\eta f_0}, \tag{9}$$

where a step of 1 Hz resulted in changing the fractional proton velocity by about 1×10^{-6}. The maximum synchrotron phase amplitude $\hat{\phi}$, and the maximum fractional momentum deviation $\hat{\delta}$ are measured at 5 s after the start of an injection cycle to allow the initial transient oscillations to damp out.

As shown in our earlier reports[4,5], the amplitudes of limit cycle attractors (LCA) can also be measured by using an oscilloscope to measure the longitudinal profile from beam position monitor (BPM) sum signal. The measured amplitudes with the optimal drag force at $H_{sol} = 160$ DAC, $V_{sol} = -185$ DAC,

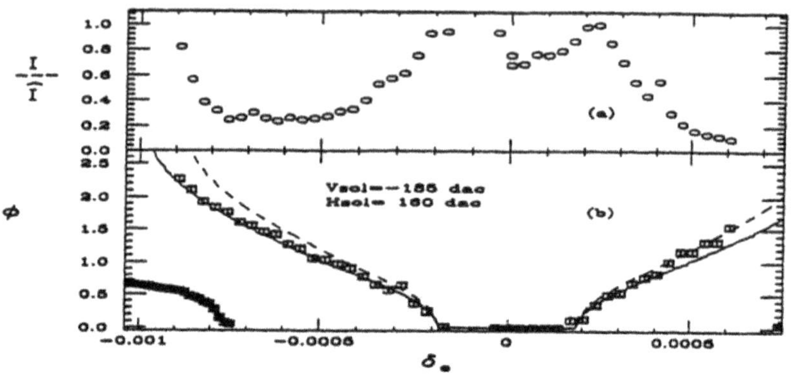

Figure 4: The measured amplitude of limit cycle attractor and the storage beam current observed vs δ_e as defined in Eq. (9) for the optimized cooling alignment, which provides a nearly symmetric Hopf bifurcation. The dashed line is obtained from numerical simulations with kinematic factor (4) with $\Delta_{e\parallel} = 1.89 \times 10^{-4}$. The tracking was done for the time equivalent of $160\tau \approx 4$ s, where the damping time is $\tau = 1/\alpha$. The solid line is the limit cycle attractor (LCA) and the shaded area in the lower left corner of the plot (b) is the region that converges to the fixed point attractor (FPA), obtained from numerical simulations using the kinematic factor (10) with $\Delta_{e\parallel} = 1.8 \times 10^{-4}, \Delta_{e\perp} = 2.2 \times 10^{-3}$, $A_\parallel = 1$, and $A_\perp = 0.5$.

and $V_{rf} = 128$ V are shown in Fig. 4b. The corresponding beam intensity normalized to the peak current is shown in Fig. 4a. Note in particular that Fig. 4 shows a nearly symmetric amplitude of the LCA vs δ_e and a close correlation between the storage beam current and the Hopf bifurcation threshold [10]. On the other hand, when the electron and proton beams were not properly aligned, we always observed asymmetric Hopf bifurcations.

To understand the nature of this instability, the dashed line shown in Fig. 4b is the result of computer simulations using Eqs. (1) and (2) with a damping force of Eqs. (3) and (4) for $\Delta_{e\parallel} = 1.89 \times 10^{-4}$, or $kT_\parallel = 8 \times 10^{-4}$ eV. Thus the temperature of the properly aligned cooling electron beam appears to be colder than that has been observed earlier [4,5]. We note also that amplitudes of limit cycle attractors are smaller than those predicted by the drag force model of Eq. (4).

When the cooling electron beam is magnetically confined by a solenoidal field, such as the IUCF Cooler Ring, the damping force can be decomposed into two components, where the magnetized cooling can enhance the effective longitudinal and transverse cooling rates (see Fig. 15 of Ref. [1]), and the longitudinal cooling force may exhibit two peaks. We are particularly interested in determining the cooling drag force for relative velocities lying between the longitudinal and transverse rms values. The measured amplitudes of limit cycle attractors enable us to study the drag force at the relative momentum deviation δ_e lying in the range of $(\Delta_{e\parallel}, \Delta_{e\perp})$.

To simulate the two component cooling force, we postulate the kinematic function for the magnetized cooling as follows:

$$g(\zeta) = A_{\parallel} g_0(\zeta) + A_{\perp} g_0(\frac{\Delta_{e\parallel}}{\Delta_{e\perp}}\zeta), \tag{10}$$

where $\zeta = (\delta - \delta_e)/\Delta_{e\parallel}$, $g_0(x)$ is given in Eq. (4), $\Delta_{e\parallel}$ and $\Delta_{e\perp}$ are respectively the longitudinal and transverse velocity spread of cooling electrons, and $A_{\parallel}$ and $A_{\perp}$ are respectively the adiabatic and the non-magnetized drag force components. Such a cooling kinematic function can be obtained with the assumption of Maxwellian distribution functions for cooling electrons. The solid line in Fig. 4b is obtained from tracking calculation with parameters $\Delta_{e\parallel} = 1.8\times10^{-4}$, $\Delta_{e\perp} = 2.2\times10^{-3}$, $A_{\parallel} = 1, A_{\perp} = 0.5$, and the synchrotron tune $\nu_s = 4.46\times10^{-4}$. The area in the lower left corner enclosed by rectangles corresponds to initial phase amplitudes which damp to a fixed point attractor at the center of the phase space. A particle with an initial synchrotron amplitude outside the shaded area will damp to the limit cycle attractor or become unstable. Note in particular that a positive slope in the kinematic factor of Eq. (10) can simultaneously generated a limit cycle attractor and a fixed point attractor for $\delta_e \leq -.00075$. The magnetized cooling at the relative velocity lying between the longitudinal velocity and the transverse velocity spreads create simultaneously a limit cycle attractor and a fixed point attractor at the center of the rf bucket. The existence of a fixed point attractor evidently provides the mechanism for a beam accumulation, which produced the higher proton beam current shown in Fig. 4 at $\delta_e < -0.0008$. Thus the electron cooling system is employing transverse velocity spread to enhance the synchrotron phase space damping when the relative velocity is larger than the longitudinal velocity spread of cooling electrons. Because of the aperture limitation of a realistic machine, such as the IUCF cooler ring, experimental measurements at $|\delta_e| > 0.0008$ were not possible.

4.2 Observation of FPA inside the LCA

Because of the dynamical aperture limitation of the off-momentum proton beam, it was difficult to attain a large relative velocity between protons and cooling electrons by changing the proton beam energy. Alternatively, one can also vary the relative velocity by changing the electron beam energy with the high voltage power supply (HVPS), i.e.

$$\delta_e = \frac{\gamma^2}{\beta c} v_{\rm rel}^{\rm lab} = \frac{1}{\beta^2} \frac{\Delta E_e}{E_e}, \tag{11}$$

where β, γ are the usual relativistic factor for the synchronous particle, $v_{\rm rel}^{\rm lab}$ is the relative velocity in the laboratory frame, E_e is the electron beam energy. Here 1 eV difference in electron energy will result in a change of $\Delta\delta_e \approx 2.1 \times 10^{-5}$. To attain proper energy resolution, we redefine the 16-bit DAC to 60 kV power supply. Thus 1 DAC corresponds to $\Delta\delta_e = 1.9 \times 10^{-5}$.

Following the experimental procedures discussed earlier by optimizing the electron beam alignment, we measured the linear and nonlinear cooling force by varying the electron beam energy. The existence of FPA inside LCA would provide a clear evidence of a double peak in the cooling drag force. Figure 5 shows a typical longitudinal beam profile (derivative type) from the signal of a medium bandwidth wall gap monitor (upper trace) after Hopf bifurcation. The rf wave signal from the pickup in the cavity is shown in the lower trace for reference. We note that the two peaks marked by vertical line correspond to the limit cycle and the central peak corresponds to a small bunch structure at the center of the bucket.

Figure 6 shows the beam current and the amplitude of the LCA as a function of δ_e. Note here that a small increase in beam current at $\delta_e \approx \pm 0.0007$ correlated with the existence of the central peak. Our experiments showed that the double peaks in the cooling force was observable for electron beam current between 0.6 to 1 A with an rf voltage less than 80 Volts. If the rf voltage is higher than 100 V, the bucket height becomes large so that the synchrotron motion would average a large range of the cooling force. The coexistence of FPA and LCA becomes less evident.

There are two characteristics in Fig. 6 worth mentioning. The first is the threshold of Hopf bifurcation, i.e. $\Delta_{e\|}$, where FPA becomes LCA. Since the bifurcation occurs at $\Delta_{e\|} = \sigma_e/\beta c$, where σ_e is the velocity spread of cooling electrons we identify the longitudinal temperature as

$$T_{\|} = \frac{1}{2} m_e \beta^2 c^2 \Delta_e^2. \tag{12}$$

The second important signature is the beam current drop to zero suddenly at $\delta_e \approx \pm .0018$. As we have learned from our model calculations that there is no possibility for beam accumulation beyond the second peak in the cooling force, we can identify the location where the beam current becomes zero as the transverse temperature, i.e.

$$T_{\perp} = \frac{1}{2} m_e \beta^2 c^2 \hat{\delta}_e. \tag{13}$$

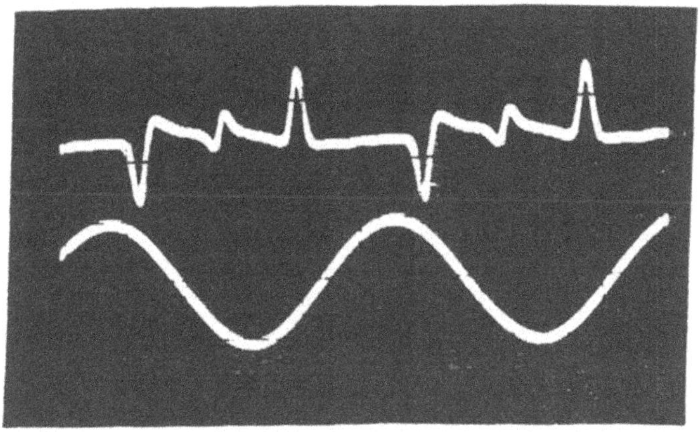

Figure 5: The upper trace shows the beam profile obtained from signals of a low bandwidth wall gap monitor on an oscilloscope. The resulting signal is the derivative of the beam profile. The lower trace shows the reference rf signal.

The left plot of Fig. 7 shows the longitudinal temperature and the right plot shows the transverse temperature. The solid line on the left plot shows the theoretical prediction given by

$$T_{\parallel th} = \frac{1}{8} m_e \beta^2 c^2 \left(\frac{\sigma_U}{U_{cath}} + \frac{kT_{cath}}{E_0}\right)^2 + 2 m_e c^2 \frac{r_e}{r_{av}}, \qquad (14)$$

where $\sigma_U \approx 3$ V is the power supply ripple of the HVPS, $kT_{cath} \approx 0.11eV$, and r_{av} is the average distance between cooling electrons. Note that the measured temperature is about 5 times higher than that predicted by the theory. The discrepancy can arise from the non-uniformity of the solenoidal field in the cooling section so that protons see electrons with different energies due mainly to space charge depression.

On the other hand, the effective transverse temperature is smaller than the 0.11 eV of the cathode. This can result from the fact that the outer peak move inwards when the two components of cooling force add together.

5 Conclusion

From our experimental data and extensive numerical simulations, we summarize our findings as follows. The Hopf bifurcation occurs when the slope $S_1 \leq 0$.

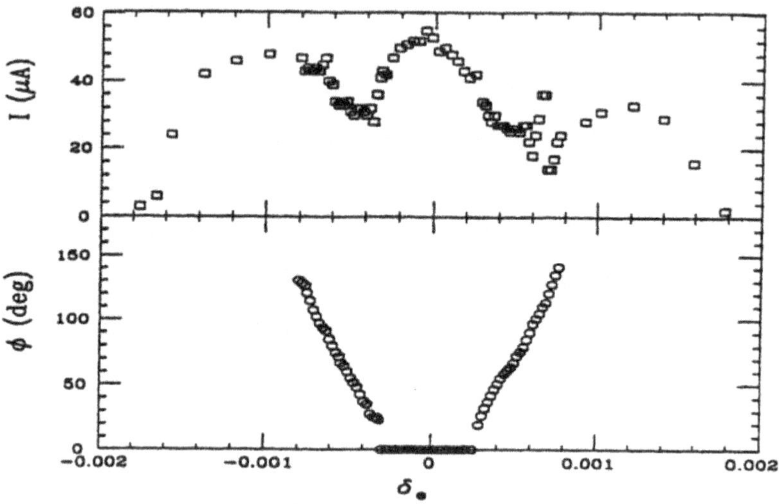

Figure 6: The upper plot shows the beam current as a function of δ_e and the lower plot shows the amplitude of limit cycle attractor.

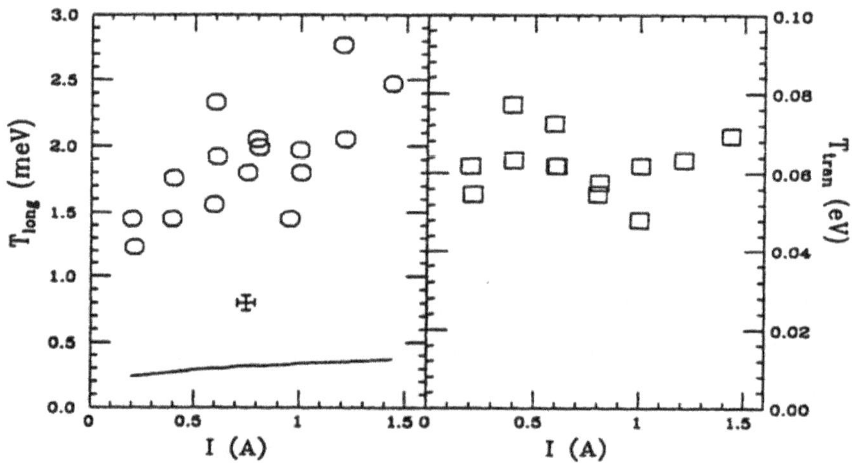

Figure 7: The left plot shows the longitudinal temperature, which is obtained from the threshold of the Hopf bifurcation, where circles are results from present experiments and the cross corresponds to the temperature obtained from Fig. 4. The right plot shows the transverse temperature obtained from the maximum δ_e where the beam current becomes zero.

The center damping free region can create by a local zero slope in the damping force, i.e. Hopf bifurcation does not necessarily imply an unstable center fixed point. Furthermore, the limit cycle attractor can co-exist with a fixed point attractor (see Fig. 4b).

In conclusion, we have studied the effect of electron beam alignment on the Hopf bifurcation. When the electron and proton beams were carefully aligned, the amplitudes of limit cycle attractors became symmetric with respect to the relative velocity between the cooling electrons and the synchronous proton. The non-magnetized drag force model with the kinematic factor of Eq. (4) fails to fit the data when the amplitude of the limit cycle attractor is large. The physics of the limit cycle attractor has been employed to determine the drag force in the region $\delta_e \in (\Delta_{e\parallel}, \Delta_{e\perp})$. By changing the electron beam energy, the fixed point attractor was observed to coexist with the limit cycle attractor confirming some essential characteristics of the magnetized cooling. The transverse velocity spread of cooling electrons plays a key role in beam accumulation when $\delta_e \in (\Delta_{e\parallel}, \Delta_{e\perp})$. Extensive numerical simulations have been performed to study effects of the drag force on particle motion. We find that the limit cycle attractor can coexist with a stable center core if the slope of the damping force become zero. A possible application of this type of damping force is to create a more uniformly distributed beam bunch in order to reduce space charge force in a beam at low energies.

Acknowledgments

This work was supported by grants from NSF PHY-9221402 and DOE DE-FG02-93ER40801.

References

1. H. Poth, Phys. Reports **196**, 135 (1990).
2. R.E. Pollock, Ann. Rev. Nucl. Sci. **41**, 357 (1991).
3. G.I. Budkar *et al.*, Particle Accelerators, **7**, 197 (1976);
4. D.D. Caussyn *et al.*, Phys. Rev. Lett. **73**, 2857 (1994).
5. D.D. Caussyn *et al.*, Phys. Rev. E **51**, 4947 (1995).
6. S.Y. Lee *et al.*, Phys. Rev. E **52**, to appear (1995).
7. S.Y. Lee, *et al.*, Phys. Rev. Lett. **67**, 3768 (1991); D.D. Caussyn *et al.*, Phys. Rev. **A46**, 7942 (1992); J.Y. Liu *et al.*, Phys. Rev. E **49**, 2347 (1994); M. Ellison *et al.*, Phys. Rev. E **50**, 4051 (1994).
8. M. Ellison, *et al.*, Phys. Rev. Lett. **70**, 591 (1993); H. Huang *et al.*, Phys. Rev. **E48**, 4678 (1993); D. Li, *et al.*, Phys. Rev. **E48**, R1638

(1993); M. Syphers *et al.*, Phys. Rev. Lett., **71**, 719 (1993); Y. Wang *et al.*, Phys. Rev. **E49**, 1610 (1994), J.Y. Liu *et al.*, Phys. Rev. **E50**, R3349 (1994), J.Y. Liu *et al.*, Particle Accelerators, **49**, 221 (1995), D. Li, *et al.*, Nucl. Inst. and Methods **A364**, 205 (1995).

9. The shift in frequency should be minimized to avoid the nonlinear region of the damping force. In our case, we measured the phase differences between ± 100 Hz shifts in the revolution frequency f_0.
10. The optimal cooling electron beam alignment depends on the closed orbit of proton beam. Thus the optimal setting of H_{sol} and V_{sol} DAC values can be different for each run. In our experiment, we found that symmetric Hopf bifurcation occurred only when the electron beam was aligned with that of proton beam. Thus, asymmetric Hopf bifurcation can be used as a signature for the misaligned cooling electron beam.

REVIEW OF RECENT WORK ON ELECTRON COOLING AT LEAR

Jacques BOSSER*
CERN, 1211 Geneva 23, Switzerland

ABSTRACT

The Low-Energy Antiproton Ring (LEAR) is expected to be used as an accumulator in the framework of the LHC lead-ion project. This implies at first that the electron-cooled ions have a lifetime which is longer than the stacking time. On the other hand, in order to improve the electron-cooling time we expect to make use of large electron beam intensities, of the order of 0.5 A, at relativistic parameter $\beta = v/c = 0.09$. With such constraints the electron beam density is rather large so the induced space-charge forces become detrimental to the cooling process itself. A neutralization of the electron beam, which aims to reduce the effects of the space-charge forces, is therefore desirable. This paper describes the measurements made on the cooling and lifetimes and on the neutralization technique implemented on the LEAR electron cooler.

1. Introduction

The Low-Energy Antiproton Ring (LEAR) is expected to be used as a lead-ion accumulator in the framework of the LHC project.[1] Multistacking and efficient electron cooling should allow one to obtain the required beam intensity and the requested small emittances.

The main LEAR and Electron Cooler parameters are given in Table 1.

Table 1. LEAR and Electron Cooler main parameters for the lead-ion project.

LEAR for lead ions	
Energy	4.2 MeV/u
Velocity factor β	0.094
Ion charge state	52^+, 53^+, or 54^+
Ion beam current	20–25 μAe
Corresponding number of ions	$2.3\text{–}2.8 \cdot 10^6$ ions/μs
Stacking process time [s]	$\simeq 2$
Vacuum pressure [torr]	$2 \cdot 10^{-11}$ (85% H_2 and He)
Electron Cooler	
Cooling length/circumferene = η_{ecool}	0.02
Electron beam radius a [cm]	2.5
Beam pipe radius b [cm]	5 or 7
Typical electron current I_e [A]	0.2 [0.4]
Electron density in cooling section $n_e = I_e/e\pi a^2\beta c$ [m^{-3}]	$1.13 \cdot 10^{14} \cdot I_e$
Effective electron density per turn $n_{\text{eff}} = \eta_{\text{ecool}} \cdot n_e$ [m^{-3}]	$2.25 \cdot 10^{12} \cdot I_e$
Typical longitudinal B-field in cooler B [T]	0.06
Lattice functions at the cooler $\beta_H = 1.9$ m, $\beta_V = 5.3$ m, $D = 3.6$ m	

*On behalf of the CERN-LEAR team and of the Russian Collaboration.

In the framework of our project it is fundamental to have a good understanding of:

- the ion beam lifetime with and without electron cooling,
- the dense electron beam space-charge effects on its transverse velocity.

As a consequence we have measured the lead-ion beam lifetimes, under different conditions, together with the cooling time. We have also implemented and tested an electron beam neutralization technique intended to reduce the space-charge effects (tests have mainly been done with protons at various energies).

The outcome of these two types of experiments will be briefly reported hereafter. For more details the reader is invited to consult References 2–6.

2. Lead-Ion Life and Cooling Times

The ion lifetime 'τ' is expressed by the following equation:

$$\frac{1}{\tau}[\mathrm{s}^{-1}] = \frac{1}{\tau_{\mathrm{vac}}} + \frac{1}{\tau_{\mathrm{rec}}}$$

where:

τ_{vac} : is the lifetime due to the effects resulting from the interactions of the ions with the residual gas molecules alone. It depends of course on the residual gas composition which has to be measured.

τ_{rec} : is the lifetime resulting from the different types of recombination of the positive lead ions with the co-moving electrons of the cooler. The recombinations may be of the radiative or dielectronic type but we were not able to distinguish between them.

2.1. Lear Vacuum Conditions

A careful analysis of the LEAR residual gas partial pressure is made at four significant locations of the ring. Table 2 shows the measurements made at the time of the actual experiments on life and cooling time. From theoretical formulae we deduce the expected lifetime which is $\tau_{\mathrm{vac}} = 17.6\ \mathrm{s} = 1/\langle t_{\mathrm{life}}\rangle = 1/0.057$. This agreement can be considered as coincidental; nevertheless the gas composition must be known with accuracy prior to any statement on the lifetime and also to compare the results of different experiments.

Table 2. Estimate of LEAR vacuum conditions and calculated lifetime.[7]

		Analyser										
		102		204[a)]		303		304			Ring averages	
		Length [m]										
		50		50		4		4				
Rest gas	Lifetime for 10^{-12} torr	P[b)]	$\langle P\rangle$[c)]	P	$\langle P\rangle$	P	$\langle P\rangle$	P	$\langle P\rangle$	$\langle P\rangle$ ring	Rel. $\langle P\rangle$ ring	$1/\langle t_{\mathrm{life}}\rangle$
H_2	625	16.00	10.19	5.10	3.25	16.00	0.81	6.10	0.31	14.56	0.73	0.023
He	624	0.60	0.38	1.30	0.83	3.10	0.16	9.40	0.48	1.85	0.09	0.003
CH_4	126	0.40	0.25	1.60	1.02	1.10	0.06	1.10	0.06	1.39	0.07	0.011
H_2O	129	0.72	0.46	1.40	0.89	0.00	0.00	1.10	0.06	1.41	0.07	0.011
N_2	92	0.00	0.00	0.00	0.00	0.00	0.00	2.00	0.10	0.10	0.01	0.001
CO	92	0.20	0.13	0.28	0.18	0.85	0.04	0.00	0.00	0.35	0.02	0.004
Ar	76	0.00	0.00	0.21	0.13	0.00	0.00	0.17	0.01	0.14	0.01	0.002
CO_2	59	0.00	0.00	0.16	0.10	0.00	0.00	0.13	0.01	0.11	0.01	0.002
Sum										**19.91**		**0.057**

a) Note: 50 m is 10 m for sector 2 and 40 m for sector 4 where vacuum is four times worse.

b) P = partial physical pressure in 10^{-12} torr.

c) $\langle P\rangle$ = contribution of this section to the average pressure of the ring.

2.2. Measurement of the Lifetime

2.2.1. Method of Measurement

Owing to the small circulating beam intensity we were not able to estimate the lifetime with the help of the classical current transformer. We therefore used the signal of a longitudinal Schottky pickup processed by a spectrum analyser. In our case $\Delta p/p$ is rather large so that the integrated Schottky noise power $S^2(t)$ is proportional the circulating unbunched ion-beam intensity. Data from the spectrum analyser are acquired by a laptop computer which does the relevant computations, namely the suppression of the electronic noise and the curve fittings.

2.2.2. Results

A plot of the inverse cooling time: $1/\tau$ [s^{-1}], as a function of the cooler current I_e [mA], is given by Fig. 1 for three different ion charge states (namely Pb^{52+}, Pb^{53+}, and Pb^{54+}).

Once the common vacuum inverse lifetime $1/\tau_{vac} = 1/20$ [s] has been subtracted one gets Fig. 2 where the three curves cross at the origin. It appears clearly that lifetimes of Pb^{52+} and Pb^{54+} are about the same and significantly larger than that of Pb^{53+} ions.

From the slope of the curves on Fig. 2 one can deduce the rate coefficient α defined by

$$\alpha = \left(\frac{1}{\tau}\right) / n_{\mathrm{eff}} \ [\mathrm{cm^3 s^{-1}}]$$

$$\alpha = \frac{1}{\tau} \cdot \frac{1}{2.25 \cdot 10^6} \cdot \frac{1}{I_e} \quad \text{for our case .}$$

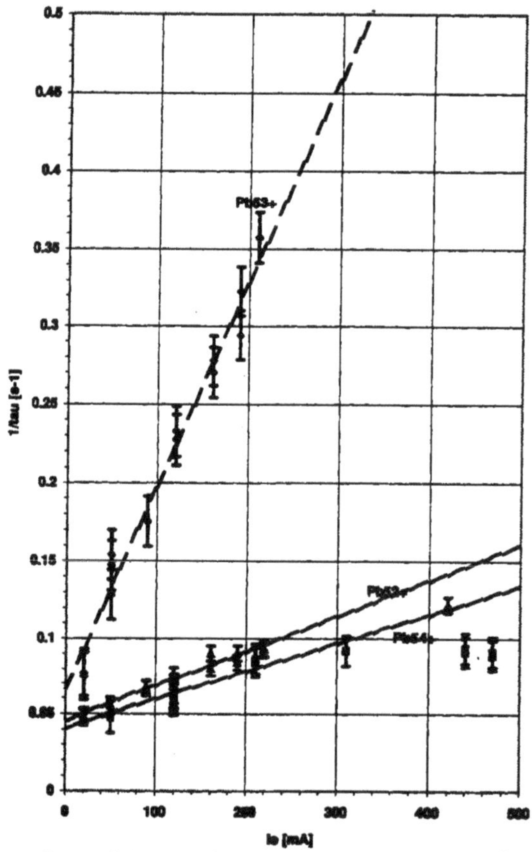

Fig. 1. Dependence of the overall inverse beam lifetime $1/\tau$ $[\mathrm{s^{-1}}]$ as a function of the electron cooling current I_e [mA], for the three lead-ion species.

According to our measurements:

$$\alpha\,[\mathrm{Pb}^{52+}] = 11 \cdot 10^{-8}, \quad \alpha\,[\mathrm{Pb}^{53+}] = 60 \cdot 10^{-8}, \quad \alpha\,[\mathrm{Pb}^{54+}] = 9 \cdot 10^{-8}\,.$$

We have no explanation for such differences of the rate coefficient with respect to the charge state number. The choice of Pb^{52+} or Pb^{54+} ions looks therefore for the time being, to be the most favourable since the linac feeding LEAR can provide the three types of ions with almost the same intensity.

Even with $I_e = 0.5$ A a lifetime of about six seconds looks quite comfortable when compared to the stacking time.

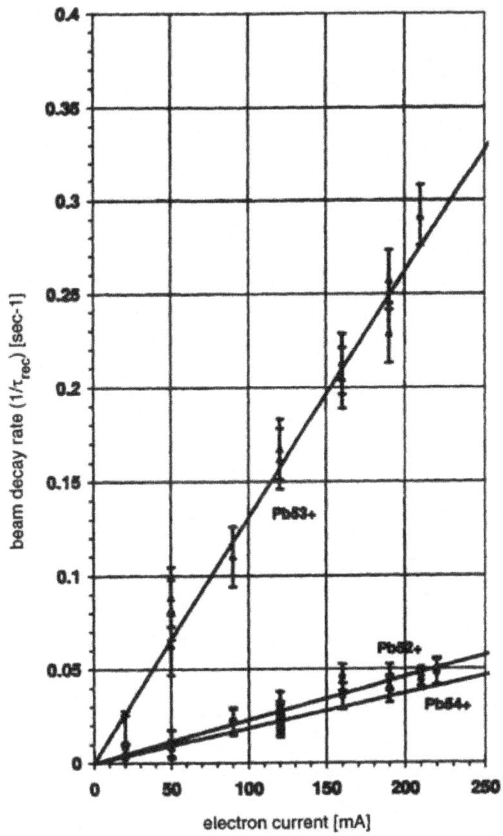

Fig. 2. Dependence of the inverse beam recombination lifetime $1/\tau_{\mathrm{rec}}$ as a function of the electron cooling current I_e [mA], for the three lead-ion species.

2.3. Cooling Time Measurement

2.3.1. Method of Measurement

The longitudinal cooling time from $\Delta p/p = 10^{-3}$ to $\Delta p/p = 10^{-4}$ is about 50 ms when $I_e = 120$ mA. It is measured, as usual, from the longitudinal Schottky pickup signal.

The transverse cooling times (H and V) are measured by horizontal and vertical ionization monitors. The electrons coming out of the micro-channel plate are collected by metallic strips distant by 1 mm one from the other and integrated at fixed time intervals. The signals are processed by a microprocessor which computes the major parameters, of the transverse distribution, versus time.

Figure 3 is an example of such a measurement in the horizontal plane for I_e = 120 mA. One determines a cooling time from a uniform ion distribution (over the machine acceptance estimated as $A_H = 150\pi$ mm · mrad), at injection, to an almost Gaussian distribution, with final r.m.s. emittance area

$$\epsilon = \pi \cdot \sigma^2/\beta_H \sim \frac{2\ \mathrm{mm}^2}{\beta_H} \cdot \pi = 2 \cdot \pi \cdot 10^{-6}\ [\mathrm{m}]\ ,$$

in about 300 ms. In the vertical plane we get almost the same results.

3. Neutralization of the Cooler Electron Beam

3.1. Aim

The electron beam space-charge potential, U_{sp}, with respect to the beam pipe, of radius b, at ground potential, is a function of the radius r, as expressed by:

$$U_{sp}(r) = -\frac{30 I_e(1-\eta)}{\beta} \cdot \begin{cases} \left(1 + 2\ln\left(\frac{b}{a}\right) - \frac{r^2}{a^2}\right) & \text{for } 0 \le r \le a \\ 2\ln\left(\frac{b}{r}\right) & \text{for } a \le r \le b \end{cases}$$

where η, the neutralization factor, is equal to $Z_i n_i/n_e$ with $Z_i n_i$ [m^{-3}] the density of the ions, of charge Z_i, which are considered uniformly stored within the electron beam volume.

The longitudinal magnetic field B is adjusted so as to, at least, limit the electron beam radius increase as a consequence of the radial electrical force $F_r = e \cdot (\partial U_{sp}/\partial r)$.

With a dense electron beam, i.e. the electron density $n_e = (I_e/e\pi a^2 \beta c)$ is large, two main issues of the space-charge effect, may be considered.

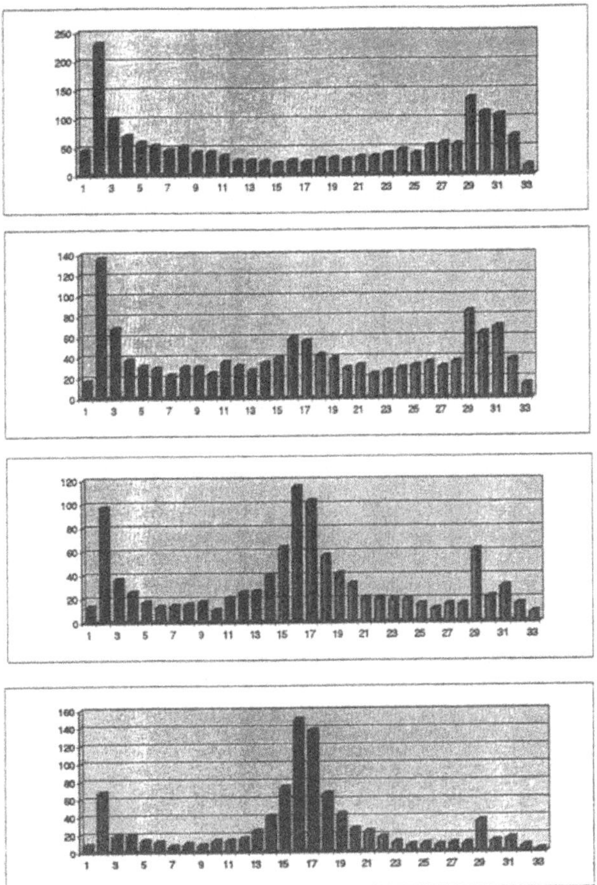

Fig. 3. Horizontal beam profile measured by the ionization profile monitor. The time interval between each measurement is 100 ms. The top measurement corresponds to 0 s while the bottom one is at $t = 300$ ms. Horizontally the distance between the strips is 1 mm.

First the electron kinetic energy E_c is a function of the radius r since:

$$\begin{aligned} E_c = (\gamma - 1)m_e c^2 &= \mathrm{e}[U_{\mathrm{cathode}} + U_{\mathrm{space-charge}}] \\ E_c(r) &= \mathrm{e}[U_c + U_{\mathrm{sp}}(r)] \simeq \frac{1}{2} m_e v_{\parallel}^2(r) \end{aligned}$$

with $v_{\parallel}$ the longitudinal component of the electron velocity. The electron beam has thus a longitudinal velocity distribution as represented by Fig. 4. Together with the electron longitudinal velocity distribution we have to consider the corresponding ion velocity distribution $v_{i\parallel}(r)$ which depends on the dispersion D [m] since $r_i = D \cdot (\Delta p/p) = D \cdot \gamma^2(\Delta v_{i\parallel}/v_{i\parallel})$ as also represented in Fig. 4 where we also show the effects of the ion transverse r.m.s. beam width $\sigma = \sqrt{\epsilon_{H,V} \cdot \beta_{H,V}}$. Depending on D the curves representing the electron on the ion velocity may intersect at a radius $r_i = a_1 \neq 0$.

The cooling force f_{cool} is proportional to $(v_e - v_i)/|v_e - v_i|^3$. Therefore an ion with radius $r_i > a_1$ such that $v_{i\parallel} < v_{e\parallel}$ will be subject to heating. Hence small $\eta\,(< 0.2)$ and large $n_e\,(> 10^{13}\,\mathrm{m}^{-3})$ may induce heating (rather than cooling) on the circulating ions. Any attempt to increase η and therefore to neutralize the electron beam, will reduce effects of this nuisance (see Fig. 4 for $\eta > 0$).

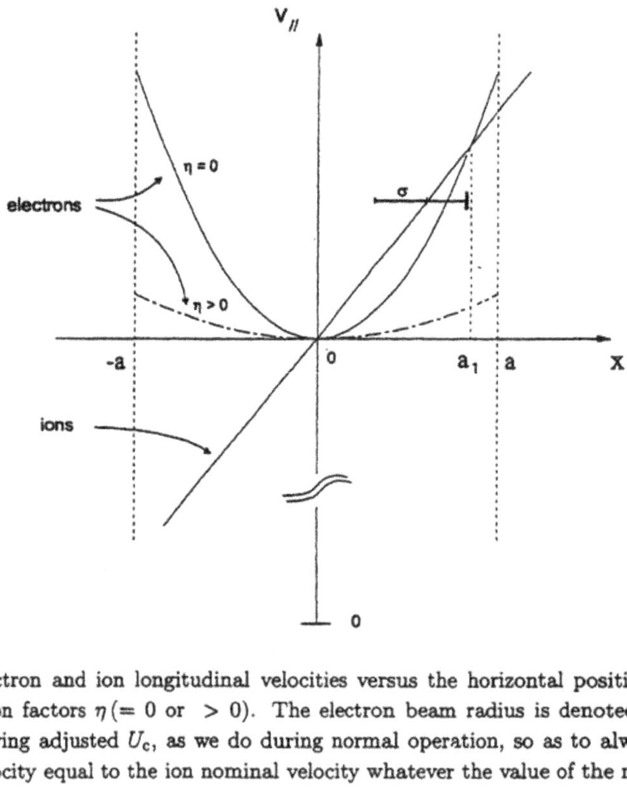

Fig. 4. Electron and ion longitudinal velocities versus the horizontal position x for two neutralization factors $\eta\,(= 0$ or $> 0)$. The electron beam radius is denoted by 'a'. We consider having adjusted U_c, as we do during normal operation, so as to always keep the electron velocity equal to the ion nominal velocity whatever the value of the neutralization coefficient (η).

As a second consequence of dense electron beams we have to consider that F_r will induce an azimuthal transverse velocity, also named drift velocity, v_d, expressed by:

$$v_d = \frac{F_r}{w_H} \quad \text{where} \quad w_H = \frac{eB}{m_e} .$$

This velocity adds quadratically to the electron nominal transverse velocity which thus, for large densities n_e, will be significantly increased. Since the electron cooling process is based on the fact that the electrons have to be very cold, a way to reduce v_d is to reduce F_r and therefore to neutralize the electron beam ($F_r(\eta = 1) = 0$).

3.2. Principle (Fig. 5)

For the purpose of neutralization we have installed two sets of electrodes, one El_g at the gun level, the other, El_c, at the collector level. Each electrode consists of two metallic half-cylinders separated by a glass insulator (small conductivity). Each metallic plate is polarized by independent positive (versus ground) voltages: $U_{\mathrm{el}\,i}(0 \le U_{\mathrm{el}\,i} \le 6$ kV, $i = 1, ..., 4)$.

On their way from the gun to the collector the energetic electrons will ionize the residual gas molecules which are almost at rest. The positively ionized ions will drift slowly, in the longitudinal magnetic field B, toward El_c or El_g were they are reflected. The slow electrons will also reach the electrodes of El_c or El_g where they are collected. The density n_i [m^{-3}] of the stored ions, between the neutralization electrodes, will thus increase with time until it reaches its asymptotic value. The neutralization factor η is defined by $\eta = (Z_i n_i / n_e)$, $(0 \le \eta \le 1)$.

It must be emphasized that, on our electron cooler the beam pipe diameter b_1, between the gun and El_g and El_c and the collector, is 100 mm while, between El_g and El_c, b_2 is equal to 140 mm. This discontinuity will induce different space-charge potentials such that, even with no potential on El_g and El_c, $\Delta U_{sp} = (30 I_e/\beta) \cdot 2 \ln(b_2/b_1)$ which has the same physical effect as that induced by an artificial neutralization by the help of the $U_{\mathrm{el}\,i}$'s. Thus these discontinuities in the beam pipe diameters induce a 'natural neutralization'.[8]

The main problem of the neutralization process is that under some conditions, namely when I_e exceeds a given threshold, η becomes unstable i.e. part or all of the stored ions are suddenly expelled. We therefore have developed some diagnostics to measure dynamically η and to observe its instability. A so-called 'shaker' has been implemented which aims to suppress the above-mentioned neutralization instabilities.

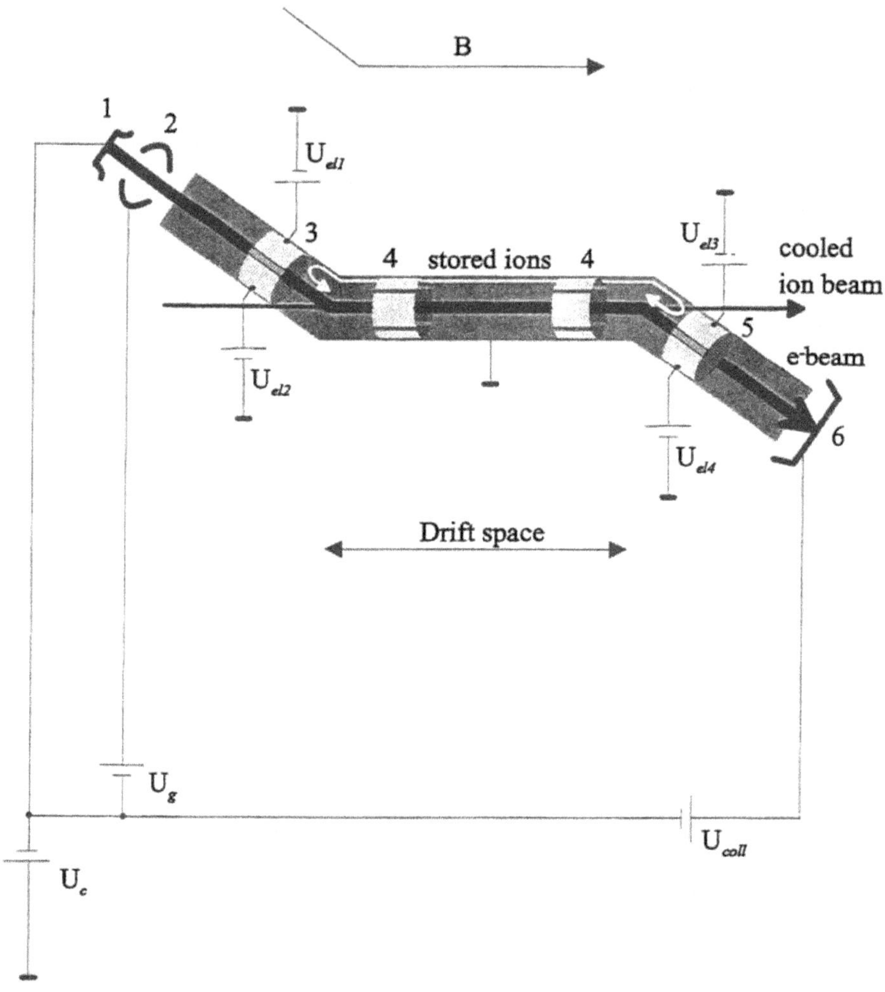

Fig. 5. Principle of neutralization.
1 Cathode + gun; 2 Control grid; 3 Neutralization 'gun' electrode El_g; 4 Horizontal and vertical pickups; 5 Neutralization 'collector' electrode El_c; 6 Collector.

3.3. Measurement of the Neutralization Factor η

We mainly use two methods, one measuring the cooled ion revolution frequency (and thereby the ion velocity) and the other making use of the electron time of flight (TOF) between El_g and El_c.

3.3.1. Revolution Frequency Measurement

The electron beam velocity at $r = 0$, which also fixes the cooled ion beam mean velocity $\langle v_i \rangle$ is given by:

$$\langle \beta_i \rangle \cdot c = \langle v_i \rangle = v_{e_\parallel}(r = 0) \simeq \left\{ \frac{2e}{m} \left(U_c - \frac{90 I_e}{\langle \beta_i \rangle}(1 - \eta) \right) \right\}^{1/2}$$

where U_c is the cathode potential (Fig. 5). An increase of η implies therefore a decrease of U_c in order to keep $\langle v_i \rangle$ or the ion revolution frequency fr at their nominal values since

$$\frac{\Delta fr}{fr} = \left(1 - \frac{\gamma^2}{\gamma_t^2}\right) \frac{\Delta\beta}{\beta} \simeq \frac{\Delta\beta}{\beta} \text{ and } \gamma_t^2 \gg 1$$

in LEAR.

As a consequence the cathode corrections ΔU_c we have to apply to the cathode in order to keep the ion revolution constant do permit us to retrieve $\Delta\eta$. At full neutralization ($\eta = 1$) any change of I_e has no influence on $\langle v_i \rangle$.

This method is used as a reference intended to calibrate the TOF technique we are going to describe now.

3.3.2. Time-Of-Flight Method (TOF)

The principle described by Fig. 6 is based on the fact that the propagation of a density modulation imposed on the electron beam depends, to the first order, primarily on the electron mean velocity and so on the neutralization factor η. For that purpose the primary electron beam longitudinal density is weakly modulated by a sinusoidal signal applied on El_g. This excitation signal is delivered by a network analyser at a frequency f close to 300 MHz. The modulated signal is detected by El_c which is at an electrical distance 'L' from El_g.

The difference in phase between the excitation and the detected signals, as measured by the analyser, is expressed by

$$\phi(El_c) - \phi(El_g) = -\frac{(2\pi f)}{v_{e_\parallel}} L$$

or

$$\Delta\phi \simeq \frac{(2\pi f)}{v_{e_\parallel}} L \cdot \frac{\Delta v_e}{v_{e_\parallel}}(\eta) \,.$$

From the measurement of $\Delta\phi$ one can easily retrieve $\Delta\eta$. An example of such a measurement is given by Fig. 7. At the time when the neutralization electrodes are polarized one observes a small but sharp change of $\Delta\phi$ followed by a much larger amplitude variation during a much longer time which corresponds to the overall neutralization time (about 3 s on Fig. 7). The short initial $\Delta\phi$ is thought to be due to the capture of stored electrons by the neutralization electrodes at the time when the positive neutralization voltages U_{el_i} are applied.

A comparison of the two methods gives comparable estimates of η within an uncertainty of 20% which can be explained by the measurement described in the next paragraph.

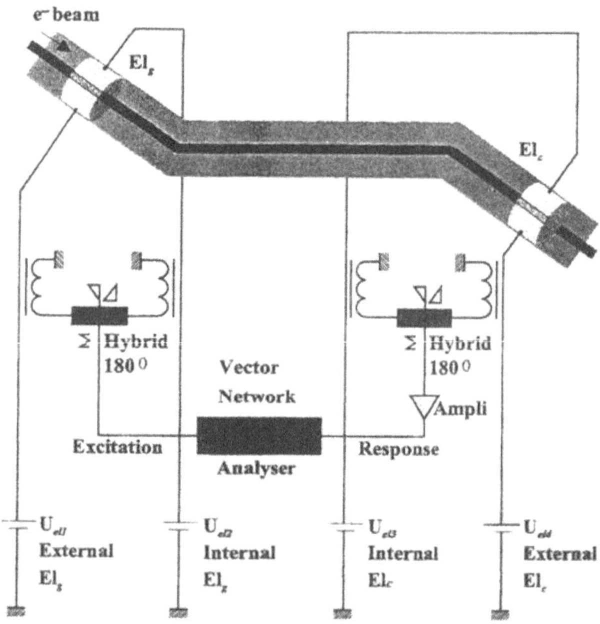

Fig. 6. Principle of the TOF method.
El_g Neutralization electrode gun; El_c Neutralization electrode collector.

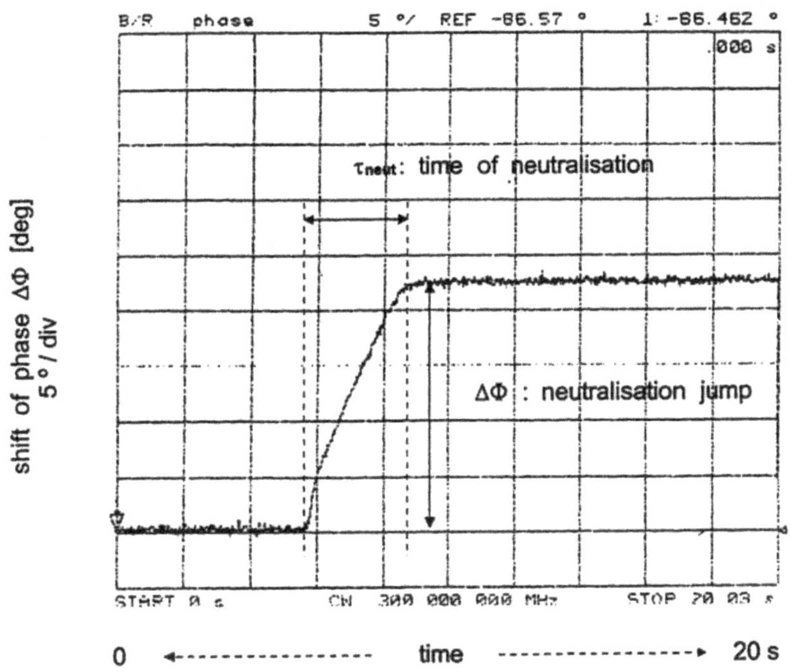

Fig. 7. TOF measurement. Plot of $\Delta\phi$ [deg] as a function of time t [s].

3.3.3. Distribution of η

The cooled ion beam has a diameter of a few millimetres and can be used as a probe of the electron beam. It is then possible to use the LEAR orbit correctors in order to displace the ion beam transversally within the electron beam and to measure the corresponding ion revolution frequency changes as a function of the ion beam radial position. Such measurements are shown in Fig. 8.

It appears clearly that, by polarizing the neutralization electrodes, the electron beam is fully neutralized over an interval of about ± 1 cm around the centre. Taking into account the finite dimension of the ion beam, the measurement for $|r| \geq 15$–20 mm must be used with care since the electron beam radius 'a' is equal to 25 mm.

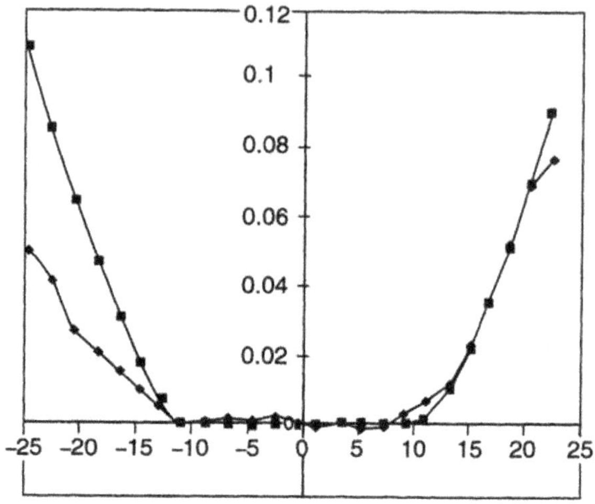

Fig. 8. Neutralization distributions. Horizontal axis: position [mm]. Vertical axis: frequency shift [kHz].

3.4. Instabilities of the Neutralization Factor: Cures

Instabilities of η are observed mainly when the primary electron beam intensity exceeds a given threshold. One observes jumps of the energy which follow from a sudden expulsion of the stored ions which participate in the neutralization process. This effect is clearly revealed by the time-of-flight method (see 3.3.2) or when looking at the signals induced on the position pickup electrodes. [6] These instabilities happen sometimes when we operate under natural neutralization conditions (see 3.2) i.e. when all the El_g and El_c electrodes are at ground potential.

One of the simplest way to suppress these instabilities is to make use of a so-called 'shaker'. The principle is shown in the lower part of Fig. 9. A low-voltage (few tenths of a volt) sinewave, with a frequency of a few 100 kHz, is applied transversally on any of the position pickup electrodes in the cooling section. This arrangement damps any instability while reducing slightly the stable value of η. Referring to the top part of Fig. 9, which shows the network analyser (measuring the TOF) display, one sees on the left the phase (see 3.3.2) or 'η' instabilities while the right part shows the stabilizing effect of the shaker for different voltage levels.

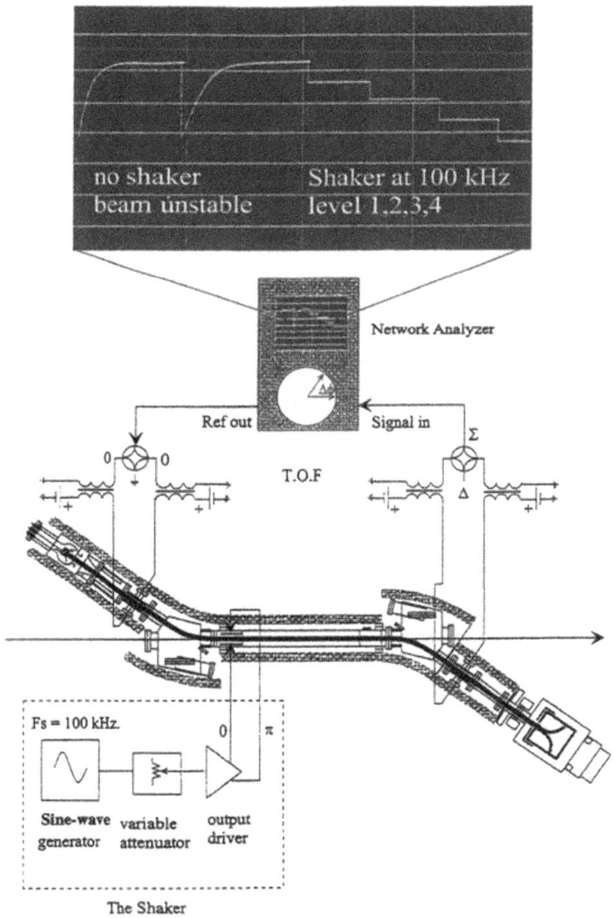

Fig. 9. Principle of the 'shaker'. Top: display of the network analyser phase (or TOF) measurement. The left part shows an unstable neutralization while the right part shows stable neutralization for different shaker voltage amplitudes.

4. Conclusions

Experiments done on LEAR showed that Pb^{52+} and Pb^{54+} are good candidates as far as the lifetime and the transverse cooling time are concerned. A future increase of the electron cooler intensity from 0.12 to 0.5 A and a lengthening of the drift space (η_{ecool} increased from 0.02 to 0.04) should permit one to lower the transverse cooling time from 0.3 to 0.1 s.

The neutralization process has been partially understood and mastered. Future improvements, like that aiming to clear the electrons which are reflected by the collector, and also the use of a quadrupolar type 'shaker' are under way. We also foresee getting rid of the natural neutralization by equalizing the beam pipe diameters along the electron beam trajectory.

References

1. P. Lefèvre and D. Möhl, Lead ion accumulation scheme for LHC, CERN/PS/DI/Note 95–09, LHC Note 257;
   P. Lefèvre and D. Möhl, A low energy accumulation ring of ions for LHC, CERN/PS/DI/Note 93–62, LHC Note 259.
2. S. Baird, J. Bosser, C. Carli, M. Chanel, P. Lefèvre, R. Ley, R. Maccaferri, S. Maury, I. Meshkov, D. Möhl, G. Molinari, F. Motsch, H. Mulder, G. Tranquille and F. Varenne, Beam lifetime tests for Pb^{52+}, Pb^{53+} and Pb^{54+} ions subject to electron cooling in LEAR (performed in June 1995), PS/AR/Note 95–12.
3. S. Baird, J. Bosser, C. Carli, M. Chanel, P. Lefèvre, R. Ley, R. Maccaferri, S. Maury, I. Meshkov, D. Möhl, G. Molinari, F. Motsch, H. Mulder, G. Tranquille and F. Varenne, Measurement of the lifetime of Pb^{52+}, Pb^{53+} and Pb^{54+} beams at 4.2 MeV per nucleon subject to electron cooling, *Phys. Lett.* **B361** (1995) 184.
4. J. Bosser, F. Caspers, M. Chanel, R. Ley, R. Maccaferri, S. Maury, I. Meshkov, G. Molinari, V. Polyakov, A. Smirnov, O. Stepashkin, E. Syresin, G. Tranquille and F. Varenne, Neutralisation of the lear electron-cooling beam: experimental results, Part. Acc. Conf., Dallas, 1995, CERN/PS 95–17 (AR).
5. J. Bosser, S. Maury, I. Meshkov, D. Möhl, E. Mustafin, E. Syresin, F. Varenne and P. Zenkerich, Stability conditions for a neutralised electron cooling beam, Part. Acc. Conf., Dallas, 1995, CERN/PS 95–16 (AR).
6. F. Varenne, Neutralisation de la charge d'espace du faisceau d'électrons du refroidisseur électronique de LEAR (in French), Thesis, Clermont-Ferrand University, November 1995.

7. B. Franzke, Vacuum requirements for heavy ion synchrotrons, Proc. Part. Acc. Conf., *IEEE Trans. Nucl. Sci.* **NS-28** (1981) 2116.
8. G. Tranquille, Suppression of the natural neutralisation of the electron beam space charge via polarised tubes, PS/OP/Note 95-47.

COOLING EXPERIMENTS AT CRYRING

HÅKAN DANARED, GUILLERMO ANDLER, and ANDERS KÄLLBERG

Manne Siegbahn Laboratory, Stockholm University
S-104 05 Stockholm, Sweden

ABSTRACT

Some experiments performed at CRYRING concerning electron cooling of bunched and costing beams are reviewed. These are cooling-force measurements with and without adiabatic beam expansion, studies of bunch lengths of cooled beams, and a test of a method for measuring cooling forces using the equilibrium between the cooling force and the rf force.

1. Introduction

CRYRING is a low-energy synchrotron and storage ring[1] equipped with electron cooling[2,3]. Highly charged ions are provided by an electron-beam ion source (EBIS), while singly charged atomic and molecular ions are produced in a plasmatron ion source. In both cases, the ions are usually accelerated to 290 keV per nucleon in an RFQ before injection into the ring. The ions can be accelerated further in the ring, up to a rigidity of 1.44 Tm, and cooled. The ring is used mainly for atomic and molecular physics. A major fraction of the experiments are studies of electron–ion recombination, where the cooler is used as an electron target. These experiments include spontaneous and laser-induced radiative recombination, dielectronic recombination, and dissociative recombination of molecular ions.

2. Cooling of coasting beams

Much of the machine development at CRYRING has been motivated by specific experimental requirements or by a general aim to increase the performance and the versatility of the machine, e.g. to extend the range of ions available in the ring. Apart from this, however, a substantial effort has been spent on studies and development of the electron cooling. Already from the first recombination experiments at CRYRING it was clear that, like at other machines, the energy resolution in spectra of dielectronic or dissociative recombination was limited by the electron temperature. At the same time, measurements of the longitudinal cooling force, i.e., the drag force that electrons exert on ions passing through the cooler, were made[2]. These showed a very good agreement with the simple model based on binary collisions between the ions and the electrons, without effects of the magnetic field in the cooler taken into account, indicating that, at CRYRING, also the cooling was limited by the electron temperature. It was therefore decided to try to expand the electron beam adiabatically and thereby reduce the transverse electron temperature[3,5].

For practical reasons, an expansion factor of 10 was chosen, obtained by using a magnetic field strength of 0.3 T at the electron gun and 0.03 T in the rest of the cooler. At the same time as the field configuration was modified, a new electron gun with a 10 times smaller cathode area was installed, ensuring that the size and current density of the expanded beam should be the same as in the original cooler, without expansion. Measurements performed with the unexpanded beam showed that it had a transverse energy spread, $kT_{e\perp}$, close to 100 meV, corresponding to the cathode temperature of about 900°C, and a longitudinal energy spread, $kT_{e\parallel}$, of approximately 0.1 meV, as expected from the internal relaxation of the electron beam[2,4]. Similar measurements with the adiabatic expansion[3,6] showed that the transverse temperature indeed had dropped by a factor 10, while the longitudinal temperature remained essentially unaffected. The longitudinal temperature should scale with the electron density (in the gun solenoid) to the power of 1/3 if it is determined by a purely longitudinal relaxation[7]. In our case the expansion would thus increase it by a factor of $10^{1/3}$. On the other hand, the transfer of energy from the transverse to the longitudinal motion becomes more strongly suppressed using the beam expansion, since it depends on the ratio between the Larmor radius of the electrons and the average distance between them. This ratio becomes smaller with increased expansion (provided that the cathode is scaled down as the expansion is increased). The longitudinal cooling force was still consistent with the binary-collision model without magnetic field. The reason why no effects of the field are seen is still not understood.

3. Cooling of bunched beams

Bunched cooled beams are of interest for studies of beam dynamics and cooling, as well as for certain types of recombination experiments where a bunched beam can give higher signal-to-background ratios due to the higher instantaneous current. The bunch length in a well-cooled beam is determined by the equilibrium between the rf potential and the beam-induced space-charge potential[8] and is given by

$$L_{\mathrm{FWHM}} = \left[\frac{3\sqrt{2}}{\pi}\frac{IC^2\ln(r_{\mathrm{v}}/r_{\mathrm{b}})}{4\pi\epsilon_0 h f_0 V_{\mathrm{rf}}}\right]^{\frac{1}{3}}. \tag{1}$$

Here, I is the ion current, C is the ring circumference, r_{v} and r_{b} are the radii of the vacuum pipe and of the beam, h is the rf harmonic, f_0 the revolution frequency of the ions, and V_{rf} is the amplitude of the rf voltage. Bunch-length measurements made at CRYRING[9] agree quite well with the theoretical predictions and with the measurements in ref 8. Since the rf system at CRYRING consists of a simple drift tube connected to a power amplifier[10] rather than a resonant cavity, it is easy to vary the rf voltage over a wide range, and the actual voltage can be measured directly on the drift tube with high accuracy. Fig. 1 shows the bunch length as a function of rf amplitude for a cooled 85 μA deuteron beam at 46 MeV and at the first harmonic of the rf frequency. The line is the theory of equation (1) using $r_v/r_b = 20$. The good agreement between experiment and theory shows that the bunch indeed is space-charge dominated.

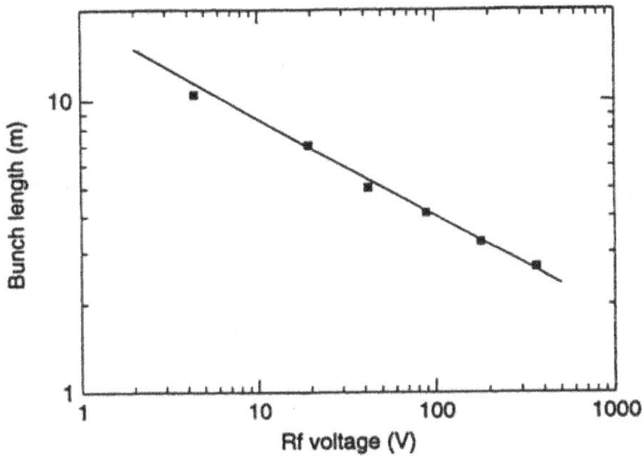

Fig. 1. The bunch length (FWHM) measured as a function of rf amplitude for a cooled 85 μA deuteron beam at 46 MeV and at the first harmonic of the rf frequency (squares). The line is the theory of equation (1).

Using very small rf voltages, equilibrium situations can be studied where the energy that an ion gains during one passage through the rf system is equal to the loss during one passage through the cooler (or vice versa). The shift of the synchronous phase that results from such an equilibrium condition can easily be detected with an oscilloscope that measures the signal induced by the bunches in an electrostatic pickup. Knowing the phase shift and the amplitude and frequency of the rf, one directly obtains the cooling force. Fig. 2 shows the result of a test of this method for cooling-force measurements that was performed recently at CRYRING using a beam of C^{3+} ions. Here, the rf was kept at a constant frequency and at an effective amplitude of 9 V, while the acceleration voltage of the cooler electrons was varied between the cooling voltage and 1.5 V above or below cooling voltage. The relative velocity between ions and electrons due to the 1.5 V change is 7600 m/s. The phase shift resulting from that voltage change corresponded to an ion energy gain of 2.2 eV in one passage through the rf system (and an equally big loss through the cooler). Scaling to an electron density of 10^{14} m^{-3} and a 1 m cooler length gives a normalized drag force $F_\parallel = 8.7$ eV/m. This is quite close to the theoretical value[3] for $kT_{e\perp} = 10$ meV, assuming that the drag force is proportional to the ion charge to the power of 1.6 at 7600 m/s relative velocity[11]. The error in this measurement was relatively large, perhaps $\pm 50\%$, mainly due to a noisy triggering signal. This can easily be improved, however. In addition, the phase shift becomes bigger for more highly charged ions, since the rf force is linear with the ion charge while the cooling force scales with a higher power of q, making this method particularly well suited for measurements with highly charged ions.

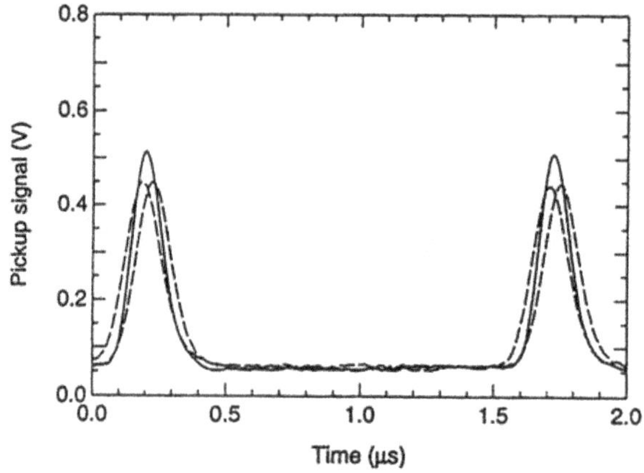

Fig. 2. Digitally averaged pickup signal from a 6 MeV per nucleon C^{3+} beam as a function of time. The two peaks correspond to two consecutive passages of the same bunch through the pickup. The full curve was taken with the cooling energy matched to the rf frequency, the dashed lines with a cooler voltage shifted by ±1.5 V from the matching condition. The time is measured relative to a fixed rf phase.

4. Acknowledgements

The authors wish to thank H. Meuth for initiating the bunch length measurements at CRYRING, and, of course, the CRYRING staff.

5. References

1. K. Abrahamsson, G. Andler, L. Bagge, E. Beebe, P. Carlé, H. Danared, S. Egnell, K. Ehrnstén, M. Engström, C. J. Herrlander, J. Hilke, J. Jeansson, A. Källberg, S. Leontein, L. Liljeby, A. Nilsson, A. Paal, K.-G. Rensfelt, U. Rosengård, A. Simonsson, A. Soltan, J. Starker, M. af Ugglas, and A. Filevich, *Nucl. Instr. Meth.* **B79** (1993) 269.
2. H. Danared, *Phys. Scr.* **48** (1993) 405.
3. H. Danared, G. Andler, L. Bagge, C. J. Herrlander, J. Hilke, J. Jeansson, A. Källberg, A. Nilsson, A. Páal, K.-G. Rensfelt, U. Rosengård, J. Starker, and M. af Ugglas, *Phys. Rev. Lett.* **72** (1994) 3775.
4. D. R. DeWitt, R. Schuch, T. Quinteros, H. Gao, W. Zong, H. Danared, M. Pajek, and N. R. Badnell, *Phys. Rev.* **A50** (1994) 1257.

5. H. Danared, *Nucl. Instr. Meth.* **A335** (1993) 397.
6. C. Biedermann, H. Gao, R. Schuch, W. Zong, S. Asp, D. R. DeWitt, and H. Kuiper, *J. Phys. B* **28** (1995) 505.
7. V.I. Kudelainen, V.A. Lebedev, I.N. Meshkov, V.V. Parkhomchuk, and B.N. Suchina, *Sov. Phys. JETP* **56** (1982) 1191.
8. T. J. P. Ellison, S. S. Nagaitsev, M. S. Ball, D. D. Caussyn, M. J. Ellison, and B. J. Hamilton, *Phys. Rev. Lett.* **70** (1993) 790.
9. H. Meuth, A. Källberg, G. Andler, and H. Danared, to be published in *Proc. 5th European Part. Acc. Conf.* (1996).
10. K. Abrahamsson, G. Andler, and C. B. Bigham, *Nucl. Instr. Meth.* **B31** (1988) 475.
11. M. Steck, K. Beckert, F. Bosch, H. Eickhoff, B. Franzke, O. Klepper, R. Moshammer, F. Nolden, P. Spädtke, and T. Winkler, *Proc. 4th European Part. Acc. Conf.*, ed. V. Suller and Ch. Petit-Jean-Genaz (World Scientific, Singapore, 1994) p. 1197.

ELECTRON COOLING AT THE ESR

M. STECK, K. BECKERT, F. BOSCH, H. EICKHOFF, B. FRANZKE, O. KLEPPER, R. MOSHAMMER, F. NOLDEN, H. REICH, B. SCHLITT, P. SPÄDTKE, T. WINKLER

Gesellschaft für Schwerionenforschung, Postfach 110552, D-64220 Darmstadt, Germany

Abstract

Electron cooling of highly charged ions has been studied systematically at the storage ring ESR. Measurements of the longitudinal cooling force as a function of the relative velocity between ions and electrons evidence that a simple q^2-dependence of the cooling force is only adequate for intermediate relative velocities. For small and large relative velocities the exponent is reduced below 2. The measured momentum spreads and emittances of electron cooled heavy ion beams can be explained as an equilibrium between cooling and intrabeam scattering. For low intensity ion beams a considerable reduction of the momentum spread is observed indicating that intrabeam scattering is suppressed. The lowest values of the momentum spread are close to the limit set by the longitudinal electron temperature. Due to the low ion beam temperature the ions can be frozen in their position which might result in a longitudinal ordering.

1. Introduction

In the operation of the storage ring ESR electron cooling is a very crucial component for beam accumulation and preparation of high quality ion beams.[1] Although it is a well established technique for protons and light ions, it comprises several new aspects when used in a storage ring which is designed for a wide range of heavy ions at various energies. For the achievement of high phase space density beams the cooling force exerted by the cold electron beam on the ion beam is the figure of merit of the cooling system. Therefore the results on longitudinal cooling forces under typical operating conditions are described with special emphasis on the scaling with the ion charge.

The phase space volume of the cooled beam is determined by an equilibrium with intrabeam scattering which constitutes the main source of heating in an ion beam of high phase space density. Since all three degrees of freedom are coupled in these internal scattering processes beam emittances as well as the longitudinal momentum spread are affected by changes of the cooling strength or variation of the ion beam intensity. For high intensity ion beams the beam temperature is orders of magnitude higher than the electron beam temperature.

By reduction of the ion beam intensity the heating rate in the ion beam is correspondingly reduced, therefore the ion beam temperature can approach the lower limit imposed by the electron beam temperature. The effect of a sudden reduction of the longitudinal ion beam temperature at low intensity indicates that intrabeam scattering is strongly suppressed in this regime. Experiments show that the longitudinal ion and electron beam temperature are of similar magnitude. The ratio of potential and thermal energy in the comoving rest frame indicates that the plasma parameter Γ approaches 1 similar to the situation which was reported from experiments at the NAP-M storage ring.[2]

2. Cooling Forces

The longitudinal cooling force has been measured as a function of the relative velocity for various species of completely stripped ions. The two applied methods, simultaneous heating with rf noise and fast stepping of the accelerating voltage of the electron beam, are complementary with respect to the relative velocity between ions and electrons.[3] Both methods allow an absolute determination of the longitudinal cooling force for a certain relative velocity with an accuracy better than 20%. The results of the cooling force measurements are shown in Fig. 1 normalized to an electron density $n_e^* = 1 \times 10^6$ cm^{-3}. The continuous curves at relative velocities $v^* \leq 10^4$ m/s are calculated from the distribution function ψ of the beam which is heated with a diffusion constant D according to $F_{\parallel} \cdot v_o = D \cdot \delta\psi/\delta E \cdot \psi^{-1}$. The data points result from fast Schottky scans which measure the drift of the ion beam due to the force of the velocity detuned electrons.

In general the velocity dependence shows the same characteristics for all ion species with a linear increase for relative velocities below 2×10^3 m/s, a broad maximum around $v^* = 1\times10^4$ m/s and an asymptotic v^{*-2}-dependence at large relative velocities. Theoretical calculations to extract the relevant electron and ion beam parameters are under way. The maximum cooling force in all these measurements at beam velocities $\beta \simeq 0.6$ - 0.7 is limited by the parallelity of the magnetic field in the cooling section. Recent magnetic field measurements proved a field quality $\langle B_\perp/B_\parallel \rangle = 3\times10^{-4}$ which will be improved to a level below 1×10^{-4} in order to gain stronger cooling. The influence of the field quality was also confirmed by a measurement of the cooling force for ion beams with $\beta = 0.33$ after deceleration which shows that the cooling force is considerably higher for slower ions.

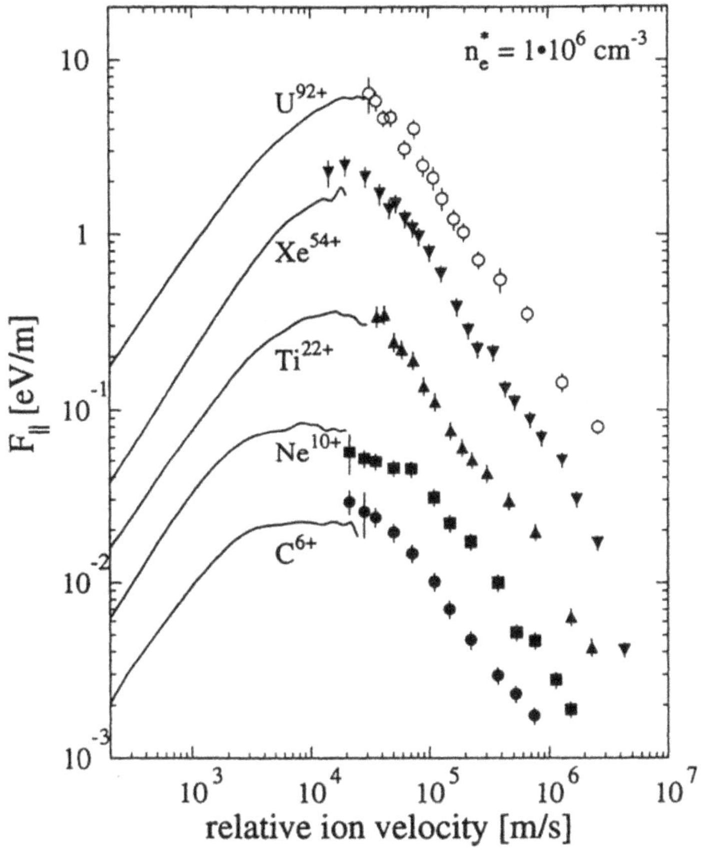

Figure 1: Longitudinal cooling force as a function of the relative velocity between ions and electrons. The continuous curves at small relative velocities were measured by rf noise heating and simultaneous cooling with an electron current of 50 mA. The discrete data points at large relative velocities were obtained by high voltage stepping with an electron current of 250 mA. All results are normalized to $n_e^* = 1 \times 10^6$ cm^{-3}.

The charge dependence of the cooling force has been determined at various relative velocities (Fig. 2) by fitting a power scaling $F_{\|} \propto q^x$ to the experimental values. From this analysis a dependence of the exponent x on the relative velocity was obtained (Fig. 3). It confirms a q^2-dependence only for intermediate velocities, but evidences significant deviations at small and large relative velocities. This is in agreement with recent experimental and theoretical studies of electron cooling at

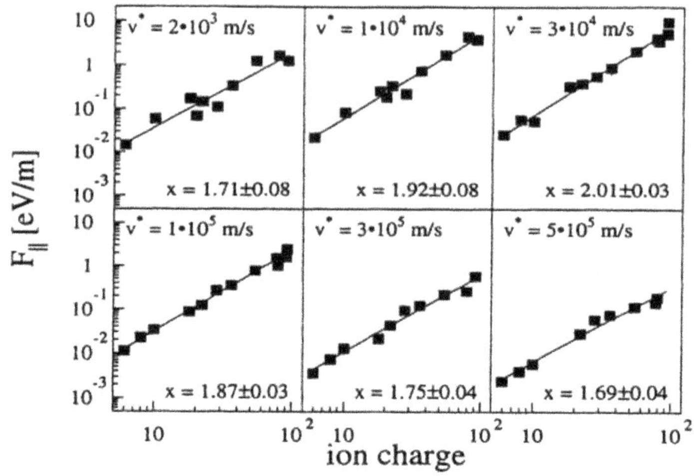

Figure 2: Cooling force increase with ion charge was fitted with a scaling $F_{\parallel} \propto q^x$. The fit value for the exponent x at certain relative velocities is given in the insert.

small relative velocities.[4] Measurements at different magnetic field strengths did not prove a strong influence of the electron beam magnetization on the cooling force, the main effect seems to be caused by a higher transverse electron temperature at low magnetic field. The effect of electron beam magnetization should be more pronounced for slow ion beams after deceleration. After the improvement of the magnetic field quality magnetization effects will be further investigated.

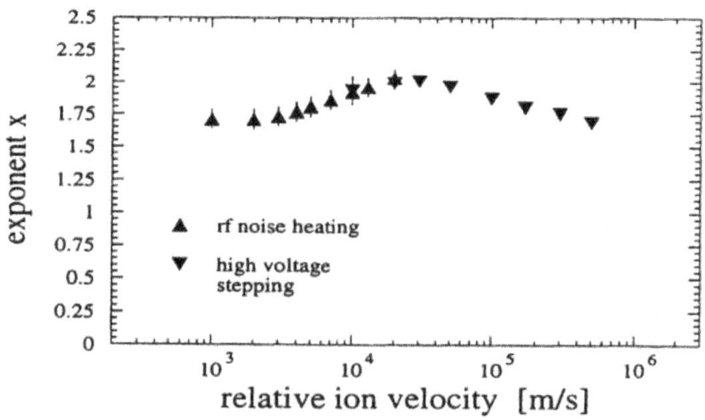

Figure 3: Exponent x in the fit $F_{\parallel} \propto q^x$ as a function of the relative velocity between ions and electrons.

3. Equilibrium between Cooling and Intrabeam Scattering

The regime of medium and high intensity heavy ion beams with electron cooling is characterized by ion beam properties which are determined by a balance between electron cooling and intrabeam scattering in the dense ion beam.[5] The heating rate by intrabeam scattering is proportional to the number of ions and inversely proportional to the phase space volume that is occupied by the ion beam. The electron cooling power increases with the electron current. Therefore the intensity of both the ion and the electron beam determine the temperature of the cooled ion beam.

The quality of a coasting ion beam in longitudinal phase space can be obtained by Schottky noise analysis of the revolution frequency distribution. The transverse emittances are measured with position sensitive particle detectors for down charged ions after the electron cooler which represent the transverse distribution of the circulating ion beam. The longitudinal and transverse beam quality is improved when a higher electron current I_e is applied for cooling (Fig. 4). The improvement of the momentum spread scales as $I_e^{-0.3}$. The transverse emittances also decrease with $I_e^{-0.3}$, but the transverse emittances are sensitive to the cooling conditions, particularly to the angular alignment of ion and electron beam and therefore the exponent is subject to variations. The gain from increased cooling strength is moderate as the intrabeam scattering rate depends on the 6-dimensional phase space density and therefore the higher cooling rate is partitioned to all subspaces.

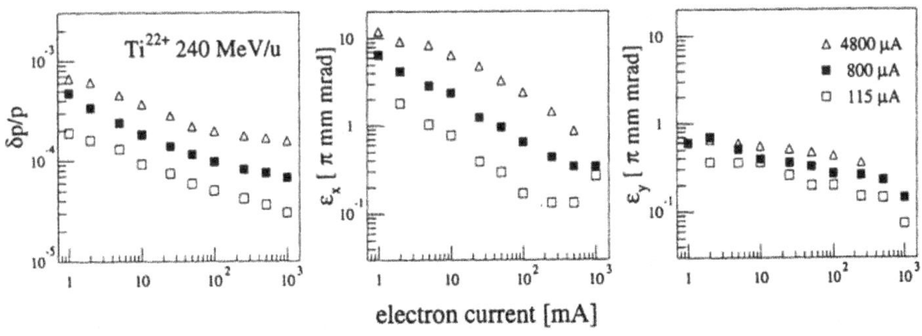

Figure 4: Reduction of the longitudinal momentum spread (FWHM) and transverse emittances (2σ) by an increase of the electron current for three currents of a Ti^{22+} beam at 240 MeV/u.

A similar behaviour is observed when the ion beam intensity is varied. With increasing particle number the phase space volume has to grow in order to keep the heating rate in balance with the constant cooling rate. Even if the electron current is constant, the cooling rate may decrease for high ion beam intensities, as the ion motion in the beam rest frame grows. The cooling force which is sensitive to all kinds of relative motion between ions and electrons will thus exceed the linear regime resulting in a reduced cooling rate.

The results of experiments with various ion species in equilibrium between cooling and intrabeam scattering as a function of the number of stored ions N are summarized in Fig. 5. The measurements evidence an increase of momentum spread and transverse emittances with the ion beam intensity. The momentum spread grows with $N^{0.3}$ for all ions, the emittances with $N^{0.5}$ on an average.[6] As for Fig. 4 again it holds true that the emittance growth is rather sensitive to the electron beam alignment. The dependence on the particle number also proves a reduced cooling rate for large particle numbers as the 6-dimensional phase space density $N/(\delta p/p \cdot \epsilon_x \cdot \epsilon_y)$ decreases for large particle numbers. On the contrary the spatial density $N/(\epsilon_x \cdot \epsilon_y)^{1/2}$ increases with the particle number N bringing the beam closer to the space charge limit. The ion charge does not affect the equilibrium values largely, the increased heating rate for higher charges is nearly balanced by the stronger cooling rate.

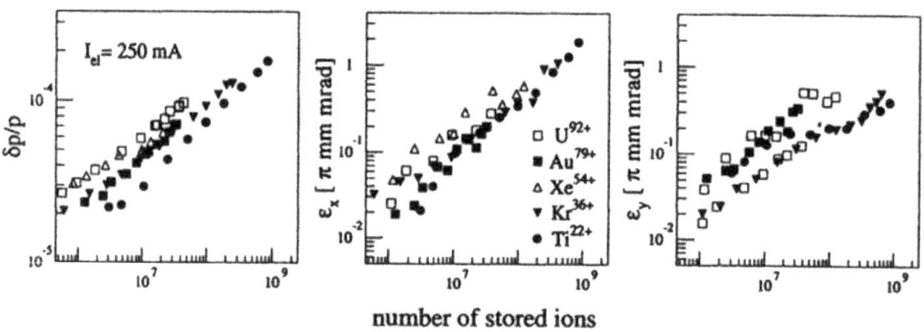

Figure 5: Summary of measurements of momentum spread and transverse emittances at typical beam velocities $\beta \simeq 0.6 - 0.7$. The ion beam intensity was varied for the ion species entered into the diagram. The electron current was 0.25 A in all measurements.

For particle numbers in excess of 10^9 the transverse beam emittance can be strongly increased by the existence of coherent transverse beam oscillations with associated particle losses at the aperture. Transverse feedback helps to damp such instabilities, but the limit for maximum beam intensity is at present around 20 mA ion current.

4. Suppression of Intrabeam Scattering at Low Intensity

The lower limit of ion beam temperatures has been investigated mainly by Schottky noise analysis.[7] Transverse diagnostics are presently limited by the resolution of the particle detectors. Some information can be obtained by destructive beam scraping.

To investigate the behaviour of the ion beam at low intensity a Au^{79+} beam at 360 MeV/u with 10^6 particles ($I_i = 24\mu A$) was injected. The initial intensity was known from a standard current transformer with an accuracy of 1 μA. The lifetime which is determined by radiative electron capture in the cooler was measured with the current transformer to $\tau = 1520$ s. During the decrease of the particle number down to nearly single ions the Schottky signal at the 35th harmonic of the revolution frequency was recorded every 10 min. The measured Schottky noise power and momentum spread are shown in Fig. 6. The momentum spread was determined from the width (FWHM) of the frequency distribution according to $\delta p/p = \eta^{-1} \cdot \delta f/f$ with $\eta = \gamma^{-2} - \gamma_t^{-2}$ and a transition energy $\gamma_t = 2.6$ of the ESR lattice.

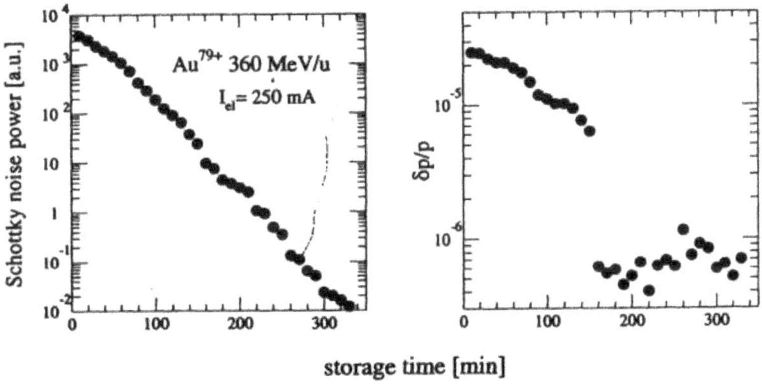

Figure 6: Reduction of the Schottky noise power and the momentum spread of a Au^{79+} beam cooled with an electron current of 0.25 A during a 5 hours' storage time. The number of stored ions N at the beginning and end are 1×10^6 and 3, respectively. Around $N \simeq 4000$ the momentum spread is reduced by a factor of ten without any significant effect in the noise power.

The Schottky noise power decreases with a lifetime $\tau = 1430$ s in good agreement with the current transformer measurement. This proves that there are no collective effects in the Schottky noise. Consequently the integrated Schottky noise power is a measure of the number of ions and the width of the Schottky signal represents the momentum spread. The momentum spread first decreases with the usual $N^{0.3}$ dependence, but at a particle number $N \simeq 4000$ the momentum spread suddenly drops by a factor 10 to a level $\delta p/p = 5 \times 10^{-7}$ on which it stays even for few stored ions. The lower level is caused by the stability of the magnetic field $\delta B/B = 4 \times 10^{-6}$ in the ESR bending magnets. The two quantities are related through $\delta p/p = \gamma_t^{-2} \cdot \delta B/B$. The actual momentum spread is likely to be smaller than the value from the frequency spread. For short record times of the spectrum analyser momentum spreads down to $\delta p/p = 2 \times 10^{-7}$ have been observed, but are normally smeared out when averaging is applied to improve the statistics of the spectra.

The importance of the power supply stability is demonstrated in Fig. 7. The momentum spread as a function of the particle number for bare gold at 360 MeV/u is compared to the values of the same ion species after deceleration to 50 MeV/u. The electron density was in both experiments 3.5×10^6 cm^{-3}. The momentum spread δp/p in the intrabeam scattering dominated regime is for the beam with reduced momentum correspondingly increased. At low intensity the momentum spread is considerably higher due to an inferior stability of the accelerating voltage of the electron beam at this energy.

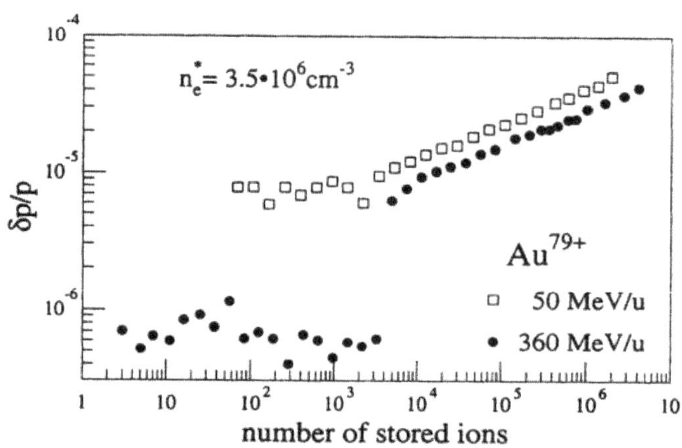

Figure 7: **Comparison of the momentum spread of a Au^{79+} beam at 360 MeV/u and at 50 MeV/u. The electron density $n_e^* = 3.5 \times 10^6$ cm^{-3} applied for cooling was equal at both energies.**

To achieve very small momentum spreads even in the low intensity limit sufficiently strong cooling is required. For reduced cooling strength the colder state of the ion beam is either achieved for smaller particle numbers or can actually not be attained (Fig. 8). The Schottky noise power is generally independent of the electron current but shows a reduction by a factor of 2 - 3 when the ion beam momentum spread drops to the lower value. Sufficiently strong cooling for otherwise equal conditions is a prerequisite to cool to very low momentum spreads. Therefore any effects of the diagnostics instrumentation can be excluded which pretend a momentum spread reduction.

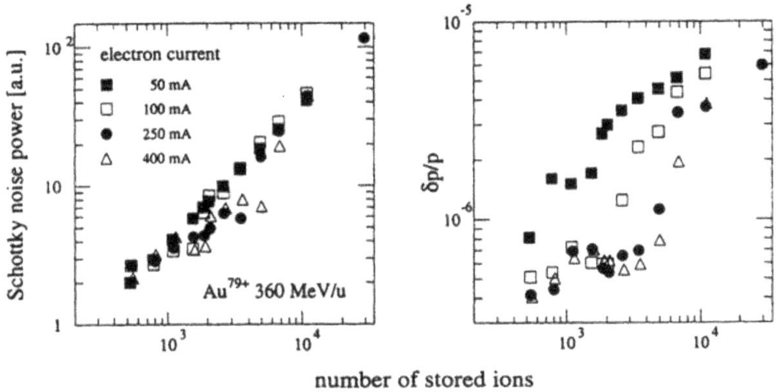

Figure 8: Schottky noise power and momentum spread for Au^{79+} at 360 MeV/u cooled with various electron currents. For particle numbers $N \lesssim 10^3$ and $N \gtrsim 10^4$ the noise power exactly agrees. The transition point to the colder state occurs at larger particle numbers when stronger cooling is applied. At the respective transition point to the colder state the Schottky noise power shows for the higher electron currents a reduction by a factor of 2 - 3.

Although the non-destructive transverse diagnostics are ruled out at low intensity information on the transverse beam size can be obtained by destructively scraping the beam. By comparison with detector measurements the beam radius from the scraper measurement is found comparable to the 3σ width at the detector location. This is reasonable as the scraper was moved to a certain position for a few seconds only and intrabeam scattering reestablished the equilibrium distribution of the remaining beam afterwards on the same time scale. The beam radius at the scraper increases continuously with the beam intensity (Fig. 9). This dependence proves that intrabeam scattering also determines the transverse ion beam properties.

Measurements with other ion species confirmed the existence of a sudden reduction of the momentum spread.[8] The reduction increases with the ion charge whereas the particle number at the transition point is nearly independent of the ion charge. For the transverse beam radius below the transition point 0.1 mm is an upper limit for all beam species. In the regime of reduced momentum spread the scraper technique is at present limited by the resolution of scraper positioning. With improved diagnostics the transverse beam properties have to be further investigated.

For the present knowledge of the transverse beam size according to the scraper measurement and its approximate equivalence to the 3σ width in the particle detector the beam emittance can be estimated. It results in a 1σ emittance of less than $1\times10^{-4}\pi$ mm mrad for 4000 particles of Au^{79+} at 360 MeV/u near the transition point.

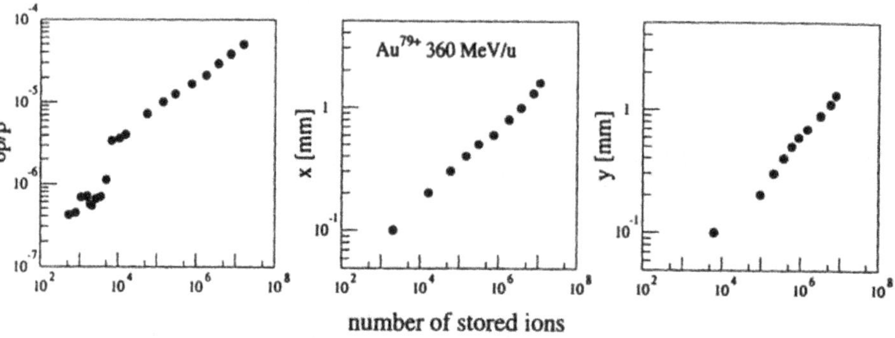

Figure 9: Momentum spread and transverse beam radius versus ion beam intensity for a Au^{79+} beam at 360 MeV/u cooled with an electron current of 0.25 A. The beam radius corresponds approximately to a 3σ value.

This corresponds to a transverse beam temperature of less than 0.4 eV. The longitudinal beam temperature above the transition point for $\delta p/p \simeq 4 \times 10^{-6}$ amounts to 0.25 eV, below the transition point the typical momentum spread $\delta p/p \simeq 5 \times 10^{-7}$ results in a longitudinal temperature of 4 meV. For lighter ions longitudinal temperatures have been found which are close to the lower limit of 5×10^{-5} eV which is determined by the longitudinal temperature of the electron beam. The longitudinal potential energy between two ions is $U = q^2e^2/(4\pi\epsilon_o\gamma d)$. For a particle number $N = 4000$ which is evenly distributed along the circumference $C = 108.3$ m the distance $d = C/N = 0.027$ m results in a potential energy $U = 2.4 \times 10^{-4}$ eV.

Another quantity of importance is the potential energy which is necessary for two neighbouring ions to overcome when passing each other. The minimum distance between the ions is on the order of the transverse betatron motion which is approximately $x \simeq 3 \times 10^{-5}$ m averaged over the storage ring circumference. Thus the particles have to overcome a potential barrier $U = q^2e^2/(4\pi\epsilon_o x)$ when changing their position. The potential energy in the particular situation is 0.35 eV and therefore close to the estimated transverse beam temperature and to the longitudinal beam temperature above the transition point. Regarding that the values are conservative estimates it can be stated that the beam is close to a situation when a change of position is hindered due to the low beam temperature. For other species of highly charged ions the situation is similar as the ratio of potential energy to beam temperature is proportional to q^2/A for the same momentum spread and beam size and therefore does not change largely with the ion charge.

In addition it should be noted that the momentum spread can be lower than the value $\delta p/p = 5 \times 10^{-7}$ determined for a revolution frequency spread which is dominated by the stability of the magnetic field. With a lower longitudinal beam temperature which is indicated by fast Schottky scans the longitudinal beam temperature could approach the longitudinal potential energy due to the interparticle distance. This would actually open the regime when the beam can be considered to be in a liquid like phase. Therefore improved longitudinal diagnostics are investigated which will allow a more accurate determination of the momentum distribution. Combined with the expected increase of cooling strength after technical improvements of the electron cooler advances in the diagnostics instrumentation are the key towards the achievement and proof of a phase transition in the cold ion beam.

References

1. B. Franzke, K. Beckert, H. Eickhoff, B. Franczak, A. Gruber, O. Klepper, F. Nolden, P. Raabe, H. Reich, P. Spädtke, M. Steck and J. Struckmeier, Proc. 1993 Part. Acc. Conf., Washington D.C. (1993) 1645-1649

2. E. N. Dementev, N. S. Dikansky, A. S. Dikansky, A. S. Medvedko, V. V. Parkhomchuk and D. V. Pestrikov, Sov. Phys. Tech. Phys. 25(8) (1980) 1001-1004 and contribution to this workshop

3. T. Winkler, K. Beckert, F. Bosch, H. Eickhoff, B. Franzke, O. Klepper, F. Nolden, H. Reich, B. Schlitt, P. Spädtke and M. Steck, Electron cooling force measurement for highly charged ions in the ESR, to be published in Hyperfine Interactions

4. Beam Cooling and Related Topics, ed. J. Bosser, CERN 94-03

5. M. Steck, K. Beckert, H. Eickhoff, B. Franzke, F. Nolden, P. Spädtke, Proc. 1993 Part. Acc. Conf., Washington D.C. (1993) 1738-1740

6. M. Steck, K. Beckert, F. Bosch, H. Eickhoff, B. Franzke, O. Klepper, R. Moshammer, F. Nolden, P. Spädtke, T. Winkler, Proc. 4th Europ. Part. Acc. Conf., London (1994) 1197-1199

7. M. Steck, K. Beckert, H. Eickhoff, B. Franzke, F. Nolden, H. Reich, P. Spädtke and T. Winkler, Suppression of intrabeam scattering in cooled heavy ion beams, to be published in Hyperfine Interactions

8. M. Steck et al., to be published in Phys. Rev. Lett.

CONCLUSIVE TALK

ARE WE APPROACHING THE ULTIMATE LIMIT OF PARTICLE ACCELERATORS?

G H REES

Rutherford Appleton Laboratory, UK

ABSTRACT

Limits of understanding, limits of performance and practical limits for particle accelerators are discussed briefly, together with limits of expectation for future project funding levels. Discussion covers only the extremes of low energy cooling rings and very high mean power particle accelerators. Advanced accelerating techniques are not discussed; they are considered to represent the future possibility for advancing the perceived existing limits.

1. Introduction

Particle accelerator concepts and designs have advanced rapidly and continuously since 1930, reaching into the fields of X-ray and atomic physics, nuclear and high energy physics, solid state physics, medicine, radiography, ion implantation, fusion, military applications, food sterilization, artificial gem production and oil and gas exploration. The driving force has been experimental high energy physics, which has progressed to ever higher particle kinetic energies. As a consequence, national funding agencies have come to reassess the relative funding for 'big' science within their overall science budgets. A slow-down in high energy physics is now evident, and it appears that future funding limitations will become more severe, making the development of new accelerating techniques a priority.

The other areas that use particle accelerators may in future become more prominent. These cover the extremes of low energy cooling rings and very high mean power accelerators for transmutation, tritium production and heavy ion fusion. In between are synchrotron light sources, spallation neutron sources and accelerators for medical and industrial applications. The extremes are selected for most comment because, on the one hand, this talk comes at the end of a workshop on Crystalline Beams, and on the other, the comparisons are of interest and the effect of space charge forces plays an important role for both. The report first discusses the limits for the high power machines and ends with some comments on special features of the cooling rings; the author comments in the context of having worked only on the former.

2. Limits for High Mean Power Accelerators

The main limit is set by the level of activation that allows hands-on-maintenance procedures. As an approximate guide for a 1 GeV accelerator, an upper limit for the total beam power losses is 1 kW, eg 1 Wm^{-1} for a 1 km linac with beam line, or 1 kW local loss at the collimators of an accumulator ring. The losses for a 100 MW accelerator must then be less than one part in 10^5.

The highest power proton accelerators built to date are the 1 MW LAMPF linac at LANL, the 0.9 MW cyclotron at PSI, Switzerland and the 0.16 MW synchrotron at RAL, UK. All three have reached their design specification, but the power levels are significantly less than those being proposed for future machines. Proton linacs with beam power up to 300 MW are being considered for transmutation of nuclear waste and also for tritium production, while neutron spallation sources based on 1 μs proton pulses are being proposed at beam powers up to 5 MW. These are considered to be upper performance limits, set mainly by beam loss levels, but also by maximum stripping foil temperatures for the charge exchange injection process of the spallation source rings.

Power increases of the order indicated suggest that the understanding of accelerators is highly advanced. This is largely true for the underlying physics issues, but the computer codes used for analysing the particle motions under space charge are still not adequate to predict with high accuracy the final performance figures. Thus, neither the halo production mechanisms in linacs, nor the evolution of the injection distributions for charge exchange injection in rings, are completely understood. Other areas where the high intensity performance of accelerator and storage rings cannot be predicted exactly are certain aspects of the stability for coherent motion. Below a ring's transition energy, the motion is found to be more stable longitudinally than is predicted by traditional coasting beam theory, and is of a different transverse form than is assumed for some higher mode head-tail patterns.

Various high mean power accelerators will now be discussed in the context of their limitations.

2.1 Cyclotrons

Studies have recently been made at PSI[1] and CERN[2] of how to achieve 10 MW of beam power from conventional separated sector cyclotrons. For both studies, the main cyclotron is designed for an injection energy of 120 MeV, an extraction energy of ~ 1 GeV, and the most critical aspect affecting beam losses is the beam size relative to the turn separation at extraction. A small beam momentum spread is advantageous, and this may be realised by the use of decelerating, third-harmonic, flat-top cavities in addition to the main cavities.

Other key items for obtaining high beam power are a high current injector and a maximising of the voltage gain per turn of the high power acceleration system. The need for the latter is related to the influence of the longitudinal space charge fields. These enhance the momenta of particles at the head of the bunches and decrease the momenta of those in the tail, resulting in a tilt of the bunches with respect to the circular orbits. The additional momentum spread due to the space charge is approximately proportional to the beam current, I, and to the square of the number of particle revolutions in the cyclotron, N. The beam size relative to the turn separation at extraction varies approximately as $I\ N^3$, so as I increases it is advantageous to decrease N. For example, a doubling of the energy gain per turn halves N and allows the current to increase by a factor of eight.

At high levels of space charge, the non-linear fields cause a distorted curvature of the tilted bunches. Coupling occurs between radial betatron and synchrotron motion and vortex motion results, leading to a round beam. The effects are largest at low energies and are thus most severe for an injector cyclotron, where it is advantageous to inject a short beam bunch and to collimate unwanted beam tails during the first and fourth beam revolutions. At the extraction energy, on the other hand, it is advisable to extend the beam bunches, using the third harmonic radio frequency system to reduce the space charge forces. Extraction losses from the 0.9 MW PSI Cyclotron are at the level of two parts in 10^4, which is about a factor of two too large for a 10 MW cyclotron. Evaluation of the mechanisms causing the beam loss and of the detailed beam dynamics in injector cyclotrons are required before the feasibility of a 10 MW cyclotron may be judged fully.

An alternative to the conventional separated sector cyclotron is the use of a SOC or Separated Orbit Cyclotron[3]. There is only one such machine in existence and it is a low energy machine, still in its commissioning phase. The transverse focussing in a SOC may be adjusted for each turn and the ring made non isochronous. An enhanced voltage gain per turn may be used and the longitudinal focusing utilised to counteract the longitudinal space charge forces. This removes one problem but introduces another, for the large voltage gain per turn increases the radial betatron-synchrotron coupling. The energies in the longitudinal and transverse phase planes are far from an equipartition state, so the coupling is expected to cause significant effects. Trinks[3] proposes the use of superconducting magnets and acceleration systems to improve the efficiencies of high power SOCs.

2.2 *Linear Accelerators*

High power linear accelerators may use room temperature or superconducting accelerating structures. At medium beam powers, room temperature cavities are pulsed, with a relatively short duty cycle for efficiency reasons. At high beam powers, continuous wave operation, cw, is used for both types of cavities, as with cyclotrons, though frequencies adopted for linac cavities are typically an order of magnitude larger than those used for cyclotrons.

There is a power level when it is advantageous to move from room temperature linacs to cw superconducting linacs, and the approximate level at which this occurs is 50 MW. For this reason, LANL now studies a superconducting option for a 100-200 MW APT facility (accelerator production of tritium). An important question for a high power cw superconducting linac is how to switch it on to full power. One possibility is to start-up with a low duty cycle and to gradually extend the pulse length to cw operation. It is, however, a difficult control problem to meet the tolerances for the amplitude and phase of the cavity fields when pulsing a superconducting linac because of the associated Lorentz force detuning. Some stiffening is needed for the cavities to reduce the detuning but this must not interfere with the main mechanical tuning. A satisfactory solution has been obtained for the nine cell superconducting cavitites of the test set-up for the TESLA e^+-e^- linear collider where the phase tolerances are $\pm 10^0$. Much tighter tolerances of $\pm 1^0$ are specified for the APT and spallation source linacs.

Losses must be kept below 1 nA m^{-1} for an APT, so that beam halo studies are a major feature of the design with three dimensional space charge codes required for both the linacs and their output beam lines. In using the codes, it is desirable to use measured output beam distributions from suitable RFQ and drift tube linacs as input data for the high energy stage of the linac. Often the experimental information is unavailable and input is assumed to have the form of constant density, ellipsoidal bunches, a form convenient for analytical use, and one which corresponds to a minimum electric field energy. It does not correspond, however, to the bunches from a linac front end which are cylindrical in shape and not ellipsoidal. Transverse emittances are approximately constant along the bunch whereas, for the ellipsoidal bunch, the particles of maximum synchrotron amplitudes have zero transverse motion. If the idealised ellipsoidal bunch is not formed at the linac input, it is difficult to see how it may evolve subsequently as the particles of large synchrotron amplitude must proceed to lose their transverse betatron energy.

APT and Spallation Source linacs have comparable peak beam currents and hence very similar beam dynamics. The 1.334 GeV coupled cavity linac[4] of the European Spallation Source, ESS, is designed for approximate equipartition (0.5) of transverse and longitudinal energies at its 70 MeV injection energy. This minimises the energy exchange through coupling, as is confirmed by multiparticle trackings. No exchange for the normalised rms emittances is found at higher energies even if the ratio of transverse to longitudinal energies is increased to 2.5 progressively along the linac. Transverse betatron phase shifts are approximately 70^0 per focusing period at injection but are reduced to 10^0 at 1.334 GeV. Tracking studies indicate a small amount of halo formation even though the normalised ring emittances stay approximately constant.

It is important that the low emittance H^- beam produced by the ESS linac does not suffer degradation in its transfer to two accumulator rings which are required for pulse compression. The most critical part of the ESS design is to achieve low loss charge exchange injection in the rings. Momentum ramping, debunching and momentum compression are used in the transfer line, and trackings have shown that the halo growth may be contained. A long achromatic collimation line is included to ensure acceptable beam parameters at the ring stripping foils.

2.3 ISIS

The 50 Hz, 800 MeV proton synchrotron of the ISIS Spallation Source has an output beam power of 160 kW, a factor 20 larger than that of any other operating synchrotron. Next generation sources require a further factor of 30 increase, so ISIS performance limitations are of interest.

Coherent beam instabilities have been of much less concern than originally anticipated. The residue of the 202 MHz linac bunch structure debunches in the ring within 30 μs, even when operating a factor of more than 10 above the predicted coasting beam threshold. Longitudinal space charge forces are relatively large at trapping but no fast longitudinal instabilities have been observed. Slower beam loading instabilities may be observed but avoided. There is no electron - proton transverse instability and there is, for most of the acceleration cycle, no other transverse instability. A higher mode, m=1, head - tail instability may develop over the 2 to 4 ms interval of the 10 ms acceleration period, but a râmping of the betatron tunes is sufficient to prevent any coherent vertical growth.

Trapping losses amount to 6% at low intensity and 11% at peak intensity but are nearly all contained on a momentum collimation system. The losses are acceptable because the protons are lost at an energy of 80 MeV or less. Peak transverse space charge tune shifts are estimated to be 0.4, which is sufficient to move the horizontal incoherent tunes of some beam particles, though not the coherent tune, across the $Q_h=4$ and $2Q_h=8$ resonances. Coherent image forces have been found to have a significant effect on the motion of some particles on crossing the resonance lines[5]. Two effects occur, one when the coherent image coefficients vary along the coherent axis of displacement, and one non-linear effect due to displacements orthogonal to the plane of particle motion. The magnitude of the effects may be estimated approximately. For higher power spallation sources, tune shifts are much less than 0.4, typically 0.1. Beam emittances and acceptances are set large enough relative to the injection emittances to keep proton foil traversals and foil temperatures at an acceptable level, and these criteria lead to the required tune shifts.

2.4 5 MW Spallation Neutron Sources

Typical specifications for pulsed, high power, spallation neutron sources are: 50 Hz repetition rate, 5 MW beam power, 1 μs maximum pulse durations, 1 to 10 GeV proton kinetic energies, and the use of two target stations, one at 10 Hz and 1 MW and the other at 50 Hz with powers up to 5 MW. The favoured designs consist of a pulsed, high power, 50 Hz, H^- linac feeding one of the following:

a) A number of accumulator rings which are fed sequentially;
b) Two 25 Hz synchrotrons, RCS, which are fed in alternate 50 Hz cycles; or
c) A 50 Hz booster, a holding ring and a 25 Hz synchrotron in series.

At low beam powers, the cheapest option is a variant of b), using a low energy H^- linac and a single 50 Hz RCS. At medium powers of up to 1 MW, the most economical options are b) and c), but at 5 MW, the option a), using a full energy linac and two accumulator rings, is the cheapest.

In the proposal for the European Spallation Source, ESS[6], the beam kinetic energy for the linac and accumulators is 1.334 GeV, the linac duty cycle is 6%, and the beam is chopped with a 60% duty factor at the 1.67 MHz rings' revolution frequency. The average H^- ion current during the unchopped intervals is 107 mA, but this is equivalent to 214 mA for space charge effects as there is low energy funnelling of 175 MHz beam bunches but a frequency of 700 MHz for the following coupled cavity linac. Funnelling is adopted because of existing performance limitations for H^- ion sources. A room temperature linac is favoured over a superconducting option because of the power levels involved in the pulsing of the latter, the complexity of the associated field controls, and the comparable costs of the options.

Injection is the most critical item for the rings, and because of this the H^- beam is collimated in the injection line. The main loss in the rings is then due to proton traversals of the stripping foil during and after injection. Every effort is made to minimise these and, in particular, the foil is made with two free unsupported edges to enhance the probability of foil avoidance. The foil is located at the centre of a special dipole magnet, whose field level is chosen to give negligible pre-stripping of H^- ions ahead of the foil, to minimise delayed stripping of H^0 atoms in the ring, and to provide a bending radius for stripped electrons large enough for direct collection. Stripping efficiencies are about 98.5%, with the remainder partially stripped H^0 atoms in a range of quantum states. The fate of these depends on their Stark state in the injection magnet. Atoms in a state of low principal quantum number n, <4, remain as H^0 and pass out of the ring for external collection. Atoms of high n, >6, strip rapidly and are accepted as protons. There remain some intermediate states, n=4 and n=5, which strip after some delay, so may be accepted or lost or become halo. Simultaneous injection painting is provided in the longitudinal and both transverse phase planes and, for a 1000 turn injection, foil traversals are restricted to seven, on average. Beam loss for the n=4 and n=5 states is then acceptable, and the dominant loss is that due to the negative momentum tail which is introduced during the foil scattering. The tail results in enhanced radial betatron motion, as the injection relies on a finite dispersion at the foil location. Downstream collectors are able to contain most of the lost beam. A low harmonic number is used for the radiofrequency containing field (h=1 and 2), and it is important to have no beam in the bunch gap. Such beam results in loss at the time of fast extraction and is also a possible cause of electron - proton instabilities.

A 2.4 GeV superconducting linac and a single 2.4 GeV accumulator ring have also been considered as an ESS option. In comparison with the 1.334 GeV option, the ring circumference is doubled, the linac peak current is approximately halved, but the linac duty cycle and chopping duty factor are unchanged. Cavity filling times are enhanced and so conditions for pulsing the superconducting cavities are less favourable. A more expensive linac results, together with a more complex accumulator. Injection criteria, for beam loss and foil temperatures, result in unchanged ring acceptances and beam spot sizes on the foil at 2.4 GeV, and the need of large acceptances seriously complicates the injection and fast extraction systems. The additional cost and complexity rule out the 2.4 GeV option, despite its reduced ion source requirements if funnelling is retained.

2.5 *Heavy Ion Fusion RF Linac and Storage Ring Drivers*

Heavy ion fusion, HIF, makes use of energetic heavy ions to produce X-rays for the implosion of pellets containing deuterium and tritium. Two types of driver are currently under consideration, one employing an induction linear accelerator and the other a radio frequency linac and an arrangement of storage rings. The main features of the latter are the use of many stages of funnelling at the front end of the linac to obtain adequate heavy ion currents, ring injection with very low losses, and large compression factors for beam bunches in both the rings and the beam lines to the reactor chamber. A pilot ignition plant[7] considers the use of ions of mass number ~200, beam kinetic energies of 10 GeV, beam energies per pulse of 3 MJ and bunch durations as small as 3 ns.

Multi-turn injection into the rings may be via charge exchange of negative heavy ions using laser stripping, or injection of positive ions via a horizontal or a corner septum unit. The first of these methods requires a high power laser beam extending over many metres, and its feasibility requires further study. The other two methods are limited in the number of turns that may be injected as the ratio of the required beam emittance in the rings to the linac output emittance is not large. The full unnormalised ring emittance (area $\div\pi$) must be ≤ 50 μ rad m and the estimated full output emittance for a 400 mA, pulsed, 10 GeV linac is approximately 4 μ rad m. The number of injected turns for loss free injection using a horizontal septum is then found to be ≤ 6, while for the corner septum, it is found to be ≤ 12. Combined horizontal and vertical injection is obtained for the corner septum by using orbit bumps, varying lattice functions and by adjusting the position and size of the incoming beam. It is also advantageous to inject into a zero dispersion straight section.

A linac-storage ring driver becomes more complicated if injection is limited to a small number of injected turns. More storage rings are needed to meet the total flux requirements, and a reduced longitudinal emittance to allow the subsequent bunch compressions. If injection is restricted to twelve turns, bunch compression to 3 ns is impractical for the predicted linac beam momentum spread of $\pm 1.4\ 10^{-4}$. It is therefore suggested that injection should be followed by longitudinal laser cooling to reduce the longitudinal emittance by an order of magnitude. The recent success with laser cooling at Århus[8] and Heidelburg[9] is thus very timely. Additional sympathetic transverse cooling[9] is not proposed, as enhanced beam loss may result. Bunched beam longitudinal laser cooling[8] is attractive to consider, however, for if it proves feasible, it would simplify considerably the subsequent bunch compressions.

2.6 *Expectations for Future Project Funding Levels*

Three categories may be defined for future projects : large international facilities, large national facilities and small to medium national facilities. In all three areas, expectations of future funding levels are falling, and approximate upper limits are now considered to be :

Large international high energy physics facilities	2 B ecu
Other large international facilities (eg ESS)	1 B ecu
Large national facilities	1 B ecu
Small to medium national facilities	0.02 to 0.2 B ecu

Future linear colliders need to set aside some of their R and D funds in an effort to study how to minimise the costs for all the main sub-components and, in addition, the operating costs. Spallation neutron sources at the 5 MW level are provisionally estimated to cost 1 B ecu and this figure may prove to be too high, so that some reduction in beam power may be proposed. At the lower end of the funding scale are the small national facilities, and here it is of interest to note that a Crystal Beam Facility is of an appropriate scale for future funding.

3. Aspects of Crystal Beams

The only common feature of the high mean power accelerators discussed in section 2 and the Crystal Beam Projects is the importance of the space charge effects. Even though beam parameters are very different for the two, the space charge fields are equally important. Crystal beam proposals rely on cooling techniques, and in particular, laser cooling, to reduce the longitudinal and transverse emittances of low intensity beams to the point where crystalline beam structures appear. The high power accelerators, on the other hand, use very intense beams but the beams are not, in most cases, in a very highly space charge dominated regime. Cooling techniques are not required, apart from the possible use in HIF storage rings.

Three dimensional space charge codes are under continuous development for studying beams in linacs and accelerators. The techniques are applicable to crystal beam studies but the codes may not be used directly for the effect of intra beam scattering must also be included. Simpler, transverse space charge, magnet lattice codes are also available and these may be used directly for evaluating the effect of linear space charge on crystal beam lattices. Important features emerge.

3.1 Magnet Lattices for Crystal Beams

The following conditions[10] have been found necessary for a crystal beam magnet lattice:

Alternating, and not constant, gradient focusing;
Operation below the ring transition energy; and
Avoidance of linear resonances between phonon modes and the lattice periodicity (ie periodicity $S > 2\sqrt{2}Q$, where Q is a betatron tune).

During the cooling process in a crystal beam ring, the beam has to pass through a wide range of operating conditions before reaching a crystalline state. The effect of the transverse space charge fields is to decrease the overall focusing, with the linear components producing a negative betatron tune shift and the non-linear components, a tune spread. Intrabeam Coulomb scattering increases as the transverse and longitudinal emittances decrease, and if there is a low cooling rate, an equilibrium beam size is reached before achieving crystallisation. To reach an ordered state, the cooling rate must be increased and instability avoided, and the very act of ordering modifies the intrabeam Coulomb scattering.

Stability for an alternating gradient magnet lattice may be expressed by a necktie diagram. At low levels of focusing, there is a narrowing of the stability region, which suggests that cooling may cause a beam to become unstable. A crystal beam lattice must be designed so that beams may be cooled to ordered states without encountering instability. The best possibility appears to be the use of symmetric focusing in the absence of space charge, that is to have approximately equal horizontal and vertical unperturbed betatron tunes. To obtain further insight, consider the use of magnet lattice codes that include the effect of linear space charge. Such codes solve the K-V envelope equations[11] for a specified magnet lattice and relevant beam parameters, finding periodic solutions for the beam envelopes over a lattice superperiod. The effect of dispersion may be included to find the modified off-momentum orbits. Equivalent betatron and dispersion parameters may be found for a chosen level of linear space charge, together with the modified value of γ_t.

Results obtained from a linear space charge lattice code are now presented for four representative lattices. The first lattice chosen, A, is that for the Legnaro Crystal Beam Proposal[12], which has been discussed at this workshop. The others are chosen to illustrate that figures of merit may be derived for crystal beam lattices. Lattice B is chosen for direct comparison with lattice A, having the same mean radius, periodicity and Q_h value, but with $Q_v \simeq Q_h$ and with γ_t increased to be greater than Q_h. The reason for the last choice will become clear later, and the mechanism for achieving it is to use a hybrid lattice cell, consisting of a horizontally focusing quadrupole and a combined function, horizontally defocusing, bending magnet in a FODO configuration. The dispersion is then at a minimum in the bending magnet, so increasing γ_t without invoking any resonance mechanism. Two further lattices, C and D, are considered and both are based on the 8.6 m FO(BD)O cell of the hybrid lattice B, but with the periodicity increased from 8 to 12 and with the betatron tunes, Q, set at ~2.85 and ~3.7 respectively, for C and D. The lattices C and D are identical apart from the different settings of the field gradients in the magnets. The betatron phase shift per cell for the lattice D is more than 90^0, which is undesirable because of envelope instability considerations[12]. However, lattice D is not realised until beams have been cooled to low temperatures in lattice C, and only then are the gradients adjusted to move lattice C to Lattice D.

Two conditions of tune depression are found for each of the lattices A, B, C and D. For the first condition, the tune depression is found which corresponds to the modified value of γ_t being depressed to reach the value of γ for the stored beam, that is $\gamma_t \simeq 1$. For the second condition, the maximum tune depressions are found for which the lattice does just not become unstable under linear space charge forces. A number of figures of merit for the lattices may be found from these conditions. Tune depressions are often expressed as the ratio Q/Q_0 where Q is the perturbed tune value under linear space charge and Q_0 is the unperturbed tune. The inverse of this ratio may be used as a figure of merit, and the ratio of the stored beam current, I, to the transverse beam emittance, ε_h or ε_v, may be used as a further figure of merit. Four such figures may be found for each lattice:

The value of Q_0/Q when $\gamma_t = 1$;
The value of Q_0/Q when just stable under linear space charge ;
The value of I/ε_h when $\gamma_t = 1$; and
The value of I/ε_v when $\gamma_t = 1$.

The justification for using these figures of merit is the following. Beams may become unstable longitudinally if γ_t, under space charge, is reduced to a value less than γ. This is the condition for the negative mass instability. Thus, the best chance to realise a crystal beam is for the modified value of γ_t to adiabatically approach γ as the cooled beam approaches the crystalline state. The condition $\gamma_t = \gamma$ is also the condition of isochronism for finite momentum beams in the lattice. High values for (Q_0/Q) and (I/ε) when $\gamma_t = 1$ indicate that a lattice allows large tune depressions and low emittances before instabilities occur. The figures of merit thus show that one type of lattice has a greater chance than another of realising a crystal beam. Results for the four lattices are tabulated and a progressive improvement may be observed from A to B to C to D.

Lattice	S	Q_{v0}	Q_{h0}	γ_t	I/ε_v	I/ε_h	Q_0/Q
A	8	1.19	1.85	1.76	2.6	4.7	1.75
B	8	1.74	1.85	2.11	5.1	5.0	2.05
C	12	2.90	2.85	3.69	8.2	8.3	3.40
D	12	3.65	3.70	5.21	15.4	15.4	4.76

Table 1. Lattices with a Space Charge Modified Value of $\gamma_t = 1$

Lattice	S	Q_{v0}	Q_{h0}	γ_t	I/ε_v	I/ε_h	Q_0/Q	γ_t
A	8	1.19	1.85	1.76	4.3	14.3	3.47	0.49
B	8	1.74	1.85	2.11	12.0	12.5	4.48	0.45
C	12	2.90	2.85	3.69	11.7	11.8	4.50	0.78
D	12	3.65	3.70	5.21	17.9	17.8	5.46	0.87

Table 2. Lattices at Limit of Stability under Space Charge.

The units for I/ε are kA (rad m)$^{-1}$, assuming 10 MeV B_e^+ ions. A further increase of the horizontal tune for lattice D from 3.70 to 3.85 increases γ_{t0} to 5.6 but the lattice is unstable under space charge before γ_t is reduced to $\gamma_t = 1$.

Too great a significance should not be read, however, into the figures of merit. Non-linear space charge forces have been neglected and these may introduce significant effects, causing a spread of both the modified betatron tunes and the modified γ_t values. Detailed considerations of a lattice must include these non-linear effects. An interesting case is the non-linear octupole lattice outlined at the workshop by P.Channell,[13] where the extreme nature of the non-linearity requires a more complex analysis for off-momentum closed orbits.

3.2 *Bunched Crystalline Beams*

Following the recent demonstrations of laser cooling of bunched beams,[8,9] the question arises whether it is possible to realize bunched crystalline beams. The general feeling is that it is possible but that it will be more difficult than for a coasting beam. The reason is that there is a greater spread of betatron tunes and modified γ_t values for a bunched beam due to the varying particle line density along the length of the bunch.

There is a need for experiments on the existing cooling rings to see what happens when a bunched beam is cooled towards the condition $\gamma_t = 1$ and beyond. Does the bunch break up if the spread of γ_t is such that part of the beam is above transition and part is below? Or do particles move adiabatically towards $\gamma_t = 1$? The higher γ_{t0} lattices discussed in section 3.1 appear to provide a greater chance of realizing a bunched crystalline beam.

3.3 *Simulations for Cooled Beams*

This topic has been discussed in many of the workshop talks and, of concern, is that some important mechanisms may have been overlooked. Thus there is a need to check whether or not simulations explain the detailed behaviour of existing cooling rings. A fruitful area for study is to see if simulations explain the mechanism of the recently discovered, sympathetic transverse laser cooling. Does the effect arise through intrabeam scattering alone, or also via the space charge forces? Transverse - longitudinal coupling may occur for both. There is a non-linear dependence of the longitudinal space charge fields on the betatron motion, and of the transverse space charge fields for the off-momentum closed orbits.

A number of questions may be posed :

1. Is a three dimensional space charge code needed? (almost certainly)
2. Should existing intrabeam scattering codes be modified for ultra cold beams? (almost certainly)
3. Are closed orbit errors and coherent image effects significant?
4. What are the effect of transient errors which are faster than the cooling times (eg high frequency power supply noise and stray fields)?
5. How do the results of simulations compare for different lattices?
6. Are the cooling mechanisms being simulated correctly?
7. What are the cooling limits of existing rings?
8. Should crystal beam rings contain electron coolers or should the beam be precooled in a separate ring?

4. Summary and Conclusions

1. Comparing very high mean power accelerators and cooling rings does not lead to any new insights, even though the beam dynamics of both is dominated by space charge effects.
2. High mean power accelerators are, at present, some way from practical beam loss limits, but next generation machines propose one to two orders of magnitude increase in beam power and their exact performance is difficult to predict.
3. Longitudinal laser cooling, developed for the cooling rings, is a technique of considerable interest for heavy ion fusion storage ring drivers, and the feasibility of such a driver may depend upon the feasibility of introducing longitudinal laser cooling.
4. Crude figures of merit may be derived for crystal beam rings from lattice codes modified for linear space charge fields, see section 3.1.
5. Hybrid lattices with enhanced γ_t are of interest for crystal beams, see section 3.1.
6. Comparisons are needed between simulations and the detailed performance of existing cooling rings.
7. As an outsider to the field, the achievement of crystal beams appears to be the ultimate machine physics experiment, but it also appears that life may never be so exciting again once the crystal beams have been obtained.

5. Acknowledgement

I am very grateful to Klaus Bongardt for many fruitful discussions on the topics of this paper.

6. References

1. Th. Stammbach et al, The Feasibility of High Power Cyclotrons, European Conference on Application of Accelerators for Research and Technology, ECAART 95, Zurich, 1995
2. N.Fiétier and P.Mandrillon, A Three-Stage Cyclotron for Driving the Energy Amplifier, CERN/AT/95-03, 1995.
3. U.Trinks, New Trends in Cyclotrons, Proceedings of the Fourth European Particle Accelerator Conference, EPAC94, 1994, pp 295-299.
4. M.Pabst and K.Bongardt, Beam Dynamics in the 1.3 GeV High Intensity ESS Coupled Cavity Linac, Proceedings of U.S. National Particle Accelerator Conference, Dallas, PAC95, 1995.
5. G.H.Rees and C.R.Prior, Image Effects on Crossing an Integer Resonance, Particle Accelerators, Vol 48, 1995, pp 251-257.
6. I.S.K.Gardner, H.Lengeler and G.H.Rees, Outline Design of the European Spallation Source, ESS 95-30-M, September 1995.
7. Heavy Ion Fusion Facility, Proposal for a European Study Group, GSI Report GSI-95-03, April 1995.
8. J.S.Hangst et al, Experience with Laser Cooling at Astrid, Proceedings of the Workshop on Crystalline Beams and Related Issues, Erice, Nov. 1995.
9. H-J.Miesner et al, Three Dimensional Cooling of a Stored Beam by Laser Light. Proceedings of the Workshop on Crystalline Beams and Related Issues, Erice, Nov. 1995.
10. J.Wei et al, Diverse Topics in Crystalline Beams, Proceedings of the Workshop on Crystalline Beams and Related Issues, Erice, Nov. 1995.
11. I.M.Kapchinsky and V.V.Vladimirsky, Proceedings of the International Conference on High Energy Accelerators and Instrumentation, CERN 1959, p.274.
12. L.Tecchio and A.Pisent, Crystal Storage Ring, Proceedings of the Workshop on Crystalline Beams and Related Issues, Erice, Nov. 1995.
13. P.J.Channell and F.R.Neri, Lattices with Minimal Space Charge Effects for Crystalline Beams, Proceedings of the Workshop on Crystalline Beams and Related Issues, Erice, Nov. 1995

LIST OF PARTICIPANTS
Workshop on Crystalline Beams and Related Issues
Erice, Sicily
11-21 November 1995

Klaus Bongardt	KFA Juelich	ess@kfa.juelich.de
Fritz Caspers	CERN	caspers@ps.msm.cern.ch
Michel Chanel	CERN	chanel@cernvm.cern.ch
Paul Channell	Los Alamos Natl. Lab	pchannell@lanl.gov
Håkan Danared	Manne Siegbahn Lab.	danared@msi.sunet.se
Daniel Dubin	UC San Diego	dubin@ssurf.ucsd.edu
Bernhard Franzke	GSI Darmstadt	FRANZKE@GSI.DE
Rudi Grimm	MPI Heidelberg	grimm@mickey.mpi-hd.mpg.de
Dietrich Habs	LMU München	habs@hobbit.mpi-hd.mpg.de
Alfons Haffmans	SUNY Stonybrook	haffmans@insti.physics.sunysb.edu
Jeffrey Hangst	Aarhus University	hangst@dfi.aau.dk
Rainer Hasse	GSI Darmstadt	R.HASSE@GSI.DE
Ingo Hofmann	GSI Darmstadt	I.Hofmann@GSI.DE
Valera Lebedev	CEBAF	LEBEDEV@CEBAF.GOV
Dusan Maletic	Brookhaven	dusan@bnlarm.bnl.gov
Igor Meshkov	Dubna	meshkov@nusun2.jinr.dubna.su
Hans-Joachim Miesner	MPI Heidelberg	miesner@trick.mpi-hd.mpg.de
Dieter Möhl	CERN	MOEHL@CERNVM.cern.ch
Hiromi Okamoto	Kyoto University	okamoto@kyticr.kuicr.kyoto-u.ac.jp

Søren Pape Møller	Aarhus University (ISA)	fyssp@dfi.aau.dk
Vasily Parkhomchuck	Novosibirsk	parkhomchuk@vxinpb.inp.nsk.su
Andrea Pisent	INFN Legnaro	pisent@lnl.infn.it
Ove Poulsen	Danish Research Ministry	opo@fsk.dk
Grahame Rees	Rutherford Laboratory	jut@uk.ac.rl.1b
Dag Reistad	Uppsala University	REISTAD@tsl.uu.se
Alessandro Ruggiero	Brookhaven	AGR@BNL.GOV
John Schiffer	Argonne National Lab.	SCHIFFER@anl.gov
Ulrich Schramm	MPI Heidelberg	schramm@mickey.mpi-md.mpg.de
Andy Sessler	Lawrence Berkeley Lab.	Andrew_Sessler@macmail.lbl.gov
Allan Sørensen	Aarhus University	ahs@dfi.aau.dk
Markus Steck	GSI Darmstadt	steck@gsi.de
Luigi Tecchio	INFN Legnaro	TECCHIO@clulnl.lnl.infn.it
Christian Toepffer	U. of Erlangen	toepffer@theorie2.physik.uni-erlangen.de
Gabriele Torelli	U. Pisa	torelli@pi.infn.it
Gerard Tranquil	CERN	tranquil@ps.msm.cern.ch
Vittorio Vaccaro	INFN Napoli	vaccaro@na.infn.it
Herbert Walther	MPI Garching	herbert.walther@mpq.mpg.de
Jie Wei	Brookhaven	jwei@pig.rhic.bnl.gov
Andreas Wolf	MPI Heidelberg	wolf@hobbit.mpi-hd.mpg.de
S. Y. Lee	Indiana Univ. (IUCF)	shylee@indiana.edu
Guenter Zwicknagel	Univ. Paris-Sud	gunter.zwicknagel@lpgp.u-psud.fr

www.ingramcontent.com/pod-product-compliance
Lightning Source LLC
LaVergne TN
LVHW060542180926
848329LV00007B/35
* 9 7 8 9 8 1 0 2 2 7 8 5 2 *